Leonard J. Brillson

**Surfaces and Interfaces of
Electronic Materials**

Related Titles

Butt, H.-J., Kappl, M.

Surface and Interfacial Forces

436 pages with 158 figures
and 17 tables
2010
Softcover
ISBN: 978-3-527-40849-8

Gerlach, G., Dotzel, W.

Introduction to Microsystem Technology
A Guide for Students

376 pages
2008
Hardcover
ISBN: 978-0-470-05861-9

Wetzig, K., Schneider, C. M. (eds.)

Metal Based Thin Films for Electronics

424 pages with 312 figures and 17 tables
2006
Hardcover
ISBN: 978-3-527-40650-0

Butt, H.-J., Graf, K., Kappl, M.

Physics and Chemistry of Interfaces

398 pages with 183 figures
2006
Softcover
ISBN: 978-3-527-40629-6

Bordo, V. G., Rubahn, H.-G.

Optics and Spectroscopy at Surfaces and Interfaces

281 pages with 144 figures and 1 tables
2005
Softcover
ISBN: 978-3-527-40560-2

Irene, E. A.

Electronic Materials Science

320 pages
2005
Hardcover
ISBN: 978-0-471-69597-4

Burshtein, A. I.

Introduction to Thermodynamics and Kinetic Theory of Matter

349 pages with 142 figures
2005
Softcover
ISBN: 978-3-527-40598-5

Leonard J. Brillson

Surfaces and Interfaces of Electronic Materials

The Author

Prof. Leonard J. Brillson
Ohio State University
Electrical & Computer Engineering
and Physics
Columbus, USA
brillson.1@osu.edu

All books published by Wiley-VCH are carefully produced. Nevertheless, authors, editors, and publisher do not warrant the information contained in these books, including this book, to be free of errors. Readers are advised to keep in mind that statements, data, illustrations, procedural details or other items may inadvertently be inaccurate.

Library of Congress Card No.: applied for

British Library Cataloguing-in-Publication Data
A catalogue record for this book is available from the British Library.

Bibliographic information published by the Deutsche Nationalbibliothek
The Deutsche Nationalbibliothek lists this publication in the Deutsche Nationalbibliografie; detailed bibliographic data are available on the Internet at <http://dnb.d-nb.de>.

© 2010 WILEY-VCH Verlag GmbH & Co. KGaA, Weinheim

All rights reserved (including those of translation into other languages). No part of this book may be reproduced in any form – by photoprinting, microfilm, or any other means – nor transmitted or translated into a machine language without written permission from the publishers. Registered names, trademarks, etc. used in this book, even when not specifically marked as such, are not to be considered unprotected by law.

Cover Spiezs design, Neu-Ulm
Typesetting Laserwords Private Limited, Chennai, India
Printing and Binding Strauss GmbH, Mörlenbach

Printed in the Federal Republic of Germany
Printed on acid-free paper

ISBN: 978-3-527-40915-0

Contents

Preface *XVII*

1 Introduction *1*
1.1 Surface and Interfaces in Everyday Life *1*
1.2 Surfaces and Interfaces in Electronics Technology *2*
 Problems *7*
 References *8*

2 Historical Background *9*
2.1 Contact Electrification and the Development of Solid-State Concepts *9*
2.2 High-Purity Semiconductor Crystals *10*
2.3 Development of the Transistor *10*
2.4 The Surface Science Era *12*
2.5 Advances in Crystal Growth Techniques *13*
2.6 Future Electronics *15*
 Problems *15*
 References *16*

3 Electrical Measurements *19*
3.1 Schottky Barrier Overview *19*
3.2 Ideal Schottky Barriers *20*
3.3 Real Schottky Barriers *22*
3.4 Schottky Barrier Height Measurements *25*
3.4.1 Current–Voltage (J–V) Technique *25*
3.4.2 Capacitance–Voltage (C–V) Technique *28*
3.4.3 Internal Photoemission Spectroscopy (IPS) *29*
3.5 Summary *33*
 Problems *33*
 References *34*

4 Interface States 37

- 4.1 Interface State Models 37
- 4.2 Simple Model Calculation of Electronic Surface States 39
- 4.3 Intrinsic Surface States 42
- 4.3.1 Experimental Approaches 42
- 4.3.2 Theoretical Approaches 43
- 4.3.3 Intrinsic Surface-State Models 44
- 4.3.4 Intrinsic Surface States of Silicon 45
- 4.3.5 Intrinsic Surface States of Compound Semiconductors 45
- 4.3.6 Dependence on Surface Reconstruction 48
- 4.3.7 Intrinsic Surface-State Summary 52
- 4.4 Extrinsic Surface States 52
- 4.4.1 Weakly Interacting Metal–Semiconductor Interfaces 52
- 4.4.2 Extrinsic Features 55
- 4.4.3 Schottky Barrier Formation and Thermodynamics 55
- 4.4.4 Extrinsic Surface-State Summary 62
- 4.5 Chapter Summary 62
- Problems 63
- References 63

5 Ultrahigh Vacuum Technology 67

- 5.1 Ultrahigh Vacuum Vessels 67
- 5.1.1 Ultrahigh Vacuum Pressures 67
- 5.1.2 Stainless Steel UHV Chambers 69
- 5.2 Pumps 70
- 5.3 Specimen Manipulators 76
- 5.4 Gauges 76
- 5.5 Deposition Sources 77
- 5.5.1 Metallization Sources 77
- 5.5.2 Crystal Growth Sources 78
- 5.6 Deposition Monitors 79
- 5.7 Summary 80
- Problems 81
- References 81
- Further Reading 82

6 Surface and Interface Analysis 83

- 6.1 Surface and Interface Techniques 83
- 6.2 Excited Electron Spectroscopies 85
- 6.3 Principles of Surface Sensitivity 88
- 6.4 Surface Analytic and Processing Chambers 89
- 6.5 Summary 92
- References 92
- Further Reading 92

7	**Photoemission Spectroscopy** 93	
7.1	The Photoelectric Effect 93	
7.2	The Optical Excitation Process 95	
7.3	Photoionization Cross Section 95	
7.4	Density of States 96	
7.5	Experimental Spectrum 96	
7.6	Experimental Energy Distribution Curves 97	
7.7	Measured Photoionization Cross Sections 100	
7.8	Principles of X-ray Photoelectron Spectroscopy 112	
7.8.1	Chemical Species Identification 112	
7.8.2	Chemical Shifts in Binding 114	
7.8.3	Distinction between Near- and Subsurface Species 114	
7.8.4	Charging and Band Bending 115	
7.9	Excitation Sources 119	
7.10	Electron Energy Analyzers 122	
7.11	Summary 125	
	Problems 125	
	References 126	

8	**Photoemission with Soft X-rays** 129	
8.1	Soft X-ray Spectroscopy Techniques 129	
8.2	Synchrotron Radiation Sources 129	
8.3	Soft X-Ray Photoemission Spectroscopy 132	
8.3.1	Basic Surface and Interface Techniques 132	
8.3.2	Advanced Surface and Interface Techniques 137	
8.3.2.1	Angular Resolved Photoemission Spectroscopy 138	
8.3.2.2	Polarization-Dependent Photoemission Spectroscopy 139	
8.3.2.3	Constant Final State Spectroscopy 140	
8.3.2.4	Constant Initial State Spectroscopy 141	
8.4	Related Soft X-ray Techniques 141	
8.5	Summary 143	
	Problems 144	
	References 144	

9	**Particle–Solid Scattering** 147	
9.1	Overview 147	
9.2	Scattering Cross Section 147	
9.2.1	Impact Parameter 147	
9.2.2	Electron–Electron Collisions 149	
9.2.3	Electron Impact Cross Section 150	
9.3	Electron Beam Spectroscopies 151	
9.4	Auger Electron Spectroscopy 153	
9.4.1	Auger Transition Probability 153	
9.4.2	Auger versus X-ray Yields 154	
9.4.3	Auger Excitation Process 156	

9.4.4	Auger Electron Energies	*158*
9.4.5	Quantitative Elemental Identification	*160*
9.5	Auger Depth Profiling	*163*
9.5	Summary	*165*
	Problems	*167*
	References	*168*

10 Electron Energy Loss Spectroscopy *169*

10.1	Overview	*169*
10.2	Dielectric Response Theory	*171*
10.3	Surface Phonon Scattering	*172*
10.4	Bulk and Surface Plasmon Scattering	*174*
10.5	Interface Electronic Transitions	*177*
10.6	Atomic-Scale Electron Energy Loss Spectroscopy	*180*
10.7	Summary	*181*
	References	*182*

11 Rutherford Backscattering Spectrometry *183*

11.1	Overview	*183*
11.2	Theory of Rutherford Backscattering	*184*
11.3	Depth Profiling	*187*
11.4	Channeling and Blocking	*190*
11.5	Interface Studies	*192*
11.6	Summary	*195*
	Problems	*195*
	References	*195*

12 Secondary Ion Mass Spectrometry *197*

12.1	Overview	*197*
12.2	Principles	*197*
12.3	SIMS Equipment	*199*
12.4	Secondary Ion Yields	*203*
12.5	Imaging	*206*
12.6	Dynamic SIMS	*207*
12.7	Organic and Biological Species	*211*
12.8	Summary	*211*
	Problems	*212*
	References	*212*

13 Electron Diffraction *213*

13.1	Overview	*213*
13.2	Principles of Low-Energy Electron Diffraction	*213*
13.3	LEED Equipment	*215*
13.4	LEED Kinematics	*216*
13.5	Surface Reconstruction	*217*

13.6	Surface Lattices and Superstructures	219
13.7	Silicon Reconstructions	221
13.8	III–V Compound Semiconductor Reconstructions	223
13.9	Reflection High-Energy Electron Diffraction	227
13.8.1	RHEED Oscillations	232
13.9	Summary	233
	Problems	234
	References	234
14	**Scanning Tunneling Microscopy**	237
14.1	Overview	237
14.2	Tunneling Theory	239
14.3	Surface Structure	244
14.4	Atomic Force Microscopy	246
14.5	Ballistic Electron Emission Microscopy	249
14.6	Atomic Positioning	252
14.7	Summary	253
	Problems	254
	References	254
15	**Optical Spectroscopies**	257
15.1	Overview	257
15.2	Optical Absorption	257
15.3	Modulation Techniques	260
15.4	Multiple Surface Interaction Techniques	262
15.5	Spectroscopic Ellipsometry	263
15.6	Surface-Enhanced Raman Spectroscopy	264
15.7	Surface Photoconductivity	267
15.8	Surface Photovoltage Spectroscopy	268
15.8.1	Theory of Surface Photovoltage Spectroscopy	268
15.8.2	Surface Photovoltage Spectroscopy Equipment	270
15.8.3	Surface Photovoltage Spectra and Photovoltage Transients	271
15.8.4	Surface Photovoltage Spectroscopy of Metal-Induced Surface States	274
15.9	Summary	276
	Problems	277
	References	277
	Further Reading	278
16	**Cathodoluminescence Spectroscopy**	279
16.1	Overview	279
16.2	Theory	281
16.2.1	Scattering Cross Section	281
16.2.2	Stopping Power	282
16.2.3	Plasmon Energy Loss	283

16.2.4	Electron Scattering Length	285
16.2.5	Semiconductor Ionization Energies	286
16.2.6	Universal Range-Energy Relation	288
16.3	Monte Carlo Simulations	291
16.4	Depth-Resolved Cathodoluminescence Spectroscopy	293
16.4.1	Surface Electronic States	295
16.4.2	Interface Electronic States	295
16.4.3	Localized CLS in Three Dimensions	297
16.4.3.1	Wafer-Scale Analysis of 2-DEG Layers	298
16.4.3.2	Schottky Barriers	300
16.4.3.3	Electronic Devices	301
16.5	Summary	302
	Problems	303
	References	304
	Further Reading	304

17 Electronic Materials' Surfaces 305

17.1	Overview	305
17.2	Geometric Structure	305
17.2.1	Surface Relaxation and Reconstruction	305
17.2.2	Extended Geometric Structures	306
17.2.2.1	Domains	306
17.2.2.2	Steps	306
17.2.2.3	Defects	310
17.3	Chemical Structure	311
17.3.1	Crystal Growth	311
17.3.1.1	Bulk Crystal Growth	311
17.3.1.2	Epitaxial Layer Crystal Growth	314
17.3.2	Kinetics of Growth, Diffusion, and Evaporation	317
17.4	Etching	318
17.4.1	Etch Processes	318
17.4.2	Wet Chemical Etching	318
17.4.3	Orientation Effects on Wet Chemical Etching	319
17.4.4	Impurities, Doping, and Light	321
17.5	Electronic Implications	323
17.6	Summary	323
	Problems	324
	References	324
	Additional Reading	326

18 Adsorbates on Electronic Materials' Surfaces 327

18.1	Overview	327
18.2	Geometric Structure	327
18.2.1	Site Specificity	328
18.2.2	Metal Adsorbates on Si	329

18.2.3	Metal Adsorbates on GaAs	*329*
18.2.4	Epitaxical Overlayers	*331*
18.2.4.1	Elemental Metal Overlayers	*331*
18.2.4.2	Metal Silicides on Si	*331*
18.2.4.3	Metal Epitaxy on Compound Semiconductors	*333*
18.2.4.4	Epitaxical Metal–Semiconductor Applications	*335*
18.3	Chemical Properties	*336*
18.3.1	Metal Overlayers on Semiconductors	*336*
18.3.1.1	Overlayer Growth Modes	*337*
18.3.1.2	Thermodynamic Factors	*337*
18.3.1.3	Nonequilibrium Energy Processing	*338*
18.3.2	Macroscopic Interface Reaction Kinetics	*339*
18.3.2.1	Silicide Phase Formation	*339*
18.3.2.2	Thin Film versus Bulk Diffusion	*341*
18.3.2.3	Mechanical and Morphological Effects	*343*
18.3.2.4	Diffusion Barriers	*343*
18.3.2.5	Atomic-Scale Metal-Si Reactions	*345*
18.3.3	Compound Semiconductor Reactions	*345*
18.4	Electronic Properties	*346*
18.4.1	Physisorption	*346*
18.4.2	Chemisorption	*347*
18.4.3	Work Function Effects	*350*
18.4.3.1	Charge Transfer	*350*
18.4.3.2	Dipole Formation	*353*
18.4.3.3	Ordered Adsorption	*354*
18.4.3.4	Negative Electron Affinity	*355*
18.4.3.5	Reconstruction Changes	*355*
18.5	Summary	*356*
	Problems	*359*
	References	*360*
	Additional Reading	*363*
19	**Adsorbate–Semiconductor Sensors**	*365*
19.1	Adsorbate–Surface Charge Transfer	*365*
19.1.1	Band-Bending Effects	*365*
19.1.2	Surface Conductance	*366*
19.1.3	Self-limiting Charge Transfer	*367*
19.1.4	Transient Effects	*369*
19.1.5	Orientational Dependence	*369*
19.2	Sensors	*370*
19.2.1	Sensor Operating Principles	*370*
19.2.2	Oxide Gas Sensors	*371*
19.2.3	Granular Gas Sensors	*374*
19.2.4	Chemical and Biosensors	*375*
19.2.4.1	Sensor Selectivity	*376*

19.2.4.2	Sensor Sensitivity 377
19.2.5	Other Transducers 377
19.2.6	Electronic Materials for Sensors 379
19.3	Summary 379
	Problems 380
	References 381

20 Semiconductor Heterojunctions 383
20.1	Overview 383
20.2	Geometric Structure 383
20.2.1	Epitaxial Growth 383
20.2.2	Lattice Matching 384
20.2.2.1	Lattice Match and Alloy Composition 384
20.2.2.2	Lattice-Mismatched Interfaces 386
20.2.2.3	Dislocations and Strain 389
20.2.3	Two-Dimensional Electron Gas Heterojunctions 393
20.2.4	Strained Layer Superlattices 394
20.2.4.1	Superlattice Energy Bands 394
20.2.4.2	Strain-Induced Polarization Fields 396
20.3	Chemical Structure 397
20.3.1	Interdiffusion 397
20.3.1.1	IV–IV Interfaces 397
20.3.1.2	III–V Compound Heterojunctions 398
20.3.2	Chemical Reactions 399
20.3.3	Template Structures 399
20.3.3.1	Bridge Layers 399
20.3.3.2	Monolayer Passivation 400
20.3.3.3	Crystal Orientations 400
20.3.3.4	Monolayer Surfactants 401
20.3.3.5	Dipole Control Structures 401
20.4	Electronic Structure 402
20.4.1	Heterojunction Band Offsets 402
20.4.2	Band Offset Characterization 405
20.4.2.1	Macroscopic Electrical and Optical Methods 405
20.4.2.2	Scanned Probe Techniques 408
20.4.2.3	Photoemission Spectroscopy Techniques 410
20.4.2.4	Band Offset Results 412
20.4.3	Interface Dipoles 412
20.4.3.1	Inorganic Semiconductors 412
20.4.3.2	Organic Semiconductors 418
20.4.4	Theories of Heterojunction Band Offsets 419
20.4.4.1	Charge Neutrality Levels 420
20.4.4.2	Local Bond Approaches 421
20.4.4.3	Empirical Deep-Level Schemes 422
20.4.5	Assessment of Theory Approaches 423

20.4.6	Interface Contributions to Band Offsets	423
20.4.6.1	Growth Sequence	424
20.4.6.2	Crystallographic Orientation	426
20.4.6.3	Surface Reconstruction: Band Bending versus Offsets	427
20.4.6.4	Surface Reconstruction: Interface Bonding	428
20.4.7	Theoretical Methods in Band Offset Engineering	429
20.4.7.1	First-Principles Calculations	429
20.4.7.2	Mathematical Approach	429
20.4.7.3	Alternative Methods	432
20.4.8	Application to Heterovalent Interfaces	432
20.4.8.1	Polarity Dependence	432
20.4.8.2	Interface Atomic Mixing	433
20.4.8.3	Atomic Interlayers	434
20.4.9	Practical Band Offset Engineering	435
20.4.9.1	Spatially-Confined Nonstoichiometry	436
20.4.9.2	Chemical Stability, Cross-Doping, and Interface States	437
20.4.9.3	"Delta" Doping	438
20.5	Summary	439
	Problems	440
	References	441
	Further Reading	445
21	**Metals on Semiconductors**	**447**
21.1	Overview	447
21.2	Metal–Semiconductor Interface Dipoles	448
21.3	Interface States	449
21.3.1	Localized States	449
21.3.2	Wavefunction Tailing (Metal-Induced Gap States)	450
21.3.3	Charge Transfer, Electronegativity, and Defects	453
21.3.4	Additional Intrinsic Pinning Mechanisms	453
21.3.5	Extrinsic States	453
21.3.5.1	Surface Imperfections and Contaminants	455
21.3.5.2	Bulk States	456
21.3.6	Interface-Specific Extrinsic States	458
21.3.6.1	Interface Reaction and Diffusion	458
21.3.6.2	Atomic Structural and Geometric Effects	460
21.3.6.3	Chemisorption-Induced Effects	461
21.3.6.4	Interface Chemical Phases	462
21.3.6.5	Organic Semiconductor–Metal Dipoles	463
21.4	Self-Consistent Electrostatic Calculations	467
21.5	Fermi-Level Pinning Models	471
21.6	Experimental Schottky Barriers	471
21.6.1	Metals on Si and Ge	472
21.6.1.1	Clean Surfaces	472
21.6.1.2	Etched and Oxidized Surfaces	473

21.6.1.3	N versus P-type Barriers	*474*
21.6.1.4	Si, Ge Summary	*475*
21.6.2	Metals on III–V Compound Semiconductors	*475*
21.6.2.1	GaAs(110) Pinned Schottky Barriers	*476*
21.6.3	InP(110) Unpinned Schottky Barriers	*479*
21.6.3.1	Clean Surfaces	*479*
21.6.3.2	Macroscopic Measurements	*479*
21.6.4	GaN Schottky Barriers	*480*
21.6.4.1	Other Binary III–V Semiconductors	*481*
21.6.5	Ternary III–V Semiconductors	*482*
21.6.6	Metals on II–VI Compound Semiconductors	*484*
21.6.6.1	Sulfides, Selenides, and Tellurides	*485*
21.6.6.2	ZnO: Dependence on Metals	*485*
21.6.6.3	ZnO: Dependence on Native Point Defects	*485*
21.6.6.4	ZnO: Dependence on Polarity	*490*
21.6.7	Metals on IV–IV, IV–VI, and III–VI Compound Semiconductors	*490*
21.6.8	Compound Semiconductor Summary	*491*
21.7	Interface Passivation and Control	*492*
21.7.1	Macroscopic Methods of Contact Formation	*492*
21.7.2	Processing Contacts	*493*
21.7.2.1	Elemental Metals on GaAs	*493*
21.7.2.2	Metal Multilayers on GaAs	*501*
21.7.2.3	Useful Metallizations for III–V Compound Ohmic Contacts	*502*
21.7.3	Atomic-Scale Control	*502*
21.7.3.1	Reactive Metal Interlayers	*502*
21.7.3.2	Less-Reactive Buffer Layers	*506*
21.7.3.3	Semiconductor Interlayers	*506*
21.7.4	Wet-Chemical Treatments	*508*
21.7.4.1	Photochemical Washing	*508*
21.7.4.2	Inorganic Sulfides	*509*
21.7.4.3	Thermal Oxides and Hydrogen	*509*
21.7.5	Semiconductor Crystal Growth	*510*
21.7.5.1	Variations in Stoichiometry	*510*
21.7.5.2	Misorientation/Vicinal Surfaces	*512*
21.7.5.3	Epitaxical Growth of Binary Alloys on Compound Semiconductors	*513*
21.8	Summary	*514*
	Problems	*514*
	References	*516*
	Further Reading	*522*

22 The Future of Interfaces *523*
22.1 Current Status *523*
22.2 Current Device Applications and Challenges *525*

22.3	New Directions	*528*
22.3.1	High-K Dielectrics	*529*
22.3.2	Complex Oxides	*530*
22.3.3	Spintronics	*532*
22.3.4	Nanoscale Circuits	*534*
22.3.5	Quantum-Scale Interfaces	*534*
22.4	Synopsis	*536*
	References	*537*

Appendices *539*

Appendix 1: Glossary of Commonly Used Symbols *541*

Appendix 2: Table of Acronyms *544*

Appendix 3: Table of Physical Constants and Conversion Factors *548*

Appendix 4: Semiconductor Properties *549*

Appendix 5: Table of Preferred Work Functions *551*

Appendix 6: Derivation of Fermi's Golden Rule *552*

Appendix 7: Derivation of Photoemission Cross Section for a Square Well *555*

Index *557*

Preface

This textbook is intended for students as well as professional scientists and engineers interested in the next generation of electronics and, in particular, in the opportunities and challenges introduced by surfaces and interfaces. As electronics technology improves with higher speed, higher sensitivity, higher power, and higher functionality, surfaces and interfaces are becoming more important than ever. To achieve higher performance at the macroscopic level, one requires even more refined control of these junctions at the microscopic and, in fact, the atomic scale. With each advance, new techniques have been developed to measure and alter physical properties with increasing refinement. In turn, these studies have revealed fundamental phenomena that have stimulated designs for new device applications. This synergy between characterization, processing, and design spans several academic disciplines including physics, chemistry, materials science, and electrical engineering.

Several excellent physics-based books are available that provide extensive mathematical analyses focused on specific effects that are also described here. However, the field of electronic surfaces and interfaces encompasses a wide range of chemical and materials science phenomena that impact electronic properties. Rather than follow advanced treatments of specific effects, this book describes at an intermediate level the full range of physical phenomena at surfaces and interfaces, the variety of techniques available to measure them, and the physical issues to be addressed in order to advance electronics to the next level of performance. The author hopes to convey the excitement of this field and the intellectual challenges ahead. He also wishes to thank many of his colleagues for paving the way for this book with their valuable discoveries and insights. Particular thanks are due to Prof. Eli Burstein, who introduced him to the physics of metal–semiconductor interfaces, Dr Charles B. Duke, whose theoretical studies of semiconductor surface structure and tunneling provided a framework on which the experimental program was built, and Prof. Giorgio Margaritondo, who helped launch his soft X-ray photoemission spectroscopy work on interfaces and introduced him to the international world of synchrotron radiation science. Finally, his deepest gratitude goes to his wife, Janice, for her patience, understanding, love, and support during the year in which this book was written.

1
Introduction

1.1
Surface and Interfaces in Everyday Life

Surfaces and interfaces are all around us. Their properties are important in our daily lives and are basic to many of today's advanced technologies. This is particularly true for the semiconductor materials that are used throughout modern electronics. The aim of this book is to present the physical principles underlying the electronic, chemical, and structural properties of semiconductor interfaces and the techniques available to characterize them. Surfaces and interfaces are a cross-disciplinary field of science and engineering. As such, this book emphasizes the principles common to physics, electrical engineering, materials science, and chemistry as well as the links between fundamental and practical issues.

Surfaces and interfaces play a central role in numerous everyday phenomena. These include (i) *triboelectricity*, the transfer of charge between two materials brought into contact – such as the static electricity built up on a comb after combing one's hair; (ii) *corrosion*, the oxidation of structural materials used in, for example, buildings, bridges, and aircraft; (iii) *passivation*, the prevention of such chemical or biological processes using special protective layers; (iv) *colloid chemistry*, the wetting of surfaces and the dispersion of particles within fluids as emulsions or colloids, for example, paints and time-release capsule medicines; (v) *tribology*, the friction between sliding objects in contact and their interface lubrication; (vi) *cleaning and chemical etching*, the removal of surface layers or adsorbed species; (vii) *catalysis*, the reduction in energetic barriers to speed up or improve the yield of chemical reactions, for example, refining oil or burning coal; and (viii) *optical interference*, the rainbow of colors reflected off thin oil layers or the internal reflection of light between stacks of materials only a few wavelengths of light thick. On a much larger scale are (ix) *electromagnetic* interfaces between the earth's atmospheric layers that bounce short-wave radio signals around the world and that alter the reflection or absorption of sunlight contributing to global warming.

Surfaces and Interfaces of Electronic Materials. Leonard J. Brillson
Copyright © 2010 WILEY-VCH Verlag GmbH & Co. KGaA, Weinheim
ISBN: 978-3-527-40915-0

1.2
Surfaces and Interfaces in Electronics Technology

Surfaces and interfaces are fundamental to microelectronics. One of the most important microelectronic devices is the transistor, all functions of which depend on the boundaries between electronic materials. Figure 1.1 illustrates the three aspects of this dependence. Here, current passes from a source metal to a drain metal through a semiconductor, in this case, silicon (Si). A gate metal between the source and the drain is used to apply voltages that attract or repel the charge carriers involved in the current flow. The result is control or "gating" of the current flow by this third electrode. This basic device element is at the heart of the microelectronics industry.

The surfaces and interfaces are the key to the transistor's operation, shown in Figure 1.1. Thus, the contact between the metal and Si is a metal silicide. Barriers can form between metals and semiconductors that impede charge movement and introduce voltage drops across their interfaces. This barrier formation is a central topic of this book. Microelectronics researchers found that promoting a chemical reaction to form silicides, such as $TiSi_2$ between Ti and Si, reduces such transport barriers and the contact resistivity ρ_c at these metal–semiconductor interfaces. This is illustrated, for example, in Figure 1.2a. Such interfacial silicide layers form low resistance, planar interfaces that can be integrated into the manufacturing process. A challenge of this approach is to achieve very

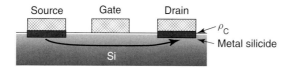

Figure 1.1 Source–gate–drain structure of a silicon transistor.

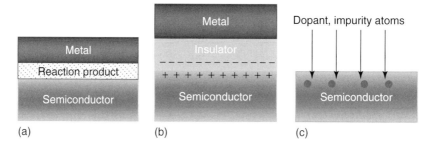

Figure 1.2 Expanded view of a (a) interface between metal and semiconductor with reacted layer, (b) gate–semiconductor interface with trapped charge in insulator and at insulator–semiconductor junction, and (c) dopant or impurity atom diffusion into semiconductor.

thin, low ρ_c contacts without allowing reactions to extend far away from the junction.

The second important interface appears at the gate–semiconductor junction, shown in Figure 1.2b. Here, the earliest transistor experiments [1] showed the presence of fixed charges at this interface that prevented control of the source–drain current. This gate interface may involve a metal in direct contact with the semiconductor or, more commonly, a stack of metal-on-insulator-on semiconductor to apply voltage bias without introducing additional current. Atomic sites within the insulator and its semiconductor interface can immobilize charge and introduce dipoles across the insulator–semiconductor interface. This localized charge produces a voltage drop that offsets applied voltages at the gate metal, opposing the gate's control of the source–drain current flow. Minimizing the formation of these localized charge sites has been one of the prime goals of the microelectronics industry since the invention of the transistor.

The third important microelectronic interface involves diffusion of atoms into and out of the semiconductor. Atomic diffusion of atoms into the semiconductor that donate or accept charge is used to control the concentration of free charge carriers within specific regions of a device. Acceleration and implantation of ionized atoms is a common process to achieve such doped layers that extend into semiconductor surfaces, here illustrated in Figure 1.2c. In addition, atomic diffusion can occur between two materials in contact that are annealed at high temperature. High-temperature annealing is often used to heal lattice damage after implantation or to promote reactions at particular device locations. However, such annealing can introduce diffusion and unintentional doping at other regions of the device. Outdiffusion of semiconductor constituents is also possible, resulting in native point defects that can also be electrically active. Balancing these effects requires careful design of materials, surface and interface preparation, thermal treatment, and device architectures.

Microelectronic circuits consist of many interfaces between semiconductors, oxides, and metals. Figure 1.3 illustrates how these interfaces form as silicon progresses from its melt-grown crystal boule to a packaged chip. The Si boule formed by pulling the crystal out of a molten bath is sectioned into wafers, which are then oxidized, diffused, or implanted with dopants, and overcoated with various metal and organic layers. Photolithography is used to pattern and etch these wafers into monolithic arrays of devices. The wafer is then diced into individual circuits that are then mounted, wire bonded, and packaged into chips.

Within each circuit element, there can be many layers of interconnected conductors, insulators, and their interfaces. Figure 1.4 illustrates the different materials and interfaces associated with a 0.18-μm transistor at the bottom of a multilayer Al–W–Si-oxide dielectric assembly [2]. Reaction, interdiffusion, and formation of localized states must all be carefully controlled at all of these interfaces during the many patterning, etching, and annealing steps involved in assembling the full structure. Figure 1.4 also shows that materials and geometries change to compensate for the otherwise increasing electrical resistance as interconnects between layers shrink into the nanoscale regime. This continuing evolution in microelectronics

1 Introduction

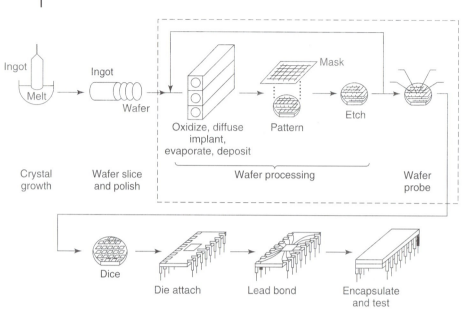

Figure 1.3 Integrated circuit manufacturing process flow.

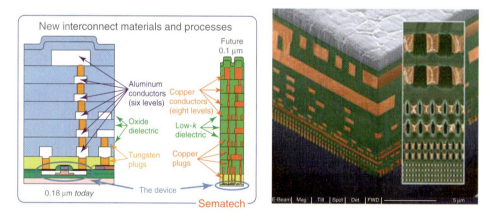

Figure 1.4 Multilayer, multimaterial interconnect architectures at the nanoscale. Feature size of interconnects at right is 45 nm [2].

underscores the importance of interfaces since the material volume associated with these interfaces becomes a larger proportion of the entire structure as circuit sizes decrease.

Many other conventional electronic devices rely on interfaces for their operation. Figure 1.5a illustrates the interface between a metal and semiconductor within a solar cell schematically as an energy E versus distance x band diagram. The Fermi

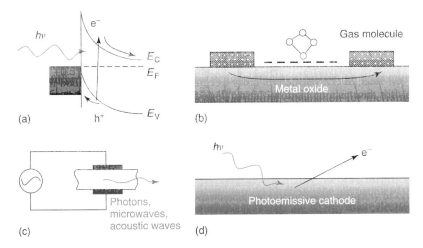

Figure 1.5 Interfaces in conventional electronics:
(a) solar cell, (b) gas sensor, (c) optoelectronic emitter, and
(d) photoemissive cathode.

levels E_F in the metal (solid line) and the semiconductor (dashed line) align at a constant energy, whereas the conduction band E_C and the valence band E_V in the semiconductor bend near the interface. Incident photons of energy $h\nu$ at this interface create electrons and holes that separate under the field set up by these bent bands. This charge separation results in photoinduced current or voltage between the metal and the semiconductor.

Figure 1.5b shows what appears to be a transistor structure except that, unlike Figure 1.1, there is no gate. Instead, molecules on this otherwise free surface adsorb on the surface, exchanging charge and inducing a field analogous to that of a gate. Figure 1.5c illustrates a circuit that generates photon, microwaves, or acoustic waves. The contacts that inject current or apply voltage to the generator layer are key to its practical operation. Unless the resistance of such contacts is low, power is lost at these contacts, reducing or totally blocking power conversion inside the semiconductor. Figure 1.5d illustrates an interface involving just a semiconductor surface that emits electrons when excited by incident photons. Chemical treatment of selected semiconductors enables these surfaces to emit multiple electrons when struck by single photons. Such surfaces are useful as electron pulse generators or photomultipliers.

Surfaces and interfaces have an even larger impact on electronics as devices move into the quantum regime. Figure 1.6 illustrates four such quantum electronic devices schematically. Figure 1.6a illustrates the energy band diagram of a quantum well, one of the basic components of optoelectronics. Here, the decrease in bandgap between E_C and E_V of one semiconductor sandwiched between layers of a larger bandgap semiconductor localizes both electrons and holes in the smaller gap material. This joint localization enhances electron–hole pair recombination and light emission. The quantum well is typically only a few atomic layers thick so

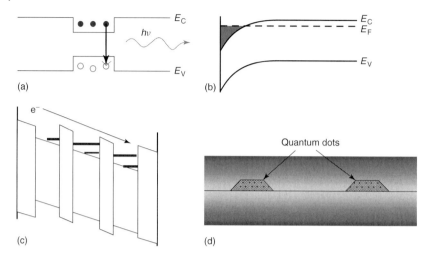

Figure 1.6 Four quantum electronic devices. (a) Charge localization, recombination, and photon emission at a quantum well; (b) carrier confinement and transport at a semiconductor inversion layer; (c) tunneling transport at an avalanche detector; and (d) carrier confinement in three dimensions at quantum dots.

that the allowed energies of electrons and holes inside the well are quantized at discrete energies. This promotes efficient carrier inversion and laser light emission. Imperfections at the interfaces of these quantum wells can reduce the quantized light emission by introducing competing channels for recombination that do not involve the quantized states in the well.

Figure 1.6b shows a schematic energy band diagram of a semiconductor with bands that bend down at the surface, allowing E_F to rise above E_C. The high concentration of electrons is confined to within a few tens of nanometers or less at the surface. This phenomenon is termed a *two-dimensional electron gas* (2DEG) *layer*. It forms a high carrier concentration, high mobility channel at the surface that is used for high-frequency, high-power devices, often termed *high electron mobility transistors* (HEMTs). Again, imperfections at the semiconductor interface can produce local electric fields that scatter charges, reduce mobility, and alter or even remove the 2DEG region.

Figure 1.6c illustrates a structure consisting of alternating high- and low-bandgap semiconductors characteristic of a *cascade laser* or of a very high frequency transistor. In either case, charge must tunnel through the ultrathin (monolayers) *"barrier"* layers into quantized energy levels. Once again, the perfection of these interfaces is crucial for the charge to tunnel efficiently between layers. Finally, Figure 1.6d represents a real-space pair of quantum dots encapsulated by other media. Such quantum dots with sizes of only a few nanometers also have quantized energy levels that yield efficient laser emission. Again, optical emission is impacted if imperfections and recombination are present at their interfaces.

The materials that comprise all electronics consist of metals, semiconductors, and/or insulators. It is instructive to realize that these three materials differ primarily in terms of the energy separation between their filled and empty electronic states. For metals, this *bandgap* is 0. For insulators, it can be quite large, typically 5 eV or more. Semiconductors lie between these two regimes with intermediate bandgaps ranging from a few hundred meV or milli-electron volts to several electron volts.

These bandgaps and Fermi levels in the energy band diagrams of Figures 1.5 and 1.6 point to another key aspect of interfaces – the alignment of energy levels between the constituents. This chapter emphasizes that contacts and charge transport between metals, semiconductor, and insulators are essential to all electronic applications. How their energy levels align is a fundamental issue that is still not well understood. The question is, does this matter? The interface band structure determines how much energy difference exists between materials and thereby what barriers exist to charge movement between them. The question is, what affects this phenomenon? There are three primary factors: (i) the constituents of the junctions, (ii) the conditions under which they form the interface, and (iii) any subsequent thermal or chemical processing. Therefore, to understand how surfaces and interfaces impact electronics, it is important to know their properties at the microscopic level and how these factors shape these properties.

Surface and interface science continues to have enormous practical value for electronics. It has helped develop the semiconductor industry into the high-performance, high-value-added industry that it is today. Surface and interface tools are essential for monitoring, controlling, and ultimately designing micro- and optoelectronic clean room processes. They are also central to controlling properties of contacts. As such, they are integral to designing the next generation of electronic device materials and fabrication processes.

Problems

1. As electronics shrinks into the nanoscale regime, the actual numbers of atoms become significant in determining the semiconductor's physical properties. Consider a 0.1-μm Si field effect transistor with a 0.1 μm × 0.1 μm cross section and a doping concentration of 10^{17} cm^{-3}. How many dopant atoms are there in the channel region? How many dopant atoms are there altogether?
2. Assume the top channel surface has 0.01 trapped electron per unit cell. How many surface charges are present? How much do they affect the channel's bulk charge density?
3. What fraction of atoms is within one lattice constant of the channel surfaces?

4. Figure 1.4 illustrates the complexity of the transistor chip structure. Describe three interface effects that could occur in these structures that could degrade electrical properties during microfabrication.
5. Name three desirable solar cell properties that could be affected by interface effects.
6. Name three interface effects that could degrade quantum well operation.

References

1. Bardeen, J. (1947) *Phys. Rev.*, **71**, 717.
2. Lammers, D. (2007) Semiconductor International, *www.semiconductor.net/article/CA6513618.html* (accessed 18 December 2007).

2
Historical Background

2.1
Contact Electrification and the Development of Solid-State Concepts

The history of electrical contacts extends far back in time past the Greco–Roman era. Contact electrification or, more precisely, triboelectricity was well known in ancient times in the form of static electricity. Thus, the rubbing of cat's fur with amber was an early example of charge transfer between solids. However, systematic observation of this electrical phenomenon began much later when, in 1874, Braun used selenium to form a rectifier, which conducted electricity unequally with positive versus negative applied voltage [1]. See Henisch's *Rectifying Semiconductor Contacts* [2] for a detailed review of early literature on semiconductor interface phenomena.

Three decades later, Einstein explained the *photoelectric effect* at metal surfaces using wave packets with discrete particle energies [3]. This established not only the particle nature of light but also the concept of work function, one of the building blocks of energy band theory. Notably, his work on the photoelectric effect, rather than his many other achievements, was the basis for Einstein's 1921 Nobel Prize. In 1931, Wilson presented a theoretical foundation for semiconductor phenomena which was based on the band theory of solids [4, 5]. Siemens *et al.* [6] and Schottky *et al.* [7] subsequently showed that the semiconductor interior was not a factor in contact rectification. In 1939, Mott [8], Davidov [9, 10], and Schottky [11–13] all published theories of rectification. Each of their models involved a band-bending region near the interface. Mott proposed an insulating region between the metal and the semiconductor that accounted for current–voltage measurements of copper oxide rectifiers. Davidov advanced the importance of thermionic work function differences in forming the band-bending region. Schottky introduced the idea that the band-bending region could arise from stable space charge in the semiconductor alone rather than requiring the presence of a chemically distinct interfacial layer. Bethe's theory of thermionic emission of carriers over the energy barrier [14], based in part on Richardson's earlier work on thermionic cathodes [15], provided a description of charge transport across the band-bending region that described the rectification process for most experimental situations. These developments laid the groundwork for describing semiconductor charge transfer and barrier formation.

Surfaces and Interfaces of Electronic Materials. Leonard J. Brillson
Copyright © 2010 WILEY-VCH Verlag GmbH & Co. KGaA, Weinheim
ISBN: 978-3-527-40915-0

See [2,16–20] for more extensive reviews of metal–semiconductor rectification, with particular emphasis on current transport and tunneling.

2.2
High-Purity Semiconductor Crystals

A major roadblock to experimental studies was the need for semiconductors with high purity. Without sufficiently high crystal perfection and purity, the intrinsic semiconductor properties were masked by the effects of impurities and lattice imperfections. By the 1940s, large high-purity semiconductor single crystals had become available using crystal-pulling and zone-refining techniques [21–26]. Two such crystal-growth techniques – Czochralski and float zone – are illustrated in Figure 2.1.

2.3
Development of the Transistor

By the mid-1940s, companies such as Bell Telephone had begun efforts aimed at producing amplifiers based on semiconductors rather than on vacuum tubes. A key event in this industrial research occurred when Bardeen, Shockley, and Pearson discovered that electric charges immobilized on the semiconductor surface were interfering with the gate modulation of the charge and current inside the semiconductor. Bardeen correctly interpreted this observation in terms of electronic

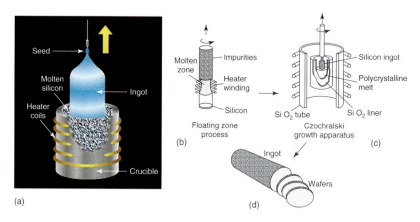

Figure 2.1 Pulling of an Si crystal from the melt (the Czochralski method): (a) silicon crystal being pulled with a seed crystal from molten silicon [25]. (b) Float-zone process with impurity segregation away from the molten zone. (c) Cutout of Czochralski crucible interior, (d) Ingot sectioning into wafers [26].

 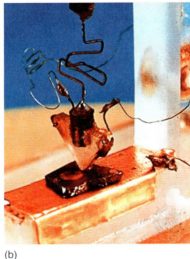

Figure 2.2 (a) Bardeen (center), Brattain (right), and Shockley (left) [28]. (b) A point contact transistor with three gold contacts on germanium crystal [29].

states that were localized at the semiconductor interface [27]. By reducing the density of these states and their effect on the field effect pictured in Figure 2.2, Bardeen, *et al.* [28] were able to demonstrate the *transistor* action [29]. For this discovery, they received the 1956 Nobel Prize in Physics. And the interface concepts inherent in the transistor continue to drive the electronics revolution.

By the early 1950s, the semiconductor employed for transistors changed from germanium to silicon, due in large part to the latter's native oxide, the dielectric separating a gate metal from a semiconductor, being more resistant to water vapor in air [22]. By 1958, Jack Kilby [30] had invented the first integrated circuits by creating arrays of transistor circuits on silicon wafers. The 1950s and early 1960s saw the development of many types of semiconductor devices such as photodiodes, sensors, photomultipliers, and light emitters. Their reliability and usefulness required not only high crystal purity but also refined methods to form their electrical contacts. This requirement led to the development of high-vacuum technology – pumps, chambers, and gauges to be discussed in following chapters.

By the mid-1960s, there was significant interest in understanding the variation of band bending and Schottky barrier formation at metal–semiconductor interfaces. The pioneering work of Mead and Spitzer resulted in barrier measurements by several techniques for metals on semiconductors. These contacts were formed in medium vacuum, $\sim 10^{-6}$ Torr and, to minimize contamination, they cleaved the semiconductor crystals in streams of evaporating metal atoms [31]. Mead and Spitzer interpreted the systematic differences they observed across classes of semiconductors in terms of different bonding between ionic versus covalent crystal lattices.

2.4
The Surface Science Era

The advent of ultrahigh vacuum technology (UHV) with chamber pressures of 10^{-10} Torr or less in the late 1960s marked the birth of *surface science* as we know it today. Researchers using surface science techniques began measuring the surface properties of semiconductors in the early 1970s. They interpreted the localized states of Bardeen [27], Tamm [32], and Shockley [33] as intrinsic surface states, that is, localized states intrinsic to the semiconductor alone and related to the bonding at the outer layer of semiconductor atoms. Kurtin *et al.* [34] pointed out a pronounced transition between ionic and covalent semiconductors that they interpreted by differences in lattice disruption between classes of compound semiconductors. Theoretical calculations of such disrupted lattices supported the appearance of localized states with energies in the semiconductor bandgap. Such gap states could, in principle, "pin" the Fermi level in a narrow range of similar energy positions. On the other hand, the examination of the atomic arrangements on semiconductor surfaces revealed that the atoms typically rearranged or reconstructed to minimize their bond energy. In general, this reconstruction moves the energies of such states out of the bandgap, where they have no effect on Fermi-level stabilization inside the gap. Early surface science experiments that had reported observing such states with energies inside semiconductor bandgaps were later reinterpreted in terms of surface artifacts. Indeed, these events highlighted the importance of surface preparation and the role of factors that were extrinsic to the semiconductor itself.

Another intrinsic state model invoked wavefunction tunneling from metals into semiconductor bandgaps. Theoretical methods based on this concept also introduced localized charge states at the metal–semiconductor interface [35–38] and could also account for differences between barriers of semiconductors with different bond ionicity [39].

By the mid-1970s, the refinement of monolithic electronic circuitry on silicon wafers had involved the material science or microelectronic metallurgy of the various electronic material interfaces. A key aspect of this metallurgy involved understanding and predicting the interfacial reactions at silicon–metal interfaces [40, 41].

Researchers involved in manufacturing microelectronics realized that small variations in interface chemical composition could introduce significant variability in electronic device properties. At this stage of microelectronic development, highly reproducible interface conditions, rather than atomically pristine ones, were paramount in satisfying manufacturing tolerances.

During this period, Andrews and Phillips found that chemical reactions at metal–silicon interfaces displayed systematic variations with Schottky barrier heights [42, 43]. Brillson demonstrated that chemical reaction and diffusion occurred on an atomic scale at interfaces between metals and compound semiconductors [44–46]. Detection of this interface chemistry became possible with the development of UHV surface science techniques that were sensitive to properties of the outer few monolayers. The systematics of Schottky barrier behavior with this interface chemistry led to the concept of a thermodynamic heat of interface reaction that could account for the qualitative transition in Schottky barriers between "reactive" and "unreactive" interfaces. This concept could be extended to interfacial layers of only a few atomic layers. Such "interlayers" could, in fact, alter the interface chemistry and, thereby, the barrier formation of a macroscopic metal–semiconductor contact [47–49].

Spicer et al. introduced another widely discussed concept of Schottky barriers, namely, the formation of defects at the semiconductor surface by the adsorption of metal or other adsorbate atoms [50, 51]. This model accounted for the nearly complete insensitivity of GaAs Schottky barriers to different metals. Technologically such effects were of high interest since GaAs and other III–V compound semiconductors possessed higher mobility than Si and were, therefore, more suitable for high-frequency applications. However, by the late 1980s, many researchers had shown that a wide range of Schottky barriers could be achieved with GaAs and other compound semiconductors [52–55], with high-quality crystalline materials and careful interface preparation.

The 1980s also saw the invention and development of the *scanning tunneling microscope* (STM), which permits the observation of atomic distributions on surfaces. Extensions of this instrument would permit local measurements of work function, capacitance, and barrier heights on a scale of nanometers. By the 1990s, one of these techniques, *atomic force microscopy* (AFM), was being used to probe insulating materials as well as semiconductors. Another STM extension, *ballistic electron energy microscopy* (BEEM), revealed heterogeneous electronic features across semiconductor surfaces that could account for multiple barrier heights within the same macroscopic contact.

These Schottky barrier developments also provided new insights into the heterojunction band offsets that are critical to micro- and optoelectronic device operation. As with metal–semiconductor contacts, researchers realized the importance of interfacial bonding and growth techniques in controlling band alignment between the two semiconductors.

2.5
Advances in Crystal Growth Techniques

Concurrent with the development of new surface science, techniques and insights were advances in crystal growth techniques. Besides the crystal pulling and float-zone methods illustrated in Figure 2.2, other techniques pictured in Figure 2.3 include horizontal Bridgman from a melt in a crucible (a and b), *liquid phase epitaxy* (LPE) (not shown) involving a melt above a single crystal substrate, and *molecular beam epitaxy* (MBE) (c) [56]. The epitaxial techniques, particularly MBE, are capable of producing atomically abrupt layers (d). Figure 2.3e illustrates the layers of a laser diode, a man-made structure enabled by atomic control of semiconductor growth.

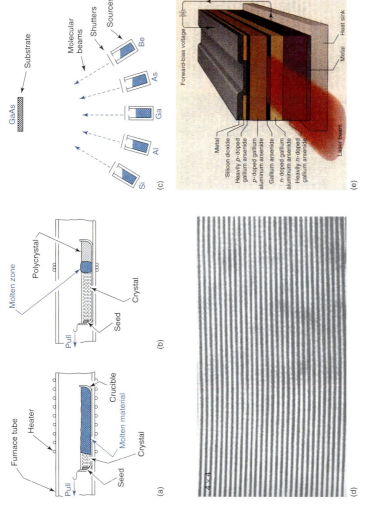

Figure 2.3 (a) Crystal growing from the melt in a crucible by solidification from one end of the melt (horizontal Bridgman method) or (b) melting and solidification in a moving zone. (c) Crystal growth by molecular beam epitaxy (MBE) by evaporation cells inside a high-vacuum chamber directing beams of Al, Ga, and As and dopants on to a GaAs substrate [56]. (d) Scanning electron micrograph of the cross section of an MBE-grown crystal having alternating layers of GaAs (dark lines) and AlGaAs (light lines). Each layer is four monolayers ($4 \times a/2 = 11.3$ Å) thick. (e) A laser diode composed of multiple GaAs and AlGaAs layers that produce coherent, monochromatic light with current injection through the layer stack. This man-made structure of atomically ordered layers is made possible by epitaxial growth techniques. (Courtesy IEEE Spectrum.)

2.6
Future Electronics

The next generations of electronics will demand even greater understanding and control of surfaces and interfaces. As devices become ultrasmall, for example, less than 100 nm dimensions, they will require ultrashallow junctions, abruptness at least below quantum-scale dimensions, lower densities of localized states, and atomic-scale optimization of electronic properties and chemical stability. Ultrahigh speed device structures will require the elimination of deep and shallow interface traps. Ultrastable contacts must function under very high temperature and possibly chemically corrosive conditions.

The history of the past six decades in microelectronics illustrates the synergy between science and technology as semiconductor discoveries led to devices, which stimulated the development of new techniques that advanced both scientific understanding and device architectures. This interplay of semiconductor growth, characterization methods, and device designs has steadily raised our understanding and control of electronic materials. There are a number of excellent books that treat the history of semiconductors, their applications, and their extension into the quantum mechanical regime at much greater depth; for example, see [57, 58].

The aim of Chapters 1 and 2 has been to motivate the study of surfaces and interfaces. Their properties alter charge densities, band structure, and electronic states on a scale of hundred of nanometers down to atomic layers. These microscopic properties manifest themselves as major changes in device features. As semiconductor structures shrink well below the micron scale, surfaces and interfaces, in fact, dominate the electronic and optical properties. In turn, atomic and nanoscale techniques are available to characterize the electronic, chemical, and geometrical properties of electronic materials structures. The chapters to follow describe the electrical and optical properties that depend on surfaces and interfaces along with the techniques developed to measure them. How these properties and techniques will contribute to the next generation of electronics comprises the concluding portion of this book.

Problems

1. (a) The first transistor was made with Ge. Why? (b) From Figure 2.2b, speculate on how low resistance contacts were made.
2. Si emerged as the material of choice for microelectronics. (a) Give five reasons why. (b) What feature of the Si/SiO_2 is the most important for device operation and why?
3. Give one example each of how surface and interface techniques have enabled modern complementary metal-oxide-semiconductor (CMOS) technology (a) structurally, (b) chemically, and (c) electronically.

4. (a) Give two reasons and two examples why GaAs, InP, and many other III–V compound semiconductors are used for quantum-scale optoelectronics. (b) Give reasons why they are advantageous for high-speed transistors.
5. (a) Which would you choose for quantum well lasers, GaAs or GaP, and why? (b) Which would you choose for high-power, high-frequency transistors, GaAs or GaN?

References

1. Braun, F. (1974) *Ann. Phys. Chem.*, **53**, 556.
2. Henisch, H.K. (1957) *Rectifying Semiconductor Contacts*, Chapter 2, Clarendon, Oxford.
3. Einstein, A. (1905) *Ann. Phys.*, **17**, 132.
4. Wilson, A.H. (1931) *Proc. R. Soc. Lond., A*, **133**, 458.
5. Wilson, A.H. (1931) *Proc. R. Soc. Lond., A*, **134**, 277.
6. Siemens, G. and Demberg, W. (1931) *Z. Phys.*, **67**, 375.
7. Schottky, W., Störmer, R., and Waibel, F. (1931) *Z. Hochfrequentztechnik*, **37**, 162.
8. Mott, N.F. (1939) *Proc. R. Soc. Lond., A*, **171**, 27.
9. Davidov, B. (1938) *J. Tech. Phys. USSR*, **5**, 87.
10. Davidov, B. (1939) *Sov. J. Phys.*, **1**, 167.
11. Schottky, W. (1939) *Z. Phys.*, **113**, 367.
12. Schottky, W. and Spenke, E. (1939) *Wiss. Veröffentl. Siemens-Werken*, **18**, 225.
13. Schottky, W. (1942) *Z. Phys.*, **118**, 539.
14. Bethe, H.A. (1942) Radiation Laboratory Report No. 43–12, Massachusetts Institute of Technology, November.
15. Richardson, O.W. (1921) *The Emission of Electricity from Hot Bodies*, Longmans-Green, Harlow, Essex.
16. Sze, S.M. (1981) *Physics of Semiconductor Devices*, 2nd edn, Chapter 3, Wiley-Interscience, New York.
17. Duke, C.B. (1969) *Tunneling in Solids*, Academic Press, New York, pp. 102–110.
18. Milnes, A.G. (1980) *Semiconductor Devices and Integrated Electronics*, Academic Press, New York, pp. 156–200.
19. Rhoderick, E.H. and Williams, R.H. (1988) *Metal-semiconductor Contacts*, Clarendon Press, Oxford.
20. Mönch, W. (1993) *Semiconductor Surfaces and Interfaces*, Springer-Verlag, Berlin.
21. Teal, G.K. and Little, J.B. (1950) *Phys. Rev.*, **78**, 647.
22. Teal, G.K. (1976) *IEEE Trans. Electron Devices*, **ED-23**, 621 and references therein.
23. Hall, R.N. (1950) *Phys. Rev.*, **78**, 645.
24. Pfann, W.G. (1952) *J. Met.*, **4**, 861.
25. www.azom.com/details.asp?ArticleID=1169.
26. www.madehow.com/Volume-1/Solar-Cell.html.
27. Bardeen, J. (1947) *Phys. Rev.*, **71**, 717.
28. http://www.corp.att.com/history/milestone_1947b.html.
29. http://www.britannica.com/ebc/art-16247/The-first-transistor-invented-by-American-physicists-John-Bardeen-Walter?articleTypeId=1.
30. http://www.ti.com/corp/docs/kilbyctr/kilby.shtml.
31. Mead, C.A. and Spitzer, W.G. (1964) *Phys. Rev.*, **134**, A713.
32. Tamm, I. (1932) *Phys. Z. Sowjetunion*, **1**, 733.
33. Schockley, W. (1939) *Phys. Rev.*, **56**, 317.
34. Kurtin, W., McGill, T.C., and Mead, C.A. (1970) *Phys. Rev. Lett.*, **22**, 1433.
35. Flores, F. and Tejedor, C. (1979) *J. Phys. C*, **12**, 731.
36. Cohen, M.L. (1980) *Adv. Electron. Electron Phys.*, **51**, 1.

37. (a) Tersoff, J. (1984) *Phys. Rev. Lett.*, **52**, 465; (b) Tersoff, J. (1984) *Phys. Rev. B*, **30**, 4875.
38. Mönch, W. (1990) *Rep. Prog. Phys.*, **53**, 221.
39. Pauling, L. (1960) *The Nature of the Chemical Bond*, 3rd edn, Cornell University Press, Ithaca.
40. Poate, J.M., Tu, K.N., and Mayer, J.W. (1978) *Thin Films – Interdiffusion and Reactions*, John Wiley & Sons, Inc., New York.
41. Mayer, J.W. and Lau, S.S. (1990) *Electronic Materials Science: for Integrated Circuits in Si and GaAs*, Macmillan, New York.
42. Andrews, J.M. and Phillips, J.C. (1975) *Phys. Rev. Lett.*, **35**, 56.
43. Andrews, J.M. and Phillips, J.C. (1975) *CRC Crit. Rev. Solid State Sci.*, **5**, 405.
44. Brillson, L.J. (1978) *Phys. Rev. Lett.*, **40**, 260.
45. Brillson, L.J. (1978) *J. Vac. Sci. Technol.*, **15**, 1378.
46. Brillson, L.J. (1978) *Phys. Rev. B*, **18**, 2431.
47. Brucker, C.F. and Brillson, L.J. (1981) *Appl. Phys. Lett.*, **39**, 67.
48. Brucker, C.F. and Brillson, L.J. (1981) *J. Vac. Sci. Technol.*, **18**, 787.
49. Brucker, C.F. and Brillson, L.J. (1981) *J. Vac. Sci. Technol.*, **19**, 617.
50. Spicer, W.E., Lindau, I., Skeath, P., and Su, C.Y. (1980) *Phys Rev. Lett.*, **44**, 420.
51. Spicer, W.E., Lilienthal-Weber, Z., Weber, E.R., Newman, N., Kendelewicz, T., Cao, R.K., McCants, C., Mahowald, P., Miyano, K., and Lindau, I. (1988) *J. Vac. Sci. Technol. B*, **6**, 1245.
52. Brillson, L.J., Viturro, R.E., Shaw, J.L., Mailhiot, C., Tache, N., McKinley, J.T., Margaritondo, G., Woodall, J.M., Kirchner, P.D., Pettit, G.D., and Wright, S.L. (1988) *J. Vac. Sci. Technol. B*, **6**, 1263.
53. Palmstrøm, C.J., Cheeks, T.L., Gilchrist, H.L., Zhu, J.G., Carter, C.B., and Nahory, R.E. (1990) in *Electronic, Optical and Device Properties of Layered Structures* (eds J.R. Hayes, M.S. Hybertson, and E.R. Weber), Materials Research Society, Pittsburgh, p. 63.
54. Offsey, S.D., Woodall, J.M., Warren, A.C., Kirchner, P.D., Chappell, T.I., and Pettit, G.D. (1986) *Appl. Phys. Lett.*, **48**, 475.
55. Waddill, G.D., Vitomirov, I.M., Aldao, C.M., and Weaver, J.H. (1989) *Phys. Rev. B*, **41**, 991.
56. Streetman, B.G. and Banerjee, S. (2000) *Solid State Electronic Devices*, 5th edn, Prentice Hall, Upper Saddle River.
57. Orton, J. (2004) *The Story of Semiconductors*, Oxford University Press, Oxford.
58. Turton, R. (1995) *The Quantum Dot, A Journey into the Future of Microelectronics*, Oxford University Press, Oxford.

3
Electrical Measurements

3.1
Schottky Barrier Overview

Electrical contacts to semiconductors are central to solid-state device performance. The potential barrier set up by band bending at the semiconductor interface is at the heart of all modern electronics. This barrier is equal to the energy difference between the Fermi level E_F of a metal relative to the band edge of the semiconductor's majority charge carrier, and it determines the "ohmic" or "rectifying" behavior of the electrical contact.

Figure 3.1a illustrates the current–voltage behavior of an "ohmic" metal–semiconductor contact. Here, the resistance $R = V/I$ is the same for both forward and reverse bias. The steep slope pictured here indicates a low resistance. Such ohmic contacts are required for devices and circuits in which the contact introduces no voltage drop. Ohmic contacts are essential in delivering all the applied voltage to the desired circuit element without loss of voltage or energy to intermediate components. Examples include contacts to (i) laser diodes to produce light emission with the lowest possible applied voltage, (ii) *metal-oxide-semiconductor field effect transistor* (MOSFET) capacitors to perform computer operations with minimum possible energy generation, and (iii) high power, high electron mobility transistors (HEMTs) for *radio frequency* (RF) wave generation to minimize resistive heating.

Figure 3.1b illustrates the current–voltage behavior of a "rectifying" metal–semiconductor contact. Here, the resistance is low in the forward direction and high in the reverse direction. This asymmetric behavior is extremely useful for active electronics since the built-in voltage of the band bending can control circuit current and voltage in numerous ways. Examples include (i) a current rectifier that blocks alternating current in one direction, (ii) a solar cell that generates voltage with illumination, (iii) a photodetector that generates current with illumination, and (iv) a laser diode that generates light with applied voltage. The utility of these bent bands is magnified by the exponential dependence of carrier density on applied voltage so that small applied voltages can have disproportionately large electrical effects.

Later chapters review many of the techniques employed to obtain low-resistance ohmic contacts. The control of rectifying, that is, Schottky barrier, contacts for particular semiconductors is a much greater challenge. The following sections

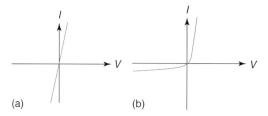

Figure 3.1 Current versus voltage plots of (a) symmetric or ohmic contact and (b) asymmetric or rectifying contact.

introduce the basic concepts of Schottky barriers and the conventional techniques used to measure these barriers in the laboratory. We then describe the difference between the electrical properties expected on the basis of this classical framework versus the actual properties observed experimentally. This difference provides the central motivation for understanding charge transfer at surfaces and interfaces on an atomic scale and thereby controlling electronic properties macroscopically.

3.2
Ideal Schottky Barriers

Ideally, the band-bending region inside the semiconductor forms as a result of charge transfer between a metal and a semiconductor. This charge transfer depends in turn on the difference in Fermi level between the two materials. Thus, the expected band bending for a given semiconductor should depend on the metal's work function. Figure 3.2 illustrates the band-bending scheme for metals on high and low work function at surfaces on n- and p-type semiconductors [1]. For the high work function metal and n-type semiconductor in Figure 3.2a, electrons flow from the semiconductor to the metal after contact, depleting a characteristic surface region in the semiconductor of electrons. With the two Fermi levels E_F^M and E_F^{SC} aligned, a double layer forms with a voltage drop of qV_B equal to the contact potential difference between the metal and the interior of the semiconductor. The double layer consists of a surface space charge region, typically 10^{-4}–10^{-6} cm thick, and an induced charge on the metal surface. The n-type depletion region in Figure 3.2a is a layer of high resistance. Thus, a voltage applied to this junction will fall mostly across the surface space charge region. The band bending depends on the difference in thermionic work function and in this simple model is expected to be

$$qV_B = \Phi_M - \Phi_{SC} \quad (3.1)$$

where Φ_M and Φ_{SC} are the metal and semiconductor *work functions*, respectively, and the n-type Schottky barrier is

$$\Phi_{SB}^n = \Phi_M - \chi_{SC} \quad (3.2)$$

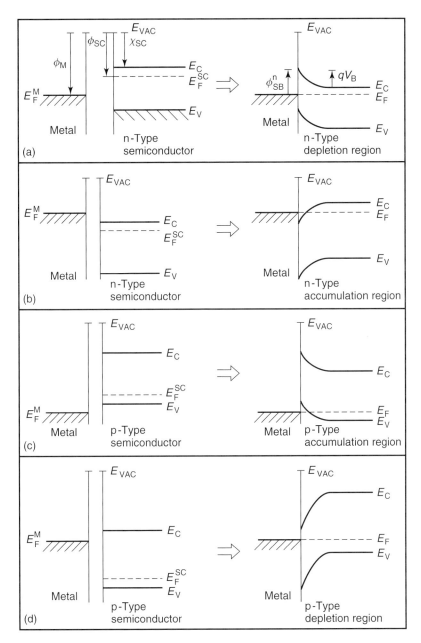

Figure 3.2 Schematic diagram of band bending before and after metal–semiconductor contact for (a) high work function metal and n-type semiconductor, (b) low work function metal and n-type semiconductor, (c) high work function and p-type semiconductor, and (d) low work function metal and p-type semiconductor [1].

where χ_{SC} is the *semiconductor electron affinity*. Note that ϕ_{SC} and χ_{SC} differ by $E_C^{bulk} - E_F$, the difference in energy between the conduction band edge E_C and the Fermi level in the semiconductor bulk so that

$$\Phi_{SB} = qV_B + E_C^{bulk} - E_F \tag{3.3}$$

For typically doped n-type semiconductors, $E_C - E_F < 0.1\,eV$. Equation 3.1 also holds for p-type semiconductors, but $\Phi_{SB}^p = \phi_M - \chi_{SC} - E_G$, where E_G is the semiconductor bandgap.

According to this band-bending model, the potential Φ within the semiconductor satisfies Poisson's equation[1]

$$\nabla^2 \Phi(x) = -\frac{\rho(x)}{\varepsilon_s} \tag{3.4}$$

In Equation 3.3, ρ is the *charge density* in the surface space charge region of width w, x denotes the coordinate axis normal to the metal–semiconductor interface, and ε_s is the *static dielectric constant* of the semiconductor. Using the abrupt approximation such that $\rho \cong qN$, the bulk concentration of ionized impurities within the surface space charge region, for $x < W$ and $\rho \cong 0$, and $d\Phi/dx = 0$ for $x > W$, one obtains [2]:

$$V = -\frac{\{qN(x-W)\}^2}{2\varepsilon_s} + V_0 \quad \text{for} \quad 0 < x < w$$
$$= V_0 \quad \text{for} \quad x > w \tag{3.5}$$

and a *depletion layer width* of

$$W = \left(\frac{2\varepsilon_s(V-V_0)}{Nq}\right)^{1/2} \tag{3.6}$$

Here $Nq = n$, the *bulk charge density*. Thus, the abrupt metal–semiconductor junction contains a parabolic band-bending region. Analogous conclusions can be drawn from the low work function metal case shown in Figure 3.2b and the high versus low work function metals on a p-type semiconductor in Figures 3.2c and d respectively. Note that Figures 3.2b and c exhibit accumulated band-bending regions, which present no barrier to majority carrier transport across the interface.

3.3
Real Schottky Barriers

In reality, the conventional or classical picture of Schottky barrier formation described by Equation 3.1 and illustrated in Figure 3.2 does not agree with experimental measurements of metal–semiconductor interfaces. Figure 3.3 illustrates this disagreement for barrier heights of different metals on Si [3–6]. The correlation between $\Phi_M - \Phi_{SC}$ (straight line) and the band bending for metals

---

1) Throughout this book, we use the rationalized MKS system with centimeters as a convenient unit of length.

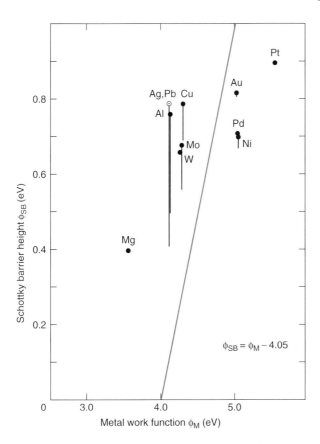

Figure 3.3 Plot of barrier heights Φ_{SB} [3–5] as a function of metal work function [7] Φ_M for metals on n-type Si. The circles are barrier heights for cleaved Si and the vertical lines cover the range of barrier values for chemically prepared surfaces after aging. The diagonal line is a plot of $\Phi_{SB} = \Phi_M - \chi_{Si}$, where $\chi_{Si} = 4.05$ eV [6]. The Φ_M are selected values obtained by photoemission techniques.

with different work functions [7] on the same semiconductor has typically not been strong. This weak dependence can be interpreted from Bardeen's work [8] to mean that localized states at the interface can accumulate charge and generate dipoles that take up much of the metal–semiconductor potential difference. These surface or interface states can have several origins.

Consider a rectification barrier with and without surface states. Figure 3.4 shows a metal and a semiconductor before and after contact. In Figure 3.4a, band bending is present even without a metal contact. This is due to negative charge that fills states localized at or near the surface at energies below the Fermi level E_F^{SC}. If the surface state density is relatively high, that is, $\sim 10^{14}$ cm^{-2}, the presence of the metal does not alter $E_F + qV_B$ appreciably. Instead, Figure 3.4b shows that most of the contact potential difference $\Phi_M - \Phi_{SC}$ falls across an atomically thin interface dipole region instead of the semiconductor space charge region. The voltage drop

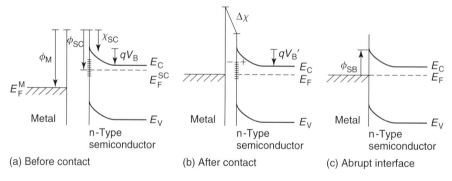

Figure 3.4 Schematic energy band diagram of a metal and a semiconductor with a high density of surface states: (a) before contact, (b) after contact, and (c) as pictured conventionally with interface dipole not shown [1].

$\Delta\chi$ across this dipole depends on the density of localized states that are filled with electrons as well as the dielectric constant of the layer across which this voltage drops. Assuming that this dipole layer is sufficiently thin that charge tunnels freely across it, the resultant abrupt interface is typically pictured without the dipole as in Figure 3.4c.

Thus $\Phi_{SB} = qV_B + E_C^{bulk} - E_F$ is not equal to $\Phi_M - \chi_{SC}$ but rather

$$\Phi_{SB} = \Phi_M - \chi_{SC} - \Delta\chi \tag{3.7}$$

For high enough densities of interface states, small movements of the Fermi level within this energy range of states produce large changes of localized state occupation. As a result, most of the potential difference between metal and semiconductor produces changes in the interface dipole rather than in the surface space charge region. The barrier height Φ_{SB} and band bending qV_B are then relatively independent of metal work function Φ_M. The Fermi level at the surface is then termed *pinned* by surface states within a narrow range of energy in the semiconductor bandgap.

The earliest direct evidence of this Fermi level "pinning" was the field effect experiment associated with the transistor development [8]. Here the presence of states at the interface reduces the effect on an applied gate voltage on the Fermi level movement within the surface space charge region, thereby preventing large changes in carrier concentration that can move within the transistor's channel.

Figure 3.5 illustrates this field effect experiment [9] and the resultant Fermi level changes with bias and surface states. In Figure 3.5a, a condenser plate applies a gate voltage V to the Si surface while current A yields conductivity of the surface space charge region. *Surface conductivity* is given by

$$\sigma = ne\mu \tag{3.8}$$

where e is *electron charge* and μ is *carrier mobility*. Electrons in the conduction band are denoted as short dashes. Without surface states or bias, the bands in Figure 3.5b are flat and E_F lies above midgap for an n-type semiconductor. With applied bias

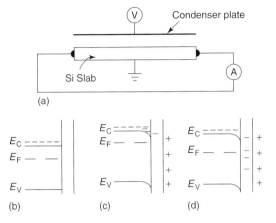

Figure 3.5 Experimental configuration for field effect measurements of Shockley and Pearson [9]. Surface conductivity of Si slab measured by current is monitored as a function of gate voltage (a) for (b) no bias, (c) bias with no surface states, and (d) bias with surface states.

and no surface states in Figure 3.5c, a positive bias increases the electron charge density inside the semiconductor surface, causing the bands to bend down as E_F moves closer to E_C. The increased charge density results in increased *conductivity* σ and *current density* $J = \sigma \mathcal{E}$, where $\mathcal{E}$ is applied electric field. With surface states present as in Figure 3.5d, this voltage-induced band bending and the addition of majority carriers are reduced. The control of carriers in the surface space charge region by an applied voltage is essential to the operation of all "active" electronic devices such as the transistor.

3.4
Schottky Barrier Height Measurements

3.4.1
Current–Voltage (J–V) Technique

There are several methods used to measure barrier heights between metals and semiconductors. They differ in (i) how difficult measurements are to perform and (ii) the factors that can complicate their interpretation. The most straightforward method is the current–voltage or J–V technique. Thermionic emission-diffusion theory yields a forward J–V characteristic given by

$$J = A^{**} T^2 \exp\left(\frac{-q\Phi_{SB}}{k_B T}\right)\left[\exp\frac{-qV}{k_B T} - 1\right] \tag{3.9}$$

where $A^{**} = f_p f_q A^* / (1 + f_p f_q v_R / v_D)$, $A^* = (m^*/m_0)(120 \text{ A/cm}^{2\circ} K^{-2})$, f_p = probability of electron emission over the semiconductor potential maximum into the metal without electron–optical phonon scattering, f_q = ratio of total J with versus without tunneling and quantum-mechanical reflection, v_R = *recombination velocity*, v_D = effective *diffusion velocity* with thermionic emission, m^* is the carrier's *effective mass*, m_0 is the free electron mass, T is *temperature*, and V is the *applied voltage*.

For $V > 2k_BT/q$ in the forward direction, A^{**} can be approximated by A^*, and ϕ_{SB} consists of a barrier height extrapolated to zero field, Φ_{SB0} minus a term $\Delta\Phi$ due to the combined effects of applied electric field and image force [10]. The effect is given by

$$\Phi_{SB} = \Phi_{SB0} - \Delta\Phi = \Phi_{SB0} - \left(\frac{q^3 \mathcal{E}}{\varepsilon_s}\right)^{1/2} \tag{3.10}$$

and illustrated in Figure 3.6. Here, the *applied electric field* $\mathcal{E} = V/W$, where W is the width of the surface space charge region and any dipole layer. Depending on N and ε_s, the contact's depletion region width W is typically <1000 Å so that voltages of only 1 V can produce field gradients $\mathcal{E}$ of 10^5 V cm^{-1} or higher. As an example, $\Delta\Phi = 0.03$ eV with $\varepsilon_s = 16\varepsilon_0$, for example, Ge, and $\mathcal{E} = 10^5$ V cm^{-1}. While the $\Delta\Phi$ term is only several tens of millivolts, it can nevertheless affect the J–V behavior for nonzero voltages.

The forward current density in Equation 3.9 extrapolated logarithmically to zero applied forward bias has an intercept at

$$J_s = A^{**}T^2 \exp\left(\frac{-q\Phi_{SB0}}{k_B T}\right) \tag{3.11}$$

so that the barrier Φ_{SB0} can be extracted from a plot of $\ln J$ versus applied forward voltage. The slope of the J–V plot in Figure 3.7 also provides the "*ideality factor*" n of the contact, defined by [10]

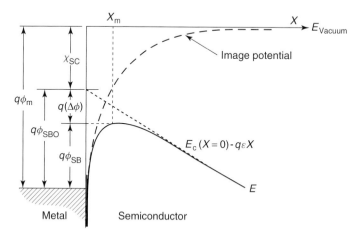

Figure 3.6 Schematic energy band diagram between a metal surface and a semiconductor. The effective barrier is lowered from Φ_{SB0} to Φ_{SB} when an electric field is applied to the interface. (After Sze [10].)

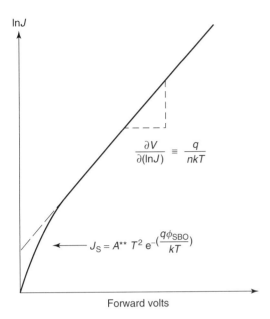

Figure 3.7 Forward current density versus applied voltage for a metal–semiconductor contact.

$$n \equiv \left(\frac{q}{k_B T}\right) \frac{\partial V}{\partial (\ln J)} = \left[1 + \left(\frac{\Delta \partial \Phi}{\partial V}\right) + \left(\frac{k_B T}{q}\right) \frac{\partial (\ln A^{**})}{\partial V}\right]^{-1} \quad (3.12)$$

In general, the second and third terms of Equation 3.12 are much less than one so that $n \approx 1$. However, a variety of physical processes can increase n. These include tunneling through the barrier [11, 12], intermediate layers with new dielectric and transport properties [12, 13], and recombination or trapping at states near the interface and within the semiconductor bandgap [13]. Later chapters discuss these processes along with the chemical and electronic interactions that produce them.

An approximation to Equation 3.6 that contains n explicitly is

$$J = A^{**} T^2 \exp\left(\frac{-q\Phi_{SB0}}{k_B T}\right) \left[\exp\left(\frac{qV}{nk_B T}\right) - 1\right] \quad (3.13)$$

However, extrapolating $\ln J$ to $V = 0$ in order to obtain Φ_{SB} is not a reliable procedure for n values that are significantly different from unity [13]. Nevertheless, Equation 3.13 can be used to extract Φ_{SB0} from the reverse current, provided that generating recombination currents within the depletion region are small compared with the Schottky emission current.

Finite voltage drops across the semiconductor bulk or its otherwise low-resistance back contact will reduce $\ln J$ with increasing forward voltage. Thus, $V_{applied} = V_{interface} + IR^{SC}_{bulk} + IR^{SC}_{back\ contact}$, where R^{SC}_{bulk} and $R^{SC}_{back\ contact}$ are parasitic semiconductor bulk and back contact resistances, respectively, and $V_{interface} < V_{applied}$. If the deviation of $\ln J$ from linearity voltage is linear with increasing applied voltage, its slope yields the total parasitic resistance $R_{\text{bulk}}^{\text{SC}} + R_{\text{back contact}}^{\text{SC}}$ and the correction term $\Delta V_{\text{applied}} = I_{\text{measured}} \left(R_{\text{bulk}}^{\text{SC}} + R_{\text{back contact}}^{\text{SC}} \right)$.

Finally, the field-lowering effect depicted in Figure 3.6 will also produce a gradual increase in current with reverse bias.

3.4.2
Capacitance–Voltage (C–V) Technique

A second common method of evaluating barrier heights is the capacitance–voltage or C–V technique. Here, the capacitance of the semiconductor depletion region is obtained by superimposing an AC voltage on the DC bias across the semiconductor. Since capacitance

$$C = \frac{\varepsilon_s A}{W} \tag{3.14}$$

where A = capacitor area and depletion width

$$W = \left[\frac{2\varepsilon_s (V - V_0)}{Nq} \right]^{1/2} \tag{3.15}$$

we have

$$C = A \left[\frac{Nq\varepsilon_s}{2(V_B - V)} \right]^{1/2} \tag{3.16}$$

As reverse bias increases, W increases and C decreases. Since d decreases with decreasing reverse voltage and vanishes at $V = V_B$, C becomes infinite and $1/C^2$ decreases to zero. Thus, the intercept of a $1/C^2$ versus V plot yields the band-bending V_B, as illustrated by Figure 3.8.

The barrier height is given by

$$\Phi_{SB}^n = qV_B + qV_n - \Delta\Phi \tag{3.17}$$

where $\Delta\Phi$ is the *image force correction* and *diffusion potential* $V_n = E_C^{\text{bulk}} - E_F$ is known from the bulk doping and m^*. Specifically,

$$n_0 = N_C \exp\left[\frac{-(E_C - E_F)}{k_B T} \right] \tag{3.18}$$

where n_0 is the *equilibrium carrier concentration* in the bulk, $N_C = 2(2\pi m_e^* k_B T/h^2)^{3/2}$ is the conduction band density of states, and h is *Planck's constant*. Furthermore, the slope of a $1/C^2$ plot versus reverse voltage

$$\frac{d(1/C^2)}{dV} = \frac{2}{(Nq\varepsilon_s A^2)} \tag{3.19}$$

yields a straight line for constant semiconductor carrier concentration N. Hence, one can use $d(1/C^2)/dV$ and $W(V)$ to obtain doping profiles of the surface space charge region.

The C–V technique has several sources of possible error, including (i) an insulating layer between the metal and semiconductor, (ii) variation of surface

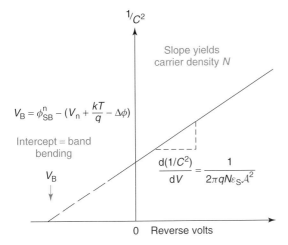

Figure 3.8 Plot of $1/C^2$ versus V for an ideal metal–semiconductor contact. Voltage intercept V_B equals the band bending within the surface space charge region of the semiconductor and is related to the Schottky barrier height $\Phi_{SB}{}^n$ as indicated.

charge (interface states) with voltage, (iii) series resistance of the junction, (iv) traps with the depletion width, and (v) variations in effective contact width with depletion layer width. The capacitance can also exhibit a frequency dependence due to trapped carriers with different capture and emission rates. Indeed one can use this frequency dependence to identify trapped charge levels and cross sections. For example, *deep-level transient spectroscopy* (DLTS) [14], *thermally stimulated capacitance* (TSCAP) [15], admittance spectroscopy [16], and photocapacitance spectroscopy [17] These techniques yield information on electronic properties of impurity and defect centers in semiconductor surface space charge regions. These properties include (i) energy levels, (ii) charge state multiplicity, (iii) thermal and optical emission rates, (iv) thermal and optical cross sections, and (v) the dependencies of the cross sections on temperature, static electric field, and photon energy [18, 19].

3.4.3
Internal Photoemission Spectroscopy (IPS)

Internal photoemission spectroscopy or IPS is perhaps the most direct and reliable macroscopic method of determining the energy barrier, which carriers must surmount in order to move across the metal–semiconductor interface. The technique employs monochromatic light to excite transitions that produce free carriers and a photocurrent in an external circuit. Figure 3.9a shows photoexcitation either through the front contact or the back, while Figure 3.9b shows photo-induced excitation of transitions from the Fermi level of the metal to the conduction

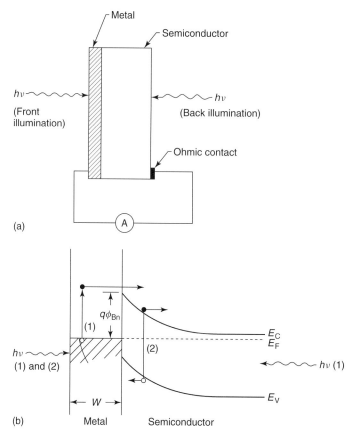

Figure 3.9 (a) Schematic diagram of internal photoemission measurement. (b) Energy band diagram for photoexcitation process. (After Sze [10].)

band edge (1) as well as band-to-band transitions (2). Light passing through the semiconductor with energy $h\nu > E_G (\equiv E_C - E_V)$ will be strongly attenuated by absorption before it reaches the metal–semiconductor interface. The free carriers so generated lie in a field-free region, recombine and do not contribute to the photoresponse. Hence, the technique either that (i) the metal film be thin enough for front illumination to be absorbed at the metal- semiconductor interface or (ii) there is back illumination through the semiconductor. For example, in back illumination, transition 1 will dominate a spectrum of photoexcited current versus incident photon energy. The photocurrent per absorbed photon per unit area is given by the Fowler theory [20] as

$$J_R \sim \left[\frac{T^2}{(E_s - h\nu)^{1/2}}\right] \left\{ \left(\frac{x^2}{2}\right) + \frac{\pi^2}{6} - \left[e^{-x} - \left(\frac{e^{-2x}}{4}\right) + \left(\frac{e^{-3x}}{9}\right) - \cdots\right]\right\} \text{ for } x > 0$$

(3.20)

where $x \equiv h(\nu-\nu_0)/k_B T$, ν = photon frequency, $h\nu_0 = q\phi_{SB}$, the barrier height, and E_S is the sum of $q\Phi_{SB}$ and the Fermi energy measured from the bottom of the semiconductor conduction band.

For $h\nu > \Phi_{SB} + 3k_B T$, photocurrent density J_R varies quadratically as a function of incident photon energies above the barrier height.

$$J_R \sim (h\nu - h\nu_0)^2 \tag{3.21}$$

Thus, a plot of $\sqrt{(J_R)}$ versus $h\nu$ yields a straight line that intersects the $h\nu$ axis at a value equal to the barrier height Φ_{SB}. Furthermore, the bias dependence of this $\sqrt{(J_R)}$ versus $h\nu$ plot yields the image-force-lowering term in Equation 3.10, from which one obtains $\varepsilon_S/\varepsilon_0$ near the interface [21]. Figure 3.10a illustrates $\sqrt{(J_R)}$ versus $h\nu$ plots for three bias conditions of Au–Si Schottky barriers. The barrier height equals 0.815 eV for zero bias. From the applied voltage and bulk carrier concentration, one can plot $\Delta\Phi$ versus electric field and obtain $\varepsilon_S/\varepsilon_0$ [21]. The presence of localized states at the metal–semiconductor interface will increase the image-force dielectric constant $\varepsilon_S/\varepsilon_0$ and the image-force lowering of the barrier [22–24].

The temperature dependence of the $\sqrt{(J_R)}$ intercept provides additional information. Temperature does not affect the functional dependence of $\sqrt{(J_R)}$ versus $h\nu$ except through changes in the semiconductor electronic structure, in particular, the

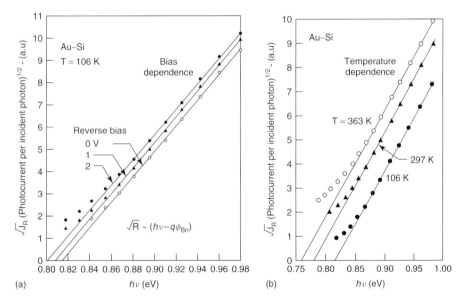

Figure 3.10 (a) Square root of photoresponse per incident photon $\sqrt{(J_R)}$ versus incident photon energy $h\nu$ for three different biasing conditions at an Au–Si interface. (After Sze et al. [10].) (b) $\sqrt{(J_R)}$ versus $h\nu$ for three semiconductor temperatures at an Au–Si interface. (After Crowell et al. [24].)

change in bandgap E_G. Changes in Φ_{SB} and in E_G can be compared to associate the Fermi level position with either the conduction or valence band edge. Figure 3.10b exhibits the temperature dependence of the Au–Si barrier height. Both the barrier height and the bandgap decrease with increasing temperature at similar rates, indicating that the Fermi level within the bandgap is linked with the valence band edge.

IPS can clearly show the dependence of Schottky barrier height on different metals. Figure 3.11 illustrates the clear difference between barriers for metals on molecular beam epitaxy (MBE)-grown GaAs [25]. Such barrier differences are much larger than those commonly reported for melt-grown GaAs and can be explained by differences in crystal quality, deposition methods, and chemically-induced defects to be discussed in later chapters. For an authoritative review of IPS measurements of barrier heights, see [26]. For a detailed description of electrical measurement techniques in general, see [27].

Each of the measurement techniques presented in this section provides a different measure of the Schottky barrier height. Each is susceptible to experimental complications, and features within the interface can introduce different types of errors. For example, J–V measurements are particularly sensitive to tunneling

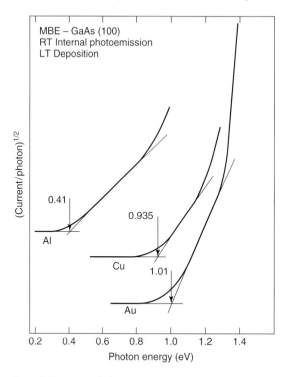

Figure 3.11 Internal photoemission spectroscopy measurements of MBE-grown GaAs barrier heights versus metal. Al, Cu, and Au overlayers yield room temperature (RT) barriers of 0.41, 0.935, and 1.01 eV, respectively [25].

through the barrier region and to trapping and recombination by deep levels within the bandgap. Artifacts of the measurement technique and deep-level trapping affect the C–V technique. Image-force lowering of the Schottky barrier affects all three techniques. In addition, measurements obtained with these techniques become difficult to interpret if the barrier is more complex than a region of parabolic band bending. Such nonparabolicity can result, for example, if electrically active sites such as dopants or traps are not uniformly distributed throughout the surface space charge region. Because of these complications, it is advisable to use more than one technique to interpret complex Schottky barrier features. Milnes [2] has compiled barrier heights for a wide variety of metal–semiconductor interfaces and compared results for the J–V, C–V, and IPS techniques. Such compilations reveal systematic differences between the results of the different techniques, a subject to be addressed in subsequent chapters.

3.5 Summary

This chapter introduced the conventional macroscopic techniques: current–voltage, capacitance–voltage, and IPS – used to measure the electrical barriers at electronic material interfaces. Each technique provides useful information but different sources of error are inherent in each. Thus, more than one technique is advisable to characterize Schottky barriers.

We have also seen that interface states can "pin" the Fermi level within the semiconductor bandgap in a narrow energy range such that the barrier height becomes insensitive to the metal work function. Such Fermi level "pinning" is due to additional electronic states near or at the semiconductor surface or interface that introduce dipoles that alter the macroscopic barriers. In general, the classical picture described here is complicated by localized charge states that arise due to both intrinsic and extrinsic phenomena. Fermi level pinning underscores the importance of controlling chemical, electronic, and geometric structures on an atomic scale. Later chapters present methods by which such control can be achieved.

Problems

1. Assume that an ideal Schottky barrier is formed by a metal with work function 4.8 eV on n-type Si with $N_d = 10^{17}$ cm^{-3}. The Fermi level position in the bandgap is obtained from the relation $n_0 = n_i e^{(E_F - E_i)kT}$, where the intrinsic carrier concentration n_i for Si is 1.5×10^{10} and E_i is approximately at the midpoint of the bandgap. Draw the intrinsic Fermi level position E_i and the equilibrium semiconductor Fermi level position E_{FS} in the bandgap.

Calculate the Schottky barrier height. What is the band bending? Draw an equilibrium band diagram as in Figure 3.2?
2. Calculate the depletion width for the Schottky diode in Problem 1. Use MKS units so that $W = \left(\frac{2\varepsilon_s(V-V_0)}{Nq}\right)^{1/2}$. What is the depletion width with a reverse bias of 0.5 V? Draw the corresponding energy band diagram.
3. Assume that a dipole forms at the Si surface due to negative surface charges with a density $n_s = 2 \times 10^{13}$ cm^{-2} separated by a distance $d = 10$ Å. What is the corresponding change in barrier height and band bending for the junction in Problem 2.
4. Calculate the capacitance for the diode in Problem 1 at zero bias with a diode diameter of 400 μm.
5. Calculate the image-force lowering of the barrier for a metal-Si diode with an applied electric field of 10^4 V cm^{-1}.
6. Calculate the forward current for a Si barrier height 0.5 eV at room temperature and $V_F = 0.3$ V forward bias for a diode with 200-μm circular diameter. Neglect image-force lowering. For Si, use $m_{e*}/m_0 = 0.26$.
7. An n-type semiconductor with $\chi = 4$ eV, $N_D = 10^{16}$ cm^{-3}, and $n_i = 10^{10}$ cm^{-3} has a room temperature bandgap $E_G = 1.2$ eV. For metals with work function $\Phi_M = 4$, 4.25, 4.5, 4.75, and 5 eV, calculate and draw Φ_{SB} versus Φ_M. Calculate and draw qV_0 versus Φ_M. For $\varepsilon/\varepsilon_0 = 15$, determine a numerical equation for the depletion region width versus Φ_M for applied voltage $V_A = 0$.
8. Consider a Si Schottky diode of area 1 cm^2 and doping N_D with a capacitance such that $1/C^2$ versus applied voltage V_A varies with slope -4.4×10^{13} F^{-2}V^{-1}. (a) Determine the semiconductor doping. (b) If $C = 168$ nF at $V = 0$, what is the band bending qV_0?
9. The bias-dependent internal photoemission curves in Figure 3.10 show a decrease in the metal-Si barrier height with increasing reverse bias. For n-type doping of 3×10^{16} cm^{-3} and a barrier height of 0.825 eV, calculate the maximum electric fields $-qN_Dd/\varepsilon$ at the interface for 0 and 2 V reverse bias. Use $n_i = 1.5 \times 10^{10}$ cm^{-3} for the intrinsic carrier concentration. What image force dielectric constant is consistent with the barrier height change?

References

1. Brillson, L.J. (1982) Surf. Sci. Rep., **2**, 123.
2. Milnes, A.G. (1980) Semiconductor Devices and Integrated Electronics, Academic Press, New York, pp. 85–136.
3. Turner, M.J. and Rhoderick, E.H. (1968) Solid-State Electron., **11**, 291.
4. Archer, R.J. and Atalla, M.M. (1963) Am. Acad. Sci. NY, **101**, 697.

5. Crowell, C.R., Spitzer, W.G., Howarth, L.E., and LaBate, E.E. (1962) *Phys. Rev.*, **127**, 2006.
6. Allen, F.G. and Gobeli, G.W. (1964) *J. Appl. Phys.*, **35**, 597.
7. Michaelson, H.B. (1977) *J. Appl. Phys.*, **48**, 4729.
8. Bardeen, J. (1947) *Phys. Rev.*, **71**, 717.
9. Shockley, W. and Pearson, G.L. (1948) *Phys. Rev.*, **74**, 232.
10. Sze, S.M. (2007) *Physics of Semiconductor Devices*, 3rd edn, Chapter 5, Wiley-Interscience, New York.
11. Duke, C.B. (1969) *Tunneling in Solids*, Academic, New York, pp. 102–110.
12. Crowell, C.R. (1965) *Solid-State Electron.*, **8**, 295.
13. Shaw, M.P. (1981) in *Handbook on Semiconductors*, Device Physics, Vol. 4, Chapter 1 (ed. C.Hilsum), North-Holland, Amsterdam.
14. Lang, D.V. (1974) *J. Appl. Phys.*, **45**, 3023.
15. (a) Carballes, J.C., Varon, J., and Ceva, T. (1971) *Solid State Commun.*, **9**, 1627; (b) Sah, C.T., Chan, W.W., Fu, H.S., and Walker, J.W. (1972) *Appl. Phys. Lett.*, **20**, 193; (c) Buehler, M.G. (1972) *Solid-Sate Electron.*, **15**, 69.
16. (a) Sah, C.T. and Walker, J.T. (1973) *Appl. Phys. Lett.*, **22**, 384; (b) Losee, D.L. (1972) *Appl. Phys.*, **21**, 54.
17. (a) Kukimoto, H., Henry, C.H., and Merritt, F.R. (1973) *Phys. Rev.*, **B7**, 2486; (b) White, A.M., Dean, P.J., and Porteous, P. (1976) *J. Appl. Phys.*, **47**, 3230.
18. Sah, C.T. (1976) *Solid-State Electron.*, **19**, 975.
19. Sah, C.T., Forbes., L., Rosier, L.L., and Tasch, A.F.Jr. (1970) *Solid-State Electron.*, **13**, 759.
20. Fowler, R.H. (1931) *Phys. Rev.*, **38**, 45.
21. Sze, S.M., Crowell, C.R., and Kahng, D. (1964) *J. Appl. Phys.*, **35**, 2534.
22. Crowell, C.R., Shore, H.B., and Labate, E.E. (1965) *J. Appl. Phys.*, **36**, 3843.
23. Parker, G.H., McGill, T.C., Mead, C.A., and Hoffman, D. (1968) *Solid-State Electron.*, **11**, 201.
24. Crowell, C.R., Sze, S.M., and Spitzer, W.G. (1964) *Appl. Phys. Lett.*, **4**, 91.
25. Chang, S., Shaw, J.L., Brillson, L.J., Kirchner, P.D., Pettit, G.D., and Woodall, J.M. (1992) *J. Vac. Sci. Technol. B*, **10**, 1932.
26. Williams, R. (1970) in *Semiconductors and Semimetals*, Vol. 6 (eds R.K. Willardson and A.C. Beer), Academic Press, New York, pp. 97–139.
27. Wieder, H.H. (1979) *Laboratory Notes on Electrical and Galvanometric Measurements*, Elsevier, Amsterdam.

4
Interface States

4.1
Interface State Models

Chapter 3 presented the concepts of interface charge transfer, the band bending induced within a space charge region below the surface, and the resultant Schottky barrier height at the metal–semiconductor junction. It also showed that electronic states localized near these interfaces alter the barrier height expected from the simple difference in metal and semiconductor work functions. Chapter 4 provides an overview of the intrinsic and extrinsic phenomena that can contribute to such localized charge states at interfaces. It also introduces the theoretical approaches used to evaluate the effects of such phenomena on Schottky barrier formation.

Localized states at interfaces fall into four general categories: (i) intrinsic surface states associated with the discontinuity of lattice potential at the semiconductor–vacuum interface; (ii) metal-induced states due to metal wavefunctions extending ("tailing") into the semiconductor at energies within the bandgap; (iii) extrinsic surface states due to imperfections of the semiconductor interface, for example, impurities or defects common to both bulk and surface; and (iv) extrinsic, interface-specific states created by interface chemical reactions, interdiffusion, or adsorbate-specific local chemical bonding. There is experimental evidence for each type of interface charge accumulation. Theoretical calculations used to infer structural information about each interface are discussed for each case along with the degree to which they can successfully account for observed spectral features. In this chapter, we first consider intrinsic surface states. After consideration of each localized state category, chapters that follow show that the interface-specific states are the dominant factor in Schottky barrier formation.

Intrinsic surface states at the semiconductor–vacuum interface have been studied longer and more extensively than any other type of interface state. Such states can arise due to lattice termination of the bulk crystal since atoms at the surface have broken bonds and fewer atomic neighbors. These structural differences can induce major changes in electronic structure versus the bulk. States at the semiconductor–vacuum interface with energies within the bandgap will be localized normal to the surface. Figure 4.1a schematically illustrates the effect of lattice termination on the wavefunction of such states. Here the wavefunction amplitude $\psi(x)$ decays exponentially with distance away from the semiconductor

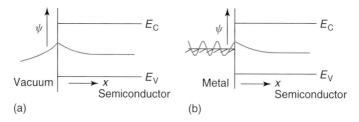

Figure 4.1 Schematic energy band diagrams of wavefunction localization due to (a) an intrinsic surface state at the semiconductor–vacuum interface. Localization is associated with decaying wavefunctions on both sides of the interface. (b) A metal-induced state at the semiconductor–metal interface. Localization is associated with the decaying wavefunction into the semiconductor.

into vacuum. The semiconductor has a forbidden energy gap due to the periodicity of its atomic lattice. Since the eigenstate for this wavefunction has an energy within the bandgap, $\psi(x)$ must decay into the semiconductor as well. The altered lattice potential at the interface accounts for the resultant localization of $\psi(x)$ and discontinuity in $\partial \psi(x)/\partial x$.

Figure 4.1b schematically illustrates the analogous wavefunction behavior at a semiconductor–metal interface. Here the wavefunction $\psi(x)$ is depicted as a propagating wave that again decays into the semiconductor's energy gap. Continuity of $\psi(x)$ at the interface again leads to wavefunction localization and charge accumulation at the interface. As with the semiconductor–vacuum interface, such states are considered to be intrinsic since they involve only the properties of the semiconductor and metal separately.

The study of intrinsic states has been motivated by the relationship between such states and Schottky barrier formation. High densities of intrinsic surface states within the semiconductor bandgap will strongly influence its Schottky barrier formation. Models of surface states must account for the variation of Schottky barrier behavior for different semiconductors. Schottky barrier measurements of different metals on a variety of vacuum-cleaved surfaces revealed that the sensitivity of a semiconductor's barrier height to different metals increased strongly with the ionicity of the semiconductor. Kurtin, McGill, and Mead defined an *index of interface behavior S* such that

$$\Phi_{SB} = S(\Phi_M - \Phi_{SC}) + C \tag{4.1}$$

where C is a constant [1]. The value of S is obtained from the slope of a qV_B versus metal work function Φ_M (or metal electronegativity χ) plot. The inset in Figure 4.2 shows a schematic energy band diagram of a metal–semiconductor interface with band bending qV_B and its difference ($\Phi_M - \Phi_{SC}$) indicated by a dipole of magnitude Δ. Interfaces with large dipoles will have low values of S and vice versa.

Figure 4.2 shows a plot of S versus a measure of the *semiconductor ionicity*, the *electronegativity difference* ΔX between anion and cation [2]. Semiconductors such as Si and GaAs exhibit S values of 0.1 or less while more ionic semiconductors

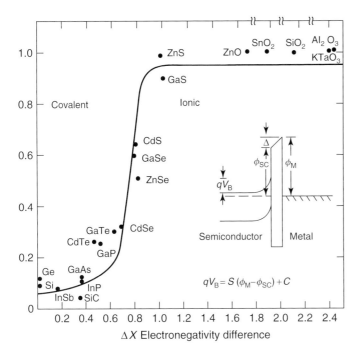

Figure 4.2 The index of interface behavior S, as defined in the inset, plotted versus electronegativity difference ΔX of the semiconductor constituents [1].

such as ZnO and SiO$_2$ have S values close to unity. This can be understood as the lattice disruption due to the lattice termination of the surface being larger for more covalent versus ionic-bonded lattices, resulting in a higher density of surface states for covalent semiconductors. This well-known curve of interface behavior was widely interpreted as evidence for intrinsic surface states.

4.2
Simple Model Calculation of Electronic Surface States

A simple model calculation [3] illustrates how quantum-mechanical boundary conditions on the semiconductor wavefunction introduce localized states at the surface. Consider a semi-infinite chain of atoms with nearly free electrons. For a periodic lattice, atoms in the surface plane have two-dimensional periodicity, but the translational symmetry of the bulk lattice normal to this plane is broken at the surface. A general one-electron wavefunction ψ_{SS} for states localized near an ideal surface consists of the product of a plane wave envelope function $\exp(i\,\mathbf{k}_\parallel \cdot \mathbf{r})$ and a periodic Bloch function $U_k(r)$ for coordinates $\mathbf{r}_\parallel = (x, y)$ parallel to the surface.

$$\psi_{SS}(\mathbf{r}_\parallel, z) = u_{k_\parallel}(\mathbf{r}_\parallel, z)\exp(i\mathbf{k}_\parallel \cdot \mathbf{r}) \tag{4.2}$$

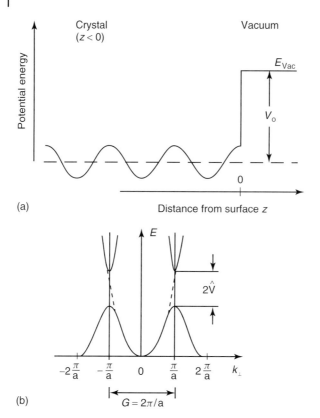

Figure 4.3 Nearly free-electron model for a cosine potential along a linear chain (z-direction). (a) Potential energy in the presence of a surface at $z = 0$. (b) Energy bands $E(k_\perp)$ for one-electron bulk states. $G = 2\pi/a$ is a reciprocal lattice vector. After Lüth [3].

where $k_\parallel = (k_x, k_y)$ is a wave vector parallel to the surface. The Bloch function has the same periodicity as the lattice potential $V(z) = 2V'\cos[2\pi z/a]$ where a is the lattice periodicity in the z-direction such that $V(z) = V(z + na)(z < 0)$.

Figure 4.3a shows this periodic potential for $V < 0$ and the potential step V_0 at the vacuum interface. Electron states at wave vectors away from the Brillouin zone (BZ) boundaries in (b) exhibit plane wave character and a parabolic dependence of energy on $k_\perp$ for $-\pi/a < k_\perp < \pi/a$ [3]. Here $k_\perp$ is a wave vector perpendicular to the surface. Near the zone boundaries $k_\perp = \pm\pi/a$, electrons scatter from a state with π/a to $-\pi/a$. Continuity of the wavefunction and its derivative at these periodic boundaries permit a solution to lowest order approximation consisting of a superposition of two plane waves.

$$\psi(z) = A\exp(ik_\perp z) + B\,\mathrm{e}\left\{i\left(k_\perp - 2\pi/a\right)z\right\} \tag{4.3}$$

4.2 Simple Model Calculation of Electronic Surface States

Substituting $\psi(z)$ and $V(z)$ into the Schrödinger equation

$$\left[-(\hbar^2/2m)(d^2/dz^2) + V(z)\right]\psi(z) = E\psi(z) \qquad (4.4)$$

yields energy eigenvalues for $k_\perp = \kappa + \pi/a$ for small κ near the zone boundary,

$$E = (\hbar^2/2m)\left[\pi/a + \kappa\right]^2$$
$$\pm V\left\{-(\hbar^2\pi\kappa)/(ma|V|) \pm \sqrt{\left[(\hbar^2\pi\kappa)/(ma|V|)^2\right] + 1}\right\} \qquad (4.5)$$

which can be approximated as

$$E \cong (\hbar^2/2m)\left[(\pi/a) + \kappa\right]^2 \pm V\{\sim 1\} \text{ for small } \kappa \qquad (4.6)$$

Hence, there are two energy solutions at the BZ boundaries such that $E_1 - E_2 = \Delta E \cong 2|V'|$. This characteristic energy splitting is equivalent to the semiconductor bandgap E_G. For example, see [4] for a discussion of plane wave solutions to the Schrödinger equation and the energy band splitting that results from lattice periodicity.

These energy eigenvalues in Equation 4.6 can now be substituted into the Schrödinger equation, Equation 4.4, to solve for prefactors A and B in Equation 4.3 $\psi(z)$ and its derivative $d\psi/dz$ must now be matched at $z = 0$ with a surface wavefunction that decays into vacuum. On the vacuum side,

$$\psi(z) = D\exp\left\{-z\sqrt{[2m(V_0 - E)/\hbar^2]}\right\}, \quad E < V_0 \qquad (4.7)$$

where V_0 is the potential difference between reference energy levels in the solid (dashed line) and the vacuum (E_{vac}).

On the semiconductor side, $\psi(z)$ can have either real or imaginary k, depending on whether the energy is outside the forbidden gap (real k) or inside (imaginary). For real k, $\Psi(z \leq 0)$ has the form of a standing Bloch wave as shown in Figure 4.4a. For imaginary $k = iq$,

$$\psi(z \leq 0) = F\exp(qz)\{\exp(i[(\pi/a)z \pm \delta])$$
$$- (\pm)\exp(-i[(\pi/a)z \pm \delta])\}\exp(-(\pm i\delta)) \qquad (4.8)$$

where $\sin(2\delta) = -(\pi\hbar^2 q)/(ma|V'|)$. Here $\Psi(z \leq 0)$ is a standing wave with an exponentially decaying amplitude as shown in Figure 4.4b. This wavefunction vanishes for $z \gg 0$, far from the surface in vacuum, and it displays a local maximum near the surface.

Matching $\psi(z)$ and $d\psi(z)/dz$ at $z = 0$ provides two equations which determine energy E and the prefactor ratio D/F. This exercise of matching exponentially decaying wavefunctions at the surface provides a real example of the localized surface state pictured schematically in Figure 4.1a, and it yields a single, well-defined energy level in the gap between bulk states. Note that such localized states are a unique feature of semiconductors since these materials possess a bandgap. This requires any wavefunctions at the surface to decay into the solid for energies inside the forbidden gap.

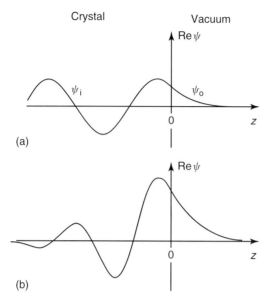

Figure 4.4 Real part of the one-electron wavefunction, Re(ψ), for (a) a standing Bloch wave (ψ_i), matched to an exponentially decaying tail (ψ_0) in the vacuum; (b) a surface-state wavefunction localized at the surface ($z = 0$). Wavelengths are on a scale of angstrom [3].

4.3
Intrinsic Surface States

4.3.1
Experimental Approaches

A number of experimental techniques have been developed to measure the properties of intrinsic states at semiconductor surfaces. Chief among these are photoemission spectroscopies that use photons to excite emission of electrons from states at the surface. These probes include *ultraviolet photoemission spectroscopy* (UPS), *soft X-ray photoemission spectroscopy* (SXPS), *angular-resolved photoelectron spectroscopy* (ARPES), *constant initial state spectroscopy* (CIS), and *constant final state spectroscopy* (CFS). Electrons incident on the semiconductor surface can also excite electron transitions that involve surface states with a technique termed *electron energy loss spectroscopy* (EELS). Several optical and electrostatic techniques such as *surface reflectance spectroscopy* (SRS), ellipsometry, and *surface photovoltage spectroscopy* (SPV) are capable of detecting surface states. Likewise, *scanning tunneling spectroscopy* (STS) can provide surface-state measurements that are localized to within individual atoms on the surface. The chapters that follow discuss all of these techniques.

4.3.2
Theoretical Approaches

Since Bardeen's classic paper, researchers have developed many theoretical approaches to describe intrinsic surface states. As Section 4.4 showed, the presence of the semiconductor surface leads to two-dimensional states and, in general, a surface band structure with distinctive differences from the bulk. Surface localization of such states within the forbidden gap is apparent from the associated wavefunctions pictured in Figure 4.4, which decay exponentially into the vacuum and into the bulk. States outside of the forbidden gap, termed *resonances*, also contribute to the surface density of states.

Theoretical calculations of electronic structure depend on the atomic structure and the correlation between electrons at the surface or interface. As a result, such calculations require either (i) a well-established surface atomic geometry or (ii) a self-consistency with both the electronic and the structural data. Until recently, well-defined surface or interface geometries have been the exception in surface or interface studies. As a result, theoretical approaches to characterize the atomic structure of semiconductor surfaces and interfaces fall into one of three categories: (i) prediction of an interface's lowest energy configuration of bonding and structure based on a number of calculational schemes; (ii) calculation and statistical comparison of low energy electron diffraction (LEED) or scanning tunneling microscopy (STM) data to provide "best-fit" models; and (iii) calculations of spectroscopic features of photoemission, electron energy loss, and other spectroscopies to be described in later chapters. See [5] for a review of theoretical techniques for the semiconductor electronic structure.

Among the various theoretical and computational methods, one can distinguish between models of ground-state properties and excited-state models that involve electrons after energy excitation [6]. Models designed to calculate one class of properties may not be appropriate for the other. An example of a ground-state model is the calculation and minimization of surface atomic structure in terms of independent structural parameters such as atomic bond lengths and angles. Here the electronic structure of the surface is involved insofar as the charge density and surface-state features contribute to the ground-state energy. Calculations based on LEED observations fall into this ground-state category. Calculations of spectroscopic features to derive the electronic structure require the use of excitation spectra obtained from various experimental techniques to be described in the following chapters. One typical approach involves comparison of the observed and calculated features for the surface-state energies and their dependence on momentum, that is, their dispersion across the BZ. Yet another involves layer-by-layer calculations of occupied valence-band densities of states.

For the elemental semiconductor Si, a wide variety of approaches have succeeded in accounting for many surface electronic features. However, despite extensive experimental study, these have yielded only an incomplete picture of the Si surface. On the other hand, several methods have successfully accounted for the changes in surface atomic structure, termed *surface reconstruction*, in zincblende compound semiconductor surfaces [6].

Electronic structure calculations of semiconductor surfaces indicate a substantial influence of the surface atomic structure. The termination of the semiconductor lattice at the surface produces both dangling bonds that project into vacuum as well as changes in back bonding that extend into the bulk. These charge distributions depend sensitively on the local coordination of atoms at the surface, for example, the formation of twofold, threefold, or multilayer bond configurations, as well as morphological features such as steps and long-range strain. Associated with these altered charge distributions are multiple energy bands, both inside and outside the forbidden bandgap. The number of such bands with the bandgap depends on the crystal orientation and the number of dangling bonds per unit cell [7]. While the full set of surface states and resonances depends on the overall charge distribution in the top semiconductor layers, it is found that changes in the coordination of the top layer alter the occupancy of these surface-state bands substantially. Thus a change in atomic coordination can change a "metallic" surface, which includes a partially filled surface-state band, to a "semiconducting" surface, which includes fully occupied and unoccupied surface-state bands [5]. While a unique correspondence does not exist in general between these surface bonds and bands, such effects serve to emphasize the first-order effect of local atomic bonding on intrinsic surface-state properties.

4.3.3
Intrinsic Surface-State Models

Intrinsic surface states can be grouped into four categories: (i) *Shockley states* [8] corresponding to the nearly free electron model used in Section 4.2; (ii) *Tamm states* [9] consisting of a linear combination of atomic eigenstates (i.e., quantum-mechanical wavefunctions of electrons orbiting a specific atom); (iii) *dangling bond states* associated with particular orbital lobes for chemical bonding, for example, sp^3 in Si; and (iv) *back-bond states* associated with surface-induced modification of chemical bonds between atomic layers. The relationship of atomic orbitals to surface states may be illustrated qualitatively by their relation to energy bands. In a solid, many atoms are brought together and, due to the Pauli exclusion principle, their atomic orbital energies spread into a continuum of energy levels termed *bands* [10]. Figure 4.5 shows two atomic energy levels A and B that form such bands at their equilibrium bond distances. These correspond to the valence and conduction bands respectively, if the lower band is fully occupied and the upper band is empty. Surface atoms are not fully coordinated with their neighbors as they would be in the bulk. Hence, in a *tight-binding model* of surface states, such atoms will have electronic energy levels that shift toward the energy levels of the free atoms. These energy levels are termed *split-off bands* and lie in the forbidden bandgap.

Surface states derived from the conduction band can accept charge and are negative when filled, whereas states derived from the valence band can donate

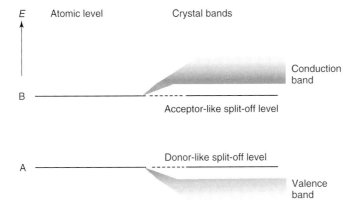

Figure 4.5 Split-off surface-state levels in a tight-binding picture. Surface atoms have fewer bonding partners than bulk atoms so that their electronic energy levels shift toward those of the free atoms. Such states can accept or donate charge depending on their valence or conduction band origin [3].

charge and are positive when empty. Following the conventional definition for shallow bulk dopant atoms, such conduction and valence band-derived states are termed *acceptors* and *donors*, respectively. In general, such states are built up from a combination of atomic orbitals and thereby valence and conduction band wavefunctions. Depending on their proximity to one or the other band, they will have more acceptor- or donor-like character.

4.3.4
Intrinsic Surface States of Silicon

The experimental and theoretical results for Si intrinsic surface states show that (i) intrinsic surface states exist near or within the Si bandgap with spectral features that vary from surface to surface and that correspond to experimental observations with varying degrees of success; (ii) the density of intrinsic surface states depends sensitively on the detailed surface reconstruction of the Si surface; (iii) complete characterization of the surface-state eigenvalue spectrum for Si surfaces requires firmer determinations of their respective surface atomic geometries plus a detailed understanding that Coulomb interactions have on deviations from one-electron models of surface electronic structure.

4.3.5
Intrinsic Surface States of Compound Semiconductors

Research on compound semiconductors shows that intrinsic surface states exist within the valence band of both cleaved (110) and noncleaved surfaces and within the bandgap on noncleaved surfaces [5, 6]. Early experimental and theoretical studies suggested that intrinsic surface states existed within the bandgap of cleaved semiconductor surfaces as well, but it is now generally accepted that the cleavage surfaces of most III–V as well as the few II–VI compound semiconductors studied have no significant intrinsic states within the forbidden gap [11, 12]. Initial observations of such bandgap states have been reinterpreted in light of (i) energy corrections due to surface core-hole relaxation and (ii) electrically-active defect phenomena such as cleavage steps [13, 14].

Calculations based on surface atoms in their bulk, that is, unreconstructed, lattice positions indicated empty or both empty and filled states within the bandgap. However, surface atoms can rearrange their positions to minimize their bond energies such that the energies of these empty and filled states move above and below the bandgap, respectively. In other words, these surface states are *swept out* of the bandgap. Refined to include such modifications, calculations could then account for the experimental result that surface states are absent within the III–V and II–VI compound semiconductor bandgaps [11]. These results demonstrate that surface states are very sensitive to the detailed atomic structure at the semiconductor surface.

Figure 4.6 provides a useful roadmap to surface-state energies and their dependence on crystal momentum along different crystallographic directions, that is, their *wave vector dispersions*. Shown here are the calculated bulk allowed energy bands of GaAs projected along various symmetry directions of the crystal [15]. In comparison, the closed circles [16] and open squares [17] indicate experimental measurements of these energies using ARPES, a technique to be described in Chapter 8 for electrons in the outer few atomic layers. Figure 4.6 shows measured surface-state bands located a few volts above and below the conduction- and valence-band edges, respectively, rather than at energies within the bandgap. States near E_C are primarily derived from cations such as Ga in GaAs, while states near E_V are derived mostly from anions such as As. States that are very near the forbidden gap are dangling bond states, analogous to those pictured in Figure 4.5. States further below the bandgap are back-bond states between surface atoms and atoms further inside the lattice. The reason that surface states lie outside rather than inside the bandgap is that a bond rotation occurs between surface atoms that lowers the total energy of occupied electronic states.

Figure 4.7 illustrates the densities of electron states versus energy near the GaAs bandgap. If the ideal GaAs crystal structure extends to the outer layer of atoms, then the surface Ga (filled circles) and As (open circles) have symmetric positions (b) and the corresponding density of states (a) has pronounced peak features at E_V and E_C that extend into the forbidden gap. However, experimental measurements reveal that there is a 27° bond rotation of Ga and As atoms within the last atomic layer of the (110) surface. As a result, these density of states peaks move into the conduction and valence bands, leaving the forbidden gap free of intrinsic surface states.

This behavior of surface states at the (110) surface is characteristic of III–V and II–VI compound semiconductors in general; that is, there are no intrinsic surface

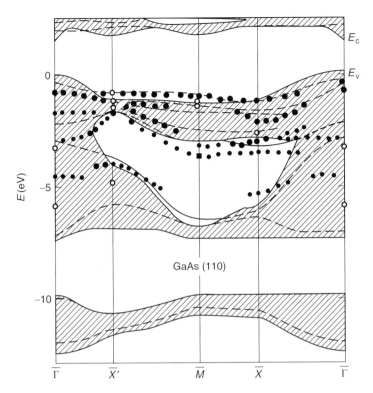

Figure 4.6 Calculated GaAs (110) surface-state dispersions (solid lines) derived from a tight-binding mode [15] compared with measured dispersions from the angle-resolved photoemission measurements (closed circles [16] and open squares [17]). Shaded areas indicate the projected bulk allowed energy bands of GaAs [6]. The relaxation sweeps the surface-state bands out of the semiconductor bandgap denoted by E_C and E_V.

states in the bandgap. Table 4.1 summarizes these observations for a wide variety of compound semiconductors. Each surface is denoted not only by its crystal orientation but also by its atomic reconstruction. Here inverse photoemission spectroscopy (IPS) yields the energies of photons emitted due to incident electrons filling unoccupied surface states [12]. In each case, surface states are present but lie at energies higher than the conduction band edge E_C. Only the elemental semiconductors Si and Ge exhibit states within the bandgap and only for certain reconstructions.

Noncleavage surface orientations of compound semiconductors can exhibit states within the bandgap. However, such surfaces have, until recently, been prepared only by polishing, ion bombardment, or vapor-phase epitaxy on oriented surfaces. Such treatments can produce different surface structures, depending on the specific process conditions and, with the exception of epitaxial growth, they can produce damage to the semiconductor lattice many layers below the surface.

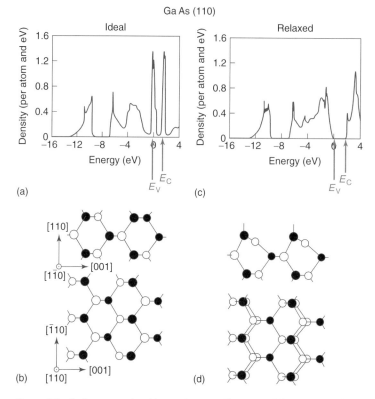

Figure 4.7 Surface-state densities and the corresponding structure models for the GaAs(110) surface. (a,c) Calculated surface-state densities for the ideal, nonreconstructed (a) and the relaxed (c) surface. The zero of the energy scale is taken to be the upper valence-band edge, $E_V = 0$. (b,d) Structure models (side and top view) for the ideal, nonreconstructed (b) and the relaxed (c) surface.

The absence of intrinsic surface states within the bandgap of cleaved compound semiconductor surfaces is significant since metal deposited on such surfaces typically results in charge transfer and Fermi level stabilization within less than a few monolayers. This behavior indicates a local electronic mechanism. The fact that intrinsic surface states play no direct role in Fermi level stabilization demonstrates that extrinsic and/or metal-induced features play a dominant role in Schottky barrier formation for many semiconductors.

4.3.6
Dependence on Surface Reconstruction

Calculations and measurements of electronic excitation spectra reveal a sensitive dependence of electronic densities of states on the surface atomic geometry. In addition to the relaxation effects such as in Figure 4.7d, surface reconstruction produces significant changes in the localized charge states and the surface electronic

Table 4.1 Energies relative to the valence-band maximum of unoccupied surface states for different semiconductors and surface reconstructions, as determined by inverse photoemission spectroscopy.[a]

Surface	Critical point	$E_c - E_v$ (eV)	$E - E_v$ (eV)
Si(111)(2 × 1)	Γ	1.1	1.3
	J	–	0.65
	J'	–	1.4
Si(111)(7 × 7)	Γ	–	1.2
Si(100)(2 × 1)	Γ	–	0.7
Ge(111)(2 × 1)	Γ	0.67	1.5
	J	–	0.15
Ge(111)c(2 × 8)	Γ	–	1.0
GaP(110)(1 × 1)	Γ	2.26	2.4
GaAs(110)(1 × 1)	Γ	1.43	2.1
	X	–	1.7
	X'	–	2.0
GaSb(110)(1 × 1)	Γ'	0.7	2.1
InP(110)(1 × 1)	Γ'	1.35	2.7
InAs(110)(1 × 1)	Γ'	0.36	1.9
InSb(110)(1 × 1)	Γ'	0.18	1.9
CdS(11–20)(1 × 1)	Γ'	2.42	3.8
	Γ'	–	5.8
CdSe(11–20)(1 × 1)	Γ'	1.73	–
CdTe(110)(1 × 1)	Γ'	1.58	2.9

[a] After Himpsel and references therein [12].

band structure. A simple example is the difference in buckling that can occur for the Si (100) surface. Electron diffraction and STM reveal that this surface can form *dimers*, that is, surface atoms that form extended arrays of bonded pairs at the surface, such that the periodicity in one crystal symmetry direction is twice that of another perpendicular to it, both in the surface plane. This (2 × 1) dimerization can yield a symmetry that can be described by either symmetric or asymmetric dimers as pictured in Figure 4.8. Both exhibit the periodicity measured experimentally but have asymmetric versus symmetric dimer atom geometries.

Calculations of the minimum surface energy for atoms based on these geometries yield the corresponding energy versus wave vector surface-state dispersions shown in Figure 4.8, where zero corresponds to the Fermi level [18, 19]. For the symmetric dimer, the occupied and empty surface states near E_V show a metallic behavior, that is, partially filled and empty energy bands. This is in contrast to the asymmetric dimer, which exhibits semiconducting behavior, that is, filled states below E_V separated by an energy gap from empty states above. This energy dependence with different dimer reconstructions shows that surface atomic geometries with the

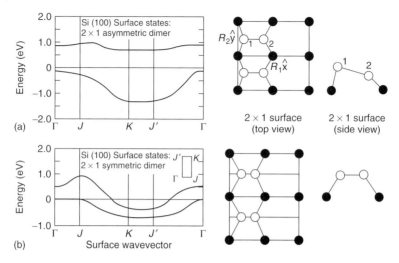

Figure 4.8 Comparison of surface-state dispersions calculated for asymmetric (a) and symmetric (b) Si(100) (2 × 1) dimer reconstructions, showing filled and empty bands near the valence-band maxima [18]. The corresponding dimer geometries (side view not to scale) appear to the right of each band diagram [19]. The partially filled and empty energy bands of the symmetric dimer indicate a metallic character, in contrast to the asymmetric dimers' semiconductor character.

same symmetry and periodicity but with different bond angles can produce different charge transfer, total energies, and semiconducting versus metallic features.

Additional reconstructions with the same geometric symmetry appear in Figure 4.9 [20]. Rather than the ideal bulk or dimer geometries shown in (a) and (b), respectively, missing rows (c) or π-bonded chains of atoms in a zigzag pattern will also exhibit the (2 × 1) symmetry. In fact, the chain model provides substantially lower total energies per surface atom than buckled geometries [21, 22]. In each case, the periodicity of the surface increases by a factor of 2. In turn, this reduces the BZ dimension by the same factor. The lattice distortion responsible for the altered symmetry can introduce an energy splitting at the new BZ boundary as pictured in Figure 4.10.

For a dangling bond state that is half filled with electrons, this distortion then represents an energy gain since the energies of filled states near the BZ edge are lower than without the perturbation. This electronic energy reduction is partially compensated by an increase in mechanical energy. Such an effect is independent of the detailed model for the surface reconstruction. The symmetry reduction splits the dangling bond's surface-state band, reducing the total energy. Molecular symmetry lowering that reduces electronic energy is common and termed a *Peierls instability* [23].

It is possible to distinguish between chainlike versus rowlike models since the former exhibits strong optically detected polarization parallel to versus perpendicular to π-chains. In general, one requires a combination of techniques to distinguish between such models.

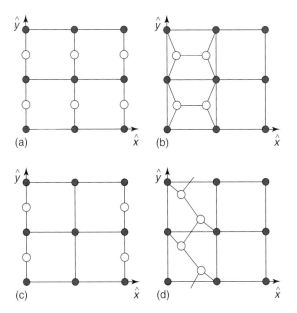

Figure 4.9 Alternative top views of the reconstructed Si(100) outermost atomic layers: (a) the ideal bulk geometry; (b) the surface dimer model; (c) the missing row model; and (d) the conjugated-chain model. Open circles denote top-layer atoms. Filled circles denote second-layer atoms [20].

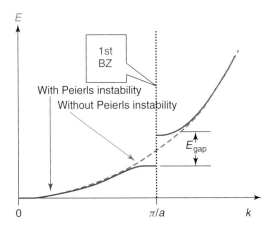

Figure 4.10 Schematic illustration of the energy gain due to the formation of a (2 × 1) surface reconstruction. The shrinking of the Brillouin-zone (BZ) dimension by a factor of 2 in one direction causes the half-filled dangling bond surface-state band (dashed line) to be folded back into the new (2 × 1) zone and split at the zone boundary, thus leading to a reduction of the total electronic energy [23].

Compared to elemental semiconductors, III–V compounds exhibit strong differences in spectroscopic features for different reconstructions, even for structures that exhibit the same diffraction pattern. Indeed, reconstructions of the GaAs(001) surface typically used in devices can be monitored using spectroscopic ellipsometry while the crystal is being grown [24]. II–VI compound reconstructions are similar to those of III–V compounds, and there is no evidence for intrinsic surface states in the bandgaps of II–VI compounds.

4.3.7
Intrinsic Surface-State Summary

Section 4.3 shows that surface states and band structure of semiconductors depend sensitively on surface atomic structure. Surface atomic structures can differ substantially from the bulk structure. These reconstructions of surface atoms can vary in many ways, each of which has distinctive electronic properties. In general, a detailed comparison of spectral and theoretical features is required to characterize surface atomic structures. While intrinsic surface states are a natural result of the lattice termination, their energies typically fall outside the forbidden energy gap. Only elemental semiconductors with particular reconstructions appear to exhibit surface states that can play a role in Schottky barrier formation.

4.4
Extrinsic Surface States

4.4.1
Weakly Interacting Metal–Semiconductor Interfaces

A second form of electronic state localization can occur at semiconductor surfaces due to the presence of a metal interface. This case is pictured in Figure 4.1b and involves the tunneling of wavefunctions from the metal into the semiconductor. Such wavefunction tunneling is not dependent on the specific atomic arrangement of atoms on the semiconductor surface or on the energy band structure of the metal. It depends instead on the bulk band structure of the semiconductor and the modified local potential at the interface due to exchange-correlation forces. Here electron–electron interactions and the microscopic details of the electronic and atomic charge densities dominate the electron dynamics [25].

The concept of wavefunction tunneling to account for localized states was first introduced by V. Heine in 1965 [26]. Since then, numerous models have been proposed that incorporate wavefunction tunneling to determine the Fermi level position inside the semiconductor bands at the metal–semiconductor interface. These models are described in chapters that follow. However, pseudopotential calculations of local charge density as a function of energy and position provide an explicit illustration of the charge redistribution that occurs at the metal–semiconductor interface [27]. They yield interface density of states information both as a function of energy and spatial position near the interface. Thus Appelbaum and Hamann's calculation of an Al–Si interface was based on a substitutional replacement of Al for the top Si(111) layer. Using a fitted model potential for the ion cores, an exact Hartree potential calculated from the valence-band charge density, and a local approximation for the exchange and correlation potential [28], they obtained several bands of surface states, including one within the Si bandgap.

Calculations of Louie and Cohen [27, 29] employed Si pseudopotentials to approximate the potentials of the ion cores for the clean Si surface [30] and a *"jellium"* model, corresponding to a "sea" of loosely bound electrons, to approximate the Al metal overlayer. The slab configuration shown in Figure 4.11 is periodically repeated to permit use of standard band structure techniques. Figure 4.11 illustrates local charge rearrangement in real space, while Figure 4.12 shows the distribution of interface states in energy. Figure 4.11a shows the contours of valence-charge density on an atomic scale in a cross section of the Al–Si(111) interface. Figure 4.11b illustrates the total valence-charge density $\rho_{total}(z)$ as a function of distance z perpendicular to the same interface. The constant $\rho_{total}(z)$ within the Al results from the jellium model assumed for the metal, while the peaks in $\rho_{total}(z)$ within the semiconductor correspond to contributions from the semiconductor bond charges.

Near the interface, $\rho_{total}(z)$ exhibits new features unlike those of either the bulk Al or the Si. Panels III and IV in Figure 4.11b show that some charge transfer takes place from the Al to the Si at the interface in this model, yielding an interface

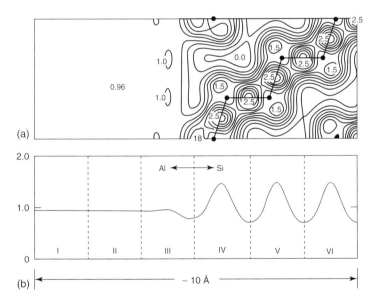

Figure 4.11 (a) Total valence-charge-density contours in a (110) plane for the Al–Si(111) interface (out of the page). Dots indicate the Si atoms. (b) Total valence-charge density averaged parallel to the interface. Spatial regions I–VI lie in the direction normal to the interface [29].

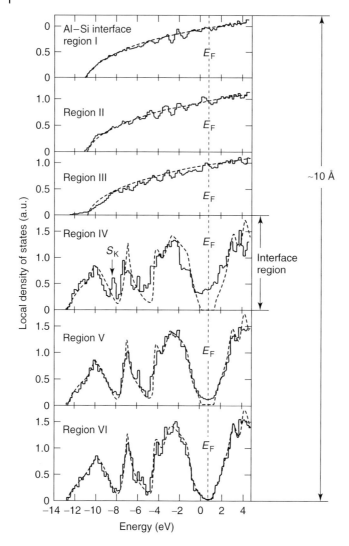

Figure 4.12 Local density of states in arbitrary units for the Al–Si(111) interface. The regions I–VI correspond to those shown in Figure 4.11. Dotted lines indicate bulk densities of Al or Si states [27].

dipole extending over atomic dimensions. The corresponding densities of states are shown in Figure 4.12 [27]. Region I displays a density of states just as in bulk Al, region VI just as in bulk Si, whereas new features appear near the plane of the Al–Si junction plane. These include interface states that are bulklike in the Si but that decay rapidly in the Al, states that are bulklike in the Al but that decay rapidly in the Si – as in Figure 4.1b, and truly localized interface states that decay away from the interface in both directions. The bulklike Al states decay into the Si whenever the range of the semiconductor bandgap is inside the metallic band. Louie and Cohen associated these *"metal-induced gap states"* or *MIGS*s with the Fermi level pinning because the energy of these states within the semiconductor bandgap agrees with experimental values [31] for Si Schottky barriers. Other calculations have refined the potential used in these calculations (32, 33). These reconfirm the disappearance of the intrinsic Si surface states and the appearance of metal-induced gap states.

The semiconductor ionicity has an effect on the metal-induced gap states. Calculations show [34] that the density of localized surface charge per unit energy decreases for more ionic semiconductors, leading to more Fermi level movement and a larger index of interface behavior S for ionic versus covalent semiconductors as in Figure 4.2. See [35] for a review of these and other techniques for calculating the electronic states at metal–semiconductor interfaces. Many of these calculations show that metal-induced interface states can produce electronic features that can account for Schottky barrier formation. However, they share the common shortcoming of employing ideal atomic geometries. The results presented earlier in this chapter emphasize that electronic structures depend sensitively on the detailed bonding arrangement at the interface. Such results underscore the importance of characterizing atomic structure near the intimate metal–semiconductor interface.

4.4.2
Extrinsic Features

The results thus far suggest the central role of metal–semiconductor interactions in charge transfer at interfaces. Further evidence emerges from correlations of experimentally measured barrier parameters with interface-specific parameters. Experimental measurements of macroscopic Schottky barriers explicitly reveal the significance of these extrinsic, interface-specific phenomena. Extrinsic surface states fall into three categories: (i) Trapped charge inherent in the semiconductor. These include bulk defects, grain boundaries, and impurities. (ii) Imperfections created with the semiconductor surface. These include cleavage steps, mechanically induced point defects, dislocations, and chemisorbed impurity atoms. (iii) Charge states created by interaction between the semiconductor and a contact material. These include interface chemical reaction, semiconductor atomic outdiffusion and/or metal indiffusion, and the associated creation of electrically active sites.

4.4.3
Schottky Barrier Formation and Thermodynamics

Indications to the effect that thermodynamics plays a role in Schottky barrier formation appeared in the early 1970s. Kipperman and Leiden [36] noticed a kink in the plots of barrier height Φ_{SB} versus metal work function Φ_M, which they attributed to a reaction product altering the effective work function of the contact metal. Data for metals on GaS and GaSe displayed this kink between metals that do or do not react with the two semiconductors. Kipperman and Leiden suggested that metal–GaS or GaSe reactions would produce a thin Ga layer at the interface

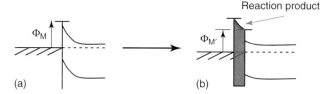

(a) (b)

Figure 4.13 Schematic [111] band diagrams of: (a) abrupt metal–semiconductor junction and metal work function Φ_M and (b) reacted metal–semiconductor interface with altered band bending due to reaction product with work function $\phi_{M'}$.

which changed the effective metal work function to that of Ga. Indeed experiments showed [37] that Al on GaS produced free Ga with heat treatment according to the reaction Al + GaS → AlS + Ga, whereas Au – which forms no reaction product with these chalcogenides – did not. For these materials, the deviation from ideal Schottky barrier formation as defined by Equation 3.3 was attributed to interface work function changes alone rather than to surface states.

A correlation between barrier heights of transition metals (TMs) on Si with the *heats of formation* ΔH_F of TM silicides first suggested that thermodynamics plays a role in Schottky barrier formation for metals on Si [38]. Figure 4.14 shows this apparently linear dependence of Φ_{SB} on ΔH_F where the barriers are among the most reproducible of those known for metal–semiconductor interfaces. The ΔH_F are those of the most stable silicide that can form and that agree with those predicted from binary phase diagrams [39].

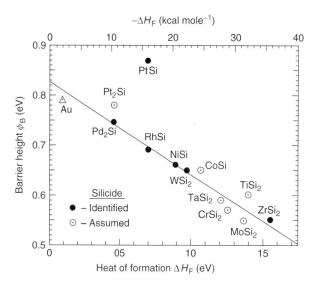

Figure 4.14 Barrier heights Φ_B of TM silicides plotted versus heats of formation ΔH_F [38].

ΔH_F is defined as $\Delta G + T\Delta S$ (at constant temperature), where $\Delta G =$ Gibbs free energy change, $\Delta S =$ enthalpy change, and ΔH_F is obtained from standard thermodynamic references. A number of tabulations list experimental values of ΔH_F for semiconductor and metal–semiconductor compounds, for example, [40, 41], and [42]. Chemical reactions between TMs and Si are well known and are indeed integral to the manufacture of microelectronics. Such reactions involve a movement of the interface away from any impurity contamination of the initial surface into the bulk of the Si [43, 44]. Hence, even though contamination may exist at the free Si surface, the interface between the metallic silicide and Si is clean.

One can interpret Figure 4.14 as the degree of hybridized bonding between TMs and Si scaling with ΔH_F such that the shift in ϕ_{SB} is proportional to the local charge transfer of the new chemical bond [38]. This work was the first to emphasize the role of microscopic chemical bonding on macroscopic electrical properties.

Subsequent work provided phenomenological evidence for a direct connection between metal–Si barriers and the bulk phase diagrams of the TM silicides. Figure 4.15 illustrates a plot of TM barrier heights with Si as a function of their lowest eutectic temperatures [45]. The eutectic chosen is that lowest on the metal side for silicides whose growth is dominated by metal diffusion and vice versa for interfaces with growth dominated by Si diffusion. Figure 4.15a shows that Φ_B increases with decreasing eutectic temperature, corresponding to the earliest phase that forms. It also reflects the strength of the interfacial bonding but encompasses more metals than included in Figure 4.14. Figure 4.15b shows a representative phase diagram Si with a TM [46]. Here the vertical lines signify compounds with well-defined compositions. The lowest melting eutectic temperature indicated is 830 °C (1103 K). Figure 4.15a is a clear demonstration that physical properties of the

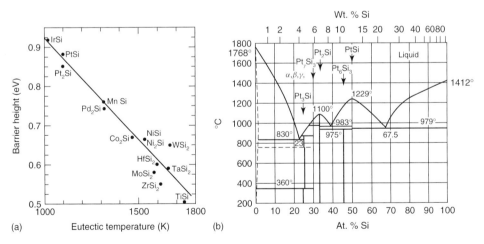

Figure 4.15 (a) Schottky barrier height Φ_{SB}^n (n-type) as a function of eutectic temperature for selected silicide-forming interfaces [45]. (b) Pt–Si alloy composition versus temperature phase diagram showing multiple eutectics and compounds [46].

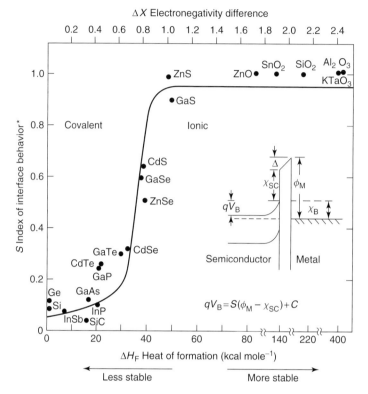

Figure 4.16 Transition in coefficient of interface behavior S between covalent and ionic semiconductors plotted versus electronegativity ΔX (upper scale) or chemical heat of formation ΔH_F (lower scale) of the semiconductor. The curve illustrates the dependence of S on the semiconductor stability against chemical reactions [50].

interface are related to an interfacial layer, although the details of this relationship are not apparent.

It is also possible to correlate barrier height with effective work functions of the TM silicide. Assuming a common limiting composition for all metal silicides of $\Phi_{M'} = (\Phi_{Si}^4 \Phi_M)^{1/5}$ at the intimate junction with Si, one obtains a roughly linear Φ_{SB}^n increase with increasing $\Phi_{M'}$ [47]. However, transmission electron microscopy (TEM) measurements of the Pd_2Si–Si [48] and $NiSi_2$ – Si [49] interfaces show abrupt interfaces with the lattice planes of the silicide and those of the Si substrates continuing virtually to intersection so that the transition from one lattice to another takes place within one lattice constant. Likewise, photoemission measurements suggest that the Pd_2Si–Si interface is metal- versus Si-rich, in contradiction to the Si-rich composition assumed. Thus an effective work function model can also account for Si barrier heights without localized states, but only with the assumption of hypothetical new interface chemical species that have not been observed.

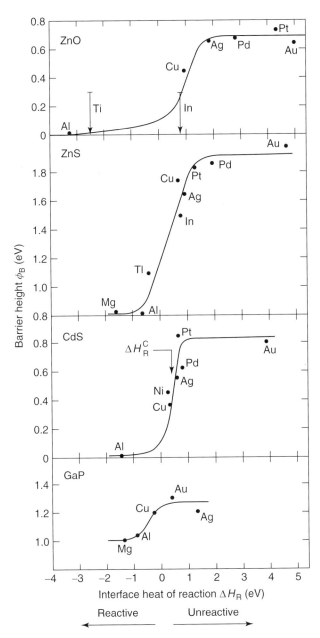

Figure 4.17 Barrier heights correlated with heats of interface chemical reaction ΔH_R for metals on ZnO, ZnS, CdS, and GaP. All of the semiconductors display the same qualitative behavior, regardless of ionicity. A critical heat of reaction $\Delta H_R^C \sim 0.5$ eV per metal atom, determined experimentally, marks the center of the transition region between reactive and unreactive interfaces. Calculated heats of reaction are per metal atom for the reaction [50].

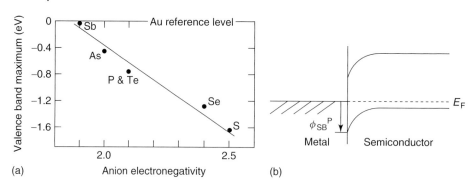

Figure 4.18 (a) Energy of the valence-band maximum relative to the Fermi level of Au contact ϕ_{SB}^P plotted versus anion electronegativity [57]. Semiconductor barrier heights for a given anion [52] are averaged. (b) Schematic energy band diagram of a p-type barrier with respect to the valence-band maximum.

The variation in the coefficient of interface behavior S defined by Equation 4.1 and presented in Figure 4.2 [1] also provides evidence for the underlying role of thermodynamics in Schottky barrier formation. Figure 4.16 presents a plot of S versus semiconductor heat of formation of the semiconductor [50]. This plot reveals a nearly identical correlation of S with ΔH_F as with electronegativity ΔX. Indeed, ΔH_F and ΔX can be related by either quadratic [2] or linear [51] expressions. This thermodynamic expression of Schottky barrier behavior for metals on compound semiconductors shows that covalent semiconductors are less stable against chemical reaction with metals than ionic semiconductors and that this relative instability correlates with narrower ranges of Schottky barriers for

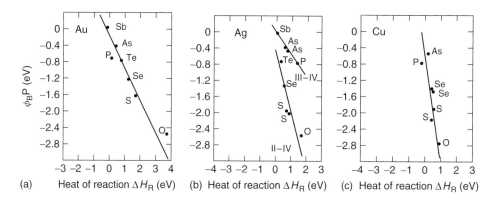

Figure 4.19 Schottky barrier height Φ_{SB}^P for (a) Au, (b) Ag, and (c) Cu plotted versus interface heat of reaction ΔH_R [11]. The same Φ_{SB}^P values plotted in Figure 4.18a are used here.

particular semiconductors. Such phenomena imply that chemical thermodynamics dominates Schottky barrier formation.

The dependence of Schottky barriers on chemical stability is even more evident when different metals on the same semiconductor are considered. The chemical stability of compound semiconductors against chemical reactions with metals can be expressed by a heat of reaction ΔH_R such that

$$\Delta H_R = (1/x)[H_F(CA) - H_F(M_X A)] \quad (4.9)$$

for the reaction $M + (1/x)CA \Rightarrow (1/x)[M_X A + C]$ where C, A, and M are the semiconductor cation, semiconductor anion, and metal, respectively, and $M_X A$ is the most stable metal–anion product normalized per metal atom [50]. Figure 4.17 illustrates plots of barrier heights measured by internal photoemission for clean interfaces [52] versus ΔH_R for the same semiconductor and different metals. In each case a qualitative change occurs near $\Delta H_R = 0$, the boundary between reactive ($\Delta H_R < 0$) and unreactive ($\Delta H_R > 0$) interfaces. Interface-specific experiments confirm that metals with $\Delta H_R < 0.5$ eV/metal atom react with CdS while metals with $\Delta H_R > 0.5$ eV/metal atom do not (50, 53).

Figure 4.16 also shows that these four semiconductors span the range from covalent to ionic, yet all four display the same qualitative change in barrier height with ΔH_R at $\Delta H_R \cong 0$. The same dependence of barrier height on ΔH_R also occurs for other semiconductors such as InP, a representative III–V compound semiconductor [54], $Zn_3 P_2$, a p-type semiconductor [55], and PbTe, a narrow-gap II–VI compound semiconductor [56]. The transition in Φ_B occurs at approximately the same $\Delta H_R \cong 0$ point in each case. The extension of Φ_B versus ΔH_R plots to other semiconductors systems appears to be limited only by the availability of electronic and thermodynamic data [11]. Figures 4.16 and 4.17 demonstrate that chemical stability factors into the metal-dependence of Schottky barrier heights.

Another barrier height correlation that emphasized the role of the anion appears in Figure 4.18. This plot shows that p-type barrier heights Φ_{SB}^P of Au on various III–V and II–VI compounds follow a systematic trend with anion electronegativity, regardless of the semiconductor [57]. These averaged Φ_{SB}^P imply that semiconductors with a common anion have approximately the same ϕ_{SB}^P, independent of the differences in bandgap. Similar plots with more scatter and some exceptions exist for Ag and Cu instead of Au [57, 58]. A roughly linear correlation between ionization potential $E_{VAC} - E_{VBM}$ and anion electronegativity is also found, emphasizing the strong anion character of the valence band [59]. These anion dependencies can be viewed at bond-producing, p-like atomic states of anion fixing the Au Fermi level and scaling with anion electronegativity. This "common-anion rule" has been verified experimentally for molecular beam epitaxy (MBE)-grown, mixed-cation layers of InGaAs [60], InGaSb [61], and $Ga_x Al_{1-x} As$ ($x < 0.3$) [62].

A similar relationship between Φ_{SB}^P and ΔH_R adds a thermodynamic perspective to this anion dependence. Figure 4.19 shows plots of the same barrier heights measured experimentally [52] versus ΔH_R for Au, Ag, and Cu [11]. As in Figure 4.18a, there is a strong correlation with anion for each metal. The nearly common ΔH_R for semiconductors with the same anion reflects the fact that semiconductors with the same anion have similar heats of formation [40] and thereby similar heats of reaction with the same metal. Such chemical reactivity plots can accommodate a larger range of semiconductors. Figures 4.17 and 4.19 provide systematic correlations of barrier height for the same semiconductor and different metals as well as the same metal and different semiconductors.

Thermodynamics can account for yet another commonly used rule of thumb, the "one-third rule," which expresses the observation that $E_F - E_{VBM} = \frac{1}{3} E_G$ for many metal–semiconductor interfaces [63, 64]. Since (i) semiconductor bandgaps are proportional to heats of formation [65, 66] and (ii) the "one-third rule" is based largely on relatively unreactive metals so that the semiconductor heat of formation dominates ΔH_R and thereby Φ_{SB}^P, both $E_F - E_{VBM}$ and E_G are proportional to H_F so that their ratio should remain relatively constant.

4.4.4
Extrinsic Surface-State Summary

This section has shown that both weakly and strongly interacting metal–semiconductor interfaces give rise to localized electronic states. Many researchers have found that chemical reactions and diffusion are quite common at metal–semiconductor interfaces such that atomically abrupt interfaces are the exception rather than the rule. These chemical interactions can occur even near room temperature and were only first detected using ultrahigh vacuum (UHV) surface science techniques capable of measuring the intimate metal–semiconductor interfaces. Various correlations of electrical barrier heights with thermodynamic parameters provide guidelines for estimating Schottky barrier heights and a starting point for understanding their underlying physical basis.

4.5
Chapter Summary

This chapter presented the various types of electronic states that can form at microscopic metal–semiconductor interfaces and that can account for the deviations of barrier height from ideal Schottky-like behavior. Intrinsic surface states at clean semiconductor surfaces form due to the lattice termination and the resultant atomic bond rearrangement. The energies of these states can lie both inside and outside the forbidden bandgap. The properties of these states depend sensitively on the detailed arrangement of the surface atoms. With the advent of UHV surface science techniques, researchers have been able to measure the unique electronic, chemical, and structural properties of microscopic metal–semiconductor interfaces as these junctions formed, monolayer by monolayer. The various correlations between barrier heights and thermodynamic parameters underscore the significance of these local electronic and chemical phenomena.

Over the past 50 years, our perspective of the solid–solid interface has changed. Measurements of the initial stages of Schottky barrier formation reveal that the interface electronic structure cannot, in general, be extracted from ideal bulk properties of abrupt metal–semiconductor interfaces. Instead the electronic structure must be understood in terms of a more complex interface involving (i) chemically reacted and/or diffused regions, (ii) local charge distribution and dipole formation, (iii) built-in potential gradients due to interdiffusion of dopants or impurities, and (iv) new dielectric properties.

Chapters 5–16 describe the wide range of techniques that have been developed to characterize and understand the properties of electronic materials' surfaces and interfaces. Chapters 17–21 show how researchers using these techniques have discovered a wide range of structural, chemical, and electronic phenomena. These findings provide the basis for understanding and controlling solid-state properties and the electronic devices that they enable.

Problems

1. (a) For a localized state in GaAs with energy level at mid-gap, calculate the decay length of its wavefunction into vacuum. (b) Assume that a metal $\psi(z)$ decays into the GaAs with an effective potential barrier $= \frac{1}{2} E_G$ and $m = m^*$. Considering only this valence band to mid-gap energy barrier, what is the corresponding decay length?
2. Compare the heats of nitride interface reaction for In, Al, and Cr (-117.2 kJ mol^{-1}) on GaN.
3. (a) For Mg on GaN, Mg$_3$N$_2$ ($\Delta H_F = -461.5$ kJ mol^{-1}) can form. Does Mg react more strongly (per metal atom) with GaN or ZnO at room temperature? Ignore entropy contributions. (b) How strongly does Ti react with GaN to form TiN (-338.1 kJ mol^{-1}) versus ZnO to form TiO$_2$ (-944 kJ mol^{-1})?
4. (a) A metal with work function $\Phi_M = 5.4$ eV in contact with Si forms a Schottky barrier $\Phi_{SB} = 0.8$ eV. Calculate the density of interface states per unit area located at $E_C - E_{SS} = 0.6$ eV, assuming a dipole length of 10 Å and a bulk $E_C - E_F = 0.1$ eV. (b) Suppose a new metal with work function $\Phi_M = 4.2$ eV forms the same density of interface states at this gap energy. What is the expected barrier height?

References

1. Kurtin, S., McGill, T.C., and Mead, C.A. (1970) *Phys. Rev. Lett.*, **22**, 1433.
2. Pauling, L. (1960) *The Nature of the Chemical Bond*, 3rd edn, Cornell University Press, Ithaca.
3. Lüth, H. (2001) *Solid Surfaces, Interfaces, and Thin Films*, 4th edn, Springer, Berlin, pp. 266–270.
4. Kittel, C. (1986) *Introduction to Solid State Physics*, Chapter 6, John Wiley & Sons, Inc., New York.
5. Bertoni, C.M., Calandra, C., Manghi, F., and Molinari, E. (1990) *Phys. Rev. B*, **27**, 1251.
6. Duke, C.B. (1988) in *Surface Properties of Electronic Materials*, The Chemical

Physics of Solid Surfaces and Heterogeneous Catalysis, Vol. 5 (D.A.King and D.P. Woodruff), Elsevier, Amsterdam, pp. 69–188.
7. Zangwill, A. (1988) *Physics at Surfaces*, Cambridge University Press, Cambridge.
8. Shockley, W. (1939) *Phys. Rev.*, **56**, 317.
9. Tamm, I. (1932) *Phys. Z. Sov. Union*, **1**, 733.
10. Streetman, B.G. and Banerjee, S. (2000) *Solid State Electronic Devices*, 5th edn, PrenticeHall, Upper Saddle River, p. 60.
11. Brillson, L.J. (1982) *Surf. Sci. Rep.*, **2**, 123.
12. Himpsel, F.J. (1990) *Surf. Sci. Rep.*, **12**, 1 and reference therein.
13. Gudat, W. and Eastman, D.E. (1976) *J. Vac. Sci. Technol.*, **13**, 831.
14. Huijser, A. and van Laar, J. (1975) *Surf. Sci.*, **52**, 202.
15. Beres, R.P., Allen, R.E., and Dow, J.D. (1983) *Solid State Commun.*, **45**, 13.
16. Williams, G.P., Smith, R.J., and Lapeyre, G.J. (1978) *J. Vac. Sci. Technol.*, **15**, 1249.
17. Huijser, A., van Laar, J., and Van Rooy, T.L. (1978) *Phys. Lett. A*, **65**, 337.
18. Chadi, D.J. (1979) *Phys. Rev. Lett.*, **43**, 43.
19. Chadi, D.J. (1979) *J. Vac. Sci. Technol.*, **16**, 1290.
20. Schlüter, M. (1988) in *Surface Properties of Electronic Materials*, The Chemical Physics of Solid Surfaces and Heterogeneous Catalysis, Vol. 5 (eds D.A.King and D.P. Woodruff), Elsevier, Amsterdam, pp. 37–68.
21. Pandey, K.C. (1982) *Phys. Rev. Lett.*, **49**, 233.
22. Northrup, J.E. and Cohen, M.L. (1982) *Phys. Rev. Lett.*, **49**, 1349.
23. Peierls, R.E. (1953) *Quantum Theory of Solids*, Oxford University Press, Oxford, p. 108.
24. Kim, S., Flock, K.L., Asar, M., Kim, I.K., and Aspnes, D.E. (2004) *J. Vac. Sci. Technol. B*, **22**, 2233.
25. Duke, C.B. (1967) *J. Vac. Sci. Technol.*, **6**, 152.
26. Heine, V. (1965) *Phys. Rev.*, **139**, A1689.
27. Louie, S.G. and Cohen, M.L. (1975) *Phys. Rev. Lett.*, **35**, 866.
28. Kittel, C. (1967) *Quantum Theory of Solids*, John Wiley & Sons, Inc., New York, pp. 86–98.
29. Louie, S.G. and Cohen, M.L. (1976) *Phys. Rev. B*, **13**, 2461.
30. (a) Chadi, D.J. (1978) *Phys. Rev. Lett.*, **41**, 1062; (b) Chadi, D.J. (1979) *Phys. Rev. B*, **19**, 2074.
31. Thanalaikas, A. (1975) *J. Phys. C (Solid State Phys.)*, **8**, 655.
32. Chelikowsky, J.R. (1977) *Phys. Rev. B*, **16**, 3618.
33. Zhang, H.I. and Schlüter, M. (1978) *Phys. Rev. B*, **15**, 1923.
34. Louie, S.G., Chelikowsky, J.R., and Cohen, M.L. (1977) *Phys. Rev. B*, **15**, 2154.
35. Herman, F. (1979) *J. Vac. Sci. Technol.*, **16**, 1101.
36. Kipperman, A.H.M. and Leiden, H.F. (1970) *Phys. Chem. Solids*, **31**, 597.
37. Van Den Vries, J.G.A.M. (1970) *J. Phys. Chem. Solids*, **31**, 597.
38. Andrews, J.M. and Phillips, J.C. (1975) *Phys. Rev. Lett.*, **35**, 56.
39. Walser, R.M. and Bene, R.M. (1976) *Appl. Phys. Lett.*, **28**, 624.
40. Wagman, D.D., Evans, W.H., Parker, V.B., Halow, I., Bailey, S.M., and Schumm, R.H. (1968–1971) National Bureau of Standards Technical Notes 270-3-270-7, US Government Printing Office, Washington, DC.
41. Mills, K.C. (1974) *Thermodynamic Data for Inorganic Sulfides, Selenides, and Tellurides*, Butterworths, London.
42. Kubaschewski, O., Evans, E.L., and Alcock, C.B. (1967) *Metallurgical Thermochemistry*, Pergamon, Oxford.
43. See for example, Lepselter, M.P. and Andrews, J.M. (1969) in *Ohmic Contacts to Semiconductors* (ed. B. Schwartz), Electrochemical Society, New York, p. 159.
44. Tu, K.N. and Mayer, J.W. (1978) in *Thin Films – Interdiffusion and Reactions* (J.M.Poate, K.N. Tu, and J.W. Mayer), Wiley-Interscience, New York, p. 359 and references therein.
45. Ottaviani, G., Tu, K.N., and Mayer, J.W. (1980) *Phys. Rev. Lett.*, **44**, 284.
46. Brandes, E.A. (1983) *Smithells Metals Reference Book*, 6th edn, Butterworths, London, pp. 11–42.

47. Freeouf, J.L. (1980) *Solid State Commun.*, **33**, 1059.
48. Schmid, P.E., Ho, P.S., Föll, H., and Rubloff, G.W. (1981) *J. Vac. Sci. Technol.*, **18**, 937.
49. Chiu, K.C.R., Poate, J.M., Rowe, J.E., Sheng, T.T., and Cullis, A.G. (1980) *Bull. Am. Phys. Soc.*, **25**, 266.
50. Brillson, L.J. (1978) *Phys. Rev. Lett.*, **40**, 260.
51. Phillips, J.C. and Van Vechten, J.A. (1970) *Phys. Rev.*, **B2**, 2147.
52. Mead, C.A. (1966) *Solid State Electron.*, **9**, 1023 and references therein.
53. Brillson, L.J. (1978) *J. Vac. Technol.*, **15**, 1378.
54. Williams, R.H., Williams, R.H., Montgomery, V., and Varma, R.R. (1978) *J. Phys. C. (Solid State Phys.)*, **11**, L735.
55. Wyether, N.C. and Catalano, A. (1980) *J. Appl. Phys.*, **51**, 2286.
56. Baars, J., Bassett, D., and Schulz, M. (1978) *Phys. Status Solidi A*, **49**, 483.
57. (a) McCauldin, J.O. (1976) *Phys. Rev. Lett.*, **36**, 56; (b) McCauldin, J.O. (1976) *J. Vac. Sci. Technol.*, **13**, 802.
58. Brillson, L.J. (1979) *J. Vac. Sci. Technol.*, **16**, 1137.
59. Swank, R.K. (1967) *Phys. Rev.*, **153**, 844.
60. Kajiyama, K., Mizushima, Y., and Sakata, S. (1973) *Appl. Phys. Lett.*, **23**, 458.
61. Keller, J.W., Roth, A.P., and Fortin, E. (1980) *Can. J. Phys.*, **58**, 63.
62. Okamoto, K., Wood, C.E.C., and Eastman, L.F. (1981) *Appl. Phys. Lett.*, **38**, 636.
63. Mead, C.A. and Spitzer, W.G. (1964) *Phys. Rev.*, **134**, A713.
64. Mead, C.A. and Spitzer, W.G. (1963) *Phys. Rev. Lett.*, **10**, 471.
65. Sirota, N.N. (1968) in *Semiconductors and Semimetals*, vol. **4** (eds R.K.Willardson and A.C. Beer), Academic Press, New York, p. 36.
66. (a) Sadagopan, V. and Gatos, H.C. (1965) *Solid-State Electron.*, **8**, 529; (b) Sadagopan, V. and Gatos, H.C. (1967) *Solid-State Electron.*, **10**, 441.

5
Ultrahigh Vacuum Technology

5.1
Ultrahigh Vacuum Vessels

5.1.1
Ultrahigh Vacuum Pressures

The study of surfaces and interfaces requires equipment to prepare and maintain clean surfaces for extended periods of time. Ambient gas pressures must be kept low enough that impinging molecules do not become a significant fraction of the surface atomic composition. Since measurements of surface and interface properties typically require hours to complete, this requirement translates to pressures in the range of 10^{-10} Torr or below, that is, *ultrahigh vacuum* (UHV). Pressure can be defined in units of Torr, where 1 atmosphere = 760 Torr. Alternate pressure units are 1 Newton/meter^2 (= 1 Pascal (Pa)) = 1×10^{-5} bar = 7.5×10^{-3} Torr.

A calculation of the rate at which molecules impinge on a surface illustrates why such low pressures are needed within these chambers. Consider residual gas molecules arriving at a rate r = number per second in a unit surface area A of 1 cm^2. From Newton's second law, the *pressure* P is defined as force per unit area $F/A = (dp/dt)/A$, the change in momentum p per unit time per unit area.

$$P = 2m <v_z> r \quad (5.1)$$

where m = the *molecular mass*, $<v_z>$ = their average *thermal velocity* normal to the surface, and $p = m<v_z>$ is the corresponding average *momentum per gas molecule*, where length z is normal to the surface. The translational kinetic energy averaged over time is

$$\frac{1}{2}m <v^2> = \left(\frac{1}{6}\right) m v_z^2 \quad (5.2)$$

since $<v^2> = v_x^2 + v_y^2 = v_z^2 = \frac{1}{3}v_z^2$. Combining the average kinetic energy with the ideal gas law

$$PV = nk_B T \quad (5.3)$$

where n is the number of molecules, k_B is the *Boltzmann constant* = 1.38×10^{-23} (Joules/Kelvin)/molecule, V is the volume, and T is temperature in degrees

Surfaces and Interfaces of Electronic Materials. Leonard J. Brillson
Copyright © 2010 WILEY-VCH Verlag GmbH & Co. KGaA, Weinheim
ISBN: 978-3-527-40915-0

Kelvin yields [1]

$$\frac{1}{2m} <v^2> = \left(\frac{3}{2}\right) k_B T \tag{5.4}$$

so that the root mean square velocity

$$<v>_{rms} = \sqrt{<v^2>} = \sqrt{3k_B \frac{T}{m}} \tag{5.5}$$

Air consists of 78% N_2 with a *molecular weight* $M_w = 28$ amu or 4.7×10^{-26} kg. At room temperature $T = 300$ K, so that $<v>_{rms} \cong 520$ m s^{-1} or nearly a third of a mile per second! Substituting Equation 5.3 into Equation 5.1 yields

$$P = \frac{6k_B Tr}{<v>_{rms}} \tag{5.6}$$

The atom density of a surface layer follows directly from the lattice dimensions. For example, a GaAs(001) surface typically used in crystal growth and device fabrication has a square mesh of lattice constant $a_o = 5.65 \times 10^{-8}$ cm, so the surface atom density $1/a_o^2 = 3.1 \times 10^{14}$ cm^{-2}. Taking this concentration as a monolayer, and for an N_2 arrival rate of 1 monolayer per second, the pressure must be $P \cong 1 \times 10^{-6}$ Torr. Thus at 10^{-6} Torr, frequently termed *high vacuum* (*HV*) conditions, the equivalent of one monolayer of atoms or molecules strikes one square centimeter each second. The probability of these atoms or molecules sticking to the surface is termed the *"sticking coefficient"* S. For metals, $S \cong 1$, while for semiconductors S can range from 1 to 10^{-6} or less, depending on the adsorbate and substrate bonding. For unity sticking coefficient $S = 1$, every impinging particle sticks to the surface. To maintain a surface clean for 1 hour, that is, 3600 seconds, such that contamination is less the 10% of a monolayer, a pressure $P \sim 1 \times 10^{-10}$ Torr is required.

Gas exposure is measured in units of *Langmuir* (L), the dosage corresponding to exposure of the surface to a gas pressure of 10^{-6} Torr for 1 second. Gas dosage is reciprocal so that 1 Langmuir is also equivalent to 100 seconds of exposure at 10^{-8} Torr, 1000 seconds at 10^{-9} Torr, and so on.

Modern equipment to enable preparation and analysis of specimens consists of (i) stainless steel chambers, (ii) pumps to remove air and other gaseous contaminants, (iii) pressure gauges to monitor the pressures, (iv) manipulators to move the sample remotely from one location to another inside the chamber, and (v) analytic equipment to measure the properties of the specimens before and after various processes. Additional equipment is available to modify and even create new materials. Chapter 2 introduced several methods of crystal growth. In addition to the preparation and analysis equipment, crystal growth facilities include (i) evaporation sources, (ii) deposition monitors, (iii) sample heaters and temperature sensors, (iv) techniques to monitor surface crystallography and stoichiometry, and (v) gas transport and/or plasma processing for sample growth or modification.

5.1.2
Stainless Steel UHV Chambers

Sealed vessels that can reach pressures of 10^{-10} Torr or lower to enable surface analysis are termed *UHV chambers*. Early versions of such chambers were fashioned from blown glass. However, the weight-bearing requirements of such chambers to hold analysis equipment rapidly gave way to chambers constructed from stainless steel. Stainless steel components of such chambers are joined together using copper gaskets sandwiched between circular flanges. When the flanges are drawn together with bolts, the gaskets deform around "knife edges" machined into the flanges, providing a seal with negligible gas leakage into the chamber. Such flanges enable one to construct systems capable of manipulating, processing, and analyzing samples inside the UHV environment.

Reaching UHV pressures requires the removal of all gases from inside the chamber. Of particular concern are water molecules that have adsorbed to the inside chamber walls. Because of the natural humidity of air, adsorbed water molecules are always present when chambers are exposed to the atmosphere. These desorb very slowly at room temperature so that UHV pressure can not be reached without continuous pumping for impractically long periods of time. Instead, UHV chambers are typically heated to temperatures of $\sim$150–180 °C for periods of 24 hours or more to accelerate this desorption process. Although higher temperatures would shorten this "bakeout" time period, care must be taken to avoid damaging elastomer valve seals that are often used with some gate valves and other components. When such non-stainless steel components are present but high-temperature anneals are required, water cooling can be used to prevent overheating vulnerable components. Bakeout hardware consists of either (i) an enclosure with resistive heater coils that serves as an oven for the UHV chamber or (ii) heater tapes that wrap around the chamber and are covered with aluminum foil to contain the heat generated.

Some elements have relatively high vapor pressures near or slightly above room temperature and present difficulties in obtaining UHV pressures. Curves of vapor pressure versus temperature for most elements are available in [2–4]. These curves show how vapor pressure increases with increasing temperature. Some high-vapor-pressure elements to avoid include Hg, Zn, Cd, S, and Se. Whether present as chamber components or as intentional evaporants, such elements can be a permanent source of residual pressure unless thoroughly removed. Small organic molecules with high vapor pressures near room temperature are of similar concern. Some elements heated to vapor can also harm components within the chamber. For example, Se vapor will react with Cu components within a chamber to form an insulating Cu–Se compound that can charge up with exposure to free electrons. Similarly, Ga vapor can form eutectics with some stainless steel welds such that chamber leaks can form at these points. It is necessary to understand what physical vapor or chemical interactions are possible for a given element or compound before introducing them to a UHV chamber.

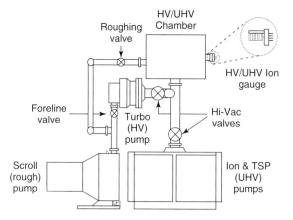

Figure 5.1 Schematic view of a UHV system. (Courtesy Varian Vacuum Technologies.)

5.2
Pumps

A combination of pumps is needed to reach UHV pressures. Different pumps operate in different ranges of pressure. Valves are used to separate these pumps since each pump operates in a different pressure range and a pump operating below its normal operating pressure will act as a leak for other pumps operating at lower pressures. Figure 5.1 illustrates a chamber with a set of pumps and valves configured for UHV. A *scroll* or *rotary* backing pump is connected to the main chamber to establish an initial vacuum before starting the ion pumps. The chamber may also contain a manipulator for positioning specimens at different locations (not shown) and an ion gauge to measure absolute pressure. The entire system can be enclosed in a shroud or oven to "*bake out*" the water vapor.

In Figure 5.1, a *rough pump* is attached to a HV/UHV chamber and to a turbomolecular pump. The rough pump removes air at atmospheric pressure through a roughing valve leading directly into the chamber and reduces the pressure to $\sim 10^{-3}$ Torr. *Sorption pumps* are an alternative to rough pumps to remove air and reduce pressure into this range. Zeolite pellets inside a stainless steel canister are cooled to liquid nitrogen temperatures. When the sorption pump is exposed to the chamber, air molecules adsorb on the large surface area of the zeolite pellets, thereby reducing the pressure to the 10^{-3} Torr range without mechanical pumping.

Once this pressure is reached, the roughing valve is closed and the foreline valve is opened to the turbopump. Once the turbopump is started, the *hi-vac valve* is opened between the turbopump and the main chamber so the turbopump can reduce the pressure further. This hi-vac valve is closed once pressures of 10^{-6} Torr or lower are reached and the lower hi-vac valve is opened to an *ion pump* and a *titanium sublimation pump* (TSP) to reduce the pressure into the UHV range. Turbopumps can also reach UHV pressures.

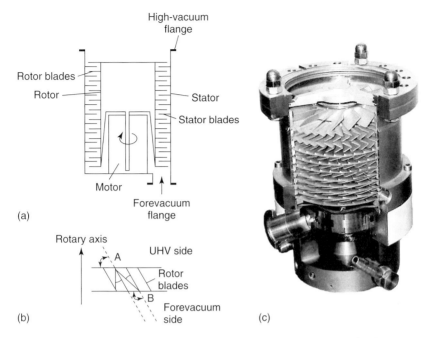

Figure 5.2 Schematic representation of a turbomolecular pump (a) general arrangement of rotor and stator. Rotor and stator blades are inclined with respect to one another [5]. (b) Qualitative view of the arrangement of the rotor blades with respect to the axis of rotation. (c) Cutout of turbopump showing tilted, rotating vanes, and drive motor. The vane angles geometrically determine the possible paths of molecules from the UHV side to the backing side and vice versa [6].

The turbopump acts as a turbine engine, pushing air through a series of vanes rotating at high speeds, typically 15 000–75 000 rpm. Depending on the pumping speed, the turbopump can reduce chamber pressures into the 10^{-9} Torr range or lower. Figure 5.2 illustrates the operating principle of the turbopump [5, 6]. Rotating vanes are tilted at an angle to produce a higher mean free path away from the UHV side such that gas molecules are pushed preferentially away from the UHV side to the backing side of the pump.

The compression ratio of the turbopump and hence its pumping speed depend on the rotor velocity as well as the molecular weight of molecules passing through. As shown in Figure 5.3, the compression ratio between the front and back of the turbopump increases approximately with the square root of molecular weight $\sqrt{M}$. For example, Figure 5.3 shows that N_2, which comprises 78% of room air, is pumped nearly 6 orders of magnitude faster than H_2, which is only a minor constituent of air. Under a high gas load, rotor speed and compression ratio decrease. Turbopumps are particularly useful for handling high gas loads since they pump most common gases efficiently and with no limiting capacity. Figure 5.3 also shows that compression ratio increases with rotor velocity. Note the supersonic velocities used in conventional turbomolecular pumps, which require

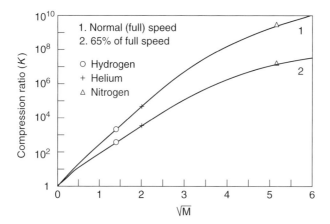

Figure 5.3 Compression ratio of a turbomolecular pump as a function of molecular weight M at normal versus 65% of full speed. (Courtesy Varian Vacuum Technologies.)

regular maintenance to avoid damage to the rotors. Of course, a rotary pump is required to remove gas at the high pressure side of the turbopump to achieve the lowest possible pressure on the low pressure side for a given compression ratio.

An *ion pump* can be used to reach UHV pressures of 10^{-10} Torr and below. After reducing the pressure in the main chamber to either 10^{-3} Torr with roughing pumps or 10^{-6} Torr with a turbopump, the hi-vac valve in Figure 5.1 from these pumps is closed and a hi-vac valve is opened between the ion pump and the main chamber. As long as the residual gas load is not too high, the ion pump can lower the pressure into the 10^{-6} Torr range in minutes and into the 10^{-10} Torr range in less than 24 hours after a chamber bakeout.

Figure 5.4 illustrates the cutout of an ion pump [5, 7] consisting of a planar array of cylindrical cells at high voltage sandwiched between two plates of Ti at ground potential. The entire package is in a magnetic field normal to the plates. Electrons spiraling around the magnetic field B between the Ti plates collide with and ionize residual gas molecules, which are then accelerated by a high voltage, typically 5 kV, and collide with either the Ti cathode plates or the auxiliary Ti cathode. The gas molecules either bond with the Ti layer and are subsequently buried or they sputter Ti atoms, which can bond with other residual gas ions.

Ion pumps have the advantage of requiring little or no maintenance for years since there are no moving mechanical parts. Furthermore, they are capable of reaching lower pressures than turbomolecular pumps. For pressures in the 10^{-5} Torr or higher range, Ti sublimation filaments can handle some of the initial gas load. These filaments operate in analogous fashion to ion pumps, creating Ti atoms that combine with gas molecules and condense on the chamber walls. The need for Ti sublimation filaments highlights a disadvantage of ion pumps, namely, their limited ability to handle high gas loads for extended periods. For such applications, turbopumps are used. Also, the high energy collisions involved in ion pumps can

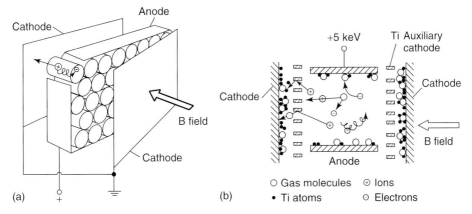

Figure 5.4 Schematic view of an ion getter pump: (a) The basic multicell arrangement. Each cell consists essentially of a tubelike anode sandwiched between at least two common cathode plates of Ti. (b) Within an individual cell, electrons spiraling around a magnetic field B hit residual gas molecules, ionizing them so they accelerate to the cathode, where they are trapped on the active cathode surface or they sputter Ti atoms from the auxiliary cathode, which, in turn, helps to trap other gas ions [5].

cause fragmentation or *cracking* of large molecules, which could produce unwanted reactions in surface chemistry studies. Ideally, ion and turbopumps can be used in tandem as shown in Figure 5.1 such that the turbopump handles the high gas loads required initially and any molecules prone to cracking, whereas the ion pump maintains UHV pressures thereafter.

Vapor pumps, also termed *diffusion pumps*, are alternatives to ion and turbopumps. This device consists of a reservoir of organic fluid that can be heated to produce vapor (Figure 5.5a). As this vapor rises, it passes through jets, cools, and condenses on to gases inside the pump. The organic settles back into the reservoir through jets that accelerate the gas molecules for removal by a forepump.

A cooled orifice or *baffle* with condensation plates above the sublimation pump and specially designed oils minimize vapor diffusing out of the pump. Since no charged particles are involved in the pumping processes, there is no molecular "cracking." However, the *"cold trap"* reduces the pumping speed. In addition, such oil-based pumping should be avoided in growth chambers where even trace impurities can degrade electronic materials' quality.

The *cryopump* represents another method of pumping by means of cooling. As shown in Figure 5.5b, this method involves no condensing fluids but rather the simple adsorption of gases on to a condenser coil connected to a source of cryogenic gas provided by an expander module. Since the cryogenic fluid resides only within a closed loop, there is no chemical interaction between the pump and the surrounding chamber. Temperatures are $\sim 70\,\mathrm{K}$ at the first stage temperatures and as low as $10\,\mathrm{K}$ at the expander module. A significant advantage of the cryopump is its ability to pump noble and very low mass gases. On the other hand, the cryopump's piston introduces vibration that can limit spatially resolved measurements. Figure 5.6

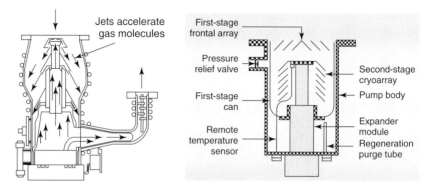

Figure 5.5 (a) Diffusion pump showing evaporation and condensation of organic fluid and gas removal by forepump. (b) Schematic diagram of cryopump showing condenser vanes and expander module used to condense residual gases inside vacuum chamber. (Courtesy Varian Vacuum Technologies.)

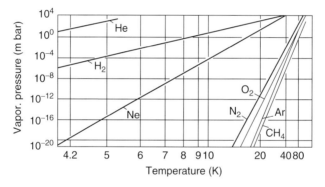

Figure 5.6 Saturation vapor pressures of various coolant materials as a function of temperature [5].

illustrates the saturation vapor pressures versus temperature for various atoms and molecules.

Figure 5.6 shows that cryogenic temperatures are capable of lowering the vapor pressure of almost all gases to below 10^{-10} Torr. For example, at a liquid helium temperature of 4.2 K, the vapor pressure of most gases is well below 10^{-10} Torr. The primary residual gas at that pressure is hydrogen. The cryopump is capable of significant gas accumulation before further gas adsorption is reduced. At that point, one must seal off the cryopump from the main chamber, allow the condenser coil to warm to room temperature, and pump away the accumulated gases. This gas removal is critical to avoid reaching potentially explosive pressures. Some gases have the potential for explosion when condensed, for example, concentrated ozone. UHV chamber with cryopumps typically include *relief valves* that open to relieve unsafe gas pressures.

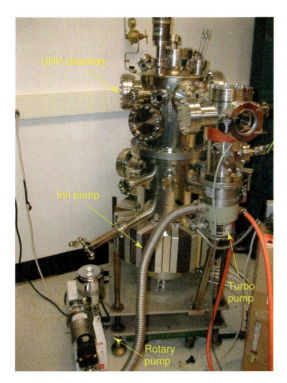

Figure 5.7 Stainless steel UHV chamber with introduction chamber connected to turbopump with backing rotary pump. A poppet valve connected to the crank shown separates the upper half of the chamber from the ion pump in the lower half. Here, the entire assembly is mounted on a platform whose height can be adjusted.

Figure 5.7 illustrates a stainless steel UHV chamber with a variety of circular flanges to which one can bolt on various manipulation, processing, and analysis hardware. The lower half of this chamber includes a ring of ion pumps with permanent magnets, a turbopump to remove gas from an introduction chamber, and a rotary pump attached to the back end of the turbopump. A crank shown actuates a *"poppet"* valve that connects the lower and upper halves of the chamber. By sealing this valve, vacuum can be preserved in the lower half while the upper half is opened to atmosphere.

The ultimate pressure that a pump can reach and the speed at which it can reach it depends on several factors including (i) the volume V to be pumped, (ii) the *desorption rate* v_d of molecules per second and per unit area A from the chamber walls, (iii) the *pumping speed* S_p, and (iv) the *conductance* C of tubes from the pump to the chamber through which gas is pumped. Pumping speed S_p is the rate at which pressure decreases with time with no gas load, expressed as volume/time, for example, liters/second. As with electric current, the conductance depends on the cross-sectional area and length. Pipe conductance also depends on the geometry of the tubes. Thus, pipe curvature can affect conductance depending on the geometric dimensions versus the mean free path of gas molecules. Conductance formulae are available for tubes and orifices that depend on cross-sectional area, tube length, bend geometries, temperature, and atomic mass unit for molecular flow [8]. Analogous to Kirchoff's electrical circuit laws, two pipes 1 and 2 in parallel have conductance $C_p = C_1 + C_2$ while the same pipes in series have conductance $1/C_s = 1/C_1 + 1/C_2$. A pump in series with pipes has an effective pumping speed S_{eff} defined as $1/S_{eff} = 1/S_p + 1/C_p$.

Example 5.1. Effective pumping speed.

What is the effective pumping speed of a system with pumping speed $S_p = 400 \, l\,s^{-1}$ and conductance $150 \, l\,s^{-1}$?

$$S_{eff} = \frac{(S_p \cdot C_p)}{(S_p + C_p)} = (400) \cdot \frac{(150)}{(550)} = 109 \, l\,s^{-1} \tag{5.7}$$

Starting from the ideal gas law $PV = nk_B T$, the rate of gas molecules being pumped out of the chamber is

$$\frac{dn}{dt} = v_d A = \frac{d}{dt}\left(\frac{PV}{k_B T}\right) = \left(\frac{1}{k_B T}\right)\left[\left(\frac{V dP}{dt}\right) + PS_p\right] \tag{5.8}$$

At steady state, $dP/dt = 0$ so that $P = k_B T \, v_d A / S_p$. This equation expresses the intuitive result that base pressure decreases with decreasing temperature, decreasing desorption rate, decreasing internal area, and increasing pumping speed.

5.3
Specimen Manipulators

Manipulators are needed to position specimens under test inside the UHV chamber. A manipulator must have three-dimensional freedom of movement plus the capability to accept samples transferred from other parts of the chamber. These degrees of freedom allow positioning the sample for various electron and photon beam excitation and signal collection. Specific requirements of the various characterization techniques appear in the following chapters. Electrical measurements in UHV of samples held by manipulators are possible for sample holders and connecting wires that are insulated from the manipulator itself. Sample holders and manipulator stages may also be fitted for heating, cooling, cleaving, line-of-sight gas dosing, and overlayer deposition.

5.4
Gauges

A number of gauges are available to monitor the pressure inside the vacuum chamber. *Diaphragm gauges* measure the volume change of a flexible diaphragm chamber due to pressure differentials in the $P > 10^{-4}$ Torr range. Similarly, *Pirani gauges* measure thermal conductivity changes of current filaments in the same pressure range. *Ionization gauges* measure pressures in the UHV range,

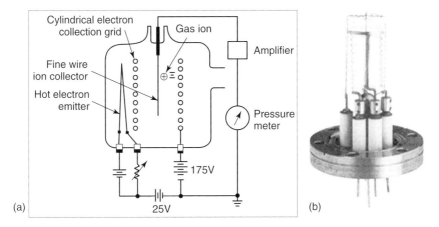

Figure 5.8 Ionization gauge and electric circuit for measuring pressure from 10^{-4} to 10^{-10} Torr and below. The electrodes can be positioned (a) with or (b) without a glass envelope. (Courtesy Varian Vacuum Technologies.)

that is, 10^{-9}–10^{-10} Torr and below. Such ion gauges operate by measuring the current of ionized molecules created by electrons from a cathode filament passing through a positively biased anode grid. At higher pressures, more molecules are ionized, giving rise to higher ion and thereby collector currents. Ionization gauges encased in glass as shown in Figure 5.8 are used when the gauge is outside the vacuum chamber. *"Nude" ion gauges* allow the gauge to be inserted inside the chamber.

5.5
Deposition Sources

5.5.1
Metallization Sources

UHV chambers provide an environment for creating well-defined metal–semiconductor or insulator interfaces by thermal evaporation and overlayer deposition on clean surfaces. A variety of techniques are available for such deposition including evaporation from high-temperature sources such as wires, boats, or crucibles in direct line of sight. Electron beam heating is effective in reaching the high temperatures required to evaporate refractory metals. Deposition is also possible by flowing source gases over the sample surface, where they can decompose and/or react to form deposited films. Yet another technique is sputtering, where a target of desired composition is bombarded with energetic ions or intense laser pulses. Excited clusters or atoms from the target

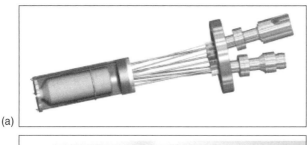

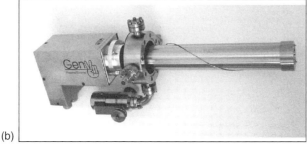

Figure 5.9 (a) Knudsen cell with tapered nozzle for epitaxial film growth. (Courtesy Veeco MBE.) (b) Oxygen plasma source. (Courtesy *tectra Physikalische Instrumenten*.)

then fly off the surface and strike the sample, where they accumulate as films of desired composition. Often background gases are needed to maintain desired stoichiometries.

5.5.2
Crystal Growth Sources

Besides metallizing surfaces, these and other sources can enable film growth of semiconductors and insulators. The vacuum growth methods described in Chapter 2 also rely on thermal evaporation sources, gas sources, or combinations. A wide variety of thermal evaporation sources are now available as a result of advances in epitaxial growth technology. Figure 5.9 illustrates two such sources used in molecular beam epitaxy (MBE). Figure 5.9a shows a *Knudsen cell* with thermocouple and heater feedthroughs passing through a knife-edge flange. The cell is tapered at its nozzle to limit heat radiating from the crucible toward the deposition substrate. Different crucible and heater designs are employed, depending on the temperature range required for evaporation.

Figure 5.9b shows a *plasma generator* that directs a beam of ionized gas at a specimen [9]. Typically, such sources employ oxygen or nitrogen for oxide or nitride growth. A leak valve controls gas flow. The ion extraction optics and nozzle are intended to maximize reactive molecules while minimizing charged species. Water lines (not shown) cool the source during operation. Gas source MBE involves the flow of precursor gases across the substrate, where they dissociate and incorporate with the growing film.

5.6 Deposition Monitors

A common method to gauge and control the deposition of thin films in UHV involves measuring thickness with *quartz crystal monitors* (QCMs). QCMs incorporate thin wafers of piezoelectric material that, when sectioned along specific planes, generate an alternating voltage at a characteristic frequency equal to the resonance frequency of the quartz crystal. Resonance frequencies in these thin films are analogous to the frequencies of a drumhead. Since these crystals have characteristic vibration frequencies that can reach megahertz frequencies, frequency counters measuring the AC input voltage from the QCM with an accuracy of 1 Hz can be used to obtain a precision of small fractions of a single atomic layer.

Figure 5.10 illustrates the parameters used to determine the relationship between deposited film thickness and vibration frequency. Essentially, changes in crystal thickness change *vibration frequency f*.

$$f = \frac{v_{tr}}{2d} = \frac{N}{d} \tag{5.9}$$

where v_{tr} is the *acoustic mode velocity* transverse to the plane of the crystal, d is the crystal thickness, and $N = v_{tr}/2$ is defined as the *crystal frequency constant*. For an "AT-cut" quartz crystal, $N = 1670$ kHz mm, while a "BT-cut" crystal has $N = 2500$ kHz mm. From Equation 5.6 [10],

$$\frac{\Delta f}{f} = -\frac{\Delta d}{d} = -\frac{\Delta m_Q}{(\rho_Q F d)} = -\frac{\Delta m}{(\rho_Q F d)} \tag{5.10}$$

where $F =$ unit area, $\rho_Q =$ quartz density, and $m_Q =$ starting quartz mass. Thus

$$\frac{\Delta f}{f} = -\frac{\Delta m^* f}{(\rho_Q F N)} \tag{5.11}$$

from Equation 5.7 and

$$\Delta f = -\frac{f^2 \rho \Delta T}{(N \rho_Q)} \tag{5.12}$$

since $\Delta m/F = \rho \Delta T$. For $f = 4.99 \times 10^6$ Hz, $\rho_Q = 2.65$ g cm^{-3}, for an AT-cut crystal, $\Delta f = (4.99 \times 10^6 \text{ Hz})^2/[(1.67 \times 10^5 \text{ Hz cm})(2.65 \text{ g cm}^{-3})] \times \rho \Delta T = \rho \Delta T/1.77$, where ΔT is in angstrom units. Thus with 1 Hz precision, thickness precision can be much less than one monolayer. Table 5.1 gives representative precisions for simple metals.

Other methods of gauging deposition include (i) counting reflection high-energy electron diffraction (RHEED) oscillations characteristic of monolayer by monolayer

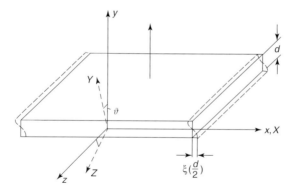

Figure 5.10 Schematic diagram of piezoelectric quartz crystal cut to exhibit a single well-defined vibration frequency.

Table 5.1 Quartz crystal monitor frequency change per monolayer of deposited metal.

Metal	Density ρ (g cm^{-3})	f per Å (Hz)	Lattice constant (Å)	f per monolayer (Hz)
Au	19.3	10.9	4.080	44.4
Al	2.69	1.53	4.050	6.20
Ga	5.91	3.33	4.510	15.02

deposition (described in Chapter 13), (ii) monitoring the vapor stream with a residual gas analyzer (RGA), and (iii) measuring the intensity of laser-induced optical luminescence from the vapor stream. Novel methods now under development include (i) the use of cathodoluminescence spectroscopy to monitor defects that are characteristic of nonstoichiometric compounds and (ii) monitoring the surface elemental stoichiometry via the X-rays emitted due to excitation by the RHEED electron beam. The measurement and control of atoms adsorbed on surfaces is of critical importance when compounds of specific elemental composition are desired.

5.7
Summary

UHV technology developed over the past six decades has provided powerful tools with which to study surfaces and interfaces. These tools include chambers, manipulators, gauges, and the combination of pumps needed to achieve UHV pressures, where surfaces can be maintained free of ambient contamination for periods of hours while their properties are studied. A variety of sources are now available to create new materials in UHV, free of contamination often encountered with growth by other techniques. Hence it is now possible to study and create structures that can extend our understanding of surfaces and interfaces on an atomic scale.

Problems

1. At atmospheric pressure, room temperature, and unity sticking coefficient, at what rate (centimeters per second) would a layer of air molecules grow on your body? Does this happen? Why or why not?
2. Calculate the room temperature pressure at which one monolayer of pure oxygen forms on the GaAs(100) surface in 1 second.
3. (a) Calculate the effective pumping speed of a system with a $400 \, \text{l s}^{-1}$ pump linked to a chamber through a pipe and orifice of conductance $100 \, \text{l s}^{-1}$.
4. Cr evaporation on to a specimen and a quartz crystal oscillator with crystal frequency constant 1670 kHz mm results in a frequency change of 250 Hz. Assuming there is uniform coverage, how many monolayers thick is the Cr layer?
5. In MBE, the flux density J (molecules per centimeter squared per second) from a Knudsen cell is given as

$$J = \frac{Ap \cdot \cos\theta}{\pi L^2 (2\pi m k_B T)^{1/2}} \tag{5.13}$$

where A is the cell aperture, p the equilibrium vapor pressure in the cell, L the distance between the cell and the substrate, m the mass of the effusing species, and T the cell temperature. For Al atom deposition on a GaAs(100) substrate with unity sticking coefficient located 40 cm from a Knudsen cell, what is the maximum deposition rate for an aperture of 1 cm^2 and a cell temperature of 1000 °C? The equilibrium vapor pressure at this temperature is 10^{-4} Torr.

References

1. Halliday, D., Resnick, R., and Walker, J. (1997) *Fundamentals of Physics Extended*, John Wiley & Sons, Inc., New York, p. 489.
2. Honig, R.E. (1962) Vapor pressure curves of the elements. *RCA Rev.*, 23, 567.
3. Glang, R. (1970) in *Handbook of Thin Film Technology* (eds L.I. Maissel and R. Glang), McGraw-Hill, New York.
4. Vossen, J.L. and Cuomo, J.J. (1978) in *Thin Film Processes* (eds J.L.Vossen and W. Kern), Academic Press, New York.
5. Lüth, H. (2001) *Solid Surfaces, Interfaces, and Thin Films*, 4th edn, Springer, New York, Panel I.
6. http://en.wikipedia.org/wiki/Vacuum_pump.
7. http://www.vakuum.cz/product_en05.htm.
8. Holkeboer, D.H., Jones, D.W., Pagano, F., and Santeler, D.J. (1993) *Vacuum*

Technology and Space Simulation, American Vacuum Society Classics, American Institute of Physics, New York.
9. *http://www.tectra.de/*.
10. Sauerbrey, G. (1959) *Z. Phys.*, **155**, 256.

Further Reading

Lafferty, J.M. (ed.) (1997) *Foundations of Vacuum Science and Technology*, Wiley Interscience, John Wiley & Sons, Inc., New York.

6
Surface and Interface Analysis

6.1
Surface and Interface Techniques

Researchers have developed a wide range of surface and interface techniques over the past few decades to measure the physical properties of surfaces and interfaces. These advances have been motivated by the explosion in materials science capabilities to create new materials. This is certainly true for electronic materials, which can now be grown by techniques with atomic layer precision and which can be tailored to improve electronic and optical performance. An additional motivation for electronic materials has been the relevance of these physical properties to the manufacture of electronics and optoelectronics.

Table 6.1 lists the vast majority of techniques developed to characterize surfaces and interfaces and the associated information they can provide. In general, Table 6.1 shows that more than one technique is available to probe a specific physical property. Furthermore, each technique can provide useful information yet each may have particular limitations. The availability of multiple techniques then allows researchers to compare results obtained by more than one technique, thereby avoiding potential artifacts of a particular technique in order to reach self-consistent conclusions.

Most of the techniques listed require UHV conditions to avoid ambient contamination during measurement as well as to allow excitation of the specimen under study. A few of the techniques require access to major instrumentation facilities such as synchrotron storage rings in order to excite their specimens with radiation not available in typical research labs. Several techniques listed can be adapted for both surface and interface measurements. For example, *static secondary ion mass spectrometry* (SSIMS), *Auger electron spectroscopy* (AES), and *cathodoluminescence spectroscopy* (CLS) can all measure features of the top few monolayers, and each has capabilities to measure similar features as a function of depth into a material's bulk.

Most spectroscopies in Table 6.1 involve excitation by photons, electrons, or ions. Likewise, photons, electrons, and ions carry the spectral information they generate. Table 6.2 illustrates this matrix of excitations and spectroscopies. Several of these techniques are available with excitation beams focused on a nanometer scale. These include confocal *photoluminescence* (PL) and *Raman scattering* (RS),

Surfaces and Interfaces of Electronic Materials. Leonard J. Brillson
Copyright © 2010 WILEY-VCH Verlag GmbH & Co. KGaA, Weinheim
ISBN: 978-3-527-40915-0

Table 6.1 Techniques and physical properties measured.

Technique	Surface information
Auger electron spectroscopy (AES)	Chemical composition, depth distribution
X-ray photoemission spectroscopy (XPS)	Chemical composition and bonding
UV photoemission spectroscopy (UPS)	E_F versus E_C and E_V, Φ_M, valence band states
Soft X-ray photoemission spectroscopy (SXPS)	Chemical composition, bonding, E_F versus E_C and E_V, Φ_M, valence band states
Constant initial state (CIS) and constant final state (CFS) spectroscopies	Empty states above E_F
Angle-resolved photoemission spectroscopy (ARPES)	Atomic bonding symmetry, Brillouin zone dispersion
Surface extended X-ray absorption fine spectroscopy (SEXAFS)	Local surface bonding coordination
Inverse photoemission spectroscopy (IPS)	Unoccupied surface state and conduction band states
Laser-excited photoemission spectroscopy (LAPS)	Bandgap state energies, symmetries
Low-energy electron diffraction (LEED) and Low-energy Positron diffraction (LEPD)	Surface atomic geometry
X-ray diffraction	Near-surface and bulk atomic geometry
Surface photovoltage spectroscopy (SPV or SPS)	Gap states energies versus E_C and E_V, Φ_M, qV_B
Infrared (IR) absorption spectroscopy	Gap state energies, atomic bonding, and coordination
Cathodoluminescence spectroscopy (CLS)	Gap state energies, compound/phase band gap
Photoluminescence spectroscopy (PLS)	Near-surface compounds, bandgap
Surface reflectance spectroscopy (SRS)	Surface dielectric response, absorption edges
Spectroscopic Ellipsometry (SE)	Composition, complex dielectric response
Surface photoconductivity spectroscopy (SPC)	Surface/near-surface gap state energies
Static secondary ion mass spectrometry (SSIMS)	Elemental composition, lateral mapping
Ion beam scattering spectrometry (ISS)	Energy transfer dynamics, charge density
Scanning tunneling microscopy (STM)	Atomic geometry, step morphology, filled and empty state geometry
Atomic force microscopy (AFM)	Morphology, electrostatic forces
Kelvin probe force microscopy (KPFM)	Φ_M, qV_B versus morphology
Magnetic force microscopy (MFM)	Magnetic moment and polarization
Scanning tunneling spectroscopy (STS)	Bandgap states, heterojunction band offsets
Field ion microscopy (FIM)	Atomic motion, atomic geometry
Low-energy electron microscopy (LEEM)	Morphology, atomic diffusion, phase Transformation, grain boundary motion
Technique	Interface information
Total external X-ray diffraction (TEXRD)	Lattice structure, strain
Low-energy electron loss spectroscopy (LEELS)	Chemical reactions, Plasmon, interband excitations

Table 6.1 *(continued)*

Technique	Surface information
Depth-resolved cathodoluminescence spectroscopy (DRCLS)	Subsurface gap states, interface compound/phase bandgap
Cross-sectional cathodoluminescence spectroscopy (XCLS)	Heterojunction band offsets, interface states, Φ_M, qV_B
Ellipsometry	Subsurface or interface dielectric response
Raman scattering spectroscopy (RSS)	Interface bonding, strain, band bending
Confocal Raman spectroscopy (CRS)	Subsurface bonding, strain
Rutherford backscattering spectroscopy (RBS)	Near-surface, bulk atomic symmetry, composition, depth distribution
Secondary ion mass spectrometry (SIMS)	Subsurface/interface elemental composition, depth distribution, lateral mapping
Ballistic electron energy microscopy (BEEM)	Barrier heights, heterojunction band offsets, barrier height lateral mapping
Scanning capacitance microscopy	Subsurface doping, capacitance
Cross-sectional kelvin probe force microscopy (XKPFM)	Φ_M, qV_B across interfaces (requires UHV)
High resolution transmission electron microscopy (HRTEM)	Interface lattice structure
Transmission electron microscopy (TEM) electron energy loss spectroscopy (EELS)	Gap state energies, atomic bonding, lattice symmetry

electron microscope-based CLS, electron energy loss spectroscopy (EELS), and AES, and focused ion beam secondary ion mass spectrometry (SIMS). The remaining techniques listed are scanned probe techniques, most of which employ atomically sharp tips as measurement devices. They detect either tunneling electric current, force-induced cantilever displacements, or capacitance displacement currents. These techniques are described in chapters devoted to photon, electron, and ion excitation, scanned probe, and related techniques with particular emphasis on those contributing significantly to our understanding of electronic surfaces and interfaces.

6.2
Excited Electron Spectroscopies

Surface science techniques have been traditionally centered on the detection of electrons. Figure 6.1 illustrates four of the major techniques that researchers have employed over the past few decades to probe surfaces.

Figure 6.1 illustrates the excitation processes involved in four of the primary spectroscopies used for measuring surface and near-surface electronic and chemical properties. The horizontal lines labeled "1," "2," and "3" in each panel correspond

Table 6.2 Representative surface and interface spectroscopies.

	Photons out	Electrons out	Ions out
Photons in	Photoluminescence Electroreflectance Ellipsometry Raman scattering	UV photoemission X-ray photoemission Soft X-ray photoemission CIS, CFS, SEXAFS	Photodesorption Desorption induced electron transition
Electrons in	Cathodoluminescence Energy dispersive X-ray Inverse photoemission	Auger electron Low-energy electron loss Low, reflection high energy electron diffraction	Field ion microscopy
Ions in	Ion-induced luminescence	Ion neutralization	Rutherford backscattering Secondary ion mass Helium backscattering Ion scattering

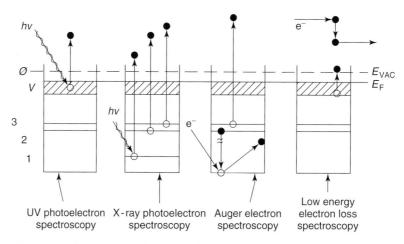

Figure 6.1 Schematic energy band level diagram illustrating the excitation processes in four major spectroscopies employing electrons.

to energies of electrons in core levels of specific atoms. The hatched regions labeled "V" refer to the energies of electrons in the higher lying valence band. Here, E_F defines a Fermi level that for simplicity resides at the top of the valence band. The dashed line represents the *vacuum level* E_{VAC}, above which electrons can escape from the solid. For ultraviolet photoelectron spectroscopy (UPS) ultraviolet photoelectron spectroscopy (UPS), a photon incident on the solid surface can excite a valence electron (filled circle) above E_{VAC}, leaving behind a hole (open circle) in the valence band. An analyzer collects the free electron and measures its energy.

The UV photon has enough energy to excite electrons only from the valence band and highest lying core levels. *X-ray photoelectron spectroscopy* (XPS) employs photons with energy sufficient to excite electrons from many deeper core levels. In the case shown, the X-ray photon has enough energy to excite electrons from core levels "1," "2," and "3," leaving behind core holes.

The kinetic energies of the excited electrons that leave the atom and the solid are characteristic of that particular atom. *Kinetic energy E_K is related to the core level binding energy E_B* according to Equation 6.1.

$$E_K = h\nu - E_B - (E_{VAC} - E_F) \tag{6.1}$$

Thus, a spectrum of electron intensity versus energy yields a set of peaks at energies equal to the difference between the incident photon energy $h\nu$ and the core level energy E_B, minus the additional energy required to leave the solid, $E_{VAC} - E_F$.

UPS and XPS involve excitation of single electrons. In AES, three electrons are involved. Here, the excitation is an energetic electron rather than a photon. The energy of this electron is sufficient to knock an electron out of a core shell, leaving behind a core hole as both electrons leave the atom. A second electron in a higher lying core level drops down into this core hole, transferring the difference in energy to a third electron, which leaves the atom and the solid. Here, the distance between levels "2" and "1" is shortened for clarity, as indicated by the vertical break.

The kinetic energy of the excited electron that leaves the atom and the solid is also quite characteristic of that particular atom. Kinetic energy $E_{123Auger}$ is related to the binding energies of the three energy levels involved in the Auger process according to Equation 6.2.

$$E_{123Auger} = E_1(Z) - E_2(Z) - E_3(Z + \Delta) - E_{VAC} \tag{6.2}$$

Here, $E_1(Z)$, $E_2(Z)$, and $E_3(Z)$ are the energies of the three core levels for elements with *atomic number Z* and is the change in core level E_3 due to the screening of the atom's nucleus by one less electron.

While the Auger process is more complex than photoemission, it nevertheless has several virtues. (i) AES yields high signals since incident electron beam fluxes are typically much higher than photon fluxes. (ii) The Auger signal has characteristic signatures that permit rapid identification. Both of these features translate into rapid signal acquisition and thus the ability to obtain information rapidly as a function of lateral position or depth. (iii) The electron beam can be focused to nanometer dimensions, permitting measurements of very small specimens or the generation of elemental maps on this nanoscale.

A fourth excited electron technique is *low-energy electron loss spectroscopy* (LEELS). Here, an incident electron impinges on a solid surface, losing some of its energy to electrons in the valence band or higher lying core levels. These filled state electrons are excited to empty states in the conduction band, empty gap states, or to levels above E_{VAC}. These excitations reduce the energy of electrons in the primary beam by the energies lost in these transitions.

In all four electron excitation processes, electrons leave the solid with energies that can be related to energy levels within the atom and the solid. These characteristic energies are reduced by any collisions or other inelastic processes that occur as the electrons travel from the excited atom to the surface of the solid. The energies of such scattered or *secondary* electrons lie in a continuous and smoothly varying range, forming a background that can be distinguished from the characteristic peak features of the unscattered electrons.

6.3
Principles of Surface Sensitivity

The excited electrons in a solid scatter over distances that depend strongly on their kinetic energies. Figure 6.2 illustrates the energy dependence of electron scattering length in a solid [1]. The individual points represent different elements measured using photoemission spectroscopy by a method to be discussed. This figure shows that scattering lengths in most solids are only a few Angstroms when kinetic energies are in the range of 50–100 eV. This means that only electrons that travel distances less than a scattering length will escape from the solid without scattering. Therefore, only these electrons within a scattering length of the free surface will contribute to the spectra with their characteristic kinetic energies unchanged. In other words, electron spectroscopies that involve measurement of excited electrons in this low-energy range are very sensitive to the top one or two monolayers of the solid. It is this feature of excited electrons that provides the basis for much of surface science.

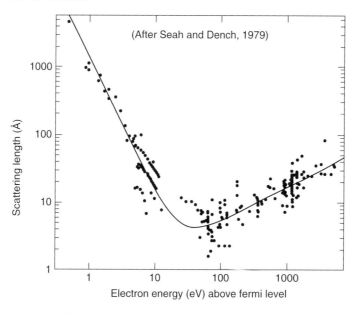

Figure 6.2 Electron scattering length in a solid. Electrons with energies in the range of 50–100 eV escape elastically only from the outer few angstroms of the surface.

Figure 6.2 also shows that the electron scattering length, also termed the *electron escape depth*, can vary over nearly 3 orders of magnitude. Thus, by adjusting the excitation to produce excited electrons at energies higher or lower than the escape depth minimum, one can vary the effective probe depth. Similarly, one may probe energy levels whose kinetic energies for a given excitation have scattering lengths close to a desired escape depth.

6.4
Surface Analytic and Processing Chambers

In general, surface and interface analysis chambers contain a variety of tools for (i) obtaining clean surfaces in UHV, (ii) characterizing their electronic, chemical, and/or structural properties, and (iii) modifying these surfaces or interfaces *in situ*, that is, without exposing the specimen being studied to air.

Figure 6.3 schematically illustrates how such tools can be arrayed within a UHV chamber. Typical surface science chambers have an intermediate pressure or load lock chamber, a port through which to insert specimens without exposing the UHV chamber to air. Translation rods mounted in bellows can move the sample across the UHV chamber to different locations using specially designed sample holders that can accommodate both rods and manipulator mounts. They enable the transfers to and from the manipulators that position the sample more precisely for analysis or processing. The sample holders themselves may have electrical contacts for resistive heating or for electrical measurements. Certainly they must enable the specimen to be electrically grounded in order to reference the measured spectrum energies to the ground reference of any energy analyzer. Spectroscopic analysis techniques that can be incorporated together in such a chamber include the electron and photon methods already discussed – AES, XPS, UPS, and LEELS – as well as optical methods such as surface photovoltage spectroscopy (SPV), PL, and CLS.

Surface analysis often involves preparing surfaces *in situ*, then measuring their properties with additional processing. Thus, high pressure processing such as *plasma oxidation* can be used to prepare clean, ordered surfaces [2] by translating the specimen through a gate valve into a side chamber with a microwave generator and ultrahigh purity gas flow. The higher pressure region can be isolated from the main chamber during this process. The residual gas is then pumped away before reintroducing the sample back into UHV. Nitrogen and hydrogen are also commonly used in plasma processing.

Other methods to control surfaces include (i) *ion milling*, that is, bombarding the sample surface with noble gas ions to remove any surface adsorbates or to etch away the outer atomic layers; (ii) *pulsed laser annealing* to raise the near-surface sample regions to high temperatures (typically above the melting point) for very short (nanosecond or less) intervals, in order to heal lattice defects or, for higher peak powers, to ablate away surface monolayers; (iii) *cleaving* to expose clean surfaces of the sample interior. Ion milling involves a focused beam of charged ions which

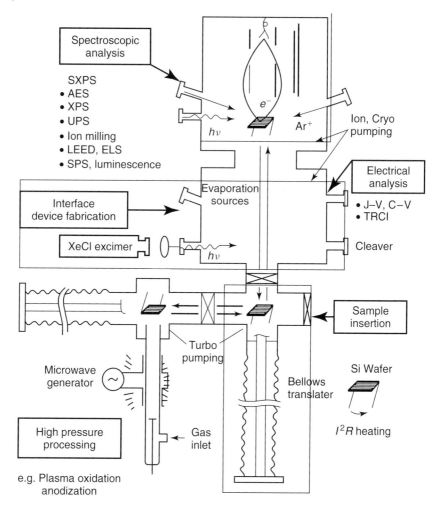

Figure 6.3 Semiconductor analysis and processing chamber with capabilities for spectroscopic analysis, device fabrication, electrical analysis, laser annealing, cleaving, and high pressure processing.

must be differentially pumped in order not to raise the UHV chamber pressure significantly. Noble ions such as Ar^+ or Xe^+ are used to avoid any chemical bonding with the specimen or other chamber components. Pulsed laser annealing typically employs an excimer laser such as XeCl to produce nanosecond pulses of deep UV light that is absorbed within shallow depths. This produces high power in very small volumes over intervals that are short compared to the time required to dissipate heat away from the irradiated zone. Sample cleaving is feasible for bulk crystals using a blade and anvil arrangement in many respects similar to the method jewelers use to fashion precious stones. Thin wafer specimens can be cleaved by applying pressure normal to the wafer plane and parallel to a cleavage plane of the clamped wafer [3].

Interface device fabrication is achieved by evaporating metals from sources in line with patterned masks positioned in front of the sample surface [4]. For metals evaporated on semiconductors, the resultant diode pattern can be analyzed electrically with *in situ* electrodes for current–voltage (J–V), capacitance–voltage (C–V), or more elaborate techniques such as *time-resolved charge injection* (TRCI) [5].

Incorporating many different analysis and processing techniques with the same UHV chamber requires careful consideration of port location and orientation. Figure 6.4 illustrates the arrangement of multiple techniques within a single chamber in more detail.

Stainless steel chambers are now readily designed using computer-aided design programs, and their standard flange and tool sizes enable researchers to design more elaborate systems that can be integrated with other facilities. Thus, surface analysis chambers are often integrated with epitaxial growth chambers to enable the study of interfaces and thin films during the initial stages of film growth. UHV chambers have been employed in pilot lines for processing state-of-the-art silicon microelectronics [6]. Such chambers can be joined with high-resolution electron microscopes to fabricate and characterize surfaces with nanometer spatial resolution [7]. Chambers have also been adapted for use at national facilities, such as storage rings for synchrotron radiation experiments or inertial confinement chambers for laser fusion.

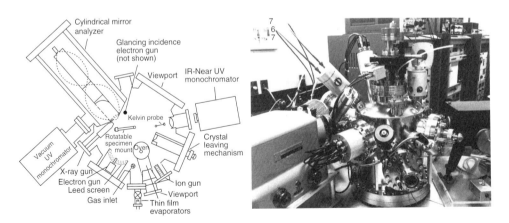

Figure 6.4 Multitechnique UHV chamber. (a) Schematic top view of a slice through the main set of horizontal ports. (b) Photo of the same chamber with the vacuum UV monochromator at left and the vibrating Kelvin probe at top right.

6.5
Summary

A wide range of techniques and tools are now available to examine surfaces and interfaces on an atomic scale. These techniques employ excitation by photons, electrons, ions, as well as scanned probe measurements of electrical transport, capacitance, and force.

Multiple techniques are available to measure the physical properties of surfaces and interfaces. Measurement of the same or related properties by more than one technique permits researchers to reach self-consistent results while avoiding potential artifacts of a particular technique.

Spectroscopies that employ excitation of electrons provide monolayer surface sensitivity because of the short electron scattering length in solids at kinetic energies in the 50–100 eV range. The energy dependence of electron scattering lengths provides a tool to vary surface sensitivity from angstroms to hundreds of nanometers.

Surface analytic and processing chambers are capable of manipulating, processing, and characterizing materials on an atomic scale. High-resolution scanning probes have now extended our ability to measure surface and interface properties laterally as well as depthwise at the nanometer scale.

References

1. Seah, M.P. and Dench, W.A. (1979) *Surf. Interface Anal.*, **1**, 2.
2. Mosbacker, H.L., Strzhemechny, Y.M., White, B.D., Smith, P.E., Look, D.C., Reynolds, D.C., Litton, C.W., and Brillson, L.J. (2005) *Appl. Phys. Lett.*, **87**, 012102.
3. Walker, D.E. Jr., Gao, M., Chen, X., Schaff, W.J., and Brillson, L.J. (2006) *J. Electron. Mater.*, **35**, 581.
4. Mosbacker, H.L., El Hage, S., Gonzalez, M., Ringel, S.A., Hetzer, M., Look, D.C., Cantwell, G., Zhang, J., Song, J.J., and Brillson, L.J. (2007) *J. Vac. Sci. Technol. B*, **25**, 1405.
5. Slowik, J.H., Brillson, L.J., and Brucker, C.F. (1982) *J. Appl. Phys.*, **53**, 550.
6. Rubloff, G.W. (1989) *J. Vac. Sci. Technol.*, **B7**, 1454.
7. Vuchic, B., Merkle, K.L., Char, K., Buchholz, D.B., Chang, R.P.H., and Marks, L.D. (1996) *J. Mater. Res.*, **11**, 2429.

Further Reading

Briggs, D. and Seah, M.P. (1990) *Practical Surface Analysis*, 2nd edn, John Wiley & Sons, Inc., New York.

7
Photoemission Spectroscopy

7.1
The Photoelectric Effect

Photoelectron spectroscopy is one of the most widely used surface science techniques. Its utility stems from the discrete or quantized nature of light, first explained by Albert Einstein in 1905 [1] and for which he received the Nobel Prize in Physics in 1921. At the turn of the century, researchers had noticed that light incident on clean metal surfaces resulted in electrons being ejected from the metal surface into vacuum. Significantly, these electrons appeared only for incident wavelengths that were shorter than a critical wavelength. Furthermore, the kinetic energies of these ejected electrons appeared to increase with decreasing wavelength. Finally, this wavelength dependence appeared to be independent of light intensity. Einstein recognized that light could deliver its energy in quantized amounts, analogous to the quantized energies of lattice vibrations in a solid that had been proposed earlier by Planck [2] to account for *black body radiation* (and for which Planck received a Nobel Prize in 1918). These quantized packets of energy, termed *photons*, could have an energy

$$E = h\nu \qquad (7.1)$$

where h is *Planck's constant*, equal to 6.626×10^{-34} J·s, and ν is the *frequency* of light, equal to the *speed of light c* divided by *wavelength* λ. Thus, as wavelength decreases, frequency ν and energy $h\nu$ increase.

Figure 7.1 illustrates these features of the *photoelectric effect*. In Figure 7.1a, electrons absorb photon energy $h\nu$ and some are ejected into vacuum. The kinetic energy of these electrons can be measured with a plate biased to retard the collection of electrons leaving the illuminated metal. Here the kinetic energy is just equal to the minimum retarding potential needed to cut off collection of these electrons. Figure 7.1b shows the dependence of the maximum kinetic energy E_{max} versus photon frequency ν. The kinetic energy is linearly proportional to ν with slope h. The minimum energy required to eject an electron is shown as the extrapolated

Surfaces and Interfaces of Electronic Materials. Leonard J. Brillson
Copyright © 2010 WILEY-VCH Verlag GmbH & Co. KGaA, Weinheim
ISBN: 978-3-527-40915-0

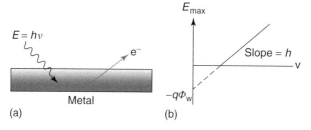

(a) (b)

Figure 7.1 The photoelectric effect. (a) Incident electrons with energy $h\nu$ are absorbed in the metal and some are ejected with finite kinetic energy. (b) Maximum kinetic energy E_{max} increases linearly with slope h as ν increases.

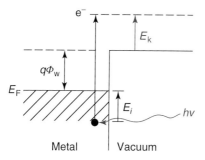

Figure 7.2 The photoexcitation process for an electron from an energy level E_i in a metal into vacuum with kinetic energy E_K by photon $h\nu$.

dashed line and defined as the *work function*, $q\Phi_W$. This dependence can be expressed as

$$E_{max} = h\nu - q\Phi_W \tag{7.2}$$

for incident photon energy $h\nu$.

An energy level diagram for this process appears in Figure 7.2. Here an electron occupying an energy level E_i inside a metal is excited by an incident photon $h\nu$ into vacuum with kinetic energy E_K. This kinetic energy is just the amount by which $h\nu$ exceeds the work function needed to exit the solid.

Exciting an electron from levels deeper in the solid requires more energy than those near the Fermi level. In general, for any energy level E_i below E_F, the photoelectron's kinetic energy obeys the following relation:

$$E_K = h\nu - E_i - q\Phi_W \tag{7.3}$$

In Figure 7.2, $E_i - E_F$ is termed the *binding energy*, and both E_i and E_F can be measured experimentally using photoexcitation spectroscopy.

7.2
The Optical Excitation Process

The photoexcitation process can be described quantum mechanically in terms of wavefunctions. For initial and final state wavefunctions $|\psi_i\rangle$ and $|\psi_f\rangle$, respectively,

$$|\psi_i\rangle - h\nu \rightarrow |\psi_f^\nu\rangle \tag{7.4}$$

or in terms of initial and final state electronic states $X(i)$ and $X(f)$, respectively,

$$h\nu + X(i) \rightarrow X^+(f) + e^- \tag{7.5}$$

Here, $X^+(f)$ is defined as $|f\rangle_{solid} = |i\rangle_{solid}$ + a hole. The electron part of $|\psi_f^\nu\rangle$ is $A_0 e^{i\mathbf{k}\cdot\mathbf{r}}$, a free electron, that is, *propagating* wavefunction, where A_0 is a constant and photon *wave vector* $\mathbf{k}$ is defined by momentum $\mathbf{p}$ as $\mathbf{p} = \hbar\mathbf{k}$, where $\hbar = h/2\pi$. The goal of photoemission spectroscopy is to gather as much information as possible on $|\psi_f^\nu\rangle$, then process it to gain information on $|\psi_i\rangle$.

7.3
Photoionization Cross Section

The photoemission process depends not only on the energy of the initial electron but also on the probability that a photon can excite it. This *cross section* for photoionization of a system in state i by a photon of energy $h\nu$ with the electron leaving the system in a final state f consisting of a photoelectron of energy E plus ion in state j is given by the quantum mechanical expression

$$\sigma_{ij}(E) = \left(\frac{4\alpha\pi a_0^2}{3 g_i}\right)(E + I_{ij})|M_{if}|^2 \tag{7.6}$$

where α = the *fine structure constant* = $e^2/\hbar c$ ($\sim 1/137$), a_0 = the Bohr radius = $\hbar^2/m_e e^2$ ($= 5.29 \times 10^{-9}$ cm), m_e = electron rest mass, g_i = a *statistical weight* (the number of degenerate, that is, same energy, sublevels of the initial discrete state), and I_{ij} = the *ionization energy* expressed in Rydbergs R_H, where $R_H = m_e e^4/2\hbar^2$ = 13.6 eV. M_{if} is a matrix element between initial and final states defined by

$$|M_{if}|^2 = \frac{4}{(E + I_{ij})^2} \sum_{i,f} |\langle f| \exp(i\mathbf{k}_\nu \cdot \mathbf{r}_u) \nabla_u |i\rangle|^2 \tag{7.7}$$

Here $\mathbf{r}_u$ = the position coordinate of the uth electron, $\mathbf{k}_\nu$ = the *wave propagation vector* of the photon, and $|\mathbf{k}_\nu| = 2\pi\nu/c$ (since $\mathbf{k}/2\pi$ is defined as $1/\lambda = \nu/c$). The matrix element $T = |\langle f| \exp(i\mathbf{k}_\nu \cdot \mathbf{r}_u) \nabla_u |i\rangle|$ in Equation 7.7 expresses the interaction between an electron and the electromagnetic field of a photon. The calculation of the matrix element $\sigma_{ij}(E)$ reduces to finding $|i\rangle$ and $|f\rangle$.

7.4
Density of States

The density of electron states at an energy E inside the solid is termed $\rho(E)$ and can be related to the photoexcitation via the *cross section* $\sigma_{ij}(E)$ according to the following relation:

$$\rho(E) = \left[\frac{1}{|T|^2}\right] \frac{\sigma(E)}{\left(\frac{\alpha}{nN}\right)} = \frac{R(E)}{|T|^2} \tag{7.8}$$

where n is the *refractive index* of the solid, N is the particle density, for example, the atom density, and R is the *rate of photoexcitation*. In other words, the rate of photoexcitation is directly proportional to the density of electron states times the transition probability, which is in turn proportional to the cross section. This relation is equivalent to a common rule of optical excitation termed *Fermi's golden rule*. Suitably normalized, this expression is

$$R = \left(\frac{2\pi}{\hbar}\right) \rho(E) |T|^2 \tag{7.9}$$

A derivation of Fermi's Golden Rule appears in Appendix 6.

7.5
Experimental Spectrum

The spectrum actually measured in a photoemission experiment involves not only the electron excitation out of the atom but also the probability that this electron can escape from the solid. This conversion from the energy distribution $N_f(E_K)$ of electrons inside the solid produced by the optical excitation to the experimental distribution of electrons measured outside the solid, termed the *energy distribution curve* (EDC), can be expressed as

$$N_\nu(E_K) \propto N_f(E_K) L_{tr}(E_K) S_{em}(E_K) \tag{7.10}$$

where $L_{tr}(E_K)$ is the distortion due to the electron transport to the surface and $S_{em}(E_K)$ is a step function that determines whether or not the electron escapes from the solid. $S_{em}(E_K) = 1$ for $E_K \geq E_{VAC}$ and $=0$ for $E_K < E_{VAC}$.

The initial distribution of excited electrons is directly proportional to $R(E_K)$ so that

$$N_\nu(E_K) \propto R(E_K) L_{tr}(E_K) S_{em}(E_K) \tag{7.11}$$

Finally, the spectrum measured in the laboratory, $N_{\nu,\text{experimental}}(E_K)$ for a given photon frequency ν, depends on the transmission and resolution of the experimental system that collects and analyzes the excited electrons.

$$N_{\nu,\text{experimental}}(E_K) \propto N_\nu(E_K) \cdot T_B(E_K) \tag{7.12}$$

where $T_B(E_K)$ is the total broadening response function of the experimental system. This response function includes the excitation linewidth as well as the energy broadening due to the analyzer resolution. Typical experimental resolutions are

0.1 eV for ultraviolet (UV) sources, ~1 eV for X-ray sources, and <0.1 eV for soft X-ray sources.

7.6 Experimental Energy Distribution Curves

Section 7.7 shows that the X-ray photoemission spectra or EDCs measured experimentally are directly proportional to the rate of photoexcitation $R(E_K)$ and hence the *densities of states* $\rho(E)$. This proportionality between EDC features and $\rho(E)$ is the attribute that establishes photoemission spectroscopy as such a powerful technique. Since $\rho(E)$ can be determined analytically from atomic wave functions and band theory and the response of the experimental system can be factored out, one can use the spectral features measured versus calculated to measure a wide range of features.

Figure 7.3 provides an illustration of the EDC that results from exciting an arbitrary initial density of states (DOS) with photon energy $h\nu$. Photoexcitation creates a replica of the density of occupied states distributed in energy below the vacuum level (dashed line), shifted up by the photon energy. Electrons that scatter inelastically are termed *secondary electrons* and form a continuum (shaded area) with the lowest density at the highest EDC energy and highest density at near-zero kinetic energy (at the sample surface), corresponding to the vacuum level cutoff. This secondary electron distribution adds to the EDC of elastic electrons to produce the composite shape illustrated schematically [3].

The secondary electron cutoff of electron emission at the vacuum level provides additional information. The width (ΔEDC) of the EDC, that is, the energy difference

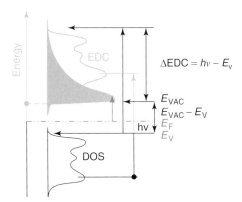

Figure 7.3 Electron density versus energy for an initial density of states and the energy distribution curve due to excitation by a photon of energy $h\nu$. The EDC is a direct indication of the DOS, modified by an additional contribution from secondary electrons that have scattered into a continuum of energies (shaded area). (Adapted from [3]).

between the highest and lowest photoelectron energies, equals $h\nu - E_V$ since electrons with the highest valence band energy are excited to the highest kinetic energy. Since E_V is defined relative to the vacuum level E_{VAC} and is equal to the electron affinity χ plus any bandgap E_G, ΔEDC yields the ionization potential $\Phi_{ionization}$ and electron affinity χ according to

$$\Delta EDC = h\nu - (E_{VAC} - E_V) = h\nu - \Phi_{ionization} = h\nu - (\chi + E_G) \quad (7.13)$$

For semiconductors with a known bandgap E_G plus the selected photon energy $h\nu$ and the measured ΔEDC, χ follows from Equation 7.13 For metals, $E_G = 0$ so that χ is simply $\Phi_{ionization}$.

The low-energy cutoff of the EDC also provides the Fermi level position E_F in the solid. This information is available from the lowest kinetic energy of electrons escaping from the solid as measured by the electron analyzer. For these electrons, $E_K = E_{VAC} - E_F^{analyzer}$ and since $E_F^{analyzer} = E_F^{solid}$ for analyzer and sample in electrical contact, $E_K = E_{VAC} - E_F^{solid}$. This is a direct measure of the work function Φ. It may be noted that if the solid has a work function lower than that of the analyzer so that $E_{VAC}^{solid} < E_{VAC}^{analyzer}$, the solid requires a negative bias in order for electrons at the solid cutoff energy to reach the analyzer.

Next, consider an example involving both valence and core level electrons. Figure 7.4 shows a representative case for a transition metal (e.g., Ni) having 3d states filled up to the Fermi level, O_{2p} states due to oxygen adsorbed on the metal surface as well as $3p_{1/2}$ and $3p_{3/2}$ metal core levels [4]. Empty s and p orbital states within the solid are shown that extend above the vacuum level. Photon energy $h\nu$ promotes

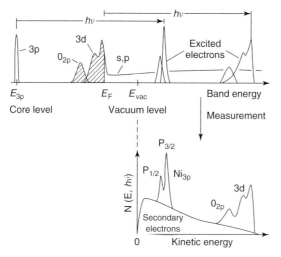

Figure 7.4 Composite energy distribution curve of photoemitted electrons from a transition metal with adsorbed oxygen. Photons excite electrons from filled valence and core level states into empty states plus a secondary electron continuum [4].

electrons from the filled states (shaded) to empty states as shown, so that the measured EDC with secondary electrons added appears as shown in the lower panel.

The spectral features in Figure 7.4 illustrate several of the most important tools that photoemission spectroscopy provides. First and foremost, peak features corresponding to core level electrons provide chemical identification of the atoms near the surface of the solid. These core level electrons occupy well-defined states and have well-defined energies. Table 7.1 displays the electronic configuration

Table 7.1 Electronic configuration for the first 36 atoms in the periodic table in their ground state.

Atomic number (Z)	Element	$n=1$, $l=0$	$n=2$, $l=0$	$l=1$	$n=3$, $l=0$	$l=1$	$l=2$	$n=4$, $l=0$	$l=1$	Shorthand notation
		1s	2s	2p	3s	3p	3d	4s	4p	
1	H	1								$1s^1$
2	He	2								$1s^2$
3	Li	\multicolumn{1}{c}{Helium core, 2 electrons}	1							$1s^2\ 2s^1$
4	Be		2							$1s^2\ 2s^2$
5	B		2	1						$1s^2\ 2s^2\ 2p^1$
6	C	Helium core,	2	2						$1s^2\ 2s^2\ 2p^2$
7	N	2 electrons	2	3						$1s^2\ 2s^2\ 2p^3$
8	O		2	4						$1s^2\ 2s^2\ 2p^4$
9	F		2	5						$1s^2\ 2s^2\ 2p^5$
10	Ne		2	6						$1s^2\ 2s^2\ 2p^6$
11	Na				1					[Ne] $3s^1$
12	Mg				2					$3s^2$
13	Al				2	1				$3s^2\ 3p^1$
14	Si	Neon core,			2	2				$3s^2\ 3p^2$
15	P	10 electrons			2	3				$3s^2\ 3p^3$
16	S				2	4				$3s^2\ 3p^4$
17	Cl				2	5				$3s^2\ 3p^5$
18	Ar				2	6				$3s^2\ 3p^6$
19	K							1		[Ar] $4s^1$
20	Ca							2		$4s^2$
21	Sc						1	2		$3d^1\ 4s^2$
22	Ti						2	2		$3d^2\ 4s^2$
23	V						3	2		$3d^3\ 4s^2$
24	Cr						5	1		$3d^5\ 4s^1$
25	Mn						5	2		$3d^5\ 4s^2$
26	Fe						6	2		$3d^6\ 4s^2$
27	Co	Argon core,					7	2		$3d^7\ 4s^2$
28	Ni	18 electrons					8	2		$3d^8\ 4s^2$
29	Cu						10	1		$3d^{10}\ 4s^1$
30	Zn						10	2		$3d^{10}\ 4s^2$
31	Ga						10	2	1	$3d^{10}\ 4s^2\ 4p^1$
32	Ge						10	2	2	$3d^{10}\ 4s^2\ 4p^2$
33	As						10	2	3	$3d^{10}\ 4s^2\ 4p^3$
34	se						10	2	4	$3d^{10}\ 4s^2\ 4p^4$
35	Br						10	2	5	$3d^{10}\ 4s^2\ 4p^5$
36	Kr						10	2	6	$3d^{10}\ 4s^2\ 4p^6$

for atoms in their ground state of the first 36 elements in the periodic table [5]. The *shell number n* and *orbital number l* for a given *atomic number Z* determine the occupancy of the individual subshells. For compactness, the electrons of each filled subshell are abbreviated as that (noble gas) atom and the number of its core electrons.

The binding energies for electrons in the various subshells of the periodic table are listed in Table 7.2 [6]. These experimental values are taken relative to the vacuum level E_{VAC}. Photoemission spectra will contain peak features corresponding to the core level energies accessible by the incident photon energy.

7.7
Measured Photoionization Cross Sections

The photoionization cross sections discussed in Section 7.3 determine the intensities of these peak features. These cross sections follow directly from the transition matrix elements in Equations 7.6 and 7.7 and depend on both the atomic orbital wavefunctions and the photon energy. For a given element, the cross section for each subshell can vary by orders of magnitude, decreasing from a maximum at the core level threshold energy for emission. The energy dependence of these cross sections is useful in identifying particular subshells.

To understand the energy dependence of such cross sections, it is useful to consider Fermi's Golden Rule in terms of electromagnetic power density. The *photoelectron cross section* σ is defined as the probability of a photoelectron event per unit photon. For a transition probability per unit time R and a *flux F* of photons [7],

$$\sigma = \frac{R}{F} = \frac{\left(\frac{\text{Transition Probability}}{\text{time}}\right)}{\left(\frac{\text{\#photons}}{\text{area} \cdot \text{time}}\right)} \quad (7.14)$$

The electromagnetic power density for electric field ε is defined as

$$\frac{c\varepsilon^2}{2} = \frac{\text{energy}}{(\text{area} \cdot \text{s})} \quad (7.15)$$

where $c =$ the speed of light so that, dividing by the energy per photon,

$$F = \frac{\text{\#photons}}{(\text{area} \cdot \text{s})} = \frac{c\varepsilon^2}{2\hbar\omega} \quad (7.16)$$

The transition probability per unit time for a density ρ is given by Fermi's Golden Rule as (see Equation A7.1)

$$R = |<\psi_f|H'|\psi_i>|^2 \frac{2\pi\rho(E)}{\hbar} = |H_{fi}|^2 \frac{2\pi\rho(E)}{\hbar} \quad (7.17)$$

To illustrate the general energy dependence of W, consider a simple square-well potential with binding energy E_B:

$$R = \left(\frac{4e^2\varepsilon^2\hbar}{m}\right)\frac{E_B^{3/2}}{E^{7/2}} \quad (7.18)$$

Table 7.2 Free atom subshell binding energies relative to E_{VAC} [6].

Z		1s	2s	2p$_{1/2}$	2p$_{3/2}$	3s	3p$_{1/2}$	3p$_{3/2}$	3d$_{3/2}$	3d$_{5/2}$	4s	4p$_{1/2}$	4p$_{3/2}$	4d$_{3/2}$	4d$_{5/2}$	4f$_{5/2}$	4f$_{7/2}$	5s	5p$_{1/2}$	5p$_{3/2}$	5d$_{3/2}$	5d$_{5/2}$	5f$_{5/2}$	5f$_{7/2}$	6s	6p$_{1/2}$	6p$_{3/2}$	6d$_{3/2}$	6d$_{5/2}$	7s
1	H	13.60																												
2	He	24.59																												
3	Li	58	5.392																											
4	Be	115	9.322																											
5	B	192	12.93	8.298	8.298																									
6	C	288	16.59	11.26	11.26																									
7	N	403	20.33	14.53	14.53																									
8	O	538	28.48	13.62	13.62																									
9	F	694	37.85	17.42	17.42																									
10	Ne	870.1	48.47	21.66	21.56																									
11	Na	1075	66	34	34	5.139																								
12	Mg	1308	92	54	54	7.646																								
13	Al	1564	121	77	77	10.62	5.986	5.986																						
14	Si	1844	154	104	104	13.46	8.151	8.151																						
15	P	2148	191	135	134	16.15	10.49	10.49																						
16	S	2476	232	170	168	20.20	10.36	10.36																						
17	Cl	2829	277	208	206	24.54	12.97	12.97																						
18	Ar	3206.3	326.5	250.6	248.5	29.24	15.94	15.76																						
19	K	3610	381	299	296	37	19	18.7			4.341																			
20	Ca	4041	441	353	349	46	28	28			6.113																			
21	Sc	4494	503	408	403	55	33	33	8	8	6.540																			
22	Ti		567	465	459	64	39	38	8	8	6.820																			
23	V		633	525	518	72	44	43	8	8	6.740																			
24	Cr		702	589	580	80	49	48	8.25	8.25	6.765																			
25	Mn		755	656	645	89	55	53	9	9	7.434																			

(continued overleaf)

Table 7.2 (continued)

Z	1s	2s	2p₁/₂	2p₃/₂	3s	3p₁/₂	3p₃/₂	3d₃/₂	3d₅/₂	4s	4p₁/₂	4p₃/₂	4d₃/₂	4d₅/₂	4f₅/₂	4f₇/₂	5s	5p₁/₂	5p₃/₂	5d₃/₂	5d₅/₂	5f₅/₂	5f₇/₂	6s	6p₁/₂	6p₃/₂	6d₃/₂	5/2	7s
26 Fe	851	726	713	98	61	59	9	9	7.870																				
27 Co	931	800	785	107	68	66	9	9	7.864																				
28 Ni	1015	877	860	117	75	73	10	10	7.635																				
29 Cu	1103	958	938	127	82	80	11	10.4	7.726																				
30 Zn	1198	1047	1024	141	94	91	12	11.2	9.394																				
31 Ga	1302	1146	1119	162	111	107	21	20	11	6.00	6.00																		
32 Ge	1413	1251	1220	184	130	125	33	32	14.3	7.90	7.90																		
33 As	1531	1362	1327	208	151	145	46	45	17	9.81	9.81																		
34 Se	1656	1479	1439	234	173	166	61	60	20.15	9.75	9.75																		
35 Br	1787	1602	1556	262	197	189	77	76	23.80	11.85	11.85																		
36 Kr	1924.6	1730.9	1678.4	292.8	222.2	214.4	95.0	93.8	27.51	14.65	14																		
37 Rb	2068	1867	1807	325	251	242	116	114	32	16	15.3					4.18													
38 Sr	2219	2010	1943	361	283	273	139	137	40	23	22					5.69													
39 Y	2375	2158	2083	397	315	304	163	161	48	30	29	6.38	6.38			6.48													
40 Zr	2536	2311	2227	434	348	335	187	185	56	35	33	8.61	8.61			6.84													
41 Nb	2702	2469	2375	472	382	367	212	209	62	40	38	7.17	7.17			6.88													
42 Mo	2872	2632	2527	511	416	399	237	234	68	45	42	8.56	8.56			7.10													
43 Tc	3048	2800	2683	551	451	432	263	259	74	49	45	8.6	8.6			7.28													
44 Ru	3230	2973	2844	592	488	466	290	286	81	53	49	8.50	8.50			7.37													
45 Rh	3418	3152	3010	634	526	501	318	313	87	58	53	9.56	9.56			7.46													
46 Pd	3611	3337	3180	677	565	537	347	342	93	63	57	8.78	8.34																
47 Ag	3812	3530	3357	724	608	577	379	373	101	69	63	11	10			7.58													
48 Cd	4022	3732	3542	775	655	621	415	408	112	78	71	14	13			8.99													
49 In	4242	3945	3735	830	707	669	455	447	126	90	82	21	20			10	5.79	5.79											
50 Sn	4469	4160	3933	888	761	719	497	489	141	102	93	29	28			12	7.34	7.34											

7.7 Measured Photoionization Cross Sections

Z	El																				
51	Sb	4385	4137		949	817	771	542	533	157	114	104	38	37			15	8.64	8.64		
52	Te		4347		1012	876	825	589	578	174	127	117	48	46			17.84	9.01	9.01		
53	I				1078	937	881	638	626	193	141	131	58	56			20.61	10.45	10.45		
54	Xe				1148.7	1002.1	940.6	689.0	676.4	213.2	157	145.5	69.5	67.5			23.39	13.43	12.13			3.89
55	Cs				1220	1068	1000	742	728	233	174	164	81	79			25	14	12.3			5.21
56	Ba				1293	1138	1063	797	782	254	193	181	94	92			31	18	16			5.58
57	La				1365	1207	1124	851	834	273	210	196	105	103			36	22	19	5.75	5.75	5.65
58	Ce				1437	1275	1184	903	885	291	225	209	114	111	6		39	25	22	6	6	5.42
59	Pr				1509	1342	1244	954	934	307	238	220	121	117	6		41	27	24			5.49
60	Nd				1580	1408	1303	1005	983	321	250	230	126	122	6		42	28	25			5.55
61	Pm				1653	1476	1362	1057	1032	335	261	240	131	127	6		43	28	25			5.63
62	Sm				1728	1546	1422	1110	1083	349	273	251	137	132	6		44	29	25			5.68
63	Eu				1805	1618	1484	1164	1135	364	286	262	143	137	6		45	30	26	6	6	6.16
64	Gd				1884	1692	1547	1220	1189	380	300	273	150	143	6		46	31	27			5.85
65	Tb				1965	1768	1612	1277	1243	398	315	285	157	150	6		48	32	28	6	6	5.93
66	Dy				2048	1846	1678	1335	1298	416	331	297	164	157	6		50	33	28			6.02
67	Ho				2133	1926	1746	1395	1354	434	348	310	172	164	6		52	34	29			6.10
68	Er				2220	2008	1815	1456	1412	452	365	323	181	172	6		54	35	30			6.18
69	Tm				2309	2092	1885	1518	1471	471	382	336	190	181	7	7	56	36	30			6.25
70	Yb				2401	2178	1956	1580	1531	490	399	349	200	190	7	7	58	37	31			7.0
71	Lu				2499	2270	2032	1647	1596	514	420	366	213	202	8	12	62	39	32	6.6	6.6	7.5
72	Hf				2604	2369	2113	1720	1665	542	444	386	229	217	13	20	68	43	35	7.0	7.0	7.9
73	Ta				2712	2472	2197	1796	1737	570	469	407	245	232	21	28	74	47	38	8.3	8.3	8.0
74	W				2823	2577	2283	1874	1811	599	495	428	261	248	30	36	80	51	41	9.0	9.0	7.9
75	Re				2937	2686	2371	1953	1887	629	522	450	278	264	38	45	86	56	45	9.6	9.6	8.5
76	Os				3054	2797	2461	2035	1964	660	551	473	295	280	47	54	92	61	49	9.6	9.6	9.1
77	Ir				3175	2912	2554	2119	2044	693	581	497	314	298	56	64	99	66	53	9.6	9.6	9.0
78	Pt				3300	3030	2649	2206	2126	727	612	522	335	318	67	75	106	71	57	9.6	9.6	9.23
79	Au				3430	3153	2748	2295	2210	764	645	548	357	339	78	87	114	76	61	12.5	11.1	10.4
80	Hg				3567	3283	2852	2390	2300	806	683	579	382	363	91	103	125	85	68	14	12	

(continued overleaf)

Table 7.2 (continued)

Z		1s	2s	2p$_{1/2}$	2p$_{3/2}$	3s	3p$_{1/2}$	3p$_{3/2}$	3d$_{3/2}$	3d$_{5/2}$	4s	4p$_{1/2}$	4p$_{3/2}$	4d$_{3/2}$	4d$_{5/2}$	4f$_{5/2}$	4f$_{7/2}$	5s	5p$_{1/2}$	5p$_{3/2}$	5d$_{3/2}$	5d$_{5/2}$	5f$_{5/2}$	5f$_{7/2}$	6s	6p$_{1/2}$	6p$_{3/2}$	6d$_{3/2}$$_{5/2}$	7s
81	Tl					3710	3420	2961	2490	2394	852	726	615	411	391	127	123	139	98	79	21	19			8	6.11	6.11		
82	Pb					3857	3560	3072	2592	2490	899	769	651	441	419	148	144	153	111	90	27	25			10	7.42	7.42		
83	Bi					4007	3704	3185	2696	2588	946	813	687	472	448	170	165	167	125	101	34	32			12	7.29	7.29		
84	Po					4161	3852	3301	2802	2687	994	858	724	503	478	193	187	181	139	112	41	38			15	8.43	8.43		
85	At					4320	4005	3420	2910	2788	1044	904	761	535	508	217	211	196	153	123	48	44			19	11	9.3		
86	Rn					4483	4162	3452	3109	2890	1096	951	798	567	538	242	235	212	167	134	55	51			24	14	10.7		
87	Fr						4324	3666	3134	2998	1153	1003	839	603	572	268	260	231	183	147	65	61			33	19	14		4.0
88	Ra						4491	3793	3254	3111	1214	1060	884	642	609	296	287	253	201	161	77	73			40	25	19		5.28
89	Ac							3921	3374	3223	1274	1116	928	680	645	322	313	274	218	174	88	83			45	29	22	5.7	6.3
90	Th							4049	3494	3335	1333	1171	970	717	679	347	338	293	233	185	97	91			50	33	25	6	6
91	Pa							4178	3613	3446	1390	1225	1011	752	712	372	362	312	248	195	104	97	6	6	50	32	24	6	6
92	U							4308	3733	3557	1446	1278	1050	785	743	396	386	329	261	203	110	101	6	6	52	34	24	6.1	6
93	Np							4440	3854	3669	1504	1331	1089	819	774	421	410	346	274	211	116	106	6	6	54	35	25	6	6
94	Pu								3977	3783	1563	1384	1128	853	805	446	434	356	287	219	122	111	6	6	53	34	23		6
95	Am								4102	3898	1623	1439	1167	887	836	467	452	355	301	220	123	112	6	6	54	44	36		6.0
96	Cm								4236	4014	1664	1493	1194	919	864	494	479	384	314	239	126	119	11	11	60	39	27	5	6
97	Bk								4366	4133	1729	1554	1236	955	898	520	504	401	329	248	142	124	12	12	63	41	27	4	6
98	Cf								4492	4247	1789	1610	1273	987	925	546	529	412	328	251	142	129	9	9	61	39	25		6
99	Es									4369	1857	1674	1316	1024	959	573	554	429	353	260	148	135	9	9	63	40	25		6
100	Fm									4498	1933	1746	1366	1068	1000	606	587	453	375	275	160	145	15	15	69	45	29	4	7
101	Md										1998	1807	1404	1100	1029	627	607	464	384	278	160	144	11	11	67	43	26		6
102	No										2071	1876	1449	1139	1064	654	633	483	400	287	166	149	14	11	69	44	26		6
103	Lw										2153	1954	1501	1186	1106	689	666	509	424	303	178	160	20	17	76	49	30	4	7
104											2237	2034	1554	1233	1149	724	700	535	448	318	190	171	26	23	82	55	33	5	8
105											2324	2117	1609	1281	1193	760	735	562	473	335	203	183	32	29	89	60	36	6	8
106											2413	2202	1664	1350	1238	797	770	590	499	350	216	194	39	35	96	65	39	7	9

(see Equation A7.2) and

$$\sigma = \frac{8e^2\hbar}{mc} \frac{E_B^{3/2}}{E^{5/2}} \tag{7.19}$$

so that σ decreases with increasing E as $E^{5/2}$ in this simple case.

A more realistic case is a hydrogenic atom. Consider an initial, known hydrogenic wavefunction $\psi_i = \psi_H$ in three-dimensional space equal to $[1/\sqrt{\pi a^3}]e^{-\rho}$ (for quantum numbers $n=1, l=0, m=0$), where ρ is Zr/a_0, $r=$ radius, $a = a_0/Z$, $Z=$ atomic number, and a_0 is the Bohr radius $= \hbar^2/me^2 = 0.529$ Å. The final state wavefunction ψ_f is that of a plane wave with momentum k and energy $E_{\text{final}} = \hbar^2 k^2/2m$, normalized to volume V such that $\psi_f = e^{ikr}/\sqrt{V}$.

The hydrogenic binding energy $E_B = Z^2 e^2/2a_0$, the energy density of states $\rho(E) = (V/2\pi^2)(2m/\hbar^2)^{3/2} E^{1/2}$ in three dimensions, and transition matrix element $H' = -e\mathcal{E}z\,e^{-i\omega t}$. Assuming $\hbar\omega \gg E_B$, one obtains a cross section $\sigma_{\text{ph}} = (288\pi/3)(e^2 \hbar/mc) E_B^{5/2}/E^{7/2}$ and with a more sophisticated description of H', $\sigma_{\text{ph}} = (128\pi/3)(e^2 \hbar/mc) E_B^{5/2}/E^{7/2}$. The quantity

$$\frac{e^2 \hbar}{mc} = 5.56 \times 10^{-2} \text{eV Å}^2 \tag{7.20}$$

so that, assuming $\hbar\omega_i \cong \hbar\omega_f$ in Equation 7.17,

$$\sigma_{\text{ph}} = \frac{7.45 \text{Å}^2}{\hbar\omega} \cdot \left(\frac{E_B}{\hbar\omega}\right)^{5/2} \tag{7.21}$$

In general, therefore, the photoionization cross section decreases with increasing energy E as $E^{5/2}$. Let us consider an example to illustrate the calculation of σ_{ph}.

Example 7.1. Calculate the photoelectron cross section for photon energy $h\nu = 6.4 \times 10^3$ eV (Fe K$_\alpha$ radiation) incident on K-shell electrons in Al. From Table 7.2, K-shell electrons in Al have binding energy $E_B^K = 1.56 \times 10^3$ eV. Therefore,

$$\sigma = \left[\frac{7.45}{(6.4 \times 10^3)}\right] \left(\frac{1.56}{6.4}\right)^{5/2} \text{Å}^2 = 3.4 \times 10^{-21} \text{ cm}^2 \tag{7.22}$$

By convention, units of cross section are given in "Barns," where 1 Barn $= 10^{-24}$ cm^2. Thus $\sigma = 3.4 \times 10^3$ Barn (b) or 3.4×10^{-3} Megabarn (Mb) where 1 Megabarn (Mb) $= 10^{-18}$ cm^2.

Table 7.3 provides some examples of photoionization cross sections σ_{nl}(Mb) for 35 commonly studied elements as a function of photon energy $\hbar\omega$ [8]. This table is a useful guide to selecting excitation energies for studying particular elements. Depending on the excitation energies available, one can select core levels to study based on the magnitude of their cross sections. For example, the Al 2p subshell with a binding energy of 73 eV has large cross sections for excitation in the range of $\hbar\omega = 120 - 160$ eV, decreasing with increasing photon energy. Similarly, the Ga 3d subshell with a binding energy 17 eV has large cross sections for excitation in the range of 40–120 eV. Note how much smaller the corresponding cross sections

Table 7.3 Atomic subshell photoionization cross sections for 23 atoms by increasing atomic number for all subshells observable for energies up to Al K$_\alpha$ radiation.

Atomic number	Atomic subshell	Binding energy (eV)	hv (eV) = 21.2	hv (eV) = 40.8	hv (eV) = 80	h (eV) = 151.4	hv (eV) = 200	hv (eV) = 300	hv (eV) = 600	hv (eV) = 800	hv (eV) = 1253.6	hv (eV) = 1486.6
1	H 1s	13.6	1.888	0.2892	0.38E − 1	0.51E − 2	0.21E − 2	0.61E − 3	0.71E − 4	0.29E − 4	0.46E − 5	0.20E − 6
6	C 1s	288						0.8591	0.1680	0.77E − 1	0.22E − 1	0.13E − 1
	C 2s	16.59	1.230	1.170	0.5440	0.1714	0.96E − 1	0.38E − 1	0.73E − 2	0.33E − 2	0.10E − 2	0.66E-3
	C 2p	8.298	6.128	1.875	0.3266	0.47E − 1	0.19E − 1	0.60E − 2	0.75E − 3	.31E − 3	0.56E − 4	0.10E-4
7	N 1s	403							0.2763	0.1329	0.39E − 1	0.24E − 1
	N 2s	20.33		1.086	0.6528	0.2383	0.1404	0.60E − 1	0.112E − 1	0.60E − 2	0.19E − 2	0.11E − 2
	N 2p	14.53	9.688	4.351	0.9605	0.1671	0.73E − 1	0.20E − 1	0.20E − 2	0.79E − 3	0.18E − 3	0.72E − 4
8	O 1s	538							0.4119	0.2044	0.63E − 1	0.40E − 1
	O 2s	28.48		0.8342	0.6901	0.2984	0.1831	0.83E − 1	0.18E − 1	0.92E − 2	0.29E − 2	0.19E − 2
	O 2p	13.62	10.67	6.816	2.054	0.4037	0.1836	0.53E − 1	0.61E − 2	0.22E − 2	0.50E − 3	0.24E − 3
13	Al 2s	121				0.5051	0.3927	0.2292	0.70E − 1	0.39E − 1	0.15E − 1	0.10E − 1
	Al 2p	77				3.841	2.176	0.8226	0.1230	0.51E − 1	0.12E − 1	0.72E − 2
	Al 3s	10.62	0.3431	0.3339	0.1572	0.58E − 1	0.37E − 1	0.19E − 1	0.50E − 2	0.29E − 2	0.11E − 2	0.78E − 3
	Al 3p	5.986	0.88E − 1	0.1230	0.65E − 1	0.20E − 1	0.11E − 1	0.46E − 2	0.64E − 3	0.36E − 3	0.72E − 4	0.59E − 4
14	Si 2s	154				0.5043	0.4241	0.2528	0.84E − 1	0.48E − 1	0.18E − 1	0.13E − 1
	Si 2p	104				4.553	2.857	1.132	0.1776	0.76E − 1	0.19E − 1	0.11E − 1
	Si 3s	13.46	0.2880	0.4232	0.2172	0.84E − 1	0.53E − 1	0.26E − 1	0.76E − 2	0.43E − 2	0.16E − 2	0.10E − 2
	Si 3p	8.151	0.3269	0.3286	0.2160	0.71E − 1	0.42E − 1	0.16E − 1	0.28E − 2	0.12E − 2	0.35E − 3	0.17E − 3
15	P 2s	191					0.4228	0.2922	0.99E − 1	0.57E − 1	0.23E − 1	0.16E − 1
	P 2p	135/134					3.489	1.501	0.2450	0.1081	0.27E − 1	0.16E − 1
	P 3s	16.15	0.1714	0.4538	0.2594	0.1087	0.69E − 1	0.35E − 1	0.10E − 1	0.56E − 2	0.22E − 2	0.14E − 2
	P 3p	10.49	1.232	0.5090	0.4275	0.1627	0.95E − 1	0.39E − 1	0.72E − 2	0.30E − 2	0.88E − 3	0.50E − 3

7.7 Measured Photoionization Cross Sections

Z	Shell											
16	S 2s	232										
	S 2p	170/168										
	S 3s	20.20		4.333				3.799		0.3134	0.1141	0.67E−1	0.27E−1	0.19E−1
	S 3p	10.36		0.4493		0.1341		0.86E−1		1.920	0.3315	0.1475	0.38E−1	0.22E−1
22	Ti 2s	567					0.2896				0.1953	0.1289	0.60E−1	0.44E−1
	Ti 2p	465/459		0.6028							1.202	0.1289	0.1746	0.1059
	Ti 3s	64			0.4308	0.2845	0.2046		0.1146	0.36E−1	0.21E−1	0.91E−2	0.64E−2
	Ti 3p	39/38				1.090	1.083	0.8011		0.4332	0.1074	0.54E−1	0.17E−1	0.11E−1
	Ti 4s	6.820	0.1951	0.1427	0.67E−1	0.26E−1	0.18E−1	0.90E−2		0.26E−2	0.14E−2	0.69E−3	0.50E−3
	Ti 3d	8	5.074	4.012	1.472		0.3318	0.1552	0.47E−1	0.49E−2	0.17E−2	0.34E−3	0.17E−3
24	Cr 2s	702										0.1465	0.72E−1	0.53E−1
	Cr 2p	589/580								1.645		0.8235	0.2450	0.1577
	Cr 3s	80					0.2961	0.2204		0.1310	0.43E−1	0.26E−1	0.11E−1	0.81E−2
	Cr 3p	49/48		0.7094		1.039		0.8462		0.5077	0.1415	0.74E−1	0.24E−1	0.15E−1
	Cr 4s	6.765	0.63E−1	0.25E−1		0.111E−1		0.70E−2		0.37E−2	0.111E−1	0.71E−3	0.31E−3	0.18E−3
	Cr 3d	8.25	9.230	8.540		4.047	1.161	0.5837		0.1900	0.21E−1	0.84E−2	0.16E−2	0.88E−3
26	Fe 2s	851											0.83E−1	0.62E−1
	Fe 2p	726/713										1.107	0.3529	0.2216
	Fe 3s	98				0.3082		0.2425		0.1495	0.52E−1	0.32E−1	0.13E−1	0.10E−1
	Fe 3p	61/59		0.6409		0.9979		0.8951		0.5869	0.1829	0.99E−1	0.34E−1	0.22E−1

(continued overleaf)

Table 7.3 (continued)

Atomic number	Atomic subshell	Binding energy (eV)	$h\nu$ (eV) = 21.2	$h\nu$ (eV) = 40.8	$h\nu$ (eV) = 80	h (eV) = 151.4	$h\nu$ (eV) = 200	$h\nu$ (eV) = 300	$h\nu$ (eV) = 600	$h\nu$ (eV) = 800	$h\nu$ (eV) = 1253.6	$h\nu$ (eV) = 1486.6
	Fe 4s	7.870	0.1349	0.1350	0.74E−1	0.32E−1	0.20E−1	0.11E−1	0.35E−2	0.21E−2	0.91E−3	0.70E−3
	Fe 3d	9	4.833	8.751	6.181	2.219	1.201	0.4247	0.54E−1	0.20E−1	0.45E−2	0.22E−2
28	Ni 2s	1015									0.93E−1	0.71E−1
	Ni 2p	877/860									0.4691	0.2998
	Ni 3s	117			0.2596	0.2498	0.1617	0.60E−1	0.37E−1	0.16E−1	0.12E−1
	Ni 3p	75/73				0.8420	0.8578	0.6313	0.2189	0.1235	0.45E−1	0.29E−1
	Ni 4s	7.635	0.93E−1	0.1299	0.75E−1	0.33E−1	0.21E−1	0.11E−1	0.38E−2	0.25E−2	0.10E−2	0.83E−3
	Ni 3d	10	3.984	8.357	7.877	3.638	2.106	0.8065	0.1142	0.49E−1	0.10E−1	0.59E−2
29	Cu 2s	1103									0.98E−1	0.75E−1
	Cu 2p	958/938									0.5345	0.3438
	Cu 3s	127				0.2812	.2424	0.1627	0.62E−1	0.39E−1	0.17E−1	0.13E−1
	Cu 3p	82/80				0.7167	0.7950	0.6257	0.2319	0.1338	0.50E−1	0.33E−1
	Cu 4s	7.726	0.36E−1	0.41E−1	0.25E−1	0.11E−1	0.83E−2	0.43E−2	0.14E−2	0.89E−3	0.42E−3	0.27E−3
	Cu 3d	11/10.4	7.553	9.934	8.712	4.441	2.595	1.040	0.1513	0.77E−1	0.21E−1	0.12E−1
30	Zn 2s	1198									0.1036	0.79E−1
	Zn 2p	1047/1024									0.6057	0.3907
	Zn 3s	141				0.2814	0.2480	0.1705	0.66E−1	0.42E−1	0.19E−1	0.14E−1
	Zn 3p	94/91				0.6509	0.7784	0.6466	0.2527	0.1485	0.56E−1	0.37E−1
	Zn 4s	9.394	0.57E−1	0.1163	0.72E−1	0.34E−1	0.23E−1	0.12E−1	0.43E−2	0.23E−2	0.11E−2	0.78E−3
	Zn 3d	12/11.2	3.572	7.292	8.895	5.059	3.237	1.380	0.2086	0.87E−1	0.21E−1	0.12E−1
31	Ga 2s	1302										0.83E−1
	Ga 2p	1146/1119									0.6851	0.4412

7.7 Measured Photoionization Cross Sections

Z	Subshell											
	Ga 3s	162										
	Ga 3p	111/107										
	Ga 4s	6.00	0.48E−1	0.1704								
	Ga 3d	21/20	3.626		0.1085	0.5886	0.2546	0.1792	0.71E−1	0.45E−1	0.21E−1	0.15E−1
32	Ge 2s	1413					0.7559	0.6645	0.2751	0.1636	0.64E−1	0.43E−1
	Ge 2p	1251/1220	8.393			0.47E−1	0.31E−1	0.17E−1	0.58E−2	0.37E−2	0.17E−2	0.12E−2
	Ge 3s	184				5.719	3.797	1.686	0.2744	0.1150	0.26E−1	0.14E−1
	Ge 3p	130/125										0.88E−1
	Ge 4s	14.3	0.22E−1	0.1989							0.7071	0.4966
	Ge 3d	33/32			0.5334		0.2584	0.1871	0.76E−1	0.48E−1	0.22E−1	0.16E−1
	Ge 4p	7.90	1.528	0.1323	0.60E−1		0.7263	0.6846	0.2958	0.1795	0.71E−1	0.48E−1
33	As 2s	1531			6.999		0.40E−1	0.22E−1	0.77E−2	0.47E−2	0.19E−2	0.15E−2
	As 2p	1362/1327			6.263		4.343	2.006	0.3438	0.1465	0.34E−1	0.19E−1
	As 3s	208		0.39E−1	0.27E−1		0.21E−1	0.14E−1	0.49E−2	0.28E−2	0.11E−2	0.74E−3
	As 3p	151/145										0.86E−1
	As 4s	17	0.24E−2	0.2085		0.4709	0.2489	0.1945	0.81E−1	0.51E−1	0.24E−1	0.18E−1
	As 3d	46/45		0.1598	0.72E−1		0.6914	0.7022	0.3186	0.1960	0.79E−1	0.54E−1
	As 4p	9.81	3.856		4.717		0.149E−1	0.27E−1	0.91E−2	0.56E−2	0.24E−2	0.18E−2
34	Se 2p	1479/1439		0.81E−1	6.578		4.871	2.357	0.4227	0.1827	0.44E−1	0.25E−1
	Se 3s	234	0.2949		0.55E−1		0.43E−1	0.28E−1	0.10E−1	0.59E−2	0.23E−2	0.17E−2
	Se 3p	173/166										0.5917
						0.6436		0.2010	0.85E−1	0.55E−1	0.26E−1	0.19E−1
								0.7139	0.3425	0.2128	0.88E−1	0.60E−1

(continued overleaf)

Table 7.3 (continued)

Atomic number	Atomic subshell	Binding energy (eV)	$h\nu$ (eV) = 21.2	$h\nu$ (eV) = 40.8	$h\nu$ (eV) = 80	h (eV) = 151.4	$h\nu$ (eV) = 200	$h\nu$ (eV) = 300	$h\nu$ (eV) = 600	$h\nu$ (eV) = 800	$h\nu$ (eV) = 1253.6	$h\nu$ (eV) = 1486.6
	Se 4s	20.15		0.2065	0.1810	0.86E − 1	0.58E − 1	0.32E − 1	0.10E − 1	0.66E − 2	0.30E − 2	0.21E − 2
	Se 3d	61/60			1.670	6.605	5.282	2.718	0.5117	0.2239	0.55E − 1	0.31E − 1
	Se 4p	9.75	8.057	0.5578	0.1353	0.94E − 1	0.74E − 1	0.48E − 1	0.17E − 1	0.10E − 1	0.38E − 2	0.26E − 2
38	Sr 3s	361							0.1071	0.70E − 1	0.34E − 1	0.25E − 1
	Sr 3p	283/273						0.7006	0.4311	0.2834	0.1252	0.89E − 1
	Sr 4s	40		0.2717			0.1044	0.57E − 1	0.19E − 1	0.11E − 1	0.53E − 2	0.39E − 2
	Sr 3d	139/137				0.4788	4.856	4.129	0.9654	0.4478	0.1186	0.69E − 1
	Sr 4s	23/22		3.307	0.4174	0.2970	0.2374	0.1546	0.56E − 1	0.35E − 1	0.14E − 1	0.10E − 1
	Sr 5s	5.69	0.1036	0.76E − 1	0.37E − 1	0.15E − 1	0.10E − 1	0.55E − 2	0.18E − 2	0.97E − 3	0.43E − 3	0.30E − 3
47	Ag 3s	724								0.1045	0.54E − 1	0.41E − 1
	Ag 3p	608/577							0.5765	0.4271	0.2225	0.1651
	Ag 4s	101				0.2222	0.1804	0.1149	0.42E − 1	0.26E − 1	0.12E − 1	0.88E − 2
	Ag 3d	379/373							2.503	1.329	0.4052	0.2474
	Ag 4p	69/63			1.057	0.4257	0.3974	0.3036	0.1322	0.84E − 1	0.38E − 1	0.27E − 1
	Ag 4d	11/10	16.62	37.48	3.218	0.4121	0.6052	0.5452	0.1896	0.1030	0.33E − 1	0.21E − 1
	Ag 5s	7.58	0.30E − 1	0.35E − 1	0.22E − 1	0.10E − 1	0.70E − 2	0.40E − 2	0.13E − 2	0.80E − 2	0.35E − 3	0.29E − 3
48	Cd 3s	775								0.1080	0.57E − 1	0.43E − 1
	Cd 3p	655/621								0.4396	0.2335	0.1740
	Cd 4s	112			0.2402	0.2289	0.1882	0.1211	0.45E − 1	0.28E − 1	0.12E − 1	0.95E − 2
	Cd 3d	415/408							2.700	1.457	0.4519	0.2776
	Cd 4p	78/71			1.183	0.4400	0.4116	0.3213	0.1432	0.91E − 1	0.41E − 1	0.30E − 1
	Cd 5s	7.58	0.65E − 1	0.1003	0.61E − 1	0.28E − 1	0.19E − 1	0.11E − 1	0.35E − 2	0.23E − 2	0.98E − 3	0.70E − 3

7.7 Measured Photoionization Cross Sections

Z	Subshell	BE (eV)										
	Cd 4d	14/13		36.91	4.856	0.4080	0.6479	0.6060	0.2238	0.1233	0.41E−1	0.26E−1
49	In 3s	830								0.1115	0.59E−1	0.45E−1
	In 3p	707/669								0.4519	0.2442	0.1830
	In 4s	126				0.2375	0.1956	0.1273	0.48E−1	0.30E−1	0.14E−1	0.10E−1
	In 3d	455/447							2.925	1.588	0.5017	0.3098
	In 4p	90/82				0.4535	0.4292	0.3404	0.1526	0.98E−1	0.45E−1	0.32E−1
	In 5s	10	0.72E−1	0.1464	0.88E−1	0.39E−1	0.26E−1	0.15E−1	0.48E−1	0.31E−2	0.13E−2	0.10E−2
	In 4d	21/20		27.20	7.176	0.4014	0.6783	0.6709	0.2614	0.1445	0.49E−1	0.31E−1
79	Au 4s	764								0.85E−1	0.45E−1	0.35E−1
	Au 4p	645/548								0.2655	0.1554	0.1223
	Au 5s	114				0.1733	0.1412	0.91E−1	0.33E−1	0.21E−1	0.10E−1	0.77E−2
	Au 4d	357/339							0.6840	0.6173	0.3629	0.2739
	Au 5p	76/61			1.389	0.3168	0.2547	0.1846	0.85E−1	0.58E−1	0.30E−1	0.22E−1
	Au 4f	91/87				0.6247	2.274	5.274	2.599	1.387	0.4176	0.2511
	Au 5d	12.5/11.1	0.28E−1	25.95	38.67	4.526	0.2330	0.1069	0.1186	0.93E−1	0.68E−1	0.35E−1
	Au 6s	9.23	0.19E−1	0.19E−1	0.28E−1	0.94E−2	0.64E−2	0.35E−2	0.13E−2	0.84E−3	0.33E−3	0.29E−3

Cross sections calculated with Hartree–Fock–Slater one-electron central potential model. J.J. Yeh and I. Lindau, Atomic Data and Nuclear Data Tables 32, 1–155 (1985). Core level energies relative to E_{VAC} [6].

for the Ga 2p subshell are in the same energy range even though these energies are near the photoionization threshold. Thus the Ga 3d subshell emission would yield the strongest signals from Ga atoms in an X-ray photoelectron spectroscopy (XPS) study. Of course, such choices presuppose the availability of various photon energies. This is the case for synchrotron radiation in the following sections. However, XPS is typically restricted to discrete energies of 1486.6 eV for Al K_α or 1253.6 eV for Mg K_α emission from conventional electron-beam-induced X-ray sources.

7.8
Principles of X-ray Photoelectron Spectroscopy

Chapter 6 introduced the concept of photon excitation of electrons and the energies required to remove electrons from the solid. XPS or electron spectroscopy for chemical analysis (ESCA) as it is sometimes termed provides sufficiently high photon energies to excite electrons from core shells of an atom. According to Equation 6.1, the kinetic energy E_K of such electrons upon leaving the solid is given as

$$E_K = h\nu - E_B - (E_{VAC} - E_F) \tag{7.23}$$

where E_B is the core level binding energy. X-ray excitation with $h\nu > 1000$ eV enables excitation of numerous core shell electrons for most elements.

Photoemission spectroscopy provides four very useful procedures for analyzing electronic materials. These are to use

1) known energies and cross sections to identify chemical species in survey spectra;
2) kinetic energy shifts, that is, "chemical shifts" to identify chemical bonding (ion) environment;
3) rigid shifts of all spectra to identify charging or band bending;
4) angle dependence or different excitations to identify near-surface species.

We now examine each of these procedures in more detail.

7.8.1
Chemical Species Identification

Figure 7.5 illustrates the binding energies and photoionization cross sections for core electron subshells of iron and uranium [9]. By definition, E_F represents zero binding energy. The distance between E_F and a particular level corresponds to the relative energy of the ion remaining after electron emission, or the binding energy of the electron. The first feature to notice is that the highest lying core electrons for these two atoms are quite different, $n = 3$ subshells for iron ($Z = 26$) and $n = 5$ and 6 for the much higher atomic number ($Z = 92$) uranium. These and the lower lying core levels permit easy distinction between such elements. Note that the various

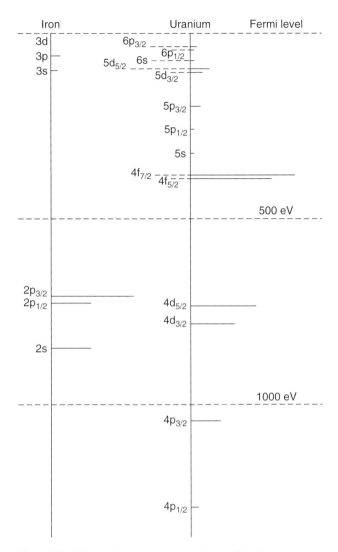

Figure 7.5 Relative ionization cross sections and ionization energies for iron and uranium. Depths below E_F are proportional to binding energy. Horizontal line lengths are proportional to photoionization cross sections [9].

subshells, for example, p, d, and f, become split in energy with ionization, with electron emission from the $p_{1/2}$, $p_{3/2}$, $d_{5/2}$, $d_{3/2}$, $f_{5/2}$, and $f_{7/2}$ subshells in the ratio 1 : 2 for p levels, 2 : 3 for d levels, and 3 : 4 for f levels according to their $2(l + 1/2)$ degeneracy in angular momentum number l.

Secondly, the lower lying core levels are at energies at the limit of photoexcitation by Mg and Al X-ray sources. Thus, according to Table 7.2, Al K_α radiation can excite both $4p_{1/2}$ and $4p_{3/2}$ electrons while Mg K_α radiation cannot. Thirdly, the photoionization cross sections are very different below $E_B = 500$ eV, with large $4f_{7/2}$ and $4f_{5/2}\sigma$'s for uranium versus much smaller σ's for iron. Thus the energy levels and their cross sections within energy ranges accessible by given photon sources provide several methods to identify specific atoms.

As an example of the photoemission signals measured in the laboratory, consider a cross section $\sigma_{ph} = 1$ Mb $= 10^{-18}$ cm^2 and flux $= 10^{12}$ photons/second such as with a commercial X-ray source. For a target with N (atoms/cm^3) $\cdot t$ (cm) per unit area normal to the beam, the number of interactions is $F \cdot \sigma \cdot N \cdot t$. For an electron escape depth of ~ 5 Å, roughly 2 ML, $N \cdot t \sim 6 \times 10^{14}$ cm^{-2}. Thus the number of interactions $= (10^{12}$ s$^{-1})(10^{-18}$ cm$^2)(6 \times 10^{14} \times$ cm$^{-2}) \cong 6 \times 10^8$ s^{-1}, of which half are emitted away from the surface and roughly 20% of those emitted from the surface are collected by the analyzer. Hence there is an order of magnitude reduction overall. Typical count rates for Mg K_α excitation with $hv = 1254.6$ eV are

σ(C1s) $= 0.022$ Mb; cps $= 1.3 \times 10^6$ cps
σ(O1s) $= 0.063$ Mb; cps $= 3.8 \times 10^6$ cps
σ(Ga3d) $= 0.026$ Mb; cps $= 1.6 \times 10^6$ cps

whereas with the same flux for synchrotron radiation at $hv = 80$ eV,

σ(Ga3d) $= 8.39$ Mb; cps $= 5.0 \times 10^9$ cps
σ(O2p) $= 2.05$ Mb; cps $= 1.2 \times 10^9$ cps

This illustrates not only typical count rates but also the value of selecting particular excitation energies and core levels to detect atomic species.

7.8.2
Chemical Shifts in Binding

Core level energies can change, depending on the bond environment of the particular atoms involved. For atoms whose electrons are withdrawn to form a bond, for example, in oxidation, the binding energy of the core electrons increases. In many cases, these core level shifts can be relatively large, for example, several volts, permitting straightforward identification of particular chemical states. Thus in Figure 7.6 [6], the formation of Cr_2O_3 shifts the Cr 3s binding energy by more than 4 eV. Similarly, oxidation of Si in fourfold coordinated (Si^{4+} charge state) SiO_2 shifts E_B by over 4 eV as well [9].

Such chemical shifts provide a tool to determine whether atoms are in their elemental state prior to inducing chemical changes or whether such changes have already occurred.

7.8.3
Distinction between Near- and Subsurface Species

The surface sensitivity of XPS is further enhanced for large take-off angles relative to the specimen's surface normal. This effect is due to the longer path length of electrons inside the solid that pass through the surface at such large angles. As a result, only electrons from proportionally shallower depths can reach the surface at

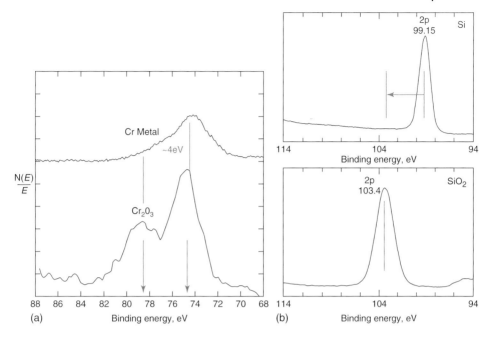

Figure 7.6 (a) Chemical shift of Cr 3s core level with oxidation. The core level shifts by approximately 4 eV to higher binding energy with formation of Cr_2O_3. (b) Chemical shift of Si 2p core level between Si in bulk Si versus in SiO_2. (Adapted from. [9]).

these angles without scattering. Figure 7.7 illustrates the dependence on take-off angle of the chemically shifted Si2p peak of a single monolayer of SiO_2 on Si [9]. The chemically shifted SiO_2 peak at 102 eV increases significantly with large take-off angle versus spectra from electrons exiting the solid normal to the solid surface. Electrons exiting normal to the surface can reach the surface from greater depths in the solid without scattering than electrons exiting at large angles. The different electron escape depths for these different take-off angles enable one to conclude that oxidized Si is present only at the top monolayer and not deeper into the specimen. The chemical shift of only ~2 eV in Figure 7.7 versus the 4 eV in Figure 7.6 is consistent with the lower coordination of Si with O in monolayer form.

7.8.4
Charging and Band Bending

Core levels can also shift if the surface potential of the solid changes. This is due to the formation of a dipole whose potential shifts the energies of the solid uniformly. Such dipole shifts of core levels may indicate sample charging. Since photoemission involves electrons leaving the solid, the solid may become positively charged unless

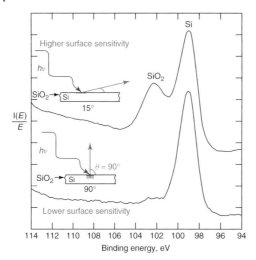

Figure 7.7 The difference in XPS spectra for normal versus large take-off angle of approximately 1 SiO$_2$ monolayer on Si. Large XPS electron take-off angles yield significantly higher surface sensitivity. (Adapted from. [9]).

electrons from the specimen's bulk can reach the surface to counterbalance this charging. However, if the surface is insulating, positive charge builds up so that a positive dipole forms between the top surface and the bulk. This effect can be offset by irradiating the surface with low-energy electrons that neutralize the positive charge.

Figure 7.8 provides an example of XPS from a surface with both neutral (thin) and charging (thicker) oxide patches. Here the spectra show a chemically shifted oxide peak that is analogous to similar oxidized peaks in Figure 7.7. These spectra consist of peaks from the metallic Al, thin Al$_2$O$_3$ patches in electrical contact with the Al, and thicker Al$_2$O$_3$ patches that are electrically insulating.

With no specimen neutralization, the top spectrum shows an additional Al$_2$O$_3$ peak that appears at an energy several volts higher binding energy than the Al$_2$O$_3$ peak of the conducting Al. With a beam of electron incident on the surface during the XPS experiment, the contribution from this insulating peak shifts to lower binding energies since the surface dipole has now changed sign. Even higher specimen neutralizer bias voltage results in an even more negative surface dipole and a further shift to lower binding energy. On the other hand, the Al metal and the Al$_2$O$_3$ patches in electrical contact with the metal remain unshifted since all excess charge is conducted away. The spectral shifts shown in Figure 7.8 demonstrate that charging is a major challenge in XPS studies of insulating samples.

Band bending at the surfaces of semiconductors also produces a dipole that shifts the bulk energies. In this case, the dipole consists of charge on the semiconductor surface and the opposite sign of charge distributed within the surface space charge region. Figure 7.9a illustrates the excitation of electrons from filled valence band states into vacuum and their collection by an electron energy analyzer. Because the

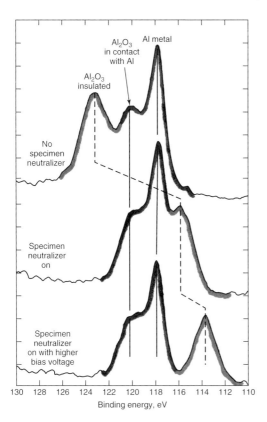

Figure 7.8 XPS spectra of Al 2s metal core levels (heavy lines) from insulating and conducting domains of Al_2O_3. Al_2O_3 peaks from insulating patches shift with surface charging, whereas Al_2O_3 peaks from Al and conducting patches of Al_2O_3 do not shift [9].

analyzer and semiconductor are grounded, they share a common Fermi level. The semiconductor bands extending away from the surface are flat in Figure 7.9a so that $E_C - E_F$ is constant. If metal atoms or other adsorbates are deposited on the semiconductor surface, charge exchange occurs. The semiconductor bands begin to bend as charge builds up on the surface, inducing a countercharge. For n-type semiconductors and negative surface charge, the charge exchange induces upward band bending toward the surface as shown in Figure 7.9b. $E_C - E_F$ is no longer constant. At the surface, $E_C - E_F$ increases and $E_F - E_V$ decreases. Since

$$h\nu = E - E_V = (E - E_F) + (E_F - E_V) \tag{7.24}$$

$E - E_F$ measured by the analyzer increases as $E_F - E_V$ decreases. With increasing charge exchange, more band bending occurs (Figure 7.9c) and $E_F - E_V$ decreases further. If $E_F - E_V$ is known for the clean surface, then XPS can monitor its position as a function of metal atom or other adsorbate coverage (Figure 7.9d).

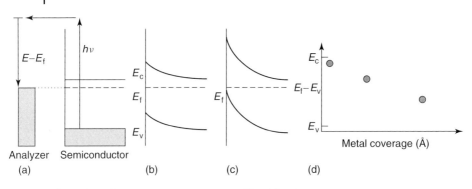

Figure 7.9 Photoemission measurement of band bending. (a) Photoexcitation from filled valence states with analyzer-measured energy $E - E_F$. (b) E_F position in gap with band bending. (c) E_F position in gap with increased band bending. (d) $E_F - E_V$ vs. metal coverage.

At surface coverages for which $E_F - E_V$ stabilizes, $E_F - E_V$ corresponds to the Schottky barrier for holes Φ_B^p, whereas Φ_B^n is its bandgap complement.

$$\Phi_B^p = E_F - E_V \tag{7.25}$$

$$\Phi_B^n = E_C - E_F = E_G - (E_F - E_V) \tag{7.26}$$

Researchers have used this XPS capability extensively to monitor the initial stages of Schottky barrier formation and charge exchange.

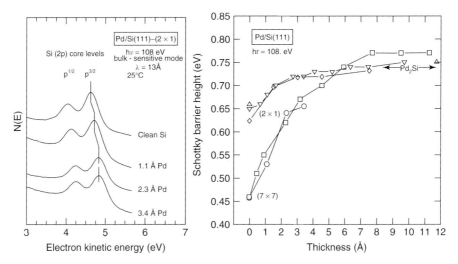

Figure 7.10 Rigid core level shifts and band bending at Pd/Si(111) interfaces. Band bending is complete within the first 10 Å of metal coverage. The final $E_F - E_V$ position corresponds to the Schottky barrier height measured macroscopically for Pd_2Si on Si [10].

Figure 7.10 illustrates (a) the rigid energy shift of Si $2p_{1/2}$ and $2p_{3/2}$ core levels with the initial deposition of Pd atoms and (b) the corresponding n-type (upward) band bending, decreasing $E_F - E_V$ and increasing n-type barrier height Φ_B^n. [10] The (2×1) and (7×7) notations signify different surface atomic geometries, termed *reconstructions*, of the same Si(111) orientation. The photon energy used produces electrons with kinetic energies of only a few volts, corresponding to a 13 Å escape depth.

The core level and $E_F - E_V$ movements are complete at coverages of only a few monolayers. This rapid $E_F - E_V$ stabilization with atomic layer coverage is characteristic of metal–semiconductor interfaces. As such, it demonstrates the microscopic character of charge exchange at interfaces and the importance of ultrahigh vacuum (UHV) surface science techniques to unravel its basic processes.

7.9
Excitation Sources

A variety of photon sources are available for photoemission spectroscopy, extending from the near-UV to the high energy X-ray region. Near-UV photons can be provided by a gas discharge lamp. Figure 7.11 pictures a cutaway drawing of such a lamp [11]. He gas is commonly used to generate photons at 21.2 eV (He I line) and 40.8 eV (He II line), the former a factor of five stronger but the latter at a higher and more useful energy to probe valence and shallow core levels with surface sensitivity. The lamp consists of a gas discharge region and a high voltage discharge in an air-cooled cavity. No windows are available to isolate the chamber from the lamp without absorbing

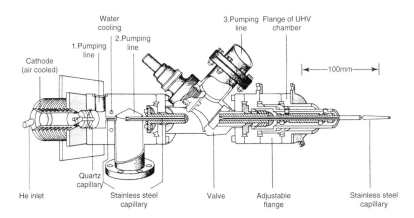

Figure 7.11 Cutaway view of a He UV discharge lamp for UV photoelectron spectroscopy. Milli-torr pressures in the He discharge cavity decrease with differential pumping to the 10^{-9} torr or lower range at the capillary's nozzle [11].

this radiation, so that pumps are required to lower the pressure from the Torr range in the cavity to UHV in the chamber. A rough pump line draws the flowing He gas through the cavity; the second and third pump lines reduce the pressure into the 10^{-10} Torr range at the nozzle of a capillary extending into the chamber. The capillary further reduces the pressure, particularly if its walls are electrically biased to promote collisions with charged ions from the discharge region. The third pump line is actually part of a straight-through gate valve that can be closed to seal off the UHV chamber. In general, this pumping scheme permits pressure differentials of almost 10 orders of magnitude.

The UV emissions result from transitions between energy levels of excited He$^+$ ions with pressures of 1 Torr for He I and 0.1 Torr for He II. Depending on the current discharge conditions, their spectral linewidths of 3 and 17 meV, respectively, enable high resolution UV spectra without the use of a UV monochromator (see Table 7.4).

Higher photon energies are available with commercial X-ray sources. Their basic design usually involves high voltage electrons accelerated into anodes of Mg or Al. This process generates X-rays as the incident electrons collide with and eject core shell electrons of the target metals and higher lying electrons emit photons as they fill the resultant core holes. Relatively strong X-ray emissions occur at 1253.6 and 1486.6 eV for Mg K$_\alpha$ and Al K$_\alpha$ lines, respectively. Figure 7.12a illustrates this basic design. The high voltage and current must pass through the UHV flanges via insulating standoffs. The heat generated by the bombarding electrons is dissipated with water cooling behind the anodes. Figure 7.12a pictures both Mg and Al targets in the same source, selectable by switching bias voltage from one anode to the other. A thin Al film covers the X-ray source nozzle to

Table 7.4 Conventional UV and X-ray sources, their energies, typical photon fluxes, and linewidths.

Source	Energy (eV)	Typical intensity at surface (photons/s)	Linewidth (meV)
He I	21.22	1×10^{12}	3
Satellites	23.09, 23.75, 24.05	$<2 \times 10^{10}$ each	–
He II	40.82	2×10^{11}	17
	48.38	2×10^{10}	–
Satellites	51.0, 52.32, 53.00	$<1 \times 10^{10}$	–
YM$_\zeta$	132.3	3×10^{11}	450
Mg K$_{\alpha 1,2}$	1253.6	1×10^{12}	680
Satellites K$_{\alpha 3}$	1262.1	9×10^{10}	–
K$_{\alpha 4}$	1263.7	5×10^{10}	–
Al K$_{\alpha 1,2}$	1486.6	1×10^{12}	830
Satellites K$_{\alpha 3}$	1496.3	7×10^{10}	–
K$_{\alpha 4}$	1498.3	3×10^{10}	–

He I and II sources are commonly used for UV excitation while Mg K$_\alpha$ and Al K$_\alpha$ sources are widely used for X-ray photoexcitation. (Adapted from [4]).

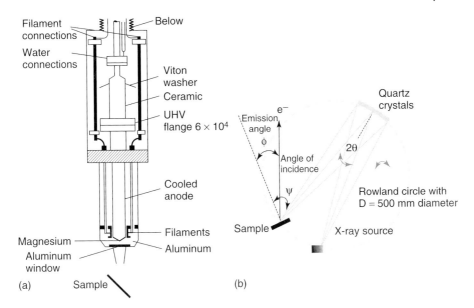

Figure 7.12 (a) Basic design of an X-ray source for XPS [4]. Filament connections to supply current and voltage pass through insulated feedthroughs to the anode region. Both Mg and Al anodes can be present in the same source. (b) Ellipsoidal geometry X-ray monochromator [12].

block ionized species and electrons from leaving the source while passing the X-rays.

The Mg and Al $K_{\alpha 1,2}$ emission lines both have lower intensity satellites that broaden the excitation linewidths to several hundred meV. Such linewidths produce corresponding broadening of spectral features, so that electronic structure or chemical shifts of less than a volt are difficult to resolve into separate peaks. X-ray monochromators reduce such linewidths by a factor of 2 using bent crystals to disperse the X-ray energies spatially. In Figure 7.12b, the X-ray source and the sample are both at the foci of the Rowland circle such that X-rays incident on the bent quartz crystal focus at the sample [12]. The bent quartz crystal causes a single wavelength of X-rays to reflect at a specific angle such that it focuses at the sample.

A much wider range of photon energies is available using synchrotron radiation. Not only does synchrotron radiation extend the range of energies from those of the sources in Table 7.4 to the soft X-ray and hard X-ray regimes, but it also provides orders of magnitude higher photon intensities and brilliance (or brightness), defined as photons per second per 0.1% band width per milliradian^2 per mm^2, as well as very high energy resolution. Figure 7.13 illustrates the photon energies and brightness of X-ray and soft X-ray sources developed over the last century [13]. This powerful source of excitation enables a wide range of

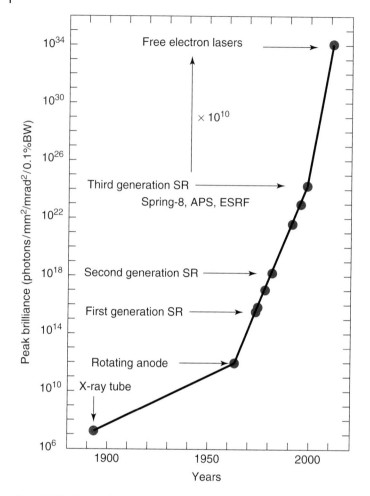

Figure 7.13 Peak brilliance (brightness) of X-ray radiation sources. X-ray sources including the European Synchrotron Radiation Facility (ESRF) and free electron lasers have increased brilliance by over 27 orders of magnitude since Roentgen's discovery.

surface and interface studies of electronic materials that are the subject of the next chapter.

7.10
Electron Energy Analyzers

Measurement of electron energies is possible using several different methods. The simplest among these is a retarding grid, whose potential can determine the maximum kinetic energies of electrons moving from the sample to the anode. A drawback of this method is that all electrons with energies greater than the retarding potential are collected. One can identify electrons within a specific energy range ΔE by applying an AC voltage to the retarding grid to discriminate between electrons collected at energy E versus $E + \Delta E$. However, the signal-to-noise ratio of this technique is relatively low. The two most common electron analyzers are the *cylindrical mirror analyzer* (CMA) and the *hemispherical analyzer* (HSA). Both the analyzers are designed for direct measurements of electrons within an energy window ΔE and hence have higher signal-to-noise characteristics. Figure 7.14 pictures a cross section of the CMA with electron trajectories and circuit for biasing the inner cylinders (ICs) [3].

The double-pass CMA, developed by Palmberg and his collaborators [14], is based on electron deflection between charged cylinders. Figure 7.14a shows photons with energy $h\nu$ causing emission of photoelectrons at sample point P. Dashed lines indicate the trajectories of the photoemitted electrons, passing between cylinders and through intermediate slits and terminated at the front end point P of an electron multiplier M. Only electrons emitted at an angle of 42.3°, determined by the physical dimensions of the CMA and the voltages applied between the IC and outer cylinder (OC), can focus at point P'. In the Physical Electronics design, only electrons with kinetic energy $E_K \cong 1.7$ eV$_p$ can be focused. Thus one collects spectra of electron multiplier signal versus E_K by monitoring collected current versus pass voltage V_p.

For photoemission spectroscopy, an additional retarding voltage at the entrance to the CMA permits one to select a constant energy resolution ΔE. In this mode, the pass energy eV_p is held constant, whereas the voltage drop between sample and pre-retardation grids V_R varies to retard electrons with energy E_K to energy $\cong 1.7$ eV$_p$ that passes through the analyzer. Thus the cylinder deflection performs the function of discriminating electron energies outside the selected ΔE range. This energy window represents the energy resolution of the photoemission measurement. Note that a reduction in V_R improves energy resolution but decreases signal quadratically owing to a reduced sample area from which electrons can pass through the CMA. To further increase signal-to-noise, one can apply a small AC voltage (oscillator O and transformer T) to the scanning pre-retardation voltage V_R. Alternatively, one can dispense with pre-retardation and constant energy resolution to achieve higher signals as in, for example, Auger electron spectroscopy. Here, one applies a modulation voltage to the cylinders themselves in order to produce an analog differentiation of the collected electron signal.

Figure 7.14b pictures an HSA with dashed lines to indicate the electron trajectory. The HSA also makes use of electron deflection to determine energy E_K. Here focusing lenses and slits determine the energy resolution. The HSA has the additional advantage of accommodating microchannel plates at its collector. Each element of these channel plate electron multipliers counts electrons passing through the HSA at a different energy. Because the HSA optics focus electrons with energies centered at a pass energy E at slightly

124 | 7 Photoemission Spectroscopy

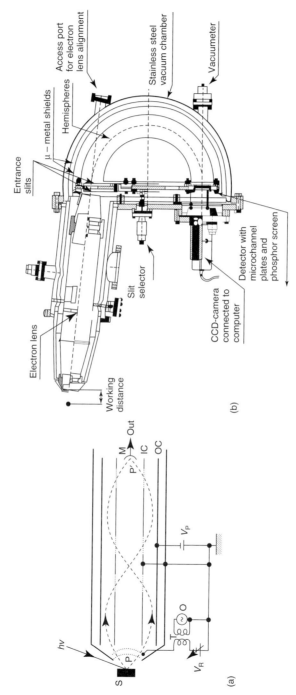

Figure 7.14 (a) Cylindrical mirror analyzer in cross section with electron trajectory; (b) hemispherical analyzer with electron trajectory.

different spatial locations, the physical spacing of the microchannel plate elements determines the energies of the electrons. Because HSA collects spectra for multiple distinct energies simultaneously, its collection efficiency is very high. As a result, the HSA permits high signal-to-noise measurements even with high energy resolution. Signal-to-noise improves further as the size of the hemispherical detector increases. While larger and more complex than a CMA, the HSA is the preferred electron energy analyzer for photoelectron spectroscopy with high signal and energy resolution. Overall, powerful photon sources and electron energy hardware are now available for photoelectron spectroscopy.

7.11
Summary

Photoemission spectroscopy is a powerful technique for surface science measurements in general and electronic material interfaces in particular. The photoemission process provides extremely high surface sensitivity for identifying chemical species from core level excitations, gauging charge transfer between atomic species from chemical binding energy shifts, and measuring Fermi level movements at the semiconductor surface from rigid core level shifts as the first few monolayers of metal form interfaces.

Photoemission spectroscopy is a quantitative technique since photoionization cross sections can be derived analytically and calculated for specific core levels and photon energies. This information yields surface chemical composition with percent level accuracy and better than 1 % surface sensitivity. The energy distribution curves of photoemitted electrons provide valence band density of states, photoionization thresholds, work functions, and Fermi level positions.

As photon sources increase in power and electron monochromators advance in detection efficiency, photoemission spectroscopy continues to yield new types of surface and interface information.

Problems

1. A clean metal film in ultrahigh vacuum is illuminated by monochromatic light. A parallel plate biased relative to the metal collects photoemitted electrons and monitors the collected current versus applied voltage. (a) For a light wavelength $\lambda = 1550$ Å incident on a Au film, what is the minimum voltage V_0 that suppresses all photoelectrons? (b) Sketch the photocurrent as a function of voltage difference between the Au and the collector plate. Sketch several curves with different light intensities.

(c) For He I photons incident on this film, what is the highest energy and velocity of a photoemitted electron?

2. For a clean single crystal Si surface illuminated by He I line emission, what is the expected room temperature width of the measured energy distribution curve?

3. In order to monitor an Al–Si system by XPS using an Al K_α source, which core levels would provide the strongest photoemission signals and why?

4. You are given a GeSi alloy whose composition you are to measure using XPS with a Mg K_α source. Identify core levels with the same symmetry and the highest photoelectron cross sections in both compounds with this excitation. Calculate and compare the photoelectron cross sections for these core level electrons.

5. From Table 7.1, what core level would you use with Al K_α X-rays to measure Ni atoms on a surface?

6. An unknown semiconductor yields Mg K_α XPS doublet peaks with kinetic energies E_{KE} relative to the vacuum level at 1103/1109, 1143/1147, 1177/1177, 1208/1209, and 1233/1234 eV. Identify this (conventional) semiconductor.

7. A ZnO surface exhibits XPS spectra with peaks at 531.15 and 533.2 eV binding energy relative to the Fermi level with deconvolved peak intensities of 4.8 and 9.2, respectively. In addition, peaks are present at 1025.35 and 283.5 eV binding energies. After oxygen plasma exposure, the low binding energy peak disappears, and peaks are now positioned at 530.4 and 1024.6 eV. Describe the surface composition, bonding, and electronic changes (a) before and (b) after surface treatment.

References

1. Einstein, A. (1905) *Ann. Phys.*, **17**, 132.
2. Planck, M. (1901) *Ann. Phys.*, **4**, 551.
3. Margaritondo, G. (1988) *Introduction to Synchrotron Radiation*, Oxford University Press, New York.
4. Lüth, H. (2001) *Solid Surfaces, Interfaces, and Thin Films*, 4th edn, Springer, Berlin, pp. 266–270.
5. Streetman, B.G. (1995) *Solid State Electronic Devices*, Prentice Hall, Englewood Cliffs.
6. Briggs, D. (ed)(1977) *Handbook of X-Ray and Ultraviolet Photoelectron Spectroscopy*, Heyden & Sons, Ltd.
7. Feldman, L.C. and Mayer, J.C. (1986) *Fundamentals of Surface and Thin Film Analysis*, Prentice Hall, Englewood Cliffs, pp. 189–195.
8. Yeh, J.J. and Lindau, I. (1985) *Atomic Data and Nuclear Data Tables*, **32**, 1.
9. Wagner, C.D., Riggs, W.M., Davis, L.E., Moulder, J.F., and Muilenberg, G.E. (1979) *Handbook of X-Ray Photoelectron Spectroscopy*, Physical Electronics Industries, Eden Prarie.
10. Rubloff, G.W. (1983) *Surf. Sci.*, **132**, 268.

11. Rowe, J.E., Christman, S.B., and Chaban, E.E. (1973) *Rev. Sci. Instrum.*, **44**, 1673.
12. *http://www.specs.de/cms/upload/bilder/IGE_11/ROWLAND_gr.gif*.
13. Shintake, T. (2007) Proceedings of PAC07, MOZBAB01, p. 89.
14. Palmberg, P.W. (1974) *J. Electron Spectrosc. Relat. Phenom.*, **5**, 69.

8
Photoemission with Soft X-rays

8.1
Soft X-ray Spectroscopy Techniques

The development of synchrotron radiation sources over the past few decades has provided powerful extensions of photoemission and related spectroscopies. This chapter describes the tools developed to harness this radiation and the techniques employed to study the surfaces and interfaces of electronic materials.

8.2
Synchrotron Radiation Sources

Today electron storage rings are the primary source of soft X-rays. These major facilities are distributed at many locations across the world. Storage rings permit electrons to move in ultrahigh vacuum (UHV) in closed orbits, bent by magnets along their path to control their trajectories and maintain their focus. Figure 8.1a provides a schematic of a storage ring [1]. Electrons are injected from an injection source (IS) into the ring, where they circulate for extended periods of time. A radio frequency (RF) cavity accelerates the electrons to compensate for energy loss due to light emission during each circuit of the ring.

The soft X-rays result from accelerating electrons in curved trajectories. Electron acceleration produces *Cerenkov radiation* [2], which becomes pronounced for electrons at relativistic velocities. The intensity of Cerenkov radiation is proportional to the electron acceleration and, hence, to its velocity and the storage ring curvature. The soft X-ray radiation is emitted tangential to the electron trajectory and is collected in beam lines L that incorporate mirrors and an X-ray monochromator to disperse the radiation in energy and direct it to the specimen under study.

Figure 8.1b illustrates the details of one section of the storage ring and a beam line. Each section of the storage ring includes bending magnets plus sets of magnetic quadrupoles to focus the electrons in transverse and vertical directions. Beam line L1 includes mirrors that focus and refocus the radiation as it passes through a monochromator that diffracts the light at different angles for

Surfaces and Interfaces of Electronic Materials. Leonard J. Brillson
Copyright © 2010 WILEY-VCH Verlag GmbH & Co. KGaA, Weinheim
ISBN: 978-3-527-40915-0

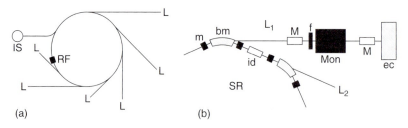

Figure 8.1 (a) Schematic top-view diagram of a synchrotron radiation facility with injector system IS, accelerating RF cavity, and beam lines that collect light emission L tangential to the curved trajectory. (b) The closed circuit includes sections with increased curvature, at which soft X-rays emanate. Bending magnets (bm), magnets (m), and insertion devices (ids) control the trajectory. X-ray photons from the ring pass through a mirror (M), a monochromator (mon), and a second mirror before reaching the sample in the experimental chamber (ec) [1].

Figure 8.2 An arc of the Tantalus synchrotron ring showing three bending magnets, ion pumps atop steel arms radiating from the ring that support beam lines with focusing mirrors, monochromators, and end stations with samples inside UHV chambers.

different wavelengths. Beam line L2 collects light emitted by an *insertion device* that undulates or wiggles the electron beam to produce radiation in the forward direction.

Figure 8.2 shows a top-view photo of the early Tantalus synchrotron storage ring to indicate the electron beam and photon trajectories. Electrons circulate in a closed orbit inside UHV tubing and bending magnets (rectangular blocks at bottom of photo). The beam lines are tangential to the ring and radiate outward and support optics to focus and diffract the synchrotron radiation. The beam line optics are typically several meters or more in length. Thus, photons with energy selected by gratings and slits pass through a monochromator and focus on a specimen inside a UHV analysis chamber at, for example, an end station at the photo's 11 o'clock position. In a photoemission spectroscopy experiment, photoemitted electrons pass into an energy analyzer. Analogous experiments, such

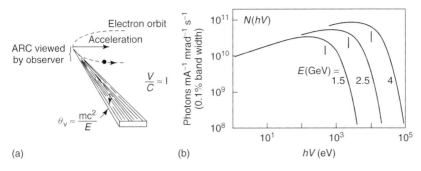

Figure 8.3 (a) Synchrotron radiation● beam emission tangential to electron orbit and (b) spectral brightness versus photon energy increases with increasing storage ring energy.

as an absorption experiment, use detectors to measure the absorbed or reflected light itself. Since any window would block most of the synchrotron radiation, the vacuum of the analysis chamber must be open to the vacuum inside the storage ring during experiments. The desk chair near the photo center provides an indication of the facility's size. Modern synchrotrons are much larger in order to produce higher energies and accommodate more beam lines.

Figure 8.3a illustrates a segment of the electron orbit as it accelerates around the curve. Light is emitted in the forward direction and subtends an angle $\theta_V \approx mc^2/E$ for $v/c \approx 1$. Thus, θ_V decreases with the increase in velocity and brightness of the source. Figure 8.3b shows the spectral distribution of photons emitted per unit current, per second, per solid angle, and per unit bandwidth [3]. Photon brightness and maximum energy increase with the increase in storage ring energy.

Increasing the electron velocity increases the X-ray intensities and also expands the range of X-ray energies. Figure 8.3b illustrates the spectral distribution of radiation emitted by a storage ring with parameters available at the Stanford Synchrotron Radiation Laboratory. The number of photons $N(h\nu)$ with energy $h\nu$ varies with photon energy as shown. $N(h\nu)$ at a given energy and the maximum energy both increase with increasing stored electron beam energy. The *photon flux*

$$N(h\nu) = 1.256 \times 10^7 \gamma\, G_1(y) \text{ photons s}^{-1} \text{ mrad}^{-1} \text{ mA}^{-1}(0.1\% \text{ bandwidth}) \quad (8.1)$$

where $\gamma = E/m_0 c^2 = [1 - v^2/c^2]^{-1/2} \cong 1957 E$ (GeV), $G_1(y) = y \int_y^\infty k^{5/3}(t)dt$, and $y = h\nu/h\nu_C$.

The photon energy $h\nu_C$ is termed the *critical photon energy* and is equal to $3hc\gamma^3/4\pi\rho$, where ρ is the radius of the curvature. As indicated by the vertical lines, $h\nu_C$ is defined as the photon energy for which half the power radiated is emitted above and below it. It increases as γ increases with the increase in velocity and as the radius of curvature decreases. Note the high energies required to achieve relativistic electron velocities and usable synchrotron radiation. Large national facilities can provide such energies, which would otherwise be difficult to achieve with table-top equipment.

Figure 7.13 showed that photon fluxes available for photoemission spectroscopy and other applications have increased significantly over the past few decades. Advances in the magnets used to focus and bend the electron beams have been contributing to this increase. Devices that produce small amplitude changes in electron trajectories, aptly termed *undulators* and *wigglers*, can boost photon intensities at specific energies by orders of magnitude over those intensities achieved by the storage ring alone. Increased flux enables extremely high sensitivity that, in principle, will enable the measurements of, for example, dopants in semiconductors as well as probes of very fast chemical processes, for example, metal–semiconductor reactions, in real time by photoemission spectroscopy.

8.3
Soft X-Ray Photoemission Spectroscopy

Soft X-ray photoemission spectroscopy or (SXPS) provides several advantages over conventional X-ray photoemission spectroscopy (XPS). These are as follows: (i) high brightness, (ii) high spectral resolution, (iii) high surface sensitivity, and (iv) variable depth sensitivity. High brightness enables sufficient signal-to-noise to detect relatively low photoemission signals from small concentrations of atoms. Brightness depends on both photon intensity and the area from which these photons emit. Increased brightness reflects both higher photon intensity and smaller spot size that a focused photon beam can achieve. Both factors contribute to higher photon flux per unit specimen area.

High spectral resolution follows from high brightness since smaller monochromator slit widths are usable while retaining adequate signal-to-noise performance. High surface sensitivity is a direct outcome of the synchrotron radiation's tunability. One can select photon energies for which the photoelectrons from a given core level emerge at kinetic energies corresponding to very short electron scattering lengths, namely, the 50–100 eV range, as shown in Figure 6.2. Variable depth sensitivity also follows from the continuous range of photon energies available. One can select photon energies for which the electrons emit at energies outside this 50–100 eV range of short scattering lengths in order to probe at different depths.

8.3.1
Basic Surface and Interface Techniques

Several basic techniques are useful in probing surfaces and interfaces of electronic materials. These are as follows: (i) chemical bonding shifts, (ii) Fermi level movements, and (iii) element-selective interdiffusion. Figure 8.4 illustrates the measurement of interface chemical reactions at a metal–semiconductor interface [4].

From Table 7.2, the metallic Al 2p binding energy is 77 eV relative to E_{VAC} (73 eV vs E_F) so that a photon energy of 140 eV excites Al 2p electrons with kinetic energies of ∼63 eV. This energy corresponds to the minimum of the escape depth

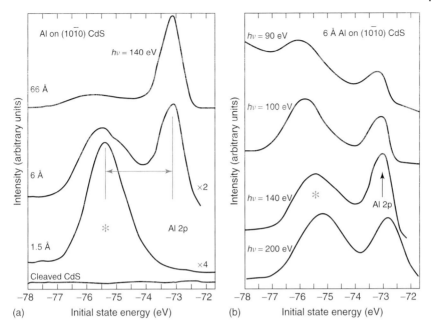

Figure 8.4 (a) Al 2p core level spectra with $h\nu = 140$ eV for cleaved CdS in UHV with increasing Al deposition. (b) Al2p core level spectra for 6 Å Al on cleaved CdS versus $h\nu$, showing the variation in ratio of reacted versus metallic Al.

curve shown in Figure 6.2. Therefore, only electrons escaping from the top 1–2 monolayers contribute to the photoemission spectrum. Figure 8.4a shows that the first 1.5 Å of Al introduces a peak feature to the otherwise featureless background spectrum of the cleaved CdS surface in this energy range. This peak appears at a binding energy of 75.5 eV. With additional Al deposition, a second Al 2p peak appears at the expected binding energy of 73 eV. The increased binding energy of the initial Al layer signifies that Al has chemically reacted with CdS. Since it is thermodynamically favorable for Al and S to react and form Al_2S_3 [5, 6], this increased binding energy indicates a transfer of electrons from Al to S in the outer semiconductor layer. The reduced electron density of the reacted Al atom results in less screening of the core hole left by the photoemitted electron and, therefore, higher binding energy. The lower binding energy of the subsequent Al layers signifies the formation of metallic Al in addition to the reacted monolayer. With an additional 60 Å of Al deposited, the metallic Al peak increases and the contribution from the reacted Al diminishes. This is consistent with the short escape depth of Al 2p photoelectrons and laminar coverage since electrons from the initial monolayer must pass through additional Al layers to escape from the film.

Figure 8.4b illustrates the effect of changing photon energy for a given thickness, in this case, 6 Å of Al as shown in Figure 8.4a. Since $h\nu = 140$ eV corresponds to

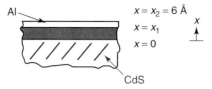

Figure 8.5 A 6 Å total Al thickness x_2 on CdS with reacted layer of unknown thickness x_1. The variation of $I_{reacted}/I_{metal}$ with x_1 for $\lambda(75\,eV) = 4$ Å yields a value of 4 Å for x_1 [4].

the escape depth minimum, higher or lower photon energies should have longer escape depths, permitting the detection of photoelectrons from below the top monolayers. Indeed, Figure 8.4b shows that the ratio of reacted to metallic Al 2p photoelectrons increases for these photon energies, consistent with the escape depth curve shown in Figure 6.2. Quantitative analysis of these peak intensities yields an effective thickness for the reacted Al layer. Assuming an unknown reacted layer thickness x_1, Figure 8.5 illustrates the total thickness of 6 Å Al on CdS [4].

The ratio of reacted-to-unreacted Al can be expressed as the ratio of two integrals over the corresponding depths of the two Al layers:

$$\frac{I_{reacted}}{I_{metal}} = \frac{\int_0^{x_1} e^{-x/\lambda(E)} dx}{\int_{x_1}^{x_2} e^{-x/\lambda(E)} dx} \tag{8.2}$$

where $x_2 = 6$ Å, $\lambda(75\,eV) = 4$ Å, and x_1 is unknown. Fitting $I_{reacted}/I_{metal}$ with x_1 yields a reacted layer thickness $\cong 4$ Å.

Overall, these spectra reveal a highly localized, reacted layer of Al on CdS and the formation of metallic Al above the reacted Al monolayer. The significance of this and analogous results for other metal–semiconductor interfaces is that charge transfer occurs between the last monolayer of the semiconductor and the first monolayer of the metal such that an electric dipole can form, which could alter the Schottky barrier formation.

Interface-sensitive SXPS can reveal the presence of interface-exchange reactions in which the deposited metal forms a bond with an anion within the top semiconductor layer. This new metal–anion bond displaces a semiconductor cation to which the anion was previously bonded. Figure 8.6 shows an example of an interface-exchange reaction [7]. Here, the SXPS spectra of all three semiconductor elements, As, Ga, and In, are shown for increasing thicknesses of Al on $In_{0.75}Ga_{0.25}As(100)$. The As 3d spectra are obtained using $h\nu = 100$ eV to achieve high surface sensitivity. From Table 7.2, $E_{VAC} - E(As\,3d_{3/2}) = 46$ eV so that kinetic energies $E_{KE} \cong h\nu - (E_{VAC} - E(As\,3d_{3/2})) = 54$ eV for elemental As are expected. Likewise, $E_{VAC} - E(In\,4d_{3/2}) = 21$ eV so that, for $h\nu = 80$ eV, $E_{KE} \cong 59$ eV for elemental In. The exact kinetic energies will depend on the Fermi level position within the semiconductor bandgap, which is discussed later.

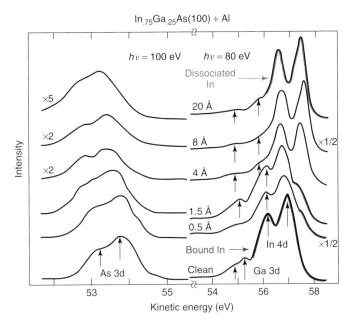

Figure 8.6 SXPS core level spectra for increasing thicknesses of Al on $In_{0.75}Ga_{0.25}As(100)$. With increasing Al deposition, dissociated In 4d core levels appear and all core levels shift to higher kinetic energy [7].

With increasing Al deposition, the spectra for all three elements change. The most significant change is the appearance of a second pair of In 4d core level peaks at almost 1 eV higher kinetic energy, that is, 1 eV lower binding energy, that correspond to the formation of dissociated, metallic In. Arrows point to the Ga $3d_{1/2}$ and $3d_{3/2}$ core levels, which merely shift to lower kinetic energies with increasing Al thickness. Similar shifts to lower kinetic energy are evident for As 3d and the chemically bound In 4d core level peaks. No significant change in As 3d core levels is expected from the chemical changes since As remains bonded to metal cations. The rigid shift of all core levels is characteristic of a Fermi level movement relative to the valence band that changes all binding energies uniformly. In this case, the uniform decreases in kinetic energies indicate an increase in $E_F - E_V$.

These SXPS core level features are significant for three reasons: (i) Al deposition causes dissociation of In from the InGaAs lattice, (ii) no dissociation of Ga takes place, and (iii) the Fermi level moves higher in the bandgap relative to the valence band. Points (i) and (ii) signify a preferential exchange reaction in which Al replaces In rather than Ga in the InGaAs. This is consistent with the higher binding energy of Al to As versus In to As as reflected by their heats of formation. $-\Delta H_{298} = 57.7$ kJ mol^{-1} for InAs versus 116.3 kJ mol^{-1} for AlAs [6].

Point (iii) indicates that the Fermi level moves more than 0.5 eV further from the valence band, causing the semiconductor band bending to become less n-type, that is, less upward bent. See Section 7.8.4 for a discussion of rigid core level shifts due to

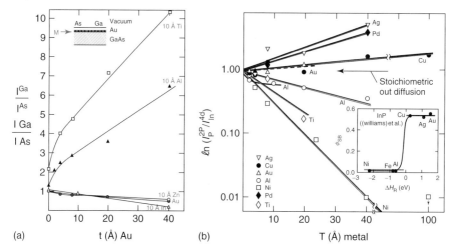

Figure 8.7 (a) Ratio of outdiffused As versus Ga atoms detected by SXPS at the free surface of Au on GaAs. $I(Ga)/I(As)$ decreases slightly for unreactive metals such as Au, In, and Zn and $I(Ga)/I(As)$ increases for reactive metals such as Al and Ti [8]. Inset shows Au/interlayer metal(M)/GaAs stack. (b) Ratio of outdiffused P versus In atoms at the free surface of the metal overlayer [9]. $I(P)/I(In)$ increases for unreactive metals and decreases for reactive metals. Inset shows the difference in Schottky barrier heights between reactive and unreactive metals [10].

Fermi level movements. Thus, both strong chemical and electronic changes occur at a metal–semiconductor interface, with only 20 Å of metal deposited. These changes occur even with room-temperature deposition of the metal. Higher temperatures would promote additional chemical reactions and diffusion. Figures 8.5 and 8.6 emphasize the local nature of metal–semiconductor interactions. Such changes would be difficult to detect without techniques such as photoemission spectroscopy that are sensitive to only a few monolayers of atoms.

An example of preferential elemental interdiffusion is the Au–GaAs interface. Photoemission studies of this interface show that both Ga and As can dissociate from their lattice and diffuse through the metal film to the free metal surface. Figure 8.7 inset is a schematic diagram of the Au/GaAs interface, showing the presence of isolated As and Ga atoms at the free surface of 20 Å Au. A layer of metal labeled M between the Au and GaAs is also shown. This intermediate layer or *interlayer* could be a relatively unreactive metal, such as Au, or a more reactive metal, such as Al. The relative concentrations of Ga to As change depending on the reactive nature of the bonding at the semiconductor–metal interface [8]. The decrease in the SXPS core level intensity ratio $I(Ga)/I(As)$, as shown in Figure 8.7a, shows that Au films deposited on UHV-clean GaAs(100) surfaces produces more As than Ga outdiffusion as the thickness of the metal increases. Similar behavior is apparent for other metals such as Zn and In, which also do not bond strongly with Ga or As at room temperature.

On the other hand, metals such as Al or Ti produce the opposite effect, namely, that $I(Ga)/I(As)$ increases strongly with the increase in the thickness of Au overlayer.

For a reactive layer that is only 10 Å thick, the $I(Ga)/I(As)$ ratio increases by over an order of magnitude relative to this ratio for unreactive interlayers. In this case, the Al or Ti interlayer binds strongly with surface As atoms, producing a reacted metal arsenide that acts as a diffusion barrier to As outdiffusion [8]. This effect becomes more pronounced with increasing interlayer metal thickness.

Figure 8.7b illustrates the analogous behavior for In and P atoms diffusing through metal overlayers to the free surfaces [9]. Unreactive metals such as Ag, Pd, Cu, and Au induce more P versus In outdiffusion, whereas more reactive metals such as Al, Ti, and Ni exhibit the reverse behavior. With 40 Å of the metal, there is nearly 3 orders of magnitude difference in $I(P)/I(In)$, depending on the metal.

The Figure 8.7b inset [10] shows a Schottky barrier plot versus the heat of reaction ΔH_R in which these reactive metals produce low barriers, whereas relatively unreactive metals produce higher barriers. This behavior suggests Fermi level movements that depend on the chemical reactivity of the metal–semiconductor interface more than on the bulk properties of the metal alone. This separation of electrical barriers into these two regimes demonstrates that the strength of interface chemical bonding determines the electrically active sites within the bandgap that are responsible for the Fermi level stabilization.

8.3.2
Advanced Surface and Interface Techniques

The unique features of soft X-rays enable a number of advanced photoemission spectroscopies that are useful in studying semiconductor surfaces. These include the following: (i) *angle-resolved photoelectron spectroscopy* (ARPES), (ii) polarization-dependent photoemission, (iii) *constant final state photoemission* (CFS), (iv) *constant initial state* (CIS) photoemission spectroscopy, and (v) absorption threshold spectroscopies.

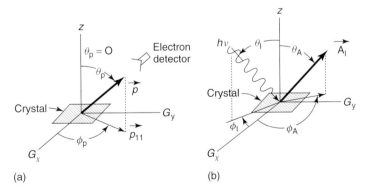

Figure 8.8 Angle-resolved photoemission measurement showing the position of electron detection and corresponding electron momentum in the surface plane of the crystal [11].

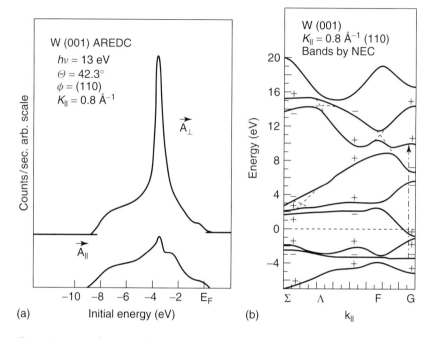

Figure 8.9 (a) Polarization-dependent and angle-resolved energy distribution curves (AREDCs) for the tungsten (001) surface with $h\nu = 13$ eV and $k_\parallel$ along the (110) direction. (b) Corresponding transition between energy bands along different symmetry directions and the band polarities. Adapted from [11].

8.3.2.1 Angular Resolved Photoemission Spectroscopy

The high brightness of synchrotron sources permits the collection of photoemitted electrons from small solid angles above the specimen. This capability enables angle-resolved photoemission spectroscopy, as shown in Figure 8.8. In the direction parallel to the specimen surface, electron momentum is conserved so that the photoemitted electron's kinetic energy E_K and *takeoff angle* θ_P relative to the surface normal define its momentum $p_\parallel = \hbar k_\parallel$ in the crystal plane according to the following relation:

$$k_\parallel = (2mE_K/\hbar^2)^{1/2} \sin\theta_P \tag{8.3}$$

To establish the momentum along a crystallographic direction (Figure 8.8a), one orients the specimen *azimuthal angle* ϕ_P with that direction in the plane defined by the photon beam $h\nu$ and the exiting electron **p** vectors (Figure 8.8b) [11].

Since $k_\parallel$ is conserved, peaks in the photoemission valence band spectrum then correspond to symmetry-allowed direct (i.e., vertical in k-space) transitions between filled and empty states. Figure 8.9 illustrates spectra with peak energy and collection angle corresponding to momentum $k_\parallel$ equal to 0.8 Å^{-1},

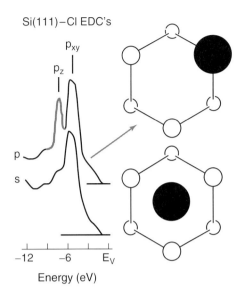

Si(111)–Cl EDC's

Figure 8.10 EDC normal emission valence band spectra for $h\nu = 32$ eV and Si(111) 2 × 1 +Cl (this surface atom notation is described in Chapter 13) with p versus s polarization. The top-view schematic indicates two alternative positions for the Cl atom on the surface Si atoms. The appearance of the additional peak expected theoretically for p_z valence band states for **A** in the plane of incidence signifies onefold coordinated Cl atoms on top of Si atoms [12].

that is, 8×10^7 cm^{-1}. The corresponding symmetry-allowed transition between filled and empty states within the Brillouin zone is also shown. The energy distribution curve (EDC) as a function of *collection angle* θ_P provides a way to map out the Brillouin zone E versus k for a given pair of bands that can be identified by their symmetry and approximate energy separation determined theoretically.

By varying θ_P and measuring energies E_K, one obtains spectral features that reflect maxima in the joint densities of initial and final states across the Brillouin zone.

8.3.2.2 Polarization-Dependent Photoemission Spectroscopy

Synchrotron-generated soft X-rays are polarized with the electric vector in the plane of the synchrotron orbit. This feature can provide additional information if the *polarization vector* **A** (Figure 8.8a) is oriented along a particular surface direction. Figure 8.10 illustrates the difference in valence band spectra of chlorine atoms on a clean Si(111) surface for polarization along within the plane of incidence (*p polarization*, finite electric vector **A** in the plane) versus perpendicular to the plane of incidence (*s polarization*, zero electric vector **A** in the plane) [12]. Chlorine has been used as an etchant for cleaning Si surfaces. The intensity of the valence band peak at 7 eV below the valence band edge E_V is present for p polarization but almost completely absent for s polarization. This signifies the directional nature of the Cl–Si bonding and favors Cl with a onefold (on top) rather than the threefold (centered) coordination as shown.

This study indicates that one can obtain useful information even without angular resolution since, for normal emission, all collected electrons have $\theta_P \equiv 0°$.

Here, polarization alone provides useful information about the chemisorption of adsorbates at semiconductor surfaces even without angular resolution.

8.3.2.3 Constant Final State Spectroscopy

CFS spectroscopy uses the tunability of synchrotron radiation to measure the density of empty states above the vacuum level into which electrons are photoexcited. First, one selects a photon energy range from slightly below to slightly above the energy difference between a relatively narrow core level E_{core} and an empty surface state at or below the semiconductor conduction band edge E_C. Second, one sets the electron analyzer to an energy E_K several eV above the cutoff energy for photoemission.

Finally, one scans $h\nu$ across this range to obtain *partial yield* spectra. These spectra exhibit peaks in the emission of secondary electrons when electrons are photoexcited from the core level to an empty state either in the conduction band or within the bandgap. The energy gained by refilling the core hole indirectly increases the emission of secondary electrons. See Figure 8.11a [13]. In Figure 8.11b, a peak in emission appears at ~31.3 eV for $E_K = 4$ eV and $E_F - E_{core}(\text{Ge } 3d_{5/2}) = 29$ eV. (Compare with $E_{VAC} - E_{core}(\text{Ge } 3d_{5/2}) = 32$ eV in Table 7.2 indicating $E_{VAC} - E_F \sim 3$ eV.)

The electron emission versus photon energy curves at a constant final kinetic energy reveal changes with surface condition that can be attributed to the formation of new surface states. Thus, in Figure 8.11b, Sb deposition on clean Ge(111) produces additional emission at a lower or split-off energy (Δ_{so}) corresponding to states $\Delta_{so} \sim 1.6$ eV below E_C for the localized gap state. However, this value must be reduced by a *lattice relaxation energy* that takes into account the exiting electron and resultant core hole. Even though complicated by this hard-to-determine correction

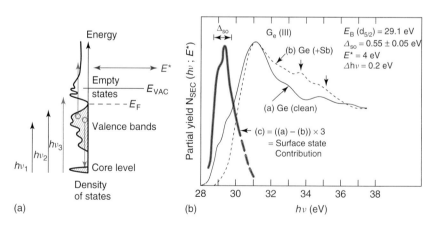

Figure 8.11 (a) Schematic density of filled and empty states and the electronic transitions induced by variable incident photon energies; (b) electron emission versus $h\nu$ at a constant (final) kinetic energy.

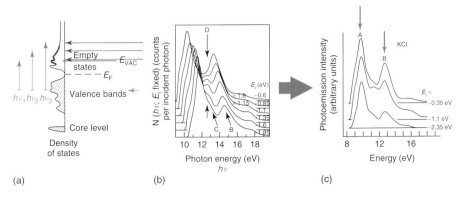

Figure 8.12 (a) Schematic density of filled and empty states and the electronic transitions induced by variable incident photon energies and (b) electron emission versus $h\nu$ at a final kinetic energy that increases with increasing $h\nu$ [14].

factor, the CFS or *partial yield spectroscopy* demonstrates the creation of new gap states as new atoms adsorb.

8.3.2.4 Constant Initial State Spectroscopy

The converse of CFS is CIS spectroscopy. In this technique, the incident photon energy is again scanned but E_K is scanned in energies to match. In other words, $E_{\text{final}} - h\nu \equiv E_i$ (initial) energy. Figure 8.12a illustrates how a given $h\nu$ produces photoelectrons above E_{VAC} [14]. Figure 8.12b shows how these spectra shift with $h\nu$, and Figure 8.12c presents the spectra with the various $h\nu$ increments subtracted such that emissions from the same initial state appear at the same energy. In this case, the two main peaks line up, indicating two maxima in the vacuum density of states.

Densities of final states are useful in determining filled state densities since the former must be known to determine the latter according to Fermi's golden rule. Note that the inverse of XPS and its variants is *inverse photoemission spectroscopy* (*IPES*), in which an energetically well-defined beam of low energy electrons loses energy with the emission of photons as the incident electrons relax into unoccupied states above the Fermi level. This technique does not require synchrotron radiation and provides useful empty state information for semiconductor and organic electronic materials. See, for example [15].

8.4
Related Soft X-ray Techniques

Soft X-rays have also been employed in absorption [16], lithography [17], microspectroscopy [18], and fluorescence measurements [19]. A particularly useful technique for determining the local atomic environment around an atom with the solid's lattice is *surface extended X-ray absorption fine structure* (*SEXAFS*). This technique

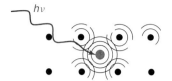

Figure 8.13 Interference of electron waves from a photoexcited atom and its nearest neighbors, which introduces fine structure in the optical absorption.

involves absorption due to excitation of an electron from a core level as a function of incident photon energy. One takes advantage of the elemental specificity of the particular core level energy by tuning $h\nu$ accordingly. Next, one selects a particular core level transition of that atom whose absorption features do not overlap those of other transitions for that or other elements in the system. Finally, one compares the absorption spectra with the calculations that take into account scattering and optical interference from neighboring atoms. These give rise to oscillations in the absorption above the threshold energy. This fine structure contains information on the distances and geometric arrangement of the probed atom relative to its atomic neighbors. In other words, it can describe the three-dimensional local coordination of the atom in the lattice.

Figure 8.13 illustrates the excitation of photoelectrons and the interference between photoelectron waves from the absorbing atom and backscattered waves from surrounding atoms. This interference produces fine structure in the oscillator strength of the absorption above its threshold value.

Figure 8.14 shows SEXAFS results for bulk Pd, Pd_2Si and ≈ 1.5 monolayers of Pd on Si(111)7×7 based on (i) the Pd L_2 $L_2M_{4,5}M_{4,5}$ Auger transition threshold yield and (ii) its Fourier transform versus the nearest neighbor distances [20]. Peak A corresponds to Pd–Si nearest neighbor distances, while peak B signifies the Pd–Pd second nearest neighbor distance. Note the correspondence between $|f(r)|$ peaks A and B in both Pd_2Si and ≈ 1.5 monolayers of Pd on Si, supporting the formation of a Pd_2Si-like compound at the intimate Pd–Si interface.

The coordination of atoms at a metal–semiconductor interface is an important ingredient in understanding localized states at Schottky barriers. SEXAFS provides the capability to probe local atomic structure of a metal on a semiconductor based on the tunability of the synchrotron photon energy over energies specific to a particular subset of surface or interface atoms. However, low signal strength from the few near-surface atoms, the high energies required for low background, and the wide energy range involved in achieving reliable results present experimental challenges for this technique.

Soft-X-ray absorption spectroscopy (SXAS) of intra- and interband transitions is a related technique that shows promise for chemical studies of electronic materials. With the high beam intensities now available at synchrotron facilities coupled with the wide energy separation of core level transitions, it is possible to identify band structure features and their changes with phase changes and chemical bonding within ultrathin films from absorption thresholds and peaks. Recent studies have focused on transition metal high-K dielectrics such as TiO_2, HfO_2, and ZrO_2 [21, 22].

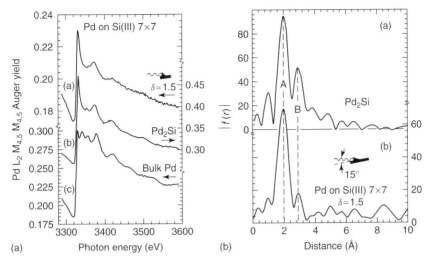

Figure 8.14 (a) Pd L_2 absorption at the $L_2M_{4,5}M_{4,5}$ Auger transition threshold energy for bulk Pd, Pd_2Si and ≈1.5 monolayers of Pd on Si(111)7×7. (b) Fourier transform $|f(r)|$ of spectra in (a) for Pd_2Si and Pd on Si(111)7×7 at a glancing incidence of 15°.

8.5 Summary

Soft X-rays enable a wide range of experimental technique based on photoexcitation. Synchrotron radiation provides high fluxes of photons with energies that span the energy range from the infrared to the hard X-ray region. This feature extends the range of atomic core levels that can be probed by photoemission spectroscopy.

The soft X-ray tunability enables highly surface-sensitive studies by selecting photon energies that yield photoelectrons with very short scattering lengths. The information obtained with SXPS from surfaces includes the following: (i) chemical bonding shifts, (ii) Fermi level movements, and (iii) element-selective interdiffusion. Such studies have revealed highly localized chemical reactions and diffusion taking place at metal–semiconductor interfaces. Furthermore, these chemical effects can be correlated with Schottky barrier formation and the resultant barrier heights.

The wide energy range and tunability of soft X-rays enables additional techniques such as ARPES, polarization-dependent photoemission, CFS, CIS photoemission spectroscopy, and absorption threshold spectroscopies that yield bulk band structure, adsorbate bonding energies and symmetries, and near-surface local bond arrangements. Many of these techniques can provide useful information on semiconductor surface cleaning, passivation, doping, local atomic environments, and interface chemical phase changes.

In subsequent chapters, we see numerous examples in which these techniques have revealed key information about electronic materials' surfaces and interfaces.

Problems

1. In an experiment to probe GaAs surfaces and interfaces with high surface sensitivity SXPS, which core levels and soft X-ray photon energies would you use?

2. In the SXPS data below, the bulk- and surface-sensitive attenuation of P 2p core levels at $h\nu = 140$ and $170\,\text{eV}$, respectively, is deconvolved to show the behavior of the substrate, a transition region, and a phosphide reaction product for Cr (top) and Fe (bottom) overlayers. Calculate the electron scattering lengths from the substrate attenuations. How do these compare with the universal escape depth curve for $h\nu = 170\,\text{eV}$ excitation of the P 2p core level?

3. Pt, Au, Mo, and Al thin films ($\Phi_M = 5.65, 5.1, 4.6$, and $4.28\,\text{eV}$, respectively) are deposited in UHV on a clean, low-defect semiconductor surface with electron affinity of $3.5\,\text{eV}$ and a bandgap of $4\,\text{eV}$. SXPS calibration of a metallic standard in electrical contact with the semiconductor places the initial Fermi level $0.1\,\text{eV}$ below the conduction band edge. Rigid core level shifts with metal deposition are 1.6, 1.5, 1.0, and $0.7\,\text{eV}$. Calculate and plot the Schottky barrier heights and compare these to that expected classically. What do you conclude about any near-surface states in the bandgap?

4. Calculate the momentum of an electron collected at angle $\theta_p = 45°$ with photon energy $h\nu = 12\,\text{eV}$ and energy level E_B $5\,\text{eV}$ below the valence band edge E_{VBM}, which is located $4\,\text{eV}$ below the vacuum level. Assume $m^* = m_0$. $\mathbf{k} = \mathbf{k}_\perp + \mathbf{k}_\parallel$.

5. In angle-resolved photoemission experiments, the momentum perpendicular to the surface is not conserved due to the barrier for electrons to escape the solid, that is, the work function Φ. Energy is still conserved, so Schroedinger's equation for initial and final wave vectors $\mathbf{k}_i$ and $\mathbf{k}_f$ can be expressed as $\hbar^2 \mathbf{k}_i^2/2m_0 - q\Phi = \hbar^2 \mathbf{k}_f^2/2m_0$. Calculate the change in $\mathbf{k}$ perpendicular to the surface for a surface with $q\Phi = 4\,\text{eV}$ and $\theta_p = 45°$.

References

1. Margaritondo, G. (1988) *Introduction to Synchrotron Radiation*, Oxford University Press, New York.
2. Cerenkov, P.A. (1934) *Dokl. Akad. Nauk SSSR*, **2**, 451.
3. Winick, H. (1980) in *Synchrotron Radiation Research*, Chapter 2 (eds H. Winick and S. Doniach), Plenum, New York.
4. Brillson, L.J., Bauer, R.S., Bachrach, R.Z., and McMenamin, J.C. (1980) *J. Vac. Sci. Technol.*, **17**, 476.
5. Wagman, D.D. et al. (1968–1971) Selected Values of Chemical Thermodynamic Properties, National Bureau of Standards Technical Notes 270-3-270-7, S. GUP, Washington, DC.

6. Kubachewski, O., Alcock, C.B., and Spencer, P.J. (1993) *Materials Thermochemistry*, 6th edn, Pergamon Press, London.
7. Brillson, L.J., Slade, M.L., Viturro, R.E., Kelly, M.K., Tache, N., Margaritondo, G., Woodall, J.M., Kirchner, P.D., Pettit, G.D., and Wright, S.L. (1986) *Appl. Phys. Lett.*, **48** (21), 1458–1460.
8. Brillson, L.J., Margaritondo, G., and Stoffel, N.G. (1980) *Phys. Rev. Lett.*, **44**, 667.
9. Brillson, L.J., Brucker, C.F., Katnani, A.D., Stoffel, N.G., and Margaritondo, G. (1981) *Appl. Phys. Lett.*, **38**, 784–786.
10. Williams, R.H., Montgomery, V., and Varma, R.R. (1978) *J. Phys. C*, **11**, L735.
11. Lapeyre, G.J., Anderson, J., and Smith, N.V. (1979) *Surf. Sci.*, **89**, 304.
12. Rowe, J.E., Margaritondo, G., and Christman, S.B. (1977) *Phys. Rev. B*, **16**, 1581.
13. Eastman, D.E. and Freeouf, J.L. (1974) *Phys. Rev. Lett.*, **33**, 1601.
14. Lapeyre, G.J., Anderson, J., Gobby, P.L., and Knapp, J.A. (1974) *Phys. Rev. Lett.*, **33**, 1290.
15. Schwieger, T., Knupfer, M., Gao, W., and Kahn, A. (2003) *Appl. Phys. Lett.*, **83**, 500.
16. Lee, P.A., Citrin, P.H., Eisenberger, P., and Kincaid, B.M. (1981) *Rev. Mod. Phys.*, **53**, 769.
17. Golovkina, V.N., Nealey, P.F., Cerrina, F., Taylor, J.W., Solak, H., David, C., and Gobrecht, J. (2004) *J. Vac. Sci. Technol. B*, **22**, 99–103.
18. Liu, P., Sun, J.Q., Guan, Y.J., Yue, W.S., Xu, L.X., Li, Y., Zhang, G.L., Hwu, Y., Je, J.H., and Margaritondo, G. (2008) *J. Synchrotron Radiat.*, **15**, 36.
19. Gonchar, A.M., Kolmogoroy, U.P., Gladkikh, E.A., Shuvaeva, O.V., Beisel, N.F., and Kolosova, N.G. (2005) *Nucl. Instrum. Methods Phys. Res. A*, **543**, 271.
20. Stöhr, J. and Yaeger, R. (1982) *J. Vac. Sci. Technol.*, **21**, 619.
21. Lucovsky, G., Hong, J.G., Futon, C.C., Zou, Y., Nemanich, R.J., and Ade, H. (2004) *J. Vac. Sci. Technol. B*, **22**, 2132.
22. Strzhemechny, Y.M., Bataiev, M., Tumakha, S.P., Goss, S.H., Hinkle, C.L., Fulton, C.C., Lucovsky, G., and Brillson, L.J. (2008) *J. Vac. Sci. Technol.*, **26**, 232.

9
Particle–Solid Scattering

9.1
Overview

Chapter 6 showed that surface and interface techniques are largely comprised of photon, electron, and ion techniques. Chapters 7 and 8 showed how cross sections for photoelectron excitation are derived and used in photoemission spectroscopy. This chapter introduces electron and ion techniques for probing electronic material surfaces and interfaces. As with photon techniques, the cross sections for particle interactions in solids are described first.

9.2
Scattering Cross Section

How do electrons and atoms scatter in solids? Unlike photon excitation, electron or ion-scattering processes involve electrostatic interactions and significant momentum changes. The energy and momentum dependence of particle–solid scattering can be used to describe many physical interactions over wide energy ranges and extending over fields as diverse as surface science to nuclear physics.

9.2.1
Impact Parameter

Consider a positively charged particle incident on a target nucleus [1]. Its trajectory is governed by energy and momentum conservation in the presence of a central force field, in this case, coulomb repulsion. Following the exposition in [1], Figure 9.1 shows the trajectory of this particle as it approaches an atomic nucleus. Here, the distance b is termed the *impact parameter*, defined as the perpendicular distance between the incident particle path and the parallel line through the target nucleus. Since the nucleus resides in a solid lattice, we take its position as stationary. A positively charged particle scattering is deflected by an angle θ off a target nucleus due to coulomb force F, where

$$F = \frac{Z_1 Z_2 e^2}{r^2} \tag{9.1}$$

Surfaces and Interfaces of Electronic Materials. Leonard J. Brillson
Copyright © 2010 WILEY-VCH Verlag GmbH & Co. KGaA, Weinheim
ISBN: 978-3-527-40915-0

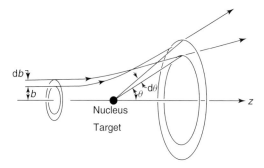

Figure 9.1 Incident particle and target nucleus showing angles of deflection θ for incident particle's impact parameter b. Cross section $\sigma(\theta)$ relates θ to b. (After. [1])

$Z_1 e$ and $Z_2 e$ are the incident and nuclear charges, respectively, and r is the distance between them.

The deflected particle conserves kinetic energy E as well as momentum p along the line through the target nucleus parallel to its incident path length (since F is symmetric for incident particle in front of or past the target nucleus). However, momentum is not conserved normal to this axis or radially with respect to the target nucleus.

Figure 9.1 shows that the incident particle is deflected at larger angles as b becomes smaller and the particle approaches closer to the target. Since F is symmetric about axis z, this scattering is symmetric about this axis. For particles incident between b and $b + db$ that scatter through angles θ to $\theta + d\theta$, the areas of the two annular rings shown in Figure 9.1 are related by $\sigma(\theta)$, the scattering cross section, according to

$$2\pi b \, db = -\sigma(\theta) \sin(\theta) d\theta \tag{9.2}$$

where the minus sign signifies the increase of θ with decreasing b. Note that $\sigma(\theta)$ has units of area. Figure 9.2a illustrates the change in momentum along an axis

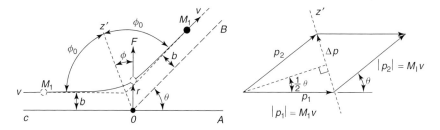

Figure 9.2 (a) Particle deflection by central force F, where angle ϕ defines the radial central force axis equidistant between incident and scattered trajectories. (b) Diagram of momentum change in particle of mass M_1 and velocity v due to central force scattering along axis z'.

z' of an incident particle of mass M_1 and velocity v scattered by an angle θ and complementary angle $2\phi_0$.

Since force F equals the change in momentum with respect to time, $\mathbf{F} = d\mathbf{p}/dt$, the change in momentum along the z' axis is given by

$$\Delta p = \int (dp)_{z'} = \int F \cos\phi \, dt = \int F \cos\phi \left(\frac{dt}{d\phi}\right) d\phi \tag{9.3}$$

Since angular momentum is conserved,

$$M_1 vb = M_1 r^2 \frac{d\phi}{dt} \tag{9.4}$$

and substituting Equations 9.1 and 9.4 into Equation 9.3,

$$\Delta p = \frac{Z_1 Z_2 e^2}{r^2} \int \cos\phi \left(\frac{r^2}{vb}\right) d\phi$$

$$= \frac{Z_1 Z_2 e^2}{vb} \int_{\phi_1}^{\phi_2} \cos\phi \, d\phi \tag{9.5}$$

where $\phi_1 = -\phi_0$, $\phi_2 = +\phi_0$, and $2\phi_0 + \theta = 180°$.

From Figure 9.2b, the change in momentum is

$$\Delta p = 2M_1 v \sin\left(\frac{\theta}{2}\right) = 2\left(\frac{Z_1 Z_2 e^2}{bv}\right) \cos\left(\frac{\theta}{2}\right) \tag{9.6}$$

from which the impact parameter b can be expressed in terms of incident particle energy E and scattering angle θ

$$b = \left(\frac{Z_1 Z_2 e^2}{M_1 v^2}\right) \cot\left(\frac{\theta}{2}\right) = \left(\frac{Z_1 Z_2 e^2}{2E}\right) \cot\left(\frac{\theta}{2}\right) \tag{9.7}$$

from which the relation

$$\sigma(\theta) = \frac{\left(\frac{Z_1 Z_2 e^2}{4E}\right)^2}{\sin^4\left(\frac{\theta}{2}\right)} \tag{9.8}$$

can be extracted using Equation 9.2 and the relation $d\cot(\theta/2) = -1/2\, d\theta/\sin^2(\theta/2)$. This *Rutherford scattering cross section* shows that scattering is inversely proportional to E^2 and $\sin^4(\theta/2)$. Experimental measurements that confirmed this relation also revealed that Z_2 was roughly proportional to one-half the atom's weight and thereby showed the concept of atomic number as opposed to atomic weight [1]. It also provides the basis for the *Rutherford backscattering* (RBS) technique to be discussed in subsequent chapters.

9.2.2
Electron–Electron Collisions

For incident particle scattering with electrons in the solid, $Z_2 = 1$ and with a small angle approximation, $\theta \cong 0$. Then, from Equation 9.6, the momentum transfer to the electron is

$$\Delta p = \frac{2Z_1 e^2}{bv} \tag{9.9}$$

and the *energy transfer* T during the collision is

$$\frac{\Delta p^2}{2m} = \frac{2Z_1^2 e^4}{b^2 m v^2} \equiv T \tag{9.10}$$

where m is the electron mass.

Taking $Z_1 = 1$ for electron–electron collisions, $\Delta p = 2e^2/bv$ so that

$$T = \frac{(\Delta p)^2}{2m} = \frac{4e^4}{2b^2 m v^2} = \frac{e^4}{b^2 E} \tag{9.11}$$

where electron energy $E = 1/2\, mv^2$.

Using Equation 9.2, we can relate a differential cross section $d\sigma(t)$ for energy transfer between T and dT to the impact parameter.

$$d\sigma(t) = -2\pi b\, db = \frac{\pi e^4}{E} \cdot \frac{dT}{T^2} \tag{9.12}$$

giving a cross section for electrons to transfer energy between lower and upper bounds $T_{\min}$ and $T_{\max}$, respectively:

$$\sigma_e = \int_{T_{\min}}^{T_{\max}} d\sigma(T) = \frac{\pi e^4}{E}\left(\frac{1}{T_{\min}} - \frac{1}{T_{\max}}\right) \tag{9.13}$$

For energetic electrons for which $E >$ a few hundred eV, $T_{\max} = E$ for $M_1 = M_2$, that is, the incident electron stops and the target electron acquires all its energy.

With $T_{\max} \gg T_{\min}$, the electron–electron scattering cross section approximates to

$$\sigma_e \approx \frac{\pi e^4}{E \cdot T_{\min}} = \frac{6.5 \times 10^{-14}}{(E \cdot T_{\min})}\, \text{cm}^2 \tag{9.14}$$

The numerical constant follows from the relations $e^2 = 14.4$ eV Å, $e^2/2r = 13.58$ eV, and radius $r = a_0 = 0.53$ Å for the hydrogen atom.

9.2.3
Electron Impact Cross Section

Finally, the parameter $T_{\min} = E_B$, the binding energy of an electron in orbit around an atom and

$$\sigma_e = \frac{\pi e^4}{E \cdot E_B} = \frac{\pi (e^2)^2}{U E_B^2} \tag{9.15}$$

where the *reduced energy* $U = E/E_B$ and $\sigma_e \equiv 0$ for $U < 1$ by definition.

Equation 9.15 represents the electron impact cross section in terms of the ratio of incident energy to electron binding energy. Defining σ_e in terms of this ratio permits us to understand general dependence of scattering cross section on incident beam energy relative to the binding energies of a particular atom's core electrons. Since the scattering removes electrons from their atomic orbits, σ_e is also termed an *ionization cross section*.

Figure 9.3 shows how this ionization cross section varies with U. In Figure 9.3a, σ_e increases with E until it reaches a peak value $\sigma_{\max}$ near $U \approx 3$–4, then decreases

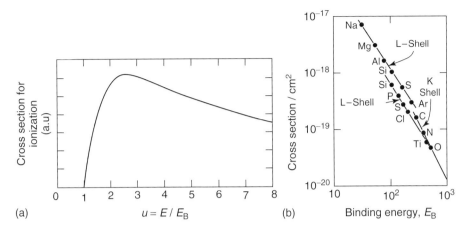

Figure 9.3 Ionization cross section versus (a) reduced energy $U = E/E_B$ that exhibits a maximum at intermediate values [2a] and (b) a quadratic decrease with increasing core-level binding energy E_B, here for $U \sim 4$ [2b].

[2]. In Figure 9.3b, σ_e decreases quadratically as E_B^{-2}, here just above σ_{max}. For $E_B = 100$ eV and $U = 4$, Equation 9.15 yields $\sigma_e = 1.6 \times 10^{-18}$ cm².

To compare the cross sections for electron excitation by incident electrons versus incident photons, let us consider the specific case of excitation of a K-shell in Al, for which $E_B = 1.56 \times 10^3$ eV from Table 7.2, and incident energy $E = 6.4 \times 10^3$ eV, corresponding to the K_α energy of Fe. (Note that X-ray photons with the latter energy would be produced if iron's K_α core electrons were excited. In turn, these photons could be used as a source of photoexcitation for Al.) Equation 9.15 yields

$$\sigma_e = \frac{\pi \cdot (14.4 \text{eV}/\text{Å})^2}{(6.4 \times 1.56 \times 10^6)} = 6.5 \times 10^{-21} \text{cm}^2 \qquad (9.16)$$

In comparison, the photoexcitation cross section for K-shell electrons in Al at this energy is $\sigma_{ph} = 3.4 \times 10^{-21}$ cm² so that $\sigma_e > \sigma_{ph}$. Thus excitation of Al $K\alpha$ core electrons by 6.4×10^3 eV electrons is more efficient than excitation with photons of the same energy. Actually, the higher flux of electrons versus photons typically available is an even stronger advantage for electron excitation as is discussed later in this chapter.

9.3
Electron Beam Spectroscopies

Electrons incident on a solid surface produce a variety of excitations that form the basis of several spectroscopies. Spectroscopies that have proved useful in studying electronic materials are (i) *Auger electron spectroscopy* (AES), (ii) *energy dispersive X-ray spectroscopy* (EDXS), (iii) *electron loss spectroscopy* (ELS), (iv) *cathodoluminescence spectroscopy* (CLS), and its low-energy counterpart (v) *depth-resolved cathodoluminescence spectroscopy* (DRCLS).

Figure 9.4 provides a schematic illustration of the excitations produced as the incident electron beam penetrates a solid surface. As the incident electrons scatter against the electrons in the solid, they produce secondary electrons. The electrons in this cascade in turn scatter both in depth and laterally, producing a characteristic pear shape inside which the electrons produce characteristic excitations. Figure 9.4 illustrates this spreading for an electron beam focused to a spot only a few tens of angstroms in diameter [3]. Such focused beams are available using a scanning electron microscope (SEM). The depth and width of this pear-shaped cascade depends on the incident beam energy and the physical properties of the excited material.

Figure 9.4 shows Auger electrons that emerge from depths of less than 5–20 Å according to the scattering length dependence in Figure 6.2. Auger electrons produced below this depth scatter before they can reach the surface, contributing to the continuum of low-energy background electrons. Secondary electrons at depths extending to the limit of the cascade can also produce X-rays that are characteristic of the atoms' core-level binding energies. Electrons that decay slowly back to their ground state are termed *fluorescent X-rays*. Since the scattering cross section for all these X-rays is much lower (see Figure 9.3a) than for electrons, they can originate from the full pear-shaped volume, typically 0.1–10 μm. This results in much lower spatial resolution than the diameter of the incident electron beam.

Chapter 10 discusses ELS. CLS rests on excitation of electron–hole pairs by the low-energy electrons produced in the final stage of the cascade. These electron–hole pairs recombine via band-to-band and transitions into and out of gap states. CLS is discussed at length in Chapter 16. A low-energy variant of CLS permits electron–hole pair excitation at depths that can be controlled on a nanometer scale with the incident electron beam energy. This technique permits studies of localized electronic states from the top semiconductor monolayers deep to tens and hundreds of nanometers below.

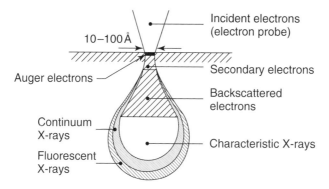

Figure 9.4 Characteristic pear-shaped volume of secondary electrons and the various excitations they produce [3].

9.4
Auger Electron Spectroscopy

9.4.1
Auger Transition Probability

Returning to Fermi's golden rule, Equation 7.9 is given by

$$W_A = \left(\frac{2\pi}{\hbar}\right) \rho(k) \left\{ \phi_f(r_1)\psi_f(r_2) \left[\frac{e^2}{|(r_1 - r_2)|}\right] \phi_i(r_1)\psi_i(r_2) dr_1 dr_2 \right\}^2 \quad (7.8)$$

where $\rho = m[V/(8\pi^3\hbar^2)]k \sin\theta d\theta d\phi$ represents the *density of states* associated with normalization in a box of volume V. To estimate the Auger transition probability in a hydrogen-like atom, consider the KLL transition shown in Figure 9.5. Here, $\phi_i(r_1)$ and $\psi_i(r_2)$ represent the two 2p electrons of the initial state, $\phi_f(r_1)$ represents the final 1s state one electron and $\psi_f(r_2)$ free electron state of the second electron.

Plugging in the known wavefunctions for hydrogenic K and L shell electrons, using the K-level energy $E_K = e^2 Z^2 / 2a_0$, and assuming $r_2 > r_1$ for 2p versus 1s electrons, one finds a transition probability [1]

$$W_A = \frac{Ce^4 m}{\hbar^3} \quad (9.17)$$

where C is a numerical constant equal to 7×10^{-3}. More precise calculations are possible using numerical techniques with more elaborate wavefunctions and an advanced description of the interaction potential [4]. However, the main result of Equation 9.17 and more detailed calculations is that the Auger transition probability is approximately independent of atomic number Z. This relative independence of W_A on Z is borne out by Table 9.1 in which W_A increases by only a factor of 4 as Z increases sevenfold.

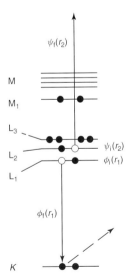

Figure 9.5 Schematic representation of KLL transition involving a two 2p electron initial state $\phi_i(r_1)\psi_i(r_2)$ and a 1s electron and free electron final state $\phi_f(r_1)\psi_f(r_2)$.

Table 9.1 Comparison of Auger transition rates and K-level X-ray emission rates (Scofield).

Atomic no.	Element	Auger (ev/$\hbar$)	K X-ray (ev/$\hbar$)
10	Ne	0.23	0.005
11	Na	0.29	0.007
12	Mg	0.36	0.010
13	Al	0.40	0.014
14	Si	0.44	0.02
15	P	0.48	0.03
16	S	0.51	0.04
17	Cl	0.54	0.05
18	Ar	0.58	0.07
20	Ca	0.65	0.12
22	Ti	0.69	0.19
24	Cr	0.72	0.28
26	Fe	0.75	0.40
28	Ni	0.78	0.55
32	Ge	0.83	1.0
36	Kr	0.89	1.69
40	Zr	0.94	2.69
46	Pd	0.99	4.94
52	Te	1.04	8.40
58	Ce	1.07	11.6
65	Tb	1.10	21.8
70	Yb	1.13	29.6

Adapted from [4].

9.4.2
Auger versus X-ray Yields

Figure 9.4 shows that the incident electron beam produces both Auger electrons and X-rays. Now, let us compare the probabilities for creating these two excitations. Again starting from Fermi's golden rule, one can show that the rate for spontaneous transition of an electron from an initial state i to a final state f is [1]

$$W_X = \left(\frac{4e^2}{3\hbar}\right)\left(\frac{\omega_{fi}}{c}\right)^3 |<\psi_f|\mathbf{r}|\psi_i>|^2 \tag{9.18}$$

where $\hbar\omega_{fi}$ is the (X-ray) energy of the emitted photon equal to the energy difference between state i and f. ψ_f and ψ_i are the wavefunctions of the final and initial states, respectively.

To a rough approximation, $<\psi_f|\mathbf{r}|\psi_i> \sim Z^{-1}$ for transition to a K-shell since the wavefunctions have finite intensity only for $|\mathbf{r}| < a_0/Z$ whereas transition energies $\hbar\omega_{fi}$ in hydrogenic atoms scale as Z^2. Thus, W_X in Equation 9.18 is proportional to

Z^4. This strong dependence of W_X on Z is borne out in Table 9.1, where K-shell X-ray emission rates increase by nearly 4 orders of magnitude over the $Z = 10$–70 range.

Now consider the competition between the Auger versus the X-ray emission processes for an excited state, that is, a core hole vacancy. The *lifetime* τ defines a rate of deexcitation $1/\tau$ that includes the sum of rates for all the decay processes.

$$\frac{1}{\tau} = W_X + W_A + W_K \tag{9.19}$$

In this general equation, W_K represents *Coster–Kronig* transitions in which an electron from the same shell fills the vacancy. For transitions to K-shell vacancies, these transitions do not occur so that the probability for X-ray emission ω_X can be written as

$$\omega_X = \frac{W_X}{(W_X + W_A)} \tag{9.20}$$

ω_X is commonly termed the *fluorescence yield*.

Given the relative independence of W_A on Z (Equation 9.17.) and the Z^4 dependence of W_X, the fluorescence yield is also proportional to Z^4.

A semiempirical expression for ω_X based on the W_X/W_A ratio is [5]

$$\omega_X = \left(\frac{W_X}{W_A}\right) \bigg/ \left(1 + \left(\frac{W_X}{W_A}\right)\right) \tag{9.21}$$

where

$$\frac{W_X}{W_A} = (-a + bZ - cZ^3)^4 \tag{9.22}$$

with the constants $a = 6.4 \times 10^{-2}$, $b = 3.4 \times 10^{-2}$, and $c = 1.03 \times 10^{-6}$. The solid curve in Figure 9.6 represents Equation 9.21 with these constants [6].

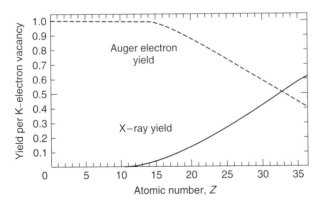

Figure 9.6 X-ray and Auger yields for deexcitation involving a K-shell core hole as a function of atomic number Z. Auger transitions dominate at low Z, while X-ray emission dominates at high Z [6].

Figure 9.6 shows that ω_X increases with Z. Auger processes are dominant for low Z while the X-ray proportion increases rapidly with Z, surpassing Auger processes above $Z = 33$ and dominating at higher Z. The *Auger electron yield* σ_e is proportional to $1 - \omega_X$. Hence, Auger transitions offer higher sensitivity to core level transitions for low Z elements, while X-ray fluorescence transitions provide higher sensitivity for high Z elements.

9.4.3
Auger Excitation Process

The Auger excitation process forms the basis for AES, one of the primary techniques for measuring surface elemental composition. With this capability, AES is used to monitor surface cleanliness in ultrahigh vacuum (UHV), thin film growth uniformity, and profiles of elemental composition as a function of depth below a surface. Figure 9.7 illustrates an experimental arrangement used for obtaining standard AES data [7]. This particular configuration includes (i) an excitation source such as a glancing incidence or coaxial electron gun with incident beam energy of 2–5 keV, (ii) an electron analyzer, in this case a cylindrical mirror analyzer (CMA) with typical energy resolution $\Delta E/E$ of $\sim 0.6\%$, (iii) an ion "sputter" gun that enables depth profiling, and (iv) electronics that introduce ramp- plus AC-modulated voltages to the cylinders in order to enhance signal-to-noise in a dN/dE analog mode of detection.

Specimens are often mounted on carousels that permit sequential and possibly automated measurements. Figure 9.8 illustrates the advantage of differentiating

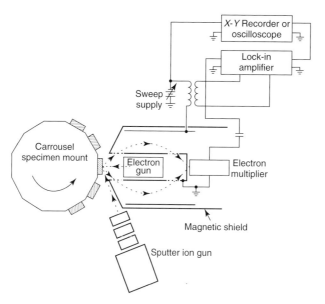

Figure 9.7 Schematic illustration of samples, electron gun, CMA, and analyzer electronics for performing AES experiments [7].

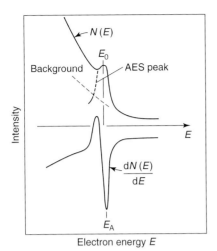

Figure 9.8 Qualitative representation of AES peak and background spectral features before and after differentiation. The resonance-like $dN(E)/dE$ feature is easily distinguished from the secondary electron background.

$N(E)$ spectra. AES peaks are typically small peaks above a large background of scattered secondary electrons. Because this background varies smoothly, differentiation flattens this spectral contribution, emphasizing only the AES peak as a $dN(E)/dE$ feature.

The Auger electrons after energy filtering reach an electron multiplier, which amplifies the signal into a current $I(V)$ that depends on CMA voltage V. Modulating V with frequency ω around a given voltage V_0 effectively differentiates the AES signal since $I(V + V_0 \sin \omega t) = I_0 + (dI/dV)V_0 \sin \omega t +$ higher order terms. Alternatively, digital techniques are available to remove the secondary electron contribution to $N(E)$ spectra, leaving only the AES peak features.

The AES spectra of most elements are dominated by a number of characteristic processes. Figure 9.9 illustrates four of the most common Auger transitions: (i) Auger KLL, (ii) LMM, (iii) LVV, and (iv) Coster–Kronig processes. KLM nomenclature refers to atomic shells while s and p notation indicates shell splitting into subshells by angular momentum and spin.

The Auger excitation process consists of three steps: (i) an incident electron (or photon) knocks a core level electron out of its orbit, leaving behind a core hole; (ii) a higher lying electron fills this hole; and (iii) the excess energy released by this decay process transfers to a third electron (or a photon) that leaves the solid. The KL_1L_2 process in Figure 9.9a involves ejection of a K-shell electron and filling of the resultant hole by an L_1 electron. An L_2 electron leaves the solid with the difference in L_1 versus K orbital energies. The $L_1M_1M_1$ process in Figure 9.9b involves an L_1 instead of a K-shell electron and M_1 electrons instead of L_1 and L_2 as in Figure 9.9a. The $L_1L_2M_1$ Coster–Kronig process in Figure 9.9c involves an intra-shell transition between $2p_{1/2}$ and 2s subshells with a higher lying 3s electron emitted. The LVV process in Figure 9.9d involves two high-lying valence electrons, one of which decays into an L_3 level while the other leaves the solid. In principle, this process can give information about the valence band density of states. The higher energy

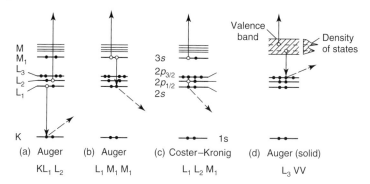

Figure 9.9 Atomic-level representation of the Auger electron processes. A primary electron beam excites a core-level electron ↘, leaving a vacancy filled by a second electron ↓, with the energy released to a third electron that exits the solid ↑. (a) KL_1L_2 Auger, (b) $L_1M_1M_1$ Auger, (c) $L_1L_2M_1$ Coster-Kronig, and (d) L_3VV Auger processes.

KLL and LMM processes in Figure 9.9 exhibit relatively strong features since they involve fewer alternative processes involving electrons within a given subshell.

9.4.4
Auger Electron Energies

AES spectral features are characteristic "fingerprints" of specific elements at specific energies. Chemical identification and composition is relatively straightforward since elements can be identified by their transition energies and these energies are strongly dependent on atomic number. See Figure 9.10 [7]. Note, however, that there are no Auger electron transitions for $Z = 1$ (H) or $Z = 2$ (He) since Auger transitions involve at least three orbital electrons.

In the kinetic energy range extending up to 2400 eV, elements with $Z < 92$ have been tabulated for KLL, LMM, and MNN transitions. Large dots signify the predominant peaks. KLL transitions have the least number of peak features, followed by LMM, and MNN. As the number of subshell avenues for recombination increases with Z, the number of Auger peaks increases. Correspondingly, the intensity of each subshell feature decreases. However, the well-defined energies of each Auger set of peaks provide an additional approach to identify particular elements.

KLL and LMM transitions for heavier elements occur at higher kinetic energies that are outside the range of this figure. Figure 9.3a showed that efficient excitation of AES transitions involves incident beam energies that are two to three times the incident beam energy. Hence, exciting these transitions for $Z > 45$ requires incident beam energies extending well above 5 keV, a common upper limit for AES electronics.

The Auger transition energies can be calculated from the difference in energies of the three electrons, corrected for a change in screening due to the ionized atom.

9.4 Auger Electron Spectroscopy

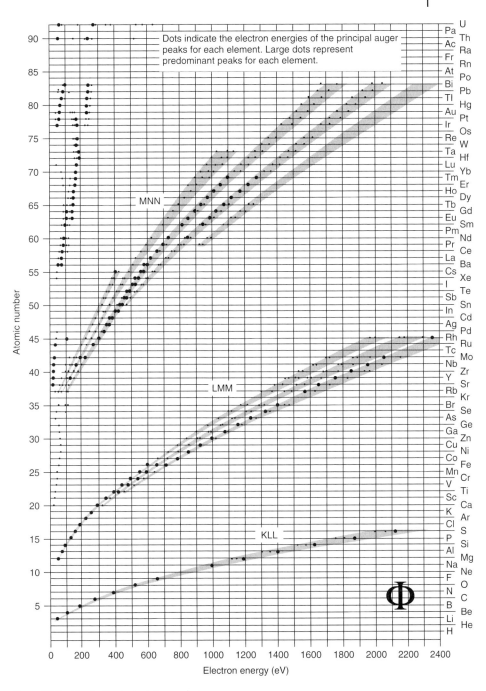

Figure 9.10 Chart of principal Auger electron energies.(L.E.Davis *et al.*, (Physical Electronics Industries, Eden Prairie, MN, 1976), with permission.)

Starting from Equation 6.2,

$$E_{123\text{Auger}} = E_1(Z) - E_2(Z) - E_3(Z+\Delta) - E_{\text{VAC}} \qquad (9.2)$$

where $E_1(Z)$, $E_2(Z)$, and $E_3(Z)$ are the energies of the three core levels for elements with atomic number Z and Δ is the change in core level E_3 due to the screening of the atom's nucleus by one less electron.

For example, in the KL_1L_2 transition process pictured in Figure 9.9a,

$$E^Z_{KL_1L_2} = E^Z_K - E^Z_{L_1} - E^Z_{L_2} - \Delta E(L_1L_2) \qquad (9.23)$$

The correction term ΔE is due to many body effects, in particular the increase of binding energy of an L_2 electron when an L_1 electron is removed, and vice versa. Thus,

$$\Delta E(L_1L_2) = 1/2 \left[E^{Z+1}_{L_2} - E^Z_{L_2} + E^{Z+1}_{L_1} - E^Z_{L_1} \right] \qquad (9.24)$$

$\Delta E(L_1L_2)$ represents the average increase in binding energy due to missing electrons. For a given Auger energy E, $\Delta E/E$ is a small correction, <1%.

Example 9.1. Consider the Auger transition energy $E^Z_{KL_1L_2}$ for iron atoms. Using Equations 6.2 and (9.23), with core level energies (relative to E_F)

for Z = 26 :   and for Z = 27 :

$E^{Fe}_K = 7114\,\text{eV}$ $E^{Co}_{E_1} = 926\,\text{eV}$

$E^{Fe}_{L_1} = 723\,\text{eV}$ $E^{Co}_{L_2} = 794\,\text{eV}$

$E^{Fe}_{L_2} = 723\,\text{eV}$

we obtain $\Delta E(L_1L_2) = 1/2\left[(794 - 723) + (926 - 846)\right] = 1/2[71 + 80] = 75.5\,\text{eV}$ so that $E^Z_{KL_1L_2} = 7114 - 846 - 723 - 75.5 = 5469.5\,\text{eV}$.

The measured $E^Z_{KL_1L_2}$ is actually 5480 eV (off the Figure 9.10 chart), within 0.2% of the calculated value.

9.4.5
Quantitative Elemental Identification

AES spectra provide $dN(E)/dE$ or $N(E)$ features whose amplitudes provide a measure of surface composition. To determine surface composition quantitatively, one requires sensitivity factors for each such feature. These sensitivity factors vary by nearly 2 orders of magnitude with Z and exhibit characteristic variations as each atomic subshell is filled. Figure 9.11 displays these variations for AES spectra obtained with a CMA and a primary beam energy of 3 keV. This follows directly from the dependence of electron scattering cross section on incident electron energy displayed in Figure 9.3a. Note the high KLL sensitivities of AES to carbon, oxygen, and nitrogen – a useful feature for surface science, where these constituents of air

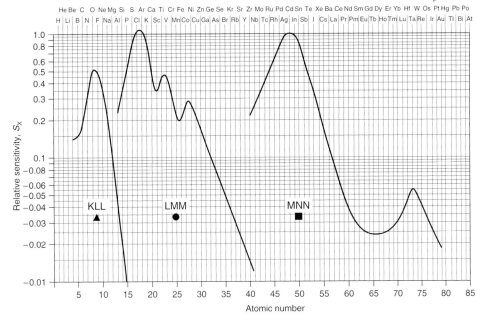

Figure 9.11 Relative AES sensitivities S_X of the elements measured with a cylindrical mirror analyzer and an incident beam energy of 3 KeV (Physcal Electronics Industries, Eden Prairie, MN, 1976, with permission.)

are usually to be avoided. Si, a basic microelectronic constituent, has a high LMM (but not KLL) sensitivity.

Sensitivity factors vary somewhat with excitation beam energy, requiring that the excitation energy match that used for the sensitivity factors. While these sensitivity factors follow general trends for all AES measurements, they are not the same for all electron analyzers. This is because the transmission function varies for different analyzers. Furthermore, $dN(E)/dE$ and $N(E)$ sensitivity factors also differ since asymmetries in the AES lineshapes affect the $dN(E)/dE$ amplitudes. Precise measures of composition require sensitivity factors for the specific analyzer used as well as the mode of data collection.

The model spectrum shown in Example 9.1 (Figure 9.12) was obtained from a silicon surface cleaned by argon ion sputtering, that is, bombardment with positively charged Ar atoms that serve to remove the topmost layers of surface atoms, exposing the underlying Si atoms. The primary features in this spectrum appear at 92 and 1619 eV, corresponding to the $Z = 14$ intersection with LMM and KLL curves in Figure 9.10. Minor features at 220, 273, and 510 eV are identified with Ar_{LMM}, C_{KLL}, and O_{KLL} Auger features, respectively.

Elemental compositions C_X are calculated by dividing dN/dE intensities of all elements by sensitivity factors S_X shown in Figure 9.11, then normalizing to 100%

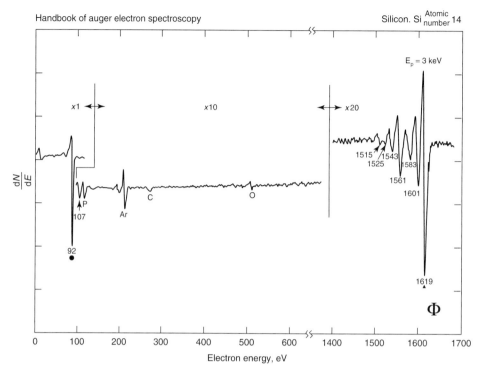

Figure 9.12 Example: Quantitative atomic composition analysis of a "clean" silicon surface with 3 KeV primary beam. (Reference [7] with permission.)

total composition.

$$C_{X1} = \frac{\left[\frac{dN(E)}{dE}\right]_1}{S_{X1}} \bigg/ \left\{ \frac{\left[\frac{dN(E)}{dE}\right]_1}{S_{X1}} + \frac{\left[\frac{dN(E)}{dE}\right]_2}{S_{X2}} + \frac{\left[\frac{dN(E)}{dE}\right]_3}{S_{X3}} + \cdots \frac{\left[\frac{dN(E)}{dE}\right]_n}{S_{Xn}} \right\} \quad (9.25)$$

where n = total number of different atomic constituents on the surface.

In this case, we have

Auger feature (eV)	Intensity	Sensitivity S_X	$1/S_X$	%
$Si_{LMM}(92)$	140	0.37	378	97.6
$Ar_{LMM}(220)$	5	1.05	4.8	1.2
$C_{KLL}(273)$	0.5	0.2	2.5	0.6
$O_{KLL}(510)$	1	0.5	2	0.5

Thus the silicon surface has less than 1% residual carbon and oxygen and ∼1% embedded Ar from the ion sputter cleaning.

9.5
Auger Depth Profiling

AES spectra acquired as surface atomic layers are removed provide atomic composition versus depth below the initial surface. Here one uses the same sensitivity and normalization of spectral features defined by Equation 6.24 for each and every spectrum acquired as the surface layers are removed. This technique is termed *Auger depth profiling*. It is useful to (i) determine changes in composition within a particular layer (ii) obtain evidence of diffusion between different layers, and (iii) measure chemical reaction near interfaces. As with XPS, the AES transitions involve core-level energies that "chemically shift" with changes in bonding environment (see Section 7.8.2). Hence, AES features can shift accordingly, so that different bonding environments can be distinguished.

Figure 9.13 illustrates an example of interfacial chemical reaction between Pd and Si, common constituents in microelectronic circuitry [8]. In this experiment, the Auger electron spectra yield not only the elements Si and Pd present but also the contributions of Si bonded to Si versus Si bonded to Pd. Figure 9.13 displays the change in composition at different depths, starting with pure Pd at the outer surface, a mixture of Pd and Si at their interface, and pure Si at depths below the interface. The depth scale is calibrated from the known thickness of deposited Pd and the crossover point between Pd and Si in the graph. Note that the rate of sputter profiling can change between layers so that this calibration is accurate only for the calibrated element.

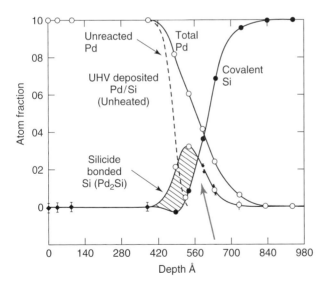

Figure 9.13 Depth profile for Pd on Si(100) obtained by AES with Ar$^+$ sputter etching and without heating. The atom fractions of silicide-bonded Si at the interface indicated the formation of Pd$_2$Si (arrow) [8].

The different curves in this figure correspond to the total Pd signal, the Si bonded covalently, and the Si bonded as the silicide Pd_2Si. The latter reaches a maximum at the Pd–Si interface, then decreases into the Si. Analysis of this depth profile leads one to conclude that (i) the interface is not abrupt; rather it is hundreds of angstroms thick even though it is formed at room temperature; (ii) a silicide forms with Pd_2Si stoichiometry; and (iii) the silicide extends into the Pd (versus the Si).

Several factors may complicate the analysis. First of all, the sputtering process used to profile the specimen can disrupt the surface. To minimize subsurface damage, one can use low ($\sim$250 eV) incident ion energies and glancing incidence of the ion beam.

Secondly, the primary electron beam used for AES can induce new bonding. One can monitor energies of spectral features as a function of time to assess how important such radiation effects may be. In principle, rapid scans at low primary beam current and voltage can minimize such effects. The sputter rates for more than one layer can be calibrated separately. Thus, AES depth profiling is a powerful tool but one requires care to interpret the results quantitatively. With such care, AES depth profiling can be effective even for ultrathin films. Figure 9.14 illustrates AES spectra of an Au film deposited on a freshly cleaved InP surface before sputtering, inside the Au film and well within the underlying InP bulk [9].

As expected, the spectrum of the 30 Å Au–InP(110) specimen after Au has been completely removed by sputtering displays just In and P features. On the other hand, the spectrum before sputtering reveals the presence of P and In even though only Au features are expected. Similarly, spectra taken at intermediate depths corresponding to the "bulk" of the Au film reveal the presence of In.

AES depth profiles of Au, In, and P for this and other Au films on InP appear in Figure 9.15. As in Figure 9.12, these graphs show the metal signal decreasing while the substrate In and P signals increase at characteristic depths, consistent with the different thicknesses of Au deposited. Also, the interfaces do not appear to be atomically abrupt despite room temperature deposition. Figure 9.15 also shows that In and P profiles are qualitatively different. P appears to diffuse through the Au film and segregate at the free Au surface. This effect becomes more apparent for thicker Au films. In atoms also appear to diffuse into the Au film but have finite concentration within the metal itself. In general, there is more In versus P out diffusion.

The spatially extended In versus P profiles are remarkable since these junctions were not annealed. Rather, the ability to measure atomic composition on a scale of monolayers reveals that finite diffusion occurs even though diffusion coefficients must be small at room temperature. The asymmetry between these profiles also confirms that the films are uniform. If bare InP were exposed through uneven thicknesses or pinholes in the metal overlayer, equal and increasing concentrations of In and P would be observed rather than the surface segregated species shown.

Perhaps most significant of all, the diffusion of semiconductor cations and anions into the metal means that vacancies of these atoms must be present within the semiconductor lattice. Such native point defects and their complexes with other defects at neighboring lattice sites are often electrically active and can thus play a role in the charge transfer involved in Schottky barrier formation.

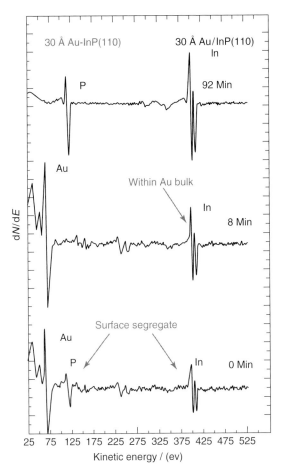

Figure 9.14 AES of 30 Å Au–InP(110) versus sputtering time. The presence of P and In at the free Au surface indicate surface segregation [9].

9.6 Summary

Electron and ion scattering in solids enable powerful techniques for probing electronic materials surfaces and interfaces. The particle–solid scattering cross section provides the basis for analyzing RBS spectra in terms of ion–ion collisions that are proportional to atomic number of the target. RBS yields elemental depth profiles inside semiconductors on a scale of nanometers to microns.

The electron–electron impact cross section yields quantitative values for rates of Auger electron transitions. Comparison between AES and photoemission cross sections shows that Auger transitionsAuger transitions and yields dominate in filling core-level holes for low Z elements while photoemission transitions and yields dominate for high Z elements, indicating the relative advantages of the AES and XPS techniques for elemental detection. Similar to photoemission, AES energies can exhibit core-level chemical shifts, although their lineshapes are more difficult to deconvolve.

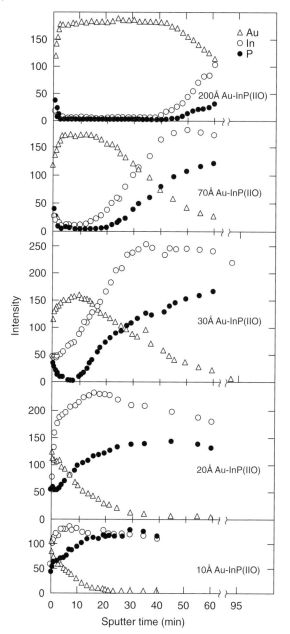

Figure 9.15 AES depth profiles for Au−InP(110) junctions with different Au overlayer thicknesses. Different In versus P out-diffusion within the Au film and at the surface is apparent [9].

AES sensitivities vary with atomic number and subshell and must be calibrated for the specific energy-dependent transmission of different electron energy analyzers.

The distinctive spectral features and energies for AES enable quantitative elemental detection and analysis. The influence of the surrounding matrix is not a major factor in AES. Chemical analysis is possible with a typical precision of 1%.

Sputter depth profiling with AES provides quantitative depth profiles of elemental composition below the free surface and through buried interfaces such as metal–semiconductor junctions. Such depth profiles can reveal interdiffusion and chemical reactions occurring on a scale of nanometers to microns, even for interfaces formed near room temperature.

Problems

1. Calculate the Bohr–Bethe range R_B of 3-keV incident electrons in GaN. Compare this result with the corresponding range in GaSb.
2. For Cu K_α radiation (E = 8.04 keV) incident on Al, calculate the photoelectric cross section σ_{ph} for the K shell of Al ($E_B = 1.56$ keV) and compare its value with the electron impact ionization cross section σ_e for 8.04 keV electrons.
3. You are given a 200 Å thick layer of Ga_xAl_{1-x} As on an InP substrate about 1 mm thick and are asked to determine the Ga to Al ratio. You can carry out XPS, AES, or EDXS analysis using 20-keV electrons and an Al K_α X-ray source. To compare the different techniques, carry out the following calculations or comparisons: (a) What is the cross section ratio σ_{Ga}/σ_{Al} for the K-shell electron impact ionization and L-shell photoeffect? (b) What is the fluorescence yield ratio $\omega_X(Ga)/\omega_X(Al)$ for the K-shell hole? (c) What is the ratio of 20-keV Ga/Al Auger yields? Photoelectron yields? X-ray yields? (d) Which technique provides the clearest Ga to Al determination?
4. Calculate the range of Auger energies for the LM_1M_2 transition of gallium. Calculate the KL_1L_2 transition energy of oxygen. How well do these calculations agree with experiment?
5. The three-3 keV Auger electron spectra show a UHV-cleaved semiconductor surface (a), after a further cleaning process (b), and after a chemical

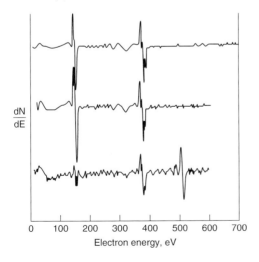

> reaction (c). Label all the elemental features in all three spectra. Identify the semiconductor, the change induced by the further cleaning process, and the type of final chemical reaction. Calculate the percent composition of each element on the chemically reacted surface.

References

1. Feldman, L.C. and Mayer, J.C. (1986) *Fundamentals of Surface and Thin Film Analysis*, Prentice Hall, p. 20.
2. (a) Chang, C.C. (1974) Analytic Auger electron spectroscopy, in *Characterization of Solid Surfaces*, Chapter 20 (eds P.F.Kane and G.R. Larrabee), Plenum Press, New York; (b) Vrakking, J.J., and Meyer, F. (1974) *Phys. Rev. A*, **9**, 1932.
3. Lüth, H. (1995) *Surface and Interfaces of Solid Materials*, 3rd edn, Springer, Berlin, p. 120.
4. Bambynek, W., Crasemann, B., Fink, R.W., Freund, H.U., Mark, H., Swift, C.D., Price, R.E., and Rao, P.V. (1972) *Rev. Mod. Phys.*, **44**, 716.
5. Burhop, E.H.S. (1955) *J. Phys. Radium*, **16**, 625.
6. Siegbahn, K., Nordling, C.N., Fahlman, A., Nordberg, R., Hamrin, K., Hedman, J., Johansson, G., Bergmark, T., Karlsson, S.E., Lindgren, I., and Lindberg, B. (1967) *ESCA, Atomic, Molecular and Solid State Structure Studied by Means of Electron Spectroscopy*, Almqvist and Wiksells, Uppsala.
7. Davis, L.E., MacDonald, N.C., Palmberg, P.W., Riach, G.E., and Weber, R.E. (1976) *Handbook of Auger Electron Spectroscopy*, 2nd edn, Physical Electronics Industries, Eden Prairie.
8. Roth, J.A. and Crowell, C.R. (1978) *J. Vac. Sci. Technol.*, **15**, 1317.
9. Shapira, Y. and Brillson, L.J. (1983) *J. Vac. Sci. Technol. B*, **1**, 618.

10
Electron Energy Loss Spectroscopy

10.1
Overview

Electron loss spectroscopy (ELS) is a second electron scattering technique that has proved useful in probing electronic materials' surfaces and interfaces. Depending on the incident electron energy, ELS encompasses a broad range of inelastic scattering mechanisms involving both short- and long-range potentials. Chapter 9 discussed the interaction of charged particles with short-range potentials such as the local atomic potential of an atom. Such interactions involve large angles and are termed *impact scattering*. Short-range or local interactions are termed *virtual excitations* since they occupy an excited state of the atom in an intermediate stage of the scattering event.

On the other hand, long-range scattering potentials include the oscillating dipole fields from collective excitations such as surface and bulk lattice vibrations, charge oscillations termed *plasmons*, and the dynamic dipole moments of vibrating adsorbed atoms or molecules. They can be distinguished in the laboratory by their angular distribution, which is near-specular, that is, having equal incoming and outgoing trajectory angles. Table 10.1 summarizes these various excitations measured by electron loss spectroscopies and the energy ranges involved.

Low-energy electrons are employed to emphasize surface excitations. As with other surface spectroscopies represented in Figure 6.1, ELS takes advantage of the short scattering length of electrons in the 50–100 eV energy range (Figure 6.2). These low energies lend themselves easily to the conventional surface science electron guns and energy analyzers used for Auger electron spectroscopy (AES) and X-ray photoelectron spectroscopy (XPS). Furthermore, scattering in this low energy range can be treated by straightforward analytic techniques.

Scattering interactions involving phonons requires high enough energy resolution to distinguish spectral features separated by only tens of meV. In this case, a monochromatic source of electrons is employed. For electronic interband transitions, interface states, plasmons, and weakly bound core levels, the electron beam provided by a conventional electron gun is sufficient. For deeper core-level and atomically resolved energy loss features, even higher excitation energies and energy filtering are required.

Surfaces and Interfaces of Electronic Materials. Leonard J. Brillson
Copyright © 2010 WILEY-VCH Verlag GmbH & Co. KGaA, Weinheim
ISBN: 978-3-527-40915-0

Table 10.1 Electron energy loss spectroscopies, the physical excitations they measure, and their relevant energy ranges.

Measure	Technique	Energy range (eV)
Surface phonons (lattice, adsorbate)	High-resolution electron energy loss spectroscopy (HREELS)	<0.01–0.1
Interface state transitions	Energy loss spectroscopy (LEELS)	0.1–10
Near-surface interband transitions	Energy loss spectroscopy (LEELS)	0.1–10
Interband transitions	Energy loss spectroscopy (ELS)	1–50
Plasmons	Energy loss spectroscopy (ELS)	10–50
Core-level transitions	Energy loss spectroscopy (ELS)	1 to >1000
Cross-sectional, atomically resolved interface transitions	High-resolution transmission electron microscopy (HRTEM) electron ELS (EELS)	>>1000

Consider a glancing incidence electron gun and cylindrical mirror analyzer (CMA) used for AES. Figure 10.1 shows an incident beam of electrons with lineshape $N(E)$ centered at *primary energy* E_P. For *high-resolution electron energy loss spectroscopy* (HREELS), an electron monochromator is required to select electron beam energies with linewidths of 1–10 meV. Achieving such a high energy resolution requires monochromators with resolving power $\Delta E/E \sim 10^{-3} - 10^{-4}$ and incident beam energies of only a few electron volts. For ELS, higher beam energies are used to excite plasmons and interband transitions, and the linewidth

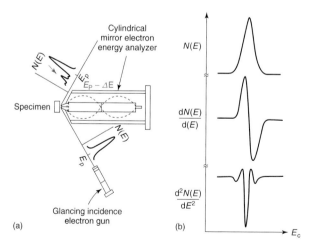

Figure 10.1 (a) Glancing incidence experimental setup for LEELS. Excitations appear as loss features (arrow) displaced to lower energies from the primary beam energy. (b) ELS $N(E)$ peak, dN/dE, and inverted d^2N/dE^2 "peak" lineshapes versus energy.

is 0.2–0.3 eV, determined by the thermal width of the electron gun filament, for example, 0.26 eV at 3000 K.

For $E_P \cong 100$ eV or less, the primary electrons penetrate the surface less than a few atomic layers. Glancing incidence further minimizes penetration of the electron cascade. The reflected or *"specular"* beam contains not only electrons with energy E_P but also electrons at energies $E_P - \Delta E$ that have lost energy due to various electronic and vibrational excitations. Since these "loss" peaks are orders of magnitude lower in intensity than the primary beam and the background of secondary electrons, differentiation of the spectra is used to minimize this background.

Differential spectra can be obtained by modulating the CMA pass voltage just as with AES spectra described in Section 9.3.3. A second-derivative spectrum generated digitally is commonly used to produce $d^2 N/dE^2$ "peaks" with the same energy but with opposite sign from peaks in the $N(E)$ spectrum.

10.2
Dielectric Response Theory

The scattering in long-range dipole fields can be treated by a quasi-continuum theory in which the dielectric properties of the solid are described by a complex dielectric function. An extended discussion of dielectric response theory is presented by Lüth [1]. The electron energy loss process can be described in terms of an *energy transfer rate* dW/dt such that

$$dWdt = \text{Re}\left\{ \int dr \mathcal{E} \cdot \frac{d\mathbf{D}}{dt} \right\} \tag{10.1}$$

where $\mathbf{D}$ is a *displacement field vector* of the moving electron, $\mathcal{E}$ is an *electric field vector* and $\mathbf{r}$ a *displacement* vector of the excitation, and $\varepsilon = \varepsilon_1 + i\varepsilon_2$ is the *complex dielectric permittivity* of the dielectric medium. Thus the energy transfer rate is the change in density of the coulomb field inside the solid, shielded by the dielectric medium.

Since both the electric field ε and the dielectric permittivity ε are dependent on both the frequency ω and the wave vector $\mathbf{q}$, $\mathbf{D}(\omega, \mathbf{q}) = \varepsilon_0 \varepsilon (\omega, \mathbf{q}) \mathcal{E} (\omega, \mathbf{q})$. Using Equation 10.1, one can show that the energy transfer rate is proportional to the imaginary part of the inverse dielectric permittivity:

$$\frac{dW}{dt} \propto \text{Im}\left\{ \frac{-1}{\varepsilon(\omega, q)} \right\} = \frac{\varepsilon_2(\omega)}{[\varepsilon_1^2(\omega) + \varepsilon_2^2(\omega)]} \tag{10.2}$$

The frequency dependence of the complex dielectric permittivity ε derives from the dielectric medium's electronic band structure, its various elementary excitations such as plasmons, excitons, and phonons, and any boundary conditions imposed by physical dimensions. Equation 10.2 states that the rate of energy loss dW/dt – the spectral response in an electron scattering event – is proportional to the well-known *bulk-loss function*. ELS spectral features appear when dW/dt increases. This occurs when $\varepsilon_2(\omega)$ exhibits maxima, for example, when $\hbar\omega$ equals the energy of an electronic band-to-band transition. This is consistent with the fact that $\varepsilon_2(\omega)$ determines the optical absorption coefficient of a solid, to be discussed in Chapter 15.

ELS peaks also occur when $\varepsilon_1(\omega) \cong 0$ and $\varepsilon_2(\omega)$ is small and monotonically varying with frequency. This occurs for longitudinal collective oscillations, for example, plasmons. Electron scattering at a solid–vacuum interface introduces a boundary condition of a semi-infinite half plane with $\varepsilon(\omega)$ bounded by a semi-infinite half plane for which $\varepsilon = \varepsilon_0 = 1$. Taking into account the continuity of tangential electrical field and normal electric displacement for the collective surface excitations, one can show [1] that this additional constraint results in

$$\frac{dW}{dt}(\text{surface}) \propto \text{Im}\left\{\frac{-1}{[\varepsilon(\omega)+1]}\right\} = \frac{\varepsilon_2(\omega)}{\{[\varepsilon_1(\omega)+1]^2 + \varepsilon_2(\omega)^2\}} \quad (10.3)$$

For this surface scattering, the ELS spectral response peaks with $\varepsilon_2(\omega)$ as for bulk scattering and, in addition, when $\varepsilon_1(\omega) \cong -1$ in frequency regions of small $\varepsilon_2(\omega)$. The latter condition determines the frequencies of surface collective excitations.

10.3
Surface Phonon Scattering

High-resolution ELS provides an effective method to probe the surface vibrational modes of solids. In particular, HREELS can identify adsorbed chemical species, adsorption sites, and bonding geometry. The key to identifying these modes is the energies of individual vibration modes obtained from infrared absorption of bulk crystal lattices or gas-phase molecules.

Figure 10.2 illustrates an HREELS spectrum for hydrogen and deuterium adsorbed on a clean GaN(0001) surface [2]. The primary beam width is sufficiently narrow that a large number of loss peaks are resolved. The highest intensity peaks labeled FK1, FK2, and FK3 are termed *Fuchs–Kliewer phonons* or surface vibrations such that $\varepsilon_1(\omega) = -1$. The loss frequencies of the lower intensity peaks correspond to the vibrational frequencies of adsorbed H with Ga or N atoms. (Note: 1 meV equals 8.065 cm^{-1}.) For adsorbates to Ga in the GaN surface, the motion is entirely in the adsorbate, whereas for adsorbates to the lighter N atoms, frequencies are corrected for center of mass motion of both adsorbate and substrate atoms. Also note the weak vibrational modes due to residual C and O on the semiconductor surface. Figure 10.2b shows the vibrational modes for deuterium. The lower adsorbate frequencies are due to deuterium's higher mass. The presence of modes involving both Ga and N shows that H and D adsorb on both Ga and N surface atoms.

Equation 10.1 shows that energy loss requires coupling between the displacement vector **D** of the moving electron with the electric dipole of the excitation. This requirement introduces dipole selection rules for surface scattering (to within the approximations used for dielectric response theory) that are useful in determining molecular orientation on surfaces.

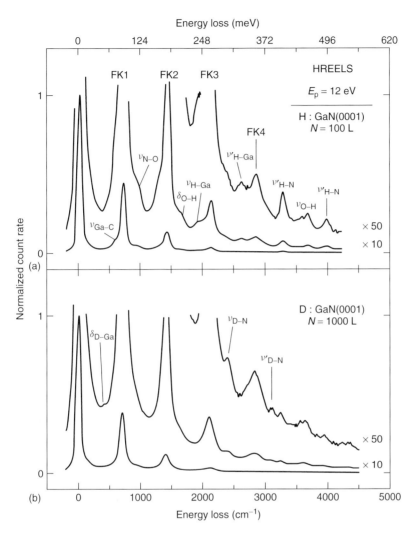

Figure 10.2 HREELS spectrum of GaN(0001) surfaces after H (a) and D (b) exposures [2].

Figure 10.3 illustrates molecular vibrations either normal or parallel to a metal surface. Vibrating molecules introduce dynamic dipoles as their bonds stretch and charge redistributes within the molecule. Such dipoles near metal surfaces induce image dipoles in the metal equidistant from the metal–vacuum interface. For molecules and their vibrations oriented normal to the surface, the image dipoles enhance the overall charge separation, whereas for vibrations parallel to the surface, the effective dipole charges are partially compensated. Thus one obtains significant energy losses for excitations with dynamic dipoles normal to the surface. Coupling to dipoles parallel to the surface is possible for oblique incident beams and off-specular data collection.

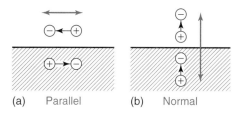

Figure 10.3 Schematic illustration of molecule vibrations normal (parallel) to a metal surface, which couple strongly (weakly) to normal incidence electron beams, and which are enhanced (partially compensated) by their image dipoles.

(a) Parallel (b) Normal

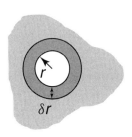

Figure 10.4 Spherical shell of radius r and thickness δr within a dielectric medium (light shading). Fluctuations in shell radius lead to a restoring force at a characteristic plasmon frequency.

10.4
Bulk and Surface Plasmon Scattering

ELS is effective in detecting plasmons, another type of collective excitation that reflects the charge density of valence and, for highly n-type crystals, conduction band electrons in the solid.

Consider the electrons orbiting atoms in a crystal. Within a shell of radius r, the number of electrons is equal to $4\pi r^2 \delta n$ for an incremental thickness δr, and electron concentration n. See Figure 10.4.

Fluctuations in charge density δn correspond to expansions or contractions in the charge at this shell radius, leading to an electric field

$$\mathcal{E} = e \cdot \frac{\delta n}{r^2} = e \cdot 4\pi r^2 \delta r \cdot \frac{n}{r^2} = 4\pi n e \, \delta r \tag{10.4}$$

The restoring force for this electric field is

$$F = -e\mathcal{E} = -4\pi n e^2 \delta r = m \frac{d^2 \delta r}{dt^2} \tag{10.5}$$

since force also equals mass times acceleration. Equation 10.5 is a second-order differential equation with the form of a harmonic oscillator. Thus

$$\frac{d^2 \delta r}{dt^2} = \frac{-4\pi n e^2 \delta r}{m} = -\omega_p^2 \delta r \tag{10.6}$$

where

$$\omega_p = \left(\frac{4\pi n e^2}{m} \right)^{1/2} \tag{10.7}$$

10.4 Bulk and Surface Plasmon Scattering

and the dielectric permittivity ε_0 is taken as 1. The solution to Equation 10.6 is $\delta r = r_0 e^{i\omega_p t}$ so that the charge density oscillates with an amplitude r_0 and a frequency ω_P.

Example 10.1. Calculate the plasmon frequency for magnesium and aluminum.

Plasmon frequency $\omega_P = (4\pi n e^2 / m)^{1/2}$

$$e^2 = 1.44 \times 10^{-7} \text{eV cm}$$
$$= 1.44 \times 10^{-7} \text{eV cm} \times (1.6 \times 10^{-12} \text{ erg eV}^{-1})$$
$$= 2.3 \times 10^{-19} \text{erg cm} \quad (10.8)$$

To calculate the electron density, we first need the number of atoms N per unit volume. To determine N, we use *Avogadro's number* $N_A = 6.023 \times 10^{23}$ atoms mol^{-1}, the density ρ, and the *atomic weight A* in the relation

$$N = \frac{N_A \rho}{A} \quad (10.9)$$

For Mg,

$$N = \frac{(6.023 \times 10^{23} \text{atoms mol}^{-1})(1.74 \text{ g cm}^{-3})}{24.31 \text{ g mol}^{-1}}$$
$$= 4.3 \times 10^{22} \text{atoms cm}^{-3} \quad (10.10)$$

Taking only two outer shell electrons per Mg atom, one estimates the electron density $n = 8.6 \times 10^{22}$ cm^{-3}. Then

$$\omega_P = \left[\frac{4\pi (8.6 \times 10^{22} \text{cm}^{-3})(2.3 \times 10^{-19} \text{erg cm})}{(9.1 \times 10^{-28} \text{g})} \right]^{1/2} = 1.65 \times 10^{16} \text{s}^{-1} \quad (10.11)$$

$$\hbar\omega_P = (1.0626 \times 10^{-27} \text{erg sec})(1.65 \times 10^{16} \text{s}^{-1}) = 1.75 \times 10^{-11} \text{erg} = 10.9 \text{ eV} \quad (10.12)$$

This compares with 10.6 eV for the zero-wave-vector value of 10.6 eV for Mg obtained from optical reflectance [3].

For Al, the electron density is calculated similarly. Here,

$$N = \frac{(6.023 \times 10^{23} \text{atoms mol}^{-1})(2.71 \text{ g cm}^{-3})}{26.98 \text{ g mol}^{-1}}$$
$$= 6.0 \times 10^{22} \text{atoms cm}^{-3} \quad (10.13)$$

With three outer shell electrons per Al atom, one estimates the electron density $n = 18.1 \times 10^{22}$ cm^{-3}. Then

$$\omega_P = \left[\frac{4\pi (18.1 \times 10^{22} \text{cm}^{-3})(2.3 \times 10^{-19} \text{erg cm})}{(9.1 \times 10^{-28} \text{g})} \right]^{1/2} = 2.39 \times 10^{16} \text{s}^{-1} \quad (10.14)$$

$$\hbar\omega_P = (1.062 \times 10^{-27} \text{erg sec})(2.39 \times 10^{16} \text{s}^{-1}) = 2.54 \times 10^{-11} \text{erg} = 15.9 \text{ eV} \tag{10.15}$$

This compares with 15.3 eV for the zero-wave-vector value for Al obtained from high energy electron energy loss [4]. Both values are close to those measured experimentally, notwithstanding the crude approximation for the electron density and a unity dielectric constant.

Example 10.1 shows that bulk plasmon energies are in the range 10–20 eV owing to the high density of outer shell electrons in these metals. For surface plasmons, the restoring force at the surface is only half that of the bulk since only a semi-infinite half plane of dielectric is present.

Since

$$F_{\text{surface}} = \frac{1}{2} F_{\text{bulk}} \tag{10.16}$$

the force given by Equation 10.6 becomes

$$\frac{d^2 \delta r}{dt^2} = \frac{-2\pi n e^2}{m \delta r} = -\omega_{SP}^2 \delta r \tag{10.17}$$

and

$$\omega_{SP} = \frac{\omega_P}{\sqrt{2}} \tag{10.18}$$

Thus, surface plasmons can be readily identified by their simple proportional relationship to their bulk counterparts.

Since metals are good electrical conductors, it is not surprising that the electron densities of simple metals are relatively high. Now consider the plasmons associated with the valence electrons in a semiconductor. For silicon, there are four atoms per unit cell on an face centered cubic (FCC) sublattice. Since there are two sublattices per unit cell over all, there are eight atoms per unit cell. Again using Equation 10.9,

$$N = (6.023 \times 10^{23} \text{atoms mol}^{-1})(2.33 \text{ g cm}^{-3})/28.09 \text{ g mol}^{-1})$$
$$= 5.0 \times 10^{22} \text{atoms cm}^{-3} \tag{10.19}$$

(Note: the unit cell dimensions provide another way to calculate N. In this case, the cubic unit cell has length $a = 5.43$ Å so that $8/a^3 = 8/(5.43 \times 10^{-8} \text{ cm})^3 = 5 \times 10^{22}$ atoms cm^{-3}).

With four electrons per atom, one obtains $n = 2 \times 10^{23}$ cm^{-3}. In other words, the electron density for silicon is approximately that of Al. The corresponding plasmon frequency is 16.7 eV, close to measured values ranging from 16.4 to 16.9 eV. Thus plasmon frequencies for both metals and semiconductors lie within an easily accessible region of energy for ELS.

We now consider conduction band electrons. Free carrier densities are typically in the range of $10^{15} - 10^{18}$ cm^{-3} so that bulk plasmon energies will be 2–3 orders of magnitude smaller. Even for a heavily degenerate semiconductor with $n \sim 10^{20}$ cm^{-3}, $\hbar\omega_P$ is only ~ 1 eV, well below the range for bulk and surface valence electron plasmons.

10.5
Interface Electronic Transitions

The ELS technique provides information on electronic transitions between filled and empty states as well as their collective excitations in bulk materials. Their energies depend on the momentum transferred during the excitation, and ELS at high incident beam energies, for example, hundreds of kilovolts, can provide enough momentum transfer to probe this *dispersion relation* across significant regions of the Brillouin zone [5]. At low excitation energies, ELS can probe the same interband and collective excitations, albeit heavily weighted toward near the Brillouin zone center (wave vector $k \cong 0$). In addition, because ELS is sensitive to surface excitations, it provides useful information about the dielectric properties of thin films and their interfaces.

The ELS sensitivity to both bulk and surface excitations enables researchers to probe interfaces as well. The procedure to accomplish this involves (i) preparing clean substrates in UHV, (ii) depositing overlayers by evaporation or by any other technique that avoids exposure to contaminants, and (iii) acquiring ELS data as a function of deposited thickness, layer by atomic layer. Measurements with atomic layer precision are needed to measure interface properties on this scale. As the photoemission results of Section 7.8.2 have shown, metal overlayers on semiconductors produce local changes that may be confined to only a few atomic layers. Beyond this thickness, the properties of the overlayer rapidly become bulklike. This is especially the case for deposition at room temperature and below. Since low energy ELS is sensitive to only a few atomic layers, it is no longer sensitive to the interface region when the overlayer thickness exceeds this range. Thus ELS spectra as a function of atomic layer thickness provide information on the bulk substrate, the interface region, and the growing bulk overlayer.

Figure 10.5 illustrates the measurement of these three regions for a metal–semiconductor interface [6]. Here a clean CdSe semiconductor surface is prepared by cleaving a bulk crystal in UHV, then exposing it to a beam of evaporated Al atoms. Analogous to Figure 8.4, Al deposited on clean CdSe bonds strongly to Se surface atoms whereas subsequent Al bonding becomes increasingly metal-like, that is, Al–Al bonds. Figure 10.5 shows a corresponding change in ELS features as a function of metal overlayer thickness. The clean CdSe surface exhibits $-d^2 N(E)/dE^2$ features associated with interband transitions at 3.0, 5.0, and 6.1 eV, a bulk plasmon at 17.2 eV, a Cd 4d-level transition to the conduction-band-density-of-states (DOS) maximum at 13.5 eV, and a 10.5 eV peak attributed to the CdSe surface plasmon. These energies agree with previous ELS measurements but their identification with surface versus bulk transitions may be more complex. Evaporation of 2 Å of Al alters the CdSe ELS features. It also removes the surface low energy electron diffraction (LEED) pattern and produces an AES LMM feature at 64 eV. At higher, metallic thicknesses, this feature shifts to 69 eV.

These ELS features are also different from those of the thicker Al film. With increasing Al deposition, new ELS peaks appear which correspond to coupled plasmons of the metal–semiconductor interface as well as the bulk Al plasmon.

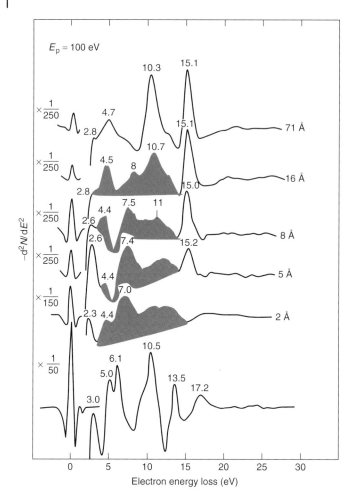

Figure 10.5 Electron energy loss spectra of Al deposited on cleaved (11 2̄0) CdSe for various thicknesses of metal deposition [6].

All CdSe features are removed by the Al overlayer. With 5 Å Al deposition, a new peak appears at 15.2 eV, which corresponds to the bulk plasmon of Al [4], along with its surface analog at 10–11 eV. The appearance of Al ELS features at this thickness indicates that the reacted interface layer is only one-to-two monolayers thick.

Thus several surface science probes indicate that a chemical reaction has taken place at the metal–semiconductor interface that is localized to only a few atomic layers. With a reacted interface layer, the metal–semiconductor junction can be viewed as a four-medium structure as shown schematically in the inset of

Figure 10.6. Here an intermediate reacted layer involving Al, Cd, and Se separates the Al metal overlayer from the CdSe substrate.

Within this scheme, the new ELS features at intermediate metal coverages in Figure 10.5 correspond to coupled plasmon modes, and a coupled mode analysis of their energy variation with thickness can provide the dielectric constant of the interface layer.

The effective dielectric constant of each medium has the form $\varepsilon_i = 1 + \omega_i^2/(\Delta_i^2 - \omega^2)$ [7], where $i =$ M (metal), I (interface), S (semiconductor), and V (vacuum). In this case, $\omega_M = 15.1$ eV and $\Delta_M = 0$ for Al, $\omega_S = 17.2$ eV and $\Delta_S = 7.85$ eV for CdSe (assuming a low-frequency dielectric constant $\varepsilon_0 = 5.8$) [8], $\omega_V = 0$ for vacuum, and ω_I and Δ_I are fitting parameters for the dielectric constant of the interface layer. From the continuity of normal electric displacement and transverse electric field across the boundaries, one obtains the following equation, valid for $k > \sim k_{\min} \gg \varepsilon_i^{1/2}\omega/c$,

$$\frac{\varepsilon_I}{\varepsilon_M} \cdot \frac{(1 - \delta e^{-2k\tau})}{(1 + \delta e^{-2k\tau})} \cdot \frac{(\gamma e^{-2kD} + 1)}{(\gamma e^{-2kD} - 1)} = 1 \tag{10.20}$$

where $\gamma = (\varepsilon_M - 1)/(\varepsilon_M + 1)$, $\delta = (\varepsilon_I - \varepsilon_S)/(\varepsilon_I + \varepsilon_S)$, D is the overlayer metal thickness, and τ is the reacted layer thickness. Since a bulk Al plasmon appears at $D + \tau = 5$ Å, τ is taken as 4Å. The solutions of Equation 10.20 can be identified with the various boundaries from roots of the equations $\varepsilon_i = -\varepsilon_j$ valid in the limit of large k, for each set i,j of semi-infinite adjoining media. This identification is possible because of the low dispersion at large wave vector. Given the form of ε_M, ε_I, and ε_S, Equation 10.20 has five solutions. These are labeled VM, (MI)$_1$, (MI)$_2$, (IS)$_1$, and (IS)$_2$ for the boundaries between vacuum, metal, reacted layer, and semiconductor. By varying only ω_i and Δ_i to fit these solutions to the experimental peak positions in Figure 10.6, one obtains $\omega_I = 14.0$ eV and $\Delta_i = 7.2$ eV. The roots of Equation 10.20 with this fit of ω_I and Δ_i yield the energy-versus-thickness dependence shown in Figure 10.6.

As shown, the ELS peak energies quickly converge to thick-film limits, characteristic of the large wave-vector scattering. The observed modes at 4.5, 7.5, and 10.3 eV show close agreement with three of the fitted curves. The IS$_1$ and IS$_2$ modes of the reacted layer are quickly attenuated by the thickening metal overlayer. The 7.5 eV mode is evidence for an interface layer with $\varepsilon_I \neq \varepsilon_S$ or ε_M since the modes calculated for a vacuum/bulk Al/CdSe structure lie at 4.8, 10.7, and 17.3 eV only. In addition, this fit is unique in that the solutions of Equation 10.18 cannot be fitted without an intermediate layer [6].

The fit of $\omega_I = 14.0$ eV and $\Delta_i = 7.2$ eV provides the low-frequency dielectric constant $\varepsilon_I(0) = 1 + \omega_I^2/\Delta_I^2 = 4.9$, significantly different from the 5.8 dielectric constant for CdSe. The lower dielectric constant is consistent with formation of a more ionic interface compound, similar to, for example, Al$_2$O$_3$ for which $\varepsilon(0) = 3.1$. Similar results are obtained for other metal–semiconductor interfaces. For example, one finds for the Al/CdS interface, $\varepsilon(0)_{\text{Al/CdS}} = 5.0$ versus $\varepsilon(0)_{\text{CdS}} = 5.5$ [6]. Likewise, Al on a clean ZnS (110) surface produces an Al–S interfacial layer and free Zn atoms that segregate to the free Al surface, analogous to Figure 9.14. This

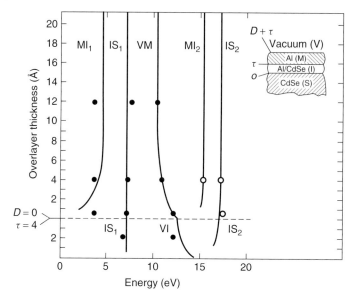

Figure 10.6 Predicted ELS energies (solid lines) with $\omega_l = 14.0$ eV and $\Delta_i = 7.2$ eV for the above four-medium interface compared with the observed ELS peak energies (●) and unresolved structure (○) as a function of metal overlayer thickness. MI, IS, and MV label mode energies associated with the metal–reacted layer, the reacted layer–semiconductor, and the metal–vacuum interfaces, respectively.

room-temperature interfacial reaction occurs over a 2–8 Å range of Al thickness that is accompanied by large chemical shifts of all three elements over the same thickness range.

The ELS technique provides evidence for room-temperature interfacial reactions at metal-III–V compound semiconductor interfaces as well. Thus the replacement reaction of Ga with Al at the Al–GaAs(110) junction produces AlAs loss features and dissociated Ga over a similar thickness range as those observed at Al/II–VI compound junctions [9].

10.6
Atomic-Scale Electron Energy Loss Spectroscopy

With the development of electron spectrometers for high energy transmission electron microscopes (TEM), it is now possible to measure energy losses of individual atoms with a resolution of several hundred meV. This microscope-based electron energy loss spectroscopy (EELS) capability enables researchers to probe electronic states within semiconductor and insulator bandgaps at interfaces on a monolayer by monolayer scale. These features can then be compared with theoretical calculations to model-specific charge distributions near the interface that are responsible for the localized state [10].

Similarly, TEM-based EELS can measure the losses near core-level threshold energies. The energy loss threshold indicates an individual atoms' charge state, providing another method of determining charge redistribution. This technique holds much promise for understanding ferroelectricity and magnetism at complex oxides interfaces, where charge, orbital nature, bond configuration, and spin are correlated.

10.7
Summary

Electron energy loss spectroscopy provides a number of powerful techniques for probing electronic and collective excitations at surfaces and interfaces. These include (i) at high energy resolution, identification of the surface bonding and vibrational modes of surface atoms and adsorbates; (ii) at intermediate energy resolution and higher energies, bulk and surface plasmon modes, interfacial dielectric constants, and interfacial reactions; and (iii) at lower energy resolution but extremely high spatial resolution, the charge state of individual atoms and the charge redistribution to form localized states.

Electrons lose energy in solids owing to transitions involving the dielectric medium's electronic band structure, its various elementary excitations such as plasmons, excitons, and phonons, and any boundary conditions imposed by physical dimensions. The energy transfer rate is proportional to the imaginary part of the solid's complex dielectric permittivity ε. This proportionality is modified at surfaces, where a half plane of $\varepsilon = 1$ is present.

HREELS with high energy resolution provides measurement of surface vibrational frequencies due to the surface atoms alone or bonded to adsorbates. Typical phonon frequencies range from tens to hundreds of meV. Higher energy loss features include bulk and surface plasmons whose characteristic frequencies depend on the density of electrons involved. Their energies lie in an energy range of a few to 20 V.

ELS has proved useful in identifying interface reactions between metals and semiconductors, detecting the presence of interfacial layers with new dielectric properties with only a few monolayers of deposited metal at room temperature. Interface electronic transitions due to coupled modes involving vacuum, metal, interfacial dielectric, and semiconductor can provide the dielectric constant of the reacted interface layer.

Electron losses measured using transmission electron microscopes can identify electronic transitions due to bandgap excitation and changes in chemical bonding on an atom-by-atom scale near interfaces.

References

1. Lüth, H. (2001) *Solid Surfaces, Interfaces, and Thin Films*, 4th edn, Chapter 4, Springer-Verlag, Heidelberg, New York.
2. Grabowski, S.P., Nienhaus, H., and Mönch, W. (2000) *Surf. Sci.*, **454–456**, 498.
3. Feuerbacker, B. and Fitton, B. (1970) *Phys. Rev. Lett.*, **24**, 499–502.
4. Gibbons, P.C., Schnatterly, S.E., Ritsko, J.J., and Fields, J.R. (1976) *Phys. Rev. B*, **13**, 2451.
5. Schnatterly, S.E. (1979) *Solid State Physics*, Vol. 34, Academic Press, Inc., New York, p. 275.
6. Brillson, L.J. (1977) *Phys. Rev. Lett.*, **38**, 245–248.
7. Barrera, R.G. and Duke, C.B. (1975) *Phys. Rev. B*, **13**, 4477.
8. Cardona, M. and Harbeke, G. (1965) *Phys. Rev.*, **137**, A1467.
9. Ludeke, R. and Esaki, L. (1975) *Surf. Sci.*, **47**, 132.
10. Windl, W., Liang, T., Lopatin, S., and Duscher, G. (2004) *Mater. Sci. Eng. B*, **156**, 114.

11
Rutherford Backscattering Spectrometry

11.1
Overview

Rutherford backscattering spectrometry (RBS) involves charged particle scattering in a Coulomb field. RBS yields chemical analysis and depth dependence of thick (micron-scale) multilayer films without ablating away the surface as in Auger electron spectroscopy (AES) depth profiling (discussed in Section 9.4) or secondary ion mass spectrometry (to be discussed in Chapter 12). The RBS technique involves a classical scattering process of high energy ions – for example, 2-MeV helium atoms – with atoms in the solid. A classical analysis is appropriate since the wavelength of the probing ions is much smaller than the range of the scattering potential. For example, according to the De Broglie relation,

$$\lambda = h/p = h/mv \tag{11.1}$$

where h is Planck's constant, m is the ion mass, and v its velocity. Velocity follows from $E_{\text{Kinetic}} = \frac{1}{2}mv^2$ and, for light ions with $E_{\text{Kinetic}} >$ MeV, yields wavelengths λ of the order of 10^{-4} Å. This is much smaller than the angstrom scale of interatomic distances and scattering potentials.

Example 11.1. Calculate the wavelength of a 2-MeV helium atom. Using $h = 6.63 \times 10^{-34}$ J-s, $m = 4 \times 1.67 \times 10^{-27}$ kg:

$$v = \left[\frac{2E_{\text{Kinetic}}}{m}\right]^{1/2} = \left[\frac{2(2 \times 10^6 \text{ eV})(1.6 \times 10^{-19} \text{ J (eV)}^{-1})}{(6.68 \times 10^{-27} \text{ kg})}\right]^{1/2}$$
$$= 9.8 \times 10^6 \text{ m s}^{-1}$$

$$\lambda = \frac{h}{mv} = \frac{6.63 \times 10^{-34} \text{ J-s}}{(6.68 \times 10^{-27} \text{ kg})(9.79 \times 10^6 \text{ m s}^{-1})}$$
$$= 1.01 \times 10^{-4} \text{ Å}$$

Surfaces and Interfaces of Electronic Materials. Leonard J. Brillson
Copyright © 2010 WILEY-VCH Verlag GmbH & Co. KGaA, Weinheim
ISBN: 978-3-527-40915-0

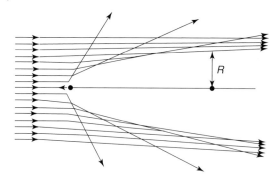

Figure 11.1 Calculated scattering trajectories of He$^+$ ions moving toward an oxygen atom from the left and the shadow cone of scattering He$^+$ ions they produce to the right of the target.

11.2
Theory of Rutherford Backscattering

Recall the theory of particle backscattering discussed in Chapter 9, which was used to describe electron scattering [1]. For Rutherford backscattering, the mass of the incoming particle is >1000, larger than the electron mass, and the positively charged ions scatter against the positive repulsion of the target nuclei. Figure 11.1 illustrates the charged particle scattering for positively charged He$^+$ ions from an oxygen atom. On the basis of an assumed Thomas–Fermi–Moliere scattering potential, He$^+$ ions approaching the target from the left produce a shadow cone of He$^+$ ions to the right of the target oxygen atom.

The closest approach of the incoming He$^+$ ion can be calculated by setting its kinetic energy $E_{kinetic}$ equal to the coulomb repulsion energy of the ion with the target oxygen atom.

$$V = \frac{(Z_1 Z_2 e^2)}{(4\varepsilon_0 \pi r_{min})} = E_{Kinetic} \tag{11.2}$$

Here Z_1 and Z_2 are the ionic charges of the ion and target. Thus the closest approach is

$$r_{min} = \frac{(Z_1 Z_2 e^2)}{(4\varepsilon_0 \pi E_{Kinetic})} \tag{11.3}$$

The higher the incident ion energy, the smaller the r_{min} and closer the approach to the atom.

Example 11.2. Calculate the closest approach of a 2-MeV helium atom to a silicon atom. Using Equation 11.3 with $Z_1 = 2$, $Z_2 = 14$, $E_{kinetic} = 2 \times 10^6$ eV, $e = 1.6 \times 10^{-19}$ C, and $\varepsilon_0 = 8.85 \times 10^{-12}$ F m^{-1}:

$$r_{min} = \frac{(2)(14)(1.6 \times 10^{-19} C)^2}{4\pi (2 \times 10^6 eV)(8.85 \times 10^{-12} F\ m^{-1})(1.6 \times 10^{-19} J\ (eV)^{-1})}$$
$$= 2.0 \times 10^{-4} Å$$

(Note: 1 Farad = 1 C V^{-1}.) This value is much smaller than a Bohr radius $a_0 = 0.53$ Å and the K-shell radius of Si, $a_0/14 \cong 3.7 \times 10^{-2}$ Å. Since the incoming atom can approach closer than the radii of electrons orbiting the target nucleus, an unscreened cross section for the Si target nucleus is justified.

Using conservation of energy and momentum, one can show that the scattered energy of the incoming particle is characteristic of the scattering atom. Because energy is conserved in the scattering process, the initial and final energies must be equal. Then

$$\frac{1}{2} M_1 v^2 = \frac{1}{2} M_1 v_1^2 + \frac{1}{2} M_2 v_2^2 \tag{11.4}$$

where M_1 and M_2 are the incident particle and target atoms, respectively, v and v_1 are the incident particle's velocity before and after scattering, respectively, and v_2 is the target atom's recoil velocity after the scattering event.

From the diagram in Figure 11.2b, conservation of momentum requires that

$$M_1 v = M_1 v_1 \cos\theta + M_2 v_2 \cos\varphi \tag{11.5a}$$
$$0 = M_1 v_1 \sin\theta + M_2 v_2 \sin\varphi \tag{11.5b}$$

From Equations 11.4 and 11.5, one can solve for the ratio of incident to final particle velocities in terms of just the scattering angle θ and masses M_1 and M_2.

$$\frac{v_1}{v} = \frac{[\pm(M_2^2 - M_1^2 \sin^2\theta)^{1/2} + M_1 \cos\theta]}{(M_1 + M_2)} \tag{11.6}$$

Likewise, the ratio of the final to initial incident particle energies (for $M_1 < M_2$) is

$$\frac{E_1}{E_0} = \left\{ \frac{[(M_2^2 - M_1^2 \sin^2\theta)^{1/2} + M_1 \cos\theta]}{(M_1 + M_2)} \right\}^2 \tag{11.7}$$

$E_1/E_0 = K$ is termed the *kinematic factor*.

A nuclear particle detector measures the scattered particle energy E_1 and velocity v_1 and at various angles θ. The detector measures energy in terms of ionized electron–hole pairs created by the incident particle within the depletion region of a reverse-biased Schottky diode. The higher the particle energy, the larger the number of electron–hole pairs and the larger the height of the pulse generated by the incident particle. The energy resolution of the particle detector is determined by statistical fluctuation in the number of electron–hole pairs, typically 10–20 keV for ^{4}He ions in the MeV energy range.

The number of pulses generated per second determines the intensity of scattering. The pulse width must be short enough to discriminate between pulses at high scattering rates. The full width at half maximum (FWHM) of such pulses will be determined by the electric field that sweeps out carriers as well as the physical thickness over which the carriers must traverse.

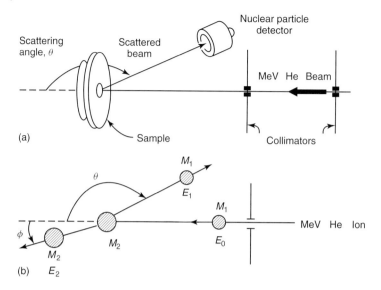

Figure 11.2 (a) Schematic diagram of RBS scattering experiment. A collimated incident beam of He atoms scatters off a sample at scattering angle θ into a nuclear particle detector. An incident particle with mass M_1, energy E_0 and velocity v collides with a target atom of mass M_2 at rest. (b) Diagram of energy and momentum conservation in a two-particle scattering event. After the collision, the incident particle scatters at angle θ, energy E_1, and velocity v_1. The target atom recoils at angle φ, energy E_2, and velocity v_2 [1].

The energy of the incident particle after scattering is determined only by the masses of the incident and target atoms and the scattering. Since M_1, v, and E_0 are known while v_1, E_1, and θ are measured, Equations 11.6 and 11.7 yield the mass M_2 of the target atom. For direct backscattering ($\theta = 180°$), the energy ratio E_1/E_0 is the lowest.

$$\frac{E_1}{E_0} = \left\{ \frac{(M_2 - M_1)}{(M_1 + M_2)} \right\}^2 \tag{11.8}$$

whereas for $\theta = 90°$,

$$\frac{E_1}{E_0} = \frac{(M_2 - M_1)}{(M_1 + M_2)} \tag{11.9}$$

A backscattering geometry produces the largest change in the scattered particle. This configuration is preferred to distinguish between target atoms of different mass M_2 since the E_1/E_0 ratio exhibits the highest possible change.

The yield ratio of the backscattered signals provides additional information. Since the scattering cross section from particle–particle scattering in Chapter 9 was derived as

$$\sigma(\theta) = \frac{(Z_1 Z_2 e^2 / 4E)^2}{\sin^4(\theta/2)} \tag{9.8}$$

the backscattered intensities for different elements in the same target material will be proportional to the ratio of their cross sections. The yield ratios of these elements are approximately proportional to these cross sections [1]. For the same incident ion, ion energy, and scattering angle θ in Equation 9.8, the scattering cross sections for two elements with atomic numbers Z_{2A} and Z_{2B} are then proportional to $(Z_{2A}/Z_{2B})^2$. The energy dependence of backscattered ions on target mass can be combined with this dependence of proportionality of scattering intensity on atomic number to characterize both the elements present and their distribution in depth within a material.

11.3
Depth Profiling

The RBS experiment involves a high energy accelerator such as that shown in Figure 11.3 that extracts specific ions, accelerates them, and focuses them on a specimen. Typically, light ions such as ionized He are used to minimize effects of electron screening of the target nuclei, maximize mass differences with target

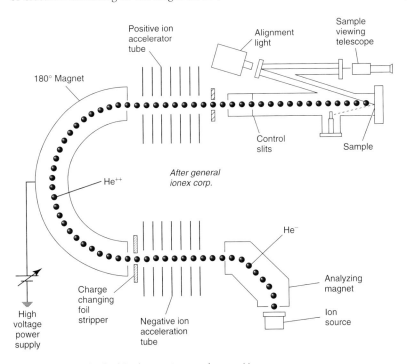

Figure 11.3 Rutherford backscattering accelerator. He atoms are ionized, charge changed, accelerated to a specific high positive voltage, and focused on a sample. The backscattered He ions are energy analyzed with a nuclear particle detector.

ions, simplify kinematics, and enable particle acceleration to high energies to increase the depth of penetration of the incident particles. Together, the energies and intensities of the backscattered ions provide concentration profiles of different elements as a function of depth from the free surface.

These primary ions lose energy continuously as they pass through a solid, mainly by electronic excitations such as plasmon creation. This continuous loss of energy along a path through a crystal provides a simple relation between scattering depth and energy loss. MeV He ions lose energy along their path at a rate $dE/dx \sim 30 - 60$ eV Å^{-1} [1]. For thin films, the total energy loss ΔE at a depth t is proportional to t.

$$\Delta E_{incident} = \int \frac{dE}{dx} \cong \frac{dE}{dx}\bigg|_{incident} \cdot t \tag{11.10}$$

Therefore, the energy of the incident particle with initial energy E_0 after reaching a given depth is

$$E(t) = E_0 - t \cdot \frac{dE}{dx}\bigg|_{incident} \tag{11.11}$$

Using the kinematic factor K defined in Equation 11.7, the particle energy after large angle scattering is $KE(t)$. The particle then loses energy along its outward path to the surface such that

$$E_1(t) = KE(t) - (t/|\cos\theta|)\frac{dE}{dx}\bigg|_{scattered}$$

$$= KE_0 - t\left(K\frac{dE}{dx}\bigg|_{incident} + (|\cos\theta|)^{-1}\frac{dE}{dx}\bigg|_{scattered}\right) \tag{11.12}$$

The energy width ΔE of the scattered particle after emerging from the solid is then

$$\Delta E = \Delta t\left(K\frac{dE}{dx}\bigg|_{incident} + (|\cos\theta|)^{-1}\frac{dE}{dx}\bigg|_{scattered}\right)$$

$$= \Delta t[S] \tag{11.13}$$

where $[S]$ is termed the *backscattering energy loss factor*. Note that the rates of energy loss dE/dx for incident and scattering particles are in general different because the particle energy changes significantly with scattering. A "surface energy approximation" that assumes constant energy losses $dE/dx|_{incident} = dE/dx|_{E_0}$ and $dE/dx|_{scattered} = dE/dx|_{KE_0}$ at E_0 and KE_0, respectively, leads to an energy width

$$\Delta E_0 = \Delta t[S_0] = \Delta t\left[K\frac{dE}{dx}\bigg|_{E_0} + (|\cos\theta|)^{-1}\frac{dE}{dx}\bigg|_{KE_0}\right] \tag{11.14}$$

This approximation is valid for thicknesses over which dE/dx does not change appreciably. It can be replaced by a mean energy approximation for longer path lengths. Monte Carlo simulations of energy loss per unit path length take such differences into account explicitly. Notwithstanding the limited range of typical dE/dx values, these approximations highlight the need for accurate experimental

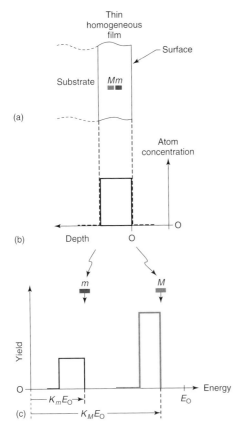

Figure 11.4 Conversion of atomic composition versus depth for a homogeneous binary film to an RBS yield versus energy spectrum. (a) Schematic of homogeneous film composed of elements A and B with atomic masses M and m. (b) Atom concentration for both masses m and M. (c) Yield versus energy showing energy separation and intensities for yields of masses m and M. $K_M E_0$ and $K_m E_0$ signify leading edges of scattering energies for masses M and m, respectively [2].

values of energy loss rates for a given particle in a particular matrix at a specific energy.

Figure 11.4 illustrates the use of the profile width to obtain thickness and the profile energy to identify mass of atoms in a given matrix [2]. Here a thin homogeneous film of a binary compound composed of elements with a heavy and a light atomic mass, M and m, respectively, is shown in Figure 11.4a. Figure 11.4b shows that both elements have equal concentration and the same depth distribution. Figure 11.4c demonstrates the separation of the M and m yields versus energy. The yield for atoms of mass M versus mass m is higher by a factor of $\cong (Z_M/Z_m)^2$ since the scattering cross section scales according to Equation 9.8. Furthermore, the energy of backscattered particles is higher for atoms of mass M versus m

according to Equation 11.8. Finally, the widths of the two yield profiles differ due to differences in the energy loss per unit path length dE/dx in the range of energies $K_M E_0$ and $K_m E_0$, where K_M and K_m are the kinematic factors from Equation 11.7 These dE/dx rates can be calibrated for a given element of known thickness in its KE_0 energy range. The continuous energy loss of a scattered particle along its path through a crystal thus provides a simple relation between scattering depth and energy loss.

11.4
Channeling and Blocking

The crystallinity of materials under study strongly affects the scattering yield of incoming particles. This is due to the regular lattice sites in crystalline materials and the large open "channels" through which particles can travel without scattering. Figure 11.5 illustrates the channels between atoms in a crystal.

Particles incident on a crystal along directions of high symmetry such as that shown here will encounter fewer target atoms as they traverse the crystal. This RBS feature provides a useful indicator of specimen crystallinity. Comparison between specimens with low densities of lattice defects and those prepared under different growth and/or processing conditions provides a measure of lattice crystallinity. Such measurements are particularly useful for microelectronic structures in which dopant atoms are implanted as one of the steps in preparing active devices. Since the implantation process involves some lattice disruption as ions pass into the solid, the lattice must be subsequently annealed to recover its prior crystallinity. Thus RBS provides a measure of this recovery process.

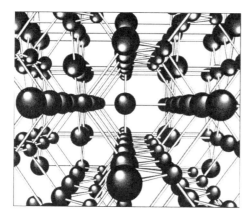

Figure 11.5 Open lattice structure of a crystalline solid and the channels between columns of atoms. Scattering yields along these symmetry directions are relatively low. (Courtesy C. Palmstrøm).

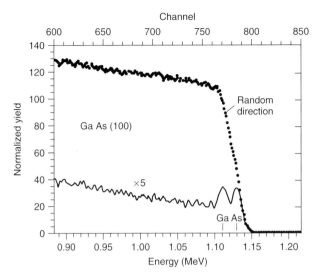

Figure 11.6 RBS yield spectra of 1.4-MeV He$^+$ for a GaAs(100) surface along a random nonchanneling direction and along the surface <100> normal direction. The channeling spectrum is an order of magnitude lower in amplitude [2].

Figure 11.6 illustrates the difference in backscattering for particles directed along a high symmetry direction versus in a random direction. Note the close proximity and intensities of the Ga and As yield peaks at their leading energy edges, consistent with their small difference in atomic mass. Besides these peaks, there is a continuous background yield due to scattering at lower depths. For high symmetry directions, the leading edge yields are well defined and strong. For a random direction, scattering takes place by atoms throughout the bulk rather than by atoms only near the surface. This results in yields that can be 2 orders of magnitude higher. High and continuous backscattered yield spectra are certainly typical for amorphous solids.

Scattering along symmetry directions affords another useful tool. As Figure 11.5 shows, ions incident along channeling directions scatter primarily from surface and near-surface atoms at the head of each column. This feature increases the surface sensitivity of RBS for studying the rearrangement of surface atoms, that is, *surface reconstruction*, as well as the position of atomic adsorbates. Such scattering experiments are performed at medium energies in order to minimize the ablation of surface atoms under study. Atoms adsorbed at preferential lattice sites "shadow" the atoms over which they sit on the surface. Hence changes in RBS yields for substrate atoms before versus after adsorption of an impurity can indicate which lattice sites the impurity atoms attach to.

RBS also indicates the position of impurity atoms *inside* the lattice. This is important since the electronic activity of dopants inside semiconductors depends on their lattice position. This requires substitution of particular lattice atoms with dopant atoms in order to increase or decrease charge valency for donors or acceptors, respectively. To begin with, scattering kinematics can separate the signals of the impurity versus the lattice atoms for impurities with a larger atomic mass than those of the host atoms. Furthermore, impurity atoms in substitutional lattice sites will scatter only minimally for incident particles along channeling directions since they are in general shadowed by lattice atoms. As the angle of incidence is varied away from a channeling direction, scattering from substitutional impurities increases. On the other hand, scattering from nonsubstitutional atoms or clusters of atoms is constant versus incident ion angle. For a detailed analysis of RBS channeling, see [1].

11.5
Interface Studies

RBS has proved to be very useful in understanding metallurgical reactions between metals and semiconductors incorporated into microelectronics. Metal contacts to semiconductors are chosen for a number of characteristics that include thermal stability at the high temperatures typically employed in the fabrication process. The final device structure must also be stable at elevated temperatures during normal operation over the life of an electronic device, typically more than a few years. Metals can react or interdiffuse at their interfaces with semiconductors. Such chemical interactions can extend over distances as long as microns. With the ever-decreasing scale of microelectronic circuits, semiconductor structures must remain unchanged on a scale of nanometers in order to preserve the electrical properties designed for a specific microelectronic device.

Figure 11.7 illustrates how RBS can measure the stability of metal–GaAs interfaces [3]. For the Au–GaAs junction, the GaAs yield profile occurs at lower energies than that of Au due to the lower masses of Ga and Au. Before annealing, only the GaAs and Au profiles are evident. After annealing, however, a new Ga peak appears at energies higher than $K_{Ga}E_0$ for the original 1500-Å depth. This feature signifies Ga diffusing out of the GaAs lattice into the Au. Likewise, the extended yield "tail" of Au atoms extending to lower energies signifies Au atoms that have diffused into the GaAs. Taken together, these features represent strong Au–Ga interdiffusion. The new Ga and Au features extend over ranges of energies that correspond to hundreds of angstroms, signifying that a macroscopic chemical interaction has taken place at elevated temperatures such that an alloy forms between the Au and GaAs.

On the other hand, the W–GaAs junction shows no sign of thermal degradation, even though the interface was annealed to a much higher temperature. The difference between these two junctions is consistent with the existence of a Au–Ga eutectic, the temperature at which all the constituents crystallize simultaneously from a molten liquid solution, at 360 °C. Thus Au and Ga have a thermodynamic

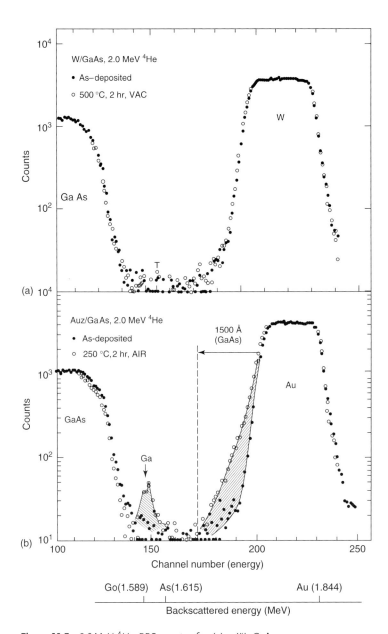

Figure 11.7 2.0 MeV ^{4}He RBS spectra for (a) a W–GaAs interface before and after annealing at 500 °C for 2 hours versus (b) a 1500 Å Au–GaAs before and after annealing at 250 °C for 2 hours. Strong interdiffusion between Au and Ga is evident for Au whereas no such thermal degradation is apparent for W even with higher annealing temperature [3].

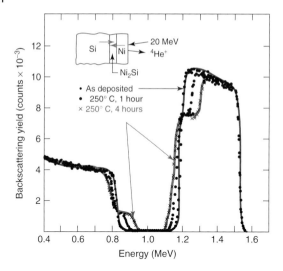

Figure 11.8 20 MeV ^{4}He$^+$ RBS profiles for a Ni film on Si and the formation of Ni$_2$Si at their as-deposited interface, with annealing at 250 °C for 1 hour, and at 250 °C for 4 hours [4].

driving force for forming alloys. However, for W with Ga or As, no such driving force exists.

Besides the extent of chemical interaction, the RBS depth profile reveals how the atoms move at the interface. Figure 11.8 illustrates the interaction between a Ni thin film and Si. Charge transport into and out of semiconductors depends on ohmic contacts with low energy loss at the metal–semiconductor interface. Ni contacts were used in Si microelectronics since they form a low-resistivity Ni silicide at the Ni–Si interface. However, Ni can form various compounds with Si with different stoichiometries and resistivities. As will be discussed in later chapters, such reaction products will depend on annealing temperature and the supply of the constituents.

RBS provides depth profiles that show how these atoms interact at different temperatures. Figure 11.7 shows how the RBS profile changes with annealing temperature at the Ni–Si interface [4]. The as-deposited Ni–Si interface displays only the Ni and Si yield profiles. With a 250 °C anneal for 1 hour, the leading edge of the Si yield profile advances to higher energies, signifying Si outdiffusion into Ni. The Ni profile extends to lower energies and decreases at energies corresponding to depths within the metal film. The relative changes in yield intensities for these two elements reveal not only that a reaction takes place at their interface but also the relative proportions of the two constituents that form the interface reaction product. In this case, there is twice as much change in Ni as in Si, indicating the formation of Ni$_2$Si.

11.6
Summary

RBS provides a relatively nondestructive tool for measuring the composition of electronic materials and the distribution of different elements as a function of depth from the free surface.

The RBS technique is based on classical kinematics so that the specific ratios of different atoms and their positions can be derived from simple calculations.

RBS is particularly useful in studying crystalline materials, where the regular positions of atoms permit channeling of the impinging particles and strong reduction in scattering yields along high symmetry directions.

The shadowing of atoms along different symmetry directions allows researchers to identify the positions of impurity atoms in the crystal lattice as well as adsorbates bonded to surface atoms.

The RBS technique has been integral in developing metal contacts to microelectronics. By monitoring yield profiles as a function of time and temperature on a scale of tens of nanometers, researchers have identified process conditions that minimize unwanted reactions or that achieve reacted species with desirable properties.

Problems

1. Calculate the velocity and wavelength of a 3-keV O atom.
2. Prove that the K-shell radius of an element with atomic number Z equals a_0/Z, where a_0 is the Bohr radius($\hbar^2/me^2$). Hint: use the balance of electrostatic and centripetal forces and the condition that momentum $mvr = n\hbar$ is quantized. What is the Bohr radius for S?
3. Calculate the closest approach of a 1-MeV helium atom to a gold atom. How does it compare to the atom's radius?
4. A Rutherford backscattering experiment involves incident 2-MeV He$^+$ ions backscattered from a SiO$_2$ target. (a) Calculate the kinematic factors for the backscattering of each element. (b) Calculate the yield ratio of Si versus O.
5. Draw the backscattering spectrum for 2-MeV He$^+$ ions from a Ge target alloyed with 10% Au.
6. For a Au overlayer of thickness 1500 Å on GaAs and an energy loss per path length of 60 eV Å^{-1}, draw the backscattering spectrum for 2-MeV He$^+$ ions. How does this compare with Figure 11.6?

References

1. Feldman, L.C. and Mayer, J.W. (1986) *Fundamentals of Surface and Thin Film Analysis*, Chapter 2, Prentice Hall, Upper Saddle River.

2. Chu, W.K., Tu, K.N., and Mayer, J.W. (1978) *Backscattering Spectrometry*, Academic Press, New York.
3. Sinha, A.K. and Poate, J.M. (1973) *Appl. Phys. Lett.*, **23**, 666.
4. Chu, W.K., Tu, K.N., and Mayer, J.W. (1975) *Thin Solid Films*, **25**, 403.

12
Secondary Ion Mass Spectrometry

12.1
Overview

The secondary ion mass spectrometry or SIMS technique is one of the most sensitive methods to detect impurities in solids. Depending on the element and its host material, the SIMS technique is capable of detecting concentrations as small as parts per billion. For a nominal atomic concentration of 10^{22} cm^{-3}, this detection limit translates to impurity levels of 10^{13}–10^{15} cm^{-3}. This concentration level is at the low end of doping densities in semiconductors, corresponding to intrinsic, that is, undoped, carrier densities in most semiconductors. Thus, SIMS provides a useful tool to determine concentration levels of dopants intentionally introduced through implantation or growth as well as unintentionally incorporated impurities that could alter the desired carrier densities. The detectivity of SIMS is several orders of magnitude better than most electron spectroscopies. However, cross sections for photoelectrons excited from atomic core levels provide a level of quantitative precision in X-ray photoemission spectroscopy (XPS), ultraviolet photoemission spectroscopy (UPS), and soft X-ray photoemission spectroscopy (SXPS) that, unlike SIMS, is relatively independent of the host matrix.

12.2
Principles

The SIMS process involves a sputtering process in which a primary ion beam with energy in the range of 1–10 keV impinges on a surface. Typical primary ions are Ar$^+$, Cs$^+$, and O$^-$. The energy of impact transfers to a secondary ion, which is ejected and analyzed by a mass spectrometer. The primary ion creates single-particle collisions with atoms adsorbed on the surface as well as atoms within the underlying solid. The resultant cascade shown in Figure 12.1 introduces lattice defects, implants ions and removes atoms from the target solid as neutral or charged secondary ions. Depending on the primary ion, adsorbate ions may emerge from the solid with negative or positive charge and either as individual atoms or in combination with target atoms at the surface. Similarly, target atoms below the surface may leave the solid with either charge, depending on the primary ion.

Surfaces and Interfaces of Electronic Materials. Leonard J. Brillson
Copyright © 2010 WILEY-VCH Verlag GmbH & Co. KGaA, Weinheim
ISBN: 978-3-527-40915-0

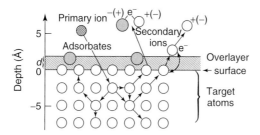

Figure 12.1 Schematic representation of the sputtering process. An initial or primary ion impinges on a surface, creating a cascade of single-particle collisions. This cascade removes surface atoms adsorbed to the surface as well as atoms in the underlying material.

In the *static* version of SIMS, the primary ion sputters atoms only from the topmost layer. In this technique, the primary ion current is limited to 10^{-9}–10^{-10} A cm^{-2}. Assuming singly charged ions, this current translates to less than 10^{10} ions/cm^2 s. Compared with a surface atom density of 10^{14}–10^{15} cm^{-2}, this ion current density is sufficiently low to remove only a fraction of surface atoms over sampling periods of minutes. Thus, the analysis is complete before significant fractions of the surface layer are disrupted. A rule of thumb is that only 0.1% of surface atomic sites should be bombarded during the measurement. Since surface atomic densities are roughly 10^{15} cm^{-2}, the primary ion dose should be below 10^{12} cm^{-2}. Detection limits can be $\sim 10^9$ cm^{-2}.

Figure 12.2 illustrates a scanning tunneling microscopy image of an atomically ordered surface of Si before and after a static secondary ion mass spectrometry (SSIMS) measurement [1]. The surface after SSIMS shows evidence for surface atom removal but still retains the ordered atomic pattern.

In the dynamic version of SIMS, higher energy ions such as isotopically pure ^{69}Ga$^+$ sputter through multiple layers into the substrate. Analysis of the spectra in this sequential removal process results in an atomic layer-by-layer depth profile

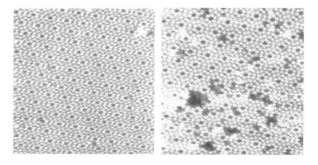

Figure 12.2 Scanning tunneling microscopy image of Si(7×7) surface before and after exposure to 2×10^{12} ions/cm^2 [1].

of the elements in the solid. Typical ion currents are 10^{-4}–10^{-5} A cm^{-2}, orders of magnitude higher than in static mode. In the dynamic secondary ion mass spectrometry (DSIMS) mode of analysis, the primary ion sputters into the sample, producing a crater. A second beam of either positive or negative ions is focused on a relatively small spot in the flat center of the crater to produce the secondary ions.

Emission of secondary ions occurs from surface areas that are proportional to their size. For atoms, this is on the order of 1 nm. For small molecules, it is 5 nm, and for large molecules, 10 nm or more.

12.3 SIMS Equipment

At a minimum, the SIMS technique requires equipment to (i) produce a focused beam of primary ions, (ii) extract, steer, and focus the resultant secondary ions, and (iii) detect and analyze the secondary ions. Figure 12.3 illustrates an example of a sector field magnetic mass separator, a type of ion mass spectrometer.

A focused beam of primary ions is produced by first ionizing neutral atoms in an ionization chamber and extraction voltage that accelerates the ionized atoms through a quadrupole lens. The lens focuses these ions, now with much higher kinetic energy, so that they travel through a magnetic sector field. Figure 12.3 shows this magnetic field deflecting the primary ions as they move toward the target, thereby removing any unintentional impurities. Once inside the ultrahigh vacuum (UHV) chamber, the ions that strike the target produce secondary ions that are separated in mass either by another magnetic sector or by another type of mass spectrometer, a quadrupole analyzer.

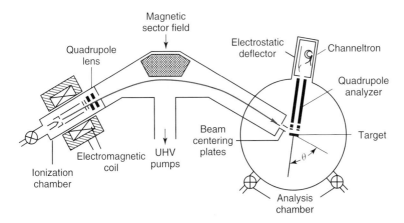

Figure 12.3 Secondary ion mass spectrometer components required for elemental analysis and trajectory of secondary ions.

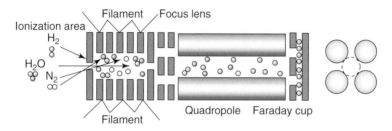

Figure 12.4 Residual gas analyzer. Filaments ionize the molecules entering the analyzer with voltages applied to the quadrupole rods filter ions with different mass-to-charge ratio. Adapted from [2].

Depending on the voltages applied to the quadrupole analyzer, only ions with a specific mass are deflected through the analyzer into a detector housing and an electron multiplier or *channeltron*, which multiplies the charge of the particle impinging on its plates. This multiplication process is analogous to that in a photomultiplier. An electrostatic mirror reflects ions into the multiplier, which is mounted off axis to minimize neutral particles or photons in direct line of sight to the target that could excite additional charge. An alternating voltage superimposed on a constant bias is applied to the quadrupole rods in order to deflect the ions. This deflection process is analogous to that of the cylindrical mirror analyzer discussed in Chapter 6 except ions rather than electrons are the particles being filtered electrostatically.

The quadrupole mass spectrometer can be used not only for SIMS, but also for residual gas analysis. Figure 12.4 illustrates the quadrupole mass spectrometer (QMS) and the ion detector for analyzing the mass of atomic or molecular species leaving the target [2]. In this case, filaments ionize the gas molecules entering the *residual gas analyzer* (RGA). A lens focuses the ions into the region between four electrodes (cylinders and circles). Depending on the voltages applied to these rods, ions with different mass-to-charge ratios are deflected along different trajectories. As the electrode voltages are swept, molecules with different mass-to-charge ratios are able to pass through the electrode region. A Faraday cup gathers the ions and generates an electric current proportional to the number of ions collected.

Figure 12.5 illustrates a typical RGA spectrum for the gases inside a vacuum chamber. The low-energy portion of this RGA spectrum shows pronounced peaks at 2, 17, 18, 28, 32, and 44 that correspond to H_2, OH, H_2O, N_2, O_2, and CO_2, respectively. These are characteristic constituents for air. Several lower intensity peaks correspond to various hydrocarbons. The high-intensity atomic mass unit (amu) 18 peak indicates the presence of residual water vapor inside the chamber. Such spectra can assist in detection of leaks using a finely focused source of helium gas around potential leak sites. This is because amu 4 is typically quite low unless He leaks into the chamber through imperfect seals.

Another type of SIMS spectrometer involves time-of-flight separation of different masses. In this technique, the primary ions arrive in pulses and generate ions

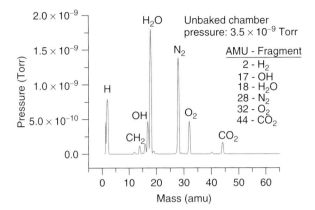

Figure 12.5 Low mass portion of a residual gas analyzer spectrum showing common molecular constituents inside an unbaked vacuum chamber at a pressure of 2×10^{-7} Torr.

that are extracted from the sample surface with a high bias voltage V_A applied to the sample. These ions enter a channel that is essentially a racetrack. A detector records the arrival of ions and the elapsed time from the initial pulse.

The secondary ion has a kinetic energy $E = \frac{1}{2}mv^2$ equal to the voltage used to extract it into the racetrack. Here, v is the secondary ion velocity. The time to traverse the distance d between sample and the detector in Figure 12.6 is simply

$$t = \frac{d}{v} = \frac{d}{(2E/m)^{1/2}} = \frac{d}{(2qV_A/m)^{1/2}} \tag{12.1}$$

This relationship shows that the elapsed time decreases with applied voltage V_A and increases with increasing mass m. Thus, $m = 2qV_A t^2/d^2$ plus any constant that is unique to any timing delay.

In practice, the instrument is calibrated using several peaks of known mass in a least squares fit to determine constants a and b in the formula $m = at^2 + b$. Figure 12.7 shows a schematic illustration of the time-of-flight secondary ion mass

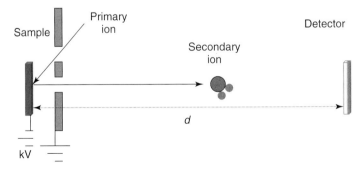

Figure 12.6 Time-of-flight method of determining particle mass in a TOF SIMS instrument.

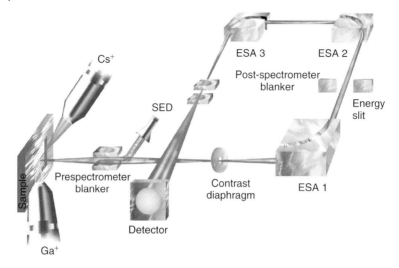

Figure 12.7 Schematic diagram of ion trajectory inside a PHI TRIFT III TOF-SIMS. Courtesy Physical Electronics Industries.

spectrometry (TOF-SIMS) apparatus. Here, a Ga$^+$ beam ablates a target surface while a Cs$^+$ beam irradiates the surface to promote formation of secondary ions. The secondary ions pass through an electrostatic blanker that stops all but surface ions generated by an initial pulse. A series of apertures and electrostatic bends force the ions around the race track and on to the detector. A secondary electron detector records the secondary electron image of electrons generated at the sample surface by an electron gun.

The TOF-SIMS and quadrupole instruments are advantaged in different ways. Compared with quadrupole and magnetic sector instruments, the TOF-SIMS instrument features high transmission by virtue of its parallel rather than sequential signal detection. This attribute accounts for the orders of magnitude higher relative sensitivity of TOF-SIMS. The practical range of the magnetic sector exceeds both quadrupole and time-of-flight instruments. However, the high mass resolution of the TOF-SIMS exceeds the $\Delta m/m$ of the others. Table 12.1 compares the three SIMS techniques [3].

Table 12.2 compares DSIMS and SSIMS with XPS and Auger electron spectroscopy (AES), alternative surface chemical techniques already discussed. All four techniques provide elemental information. XPS and AES supply chemical bonding information whereas SIMS fragments can reveal information on molecular species. The spatial resolution of conventional XPS is in the low micron range, whereas AES with focused electron beams can resolve features below 10 nm. The focused ions employed in DSIMS and SSIMS can resolve spatial features down to 50 nm. Perhaps most significantly, SIMS detection sensitivity is orders of magnitude greater than either XPS or AES. Whereas the latter can detect fractions of a percent surface coverage, SIMS detection, particularly in DSIMS can reach 10 ppb or less.

Table 12.1 Comparison of mass spectrometers for SIMS.

Mass spectrometer	Resolution	Practical mass range	Transmission	Mass detection	Relative sensitivity
Quadrupole	10^2-10^3	$<10^3$	0.01–0.1	Sequential	1
Magnetic sector	10^4	$>10^4$	0.1–0.5	Sequential	10
Time-of-flight	$>10^4$	10^3-10^4	>0.5	Parallel	10^4

After 3.

Table 12.2 Comparison of XPS, AES, DSIMS, and SSIMS.

	XPS	AES	DSIMS	SSIMS
Probe beam	Photons	Electrons	Ions	Ions
Analyzed beam	Electrons	Electrons	Ions	Ions
Sampling depth (nm)	0.5–5	0.5–5	0.1–1	0.1–1
Detection limits	1×10^{-4}	1×10^{-4}	1×10^{-9}	1×10^{-6}
Information	Elemental Chemical	Elemental Chemical	Elemental Structural	Elemental Molecular
Spatial resolution	1–30 μm	10 nm	50 nm	120 nm
Materials	All solids	Inorganics	Inorganics	All solids

Adapted with courtesy, Physical Electronics.

SSIMS and XPS can probe all solids, organic, and inorganic, whereas AES and DSIMS are hampered by sample charging.

12.4 Secondary Ion Yields

Ionization of the sputtered atoms is required so that the resultant ions move under the applied extraction voltage. This ionization depends on charge transfer between the sputtered atom and the lattice it leaves. In turn, the efficiency of this ionization depends sensitively on the electronegativity of the primary ion. This SIMS ionization efficiency is termed *ion yield* and equals the fraction of sputtered atoms that become ionized. These ion yields can vary by orders of magnitude for different elements.

Figure 12.8 displays the variation in ion yield for a wide range of elements as (i) positive ions induced by O ion sputtering and (ii) negative ions induced by Cs ion sputtering. Figure 12.8a shows that as the ionization potential (the energy required to ionize the sputtered atom) increases, the positive ion yield decreases by orders of magnitude. Similarly, as the electron affinity of the sputtered atom

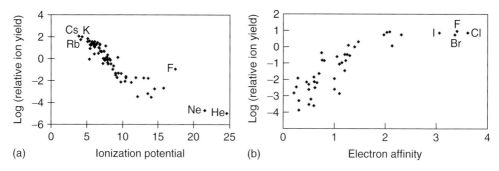

Figure 12.8 Positive (a) and negative (b) ion yields with O ion or Cs ion sputtering, respectively relative to that for Si in Si [4].

and the ease with which the ion attracts an electron increases in Figure 12.8b, the ion yield increases exponentially. Oxygen bombardment increases the ion yield of positive ions, while Cs bombardment increases the ion yield of negative ions.

From the Charles Evans and Associates web site: "Oxygen enhancement occurs as a result of metal–O bonds in an O-rich zone. When these bonds break in the ion emission process, the O becomes negatively charged because its high electron affinity favors electron capture and its high ionization potential inhibits positive charging. The metal is left with the positive charge. O beam sputtering increases the concentration of O in the surface layer." [4] Likewise, "The enhanced negative ion yields produced with Cs bombardment can be explained by work functions that are reduced by implantation of Cs into the sample surface. More secondary electrons are excited over the surface potential barrier. Increased availability of electrons leads to increased negative ion formation" [4].

Figure 12.9 reproduces a chart by Evans and Associates of secondary ion yields in terms of the primary ions most effective in ionizing the secondaries. Once the appropriate primary ion is selected, the secondary ion intensity I_E for a given element E at a concentration C_E of that element generated by the primary ion determines a relative sensitivity factor (RSF). For element E, this factor is termed RSF_E and is just the proportionality between the I_E/C_E ratio versus a reference secondary ion intensity to concentration ratio I_R/C_R. Thus,

$$I_R/C_R = RSF_E \cdot I_E/C_E \tag{12.2}$$

The lower the RSF_E is, the higher the sensitivity of SIMS to that element. If one takes the reference element R to be the host matrix element M, then Equation 12.1 can be reexpressed as

$$C_E = RSF \cdot I_E/I_M \tag{12.3}$$

This common SIMS equation yields the concentration C_E of element E as the ratio between secondary ion intensity ratio of E versus that of M multiplied by the RSF for these elements.

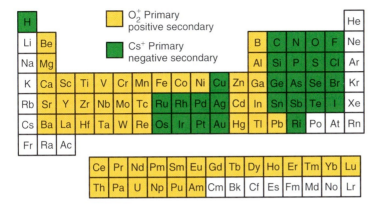

Figure 12.9 Partial table of elements color coded to indicate the primary ion likely to produce the higher ion yield [4].

Figure 12.10 presents the relative sensitivity factors for most elements with O_2^+ primary ions [5]. For example, this chart shows that Column III elements such as Al, Ga, and In as well as the Column I elements have very low RSF of 10^{21}, indicating that SIMS using O_2^+ primary ions will detect these elements with relatively high sensitivity.

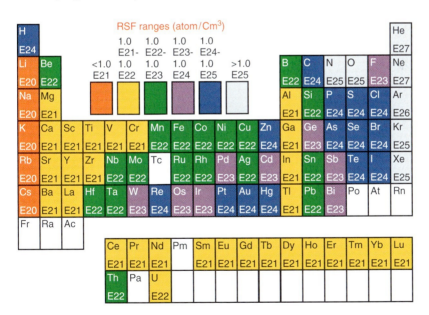

Figure 12.10 Periodic table of elements in terms of relative sensitivity factors for O_2^+ primary ions expressed as atoms per cubic centimeter. Low RSF corresponds to high sensitivity. After [5].

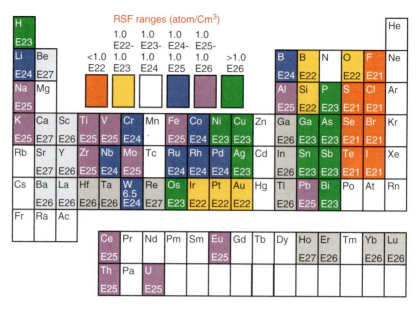

Figure 12.11 Periodic table of elements in terms of relative sensitivity factors for Cs⁺ primary ions expressed as atoms per cubic centimeter. Low RSF corresponds to high sensitivity. After [5].

Figure 12.11 presents the relative sensitivity factors for most elements with Cs⁺ primary ions. For Cs⁺ primary ions, this chart shows that Column VI elements such as S, Se, and Te as well as the Column VII elements have relatively low RSF of 10^{22}, indicating that SIMS using Cs⁺ primary ions will detect these elements with relatively high sensitivity. Figures 12.10 and 12.11 show that sensitivities can vary by orders of magnitude from one element to another. Furthermore, the ion yields depend sensitively on the matrix in which the particular atom is situated. Therefore, quantitative analysis requires standards of the same elements with known concentration in the same matrix.

12.5
Imaging

Parallel signal detection provides sufficiently high signal-to-noise spectra to permit imaging of elements across a specimen's surface, pixel by pixel, in relatively short times. Figure 12.12 illustrates the principle underlying the imaging technique. Here, each specific region of the surface provides a spectrum of intensity versus mass per charge m/z. As the focused ion beam scans across the surface in raster fashion, the data acquisition system constructs a map of secondary ion intensity versus position. One such map is the total intensity of all ions collected. See total ion image in Figure 12.12. In addition, the individual peak intensities within each

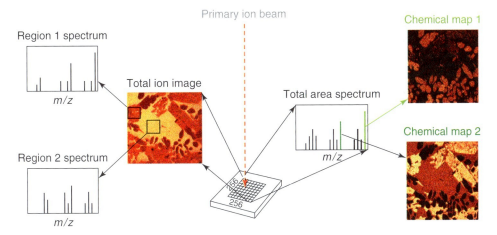

Figure 12.12 TOF-SIMS imaging for primary ion beam rastered across a specimen surface, yielding intensity versus m/z spectra at specific locations and maps of total ion as well as individual m/z versus position chemical maps [3].

intensity versus m/z spectrum provide maps of specific mass intensities across the specimen surface. For example, see chemical maps 1 and 2. Focused ion beams permit one to obtain such maps with a raster element size of 100 nm or less.

As an example of the TOF-SIMS imaging capability, consider a structural material of high interest, 304 stainless steel, in which precipitates may play a role in changing local elemental composition [6]. The first two rows of Figure 12.13 show that MnS and Mn(Na,K)O precipitates are present in the stainless steel matrix. The third row taken from a different area shows that Mn precipitates do not deplete Cr from its uniform distribution in the surrounding matrix. This conclusion is significant for understanding corrosion in stainless steel, where Cr plays an important role.

12.6
Dynamic SIMS

Dynamic SIMS provides concentrations of elements and molecular complexes as a function of depth below a free surface, information of significant importance for electronic materials. SIMS is commonly used to measure absolute concentrations of elements used as dopants in semiconductors. Such concentrations are critical to the functioning of all active electronic devices such as transistors, switching memory elements, and lasers since they determine internal potentials and band structure as a function of location. To calibrate these concentrations, researchers use ion accelerators to implant known doses of specific elements inside the semiconductors of interest. Monte Carlo simulations of the implanted ion distribution versus depth then predict the concentration profiles and provide the calibration for the SIMS profiles actually measured.

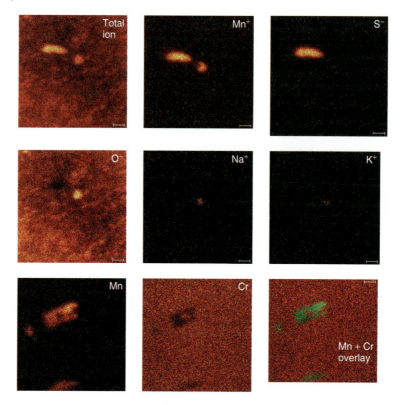

Figure 12.13 SIMS ion images of precipitates and segregation in the same surface area of 304 Stainless Steel. 10 μm × 10 μm images illustrate total positive ion image, Mn^+, S^-, O^-, Na^+, and K^+ maps. Bottom row shows maps of Mn, Cr, and their combined signals superimposed in another area [6].

SIMS depth profiles can also measure the extent of chemical diffusion or reaction at interfaces between different electronic materials. Such interactions can take place due to thermodynamic driving forces, especially at the elevated temperatures often encountered during device operation. Similarly, semiconductors can be subjected to elevated temperatures during device processing such as ohmic contact formation or annealing to eliminate lattice damage generated by ion implantation. See Figure 1.2.

Elevated temperatures are also necessary for semiconductor growth. High temperatures promote atomic movement across surfaces so that deposited atoms incorporate rapidly into regular lattice sites before more atoms reach the surface. Such movement can take place across interfaces as well. As an example of the dilemma faced by crystal growers, consider the growth of double heterostructures for which the optimum growth temperatures of each layer are different. Thus, growth of an intermediate layer at a lower temperature must be followed by growth

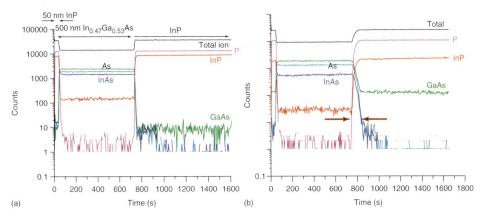

Figure 12.14 Cs⁺ SIMS depth profile of InP/500 nm In$_{0.53}$Ga$_{0.47}$As/InP/50 nm InP grown with 40 versus 170 seconds P exposure between In shutter closing and As shutter opening. Arrows indicate interface broadening [7].

of the confinement layer at a higher temperature that could degrade the previous layer.

Furthermore, details of the growth stop between layers can have an effect on the interfacial chemical (and thus the electronic) structure. As an example, Figure 12.14 illustrates SIMS depth profiles for an Molecular beam epitaxy (MBE)-grown InP/InGaAs/InP double heterostructure in which growers varied the length of time elapsed between the end of InP growth and the onset of InGaAs growth during which the P shutter remained open to prevent In clustering on the InP growth surface [7]. The 0.53 : 0.47 In : Ga ratio produces lattice matching of InGaAs to InP.

The short (40 seconds) growth stop between P and As growth in Figure 12.14a results in an abrupt interface – a transition between InP and InGaAs of less than 5–10 nm. This appears as orders of magnitude increase or decrease in the Ga and As versus In and P signals, respectively, across the interface over distances of only a few nanometers.

On the other hand, Figure 12.14b shows that a longer (170 seconds) delay in closing the P shutter after the As shutter opens results in a slower increase of As and InAs such that the interface width is now ∼100 nm. Note the abrupt transition at the outer InP/InGaAs interface, for which conditions were identical during growth. The rapid increase in In bonding to As occurs once the P source is closed. Likewise, As bonding to In slows as the Ga shutter opens.

The interface broadening and changes in film composition result in the appearance of new electronic structure localized to the junction [7, 8]. Such DSIMS spectra underscore the delicate balance among growth conditions, interface chemical composition, and electronic structure.

The identification of GaAs versus InP clusters in Figure 12.14 was made possible by the excellent mass resolution afforded by time-of-flight instruments. Thus, both

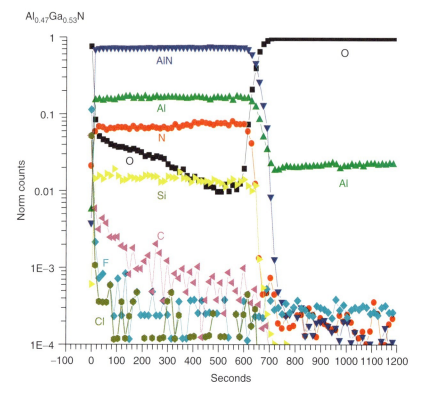

Figure 12.15 Normalized SIMS Cs$^+$ depth profile of 500 nm Al$_{0.47}$Ga$_{0.53}$N grown on Al. The O signal extends into AlGaN and residual N extends into Al$_2$O$_3$.

GaAs and InP have nominally the same atomic mass, 145.8 amu. However, with a mass resolution $> 10^4$, the 0.03 amu separation between InP and GaAs is easily resolved. This capability is also useful in separating isotopes of various elements such as D^2, O^{18}, and C^{13}.

DSIMS reveals the physical basis for unintentional dopants in semiconductors. For example, MBE growth of AlGaN on sapphire results in formation of degenerately doped layers at the AlGaN/Al$_2$O$_3$ junction. Figure 12.15 illustrates the SIMS profile for a 500 nm Al$_{0.47}$Ga$_{0.53}$N grown on Al.

Oxygen atoms serve as n-type dopants in AlGaN and other nitrides. Diffusion of O into AlGaN from sapphire (Al$_2$O$_3$) growth substrates results in concentrations that easily account for donor densities that easily exceed 10^{20} cm^{-3}.

The SIMS depth resolution degrades with depth of analysis since irregularities in the sputtered crater introduce unintentional depth variations. For depths below a micron, depth resolutions of <5–10 nm can be achieved.

The speed of depth profiling depends on the sputtering beam atom. Instead of Ga^{69}, new sputter beams with higher mass such as Au or buckyballs provide faster and more intense secondary ion generation.

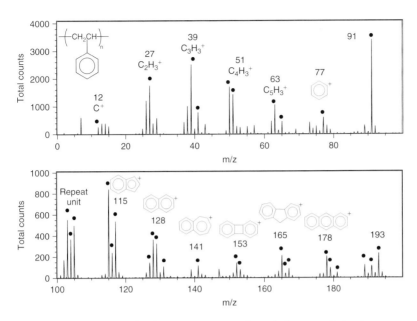

Figure 12.16 TOF-SIMS of polystyrene. Courtesy, Physical Electronics Industries.

12.7
Organic and Biological Species

Inorganic semiconductors are among the least complicated materials for SIMS to analyze. More challenging are organic and biological molecules because so many are composed of complex molecular groupings with atomic weights that can be interpreted in several ways. Figure 12.16 illustrates the numerous fragments and their atomic weights from SIMS analysis of polystyrene. Identification of organic molecules is particularly difficult since the constituents are typically C, O, H, and N and little else. To identify the parent molecule from such fragments, one has libraries of spectra available to catalog the various species generated during secondary ion emission and their relative intensities for a particular molecule. Similar libraries of fragments are becoming available for biological molecules [9]. Again, the elemental species consist of C, O, N, H, and possibly P and S. Such analyses will prove extremely valuable in evaluating conformation of proteins on surfaces.

12.8
Summary

SIMS is an important chemical analysis technique for electronic materials. Its parts-per-billion sensitivity permits quantitative analysis of dopants in semiconductors beyond the capabilities of other surface science techniques.

SSIMS permits detection of trace element concentrations at surfaces with detection limits of parts per million.

Quantitative analysis using SIMS requires standards. This requires analysis using the same equipment of the element of interest inside the same matrix as the host material under investigation.

Chemical interdiffusion and reaction can take place at electronic material interfaces associated with the growth or processing of active electronics. Such effects can actually take place near room temperature – a topic that is discussed in subsequent chapters.

SIMS can be extended to organic and biological materials in which most of the constituent atoms are common to many species. Analysis requires libraries of SIMS spectra that catalog intensity distributions across wide mass ranges.

Problems

1. Assuming a primary O^- current of 10^{-9} A cm^{-2} impinging on a GaAs(100) surface and one atom per incident ion removed by a primary ion, what is the maximum time available to acquire a static SIMS spectrum without significantly disrupting the surface?
2. The time-of-flight mass spectrometer is capable of $>10^4$ mass resolution $\Delta m/m$. For mass 100 amu, 2 keV applied voltage, and a flight path equal to 2 m, what clock speed of the detection electronics is required in order to achieve this resolution?
3. Which ionization source(s) would you use to probe for C, O, Mg, Cl, and Co using SIMS with host semiconductor of GaAs?
4. Which ionization source(s) would you use to probe for interdiffusion across (a) a CdS/Si interface? and across (b) a GaAs/InP heterojunction?

References

1. Zandvliet, H.J.W., Elswijk, H.B., van Loenen, E.J., and Tsong, I.S.T. (1992) *Proceedings of the 8th International Conference on SIMS*, (eds A. Benninghoven, J. Tjuer and H.W. Werner), John Wiley & Sons, Ltd, Chichester, p. 3–9.
2. http://www.stechoriba.com/rga%20principle.jpg.
3. Courtesy, Physical Electronics Industries, Chanhassen, MN, http://www.phi.com.
4. http://www.eaglabs.com/training/tutorials/sims_theory_tutorial/elemeff.php.
5. Wilson, R.G. (1995) *Int. J. Mass Spectrom. Ion Process.*, **143**, 43.
6. Meng, Q., Frankel, G.S., Colijn, H.O., and Goss, S.H. (2003) *Nature*, **424**, 389.
7. Smith, P., Bradley, S.T., Goss, S.H., Hudait, M., Lin, Y., Ringel, S.A., and Brillson, L.J. (2004) *J. Vac. Sci. Technol. B*, **22**, 554.
8. Smith, P.E., Goss, S.H., Gao, M., Hudait, M.K., Lin, Y., Ringel, S.A., and Brillson, L.J. (2005) *J. Vac. Sci. Technol. B*, **23**, 1832–1837.
9. Belu, A., Graham, D.J., and Castner, D.G. (2003) *Biomaterials*, **24**, 3635.

13
Electron Diffraction

13.1
Overview

Low-energy electron diffraction (LEED) is a standard surface analysis technique for obtaining structural information on surfaces. LEED has been an integral component of ultrahigh vacuum (UHV) surface science since UHV chambers became available. It has provided the vast majority of surface structure data. In the past two decades, it has been supplemented by scanning tunneling microscopy (STM) to be discussed in Chapter 14. LEED and STM have revealed the complex and often beautiful organization of atoms on semiconducting and metallic surfaces, allowing researchers insight into their electronic and chemical properties.

13.2
Principles of Low-Energy Electron Diffraction

Davisson and Germer were the first to observe the coherent scattering of electrons from solids [1]. They directed electrons with energies in the range of $10 < E < 1000\,eV$ in UHV at clean nickel surfaces. The diffraction pattern they observed proved for the first time that electrons can diffract just as light waves do. They received the 1927 Nobel Prize for directly demonstrating that particles can act as waves, one of the fundamental tenets of quantum mechanics.

One can view a crystal as a set of geometrically equivalent planes parallel to a surface. Within each plane, atoms comprise a regular two-dimensional array that acts as a diffraction grating. In other words, electrons diffract from surfaces because the regular array of atoms represents a diffraction grating with "grooves" that are separated by atomic bond lengths. Figure 13.1 illustrates the geometry of an LEED experiment. Here, an incident electron beam impinges at near-normal incidence on a crystal surface. The diffracted electrons project backward toward a screen with a fluorescent coating. A high voltage is applied between this screen and a grounded mesh in order to accelerate the electrons into the screen and create

Surfaces and Interfaces of Electronic Materials. Leonard J. Brillson
Copyright © 2010 WILEY-VCH Verlag GmbH & Co. KGaA, Weinheim
ISBN: 978-3-527-40915-0

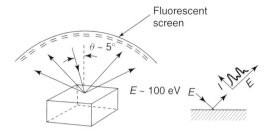

Figure 13.1 Schematic illustration of LEED apparatus (left) with ~100 eV electrons diffracted toward fluorescent screen and (right) dynamical LEED spectrum illustrating intensity versus incident beam energy.

a visible emission pattern. The resultant diffraction pattern is a direct measure of (i) the two-dimensional surface geometry, (ii) the lattice spacing of surface atoms, and (iii) the periodic spacing of groups of atoms that have reorganized.

A typical LEED pattern appears in Figure 13.2. The arrangement of spots indicates periodicity that is twice as large in the vertical versus the horizontal direction. The central region is shadowed by the electron gun supplying the incident beam. Similar to X-ray diffraction of bulk crystals, the spacing of diffraction spots is directly related to the wavelength of the incident beam of electrons as well as the spacing of atoms. The wavelength of incident electrons is governed by the de Broglie relation

$$\lambda = \frac{h}{p} \qquad (13.1)$$

Figure 13.2 Photograph of reverse-view LEED pattern for 92 eV electrons bombarding a (2 × 1) Si(100) surface.

where the electron momentum $p = mv$. For electron mass m and kinetic energy $E = \frac{1}{2} mv^2$,

$$\lambda = \sqrt{\left(\frac{150.4}{E}\right)} \tag{13.2}$$

where E is in units of eV. Thus, an electron with energy $E = 150.4$ eV, has a wavelength $\lambda = 1$ Å. These low energies are required in order to produce diffraction patterns whose spots are visibly separated. As with X-ray diffraction, the relation between lattice spacing and diffracted wavelength is given by

$$n\lambda = d \sin \theta \tag{13.3}$$

where θ is the angle between the incident and diffracted beams and n is the diffraction order. Hence, the diffracted wavelength must be comparable to the lattice spacing in order to observe diffraction patterns with measurable diffraction angles θ between spots.

Furthermore, at energies of a few hundred electron volts or less, the incident electrons scatter primarily in the outer few angstroms of the crystal surface. See Figure 6.2. Therefore, the diffraction pattern of incident electrons in this energy range must be characteristic of the top few layers of crystal surface atoms. For a clean uniform surface, one can then determine (i) the periodicity of the systems – the surface *unit mesh* by measuring the symmetry and separation of spots, and (ii) the location of atoms within the unit mesh by performing detailed measurements of diffracted intensities versus energy as illustrated schematically in Figure 13.1.

13.3
LEED Equipment

The LEED equipment consists of a set of retarding and accelerating grids that act as filters to pass and display electrons as a function of energy. As shown in Figure 13.3, the innermost pair of grids permit retardation of the diffracted electrons so that only elastically scattered electrons at the incident beam energy pass through [2]. The second pair of grids accelerate the electrons toward the fluorescent screen. The electron beam spot size is ~ 1 mm; the current is ~ 1 µA with an energy beam spread $\Delta E \sim 0.5$ eV due to thermal broadening at the high temperatures of the electron gun filament. Advanced LEED equipment includes channel plates instead of fluorescent screens, digitized images, computer-controlled beam tracking, and lower beam currents of only $\sim 10^{-10}$ A. The LEED "optics" can also be used for Auger electron spectroscopy since the retarding grids provide a means of determining the kinetic energies of the scattered electrons. However, this retarding grid method is less efficient than using cylindrical mirror or hemispherical energy analyzers.

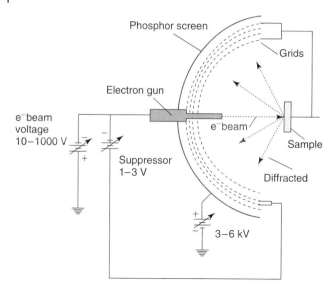

Figure 13.3 LEED equipment including electron gun, fluorescent screen at high voltage, along with variable-voltage suppressor, acceleration, and grounded screens to control kinetic energy of diffracted electrons [2].

13.4
LEED Kinematics

The spatial distribution of diffracted beams relative to the incident beam depends on the translational symmetry of atoms in the outermost crystal layer. The momenta parallel to the surface of the incident and diffracted electrons are related by the conservation relation

$$\mathbf{k}_{i,\|} = \mathbf{k}_{f,\|} + \mathbf{g} \tag{13.4}$$

where $\mathbf{k}_{i,\|}$ = vector momentum of incident electron parallel to the surface, $\mathbf{k}_{f,\|}$ = vector momentum of the diffracted electron parallel to the surface, and $\mathbf{g}$ = the vector of the two-dimensional reciprocal lattice (termed the *Bravais lattice*) associated with the surface plane. Since $E = \hbar^2 k^2 / 2m$, the momentum parallel to the surface

$$k_{i,\|} = (2mE/\hbar^2)^{1/2} \sin \theta \tag{13.5}$$

and $\mathbf{g} = h\mathbf{A} + k\mathbf{B}$, where $\mathbf{A}$ and $\mathbf{B}$ are *primitive vectors* of the reciprocal lattice and h and k are integers.

According to Equations 13.2 and 13.3, as the incident electron beam energy decreases, the wavelength λ and diffraction angle θ increase so that the spacing between spots expands. The primitive vectors allow one to extract the dimensions of the unit mesh from the energy dependence of θ along symmetry directions.

A geometric representation of momentum conservation in terms of $\mathbf{k}_{i,\|}$, $\mathbf{k}_{f,\|}$, and the primitive vectors – termed the *Ewald sphere* – appears in Figure 13.4b

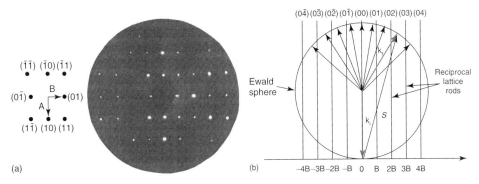

Figure 13.4 (a) Photograph of normal incident LEED pattern of the cleaved GaAs(100) surface using 150 eV incident electron beam energy. Each spot is indexed according to the primitive vectors shown to the left of the photo. (b) Ewald sphere with rods of the two-dimensional reciprocal lattice, incident wave vector k_i, diffracted wave vector k_f, and momentum transfer **S** along the **B** direction, shown here for the (03) beam. (After Kahn [3].)

along with a LEED pattern of the GaAs(110) surface with the primitive vectors **A** and **B** illustrated in Figure 13.4a [3]. Here, the rectangular diffraction pattern indicates different periodic distances along the <100> and <010> directions. The lines parallel to the incident beam momentum k_i are termed as *rods*. The intersection of these rods with the Ewald sphere determines the direction of the diffracted beam. The (h,k) notation of the spots illustrated in Figure 13.4a corresponds to the combination of primitive vectors required to produce the particular spots in the diffraction pattern. Figure 13.4b illustrates the change in momenta allowed for diffracted electrons along the **B** direction of the reciprocal lattice.

The symmetry of the spot pattern can also provide (i) adsorbate atom site symmetry and (ii) evidence for surface domains and steps.

13.5
Surface Reconstruction

Atoms near the surface can change their bonding because of the lattice termination. The presence of the surface destroys the translational invariance of the crystal perpendicular to the surface. Thus, the normal geometry of diffracted beams alone cannot give sufficient information to describe the atomic spacings in the direction normal to the surface. However, one can analyze the diffracted beam intensities versus energy to obtain the three-dimensional unit cell structure within each layer and their geometry with respect to each other.

Multiple scattering effects prevent extension of X-ray diffraction theory to low energies. Instead one must use the strong interaction of low-energy electron beams with the first few layers of solid to extract the atomic geometry in the direction normal to the surface. The strong near-surface interaction requires

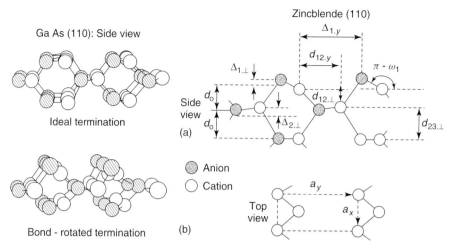

Figure 13.5 Schematic illustration surface atoms before (a) and after (b) surface reconstruction for (110) surfaces of compound semiconductors with the zincblende structure. The schematic of atomic angles and bond lengths (right) represents the independent structural variables that describe the reconstructed atomic geometries. (After Duke [4].)

multiple scattering or "dynamical" models to interpret the energy and temperature dependence of LEED spot intensities.

Figure 13.5 illustrates the outer atomic layers of GaAs – arguably the most important compound semiconductor to date – for the (110) surface. This crystal face is nonpolar; that is, it has equal numbers of Ga and As atoms. Figure 13.4 shows the LEED pattern for this surface. Figure 13.5a illustrates the ideal termination of this surface assuming no change in the crystal lattice from the bulk. The side view shows that Ga and As are equally present in the outermost atomic plane of atoms.

Compound semiconductors with the zincblende structure such as GaAs exhibit reconstruction of the outer atomic layers due to the lattice termination and the redistribution of bond charge. Thus, the *bond-rotated* termination (Figure 13.5b) illustrates a relaxation of the cation atoms (in this case, Ga) into the surface by an angle $\omega_1 \equiv \tan^{-1}[\Delta_{1,\perp}/(a_y - \Delta_{1,y})]$ [4]. This rotation reflects the sp^3 hybridization for As surface atoms and sp^2 hybridization for Ga atoms [5], consistent with minimizing the energy of the As atoms' lone pair electrons.

One arrives at the bond angles and bond lengths by simulating beam profiles iteratively and finding the best match to experimental profiles. The accurate calculation of the GaAs surface reconstruction represents an important success in surface science. This has been confirmed by other theoretical approaches and complementary experiments. The results showed that bond length changes of surface atoms are small. Reconstruction is due primarily to bond rotation. Subsequently, this result has been extended to show that surface bond rotation is

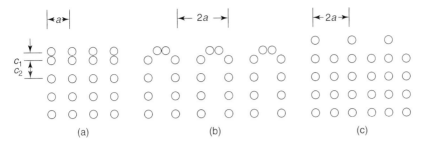

Figure 13.6 Side views of (a) atomic relaxation, (b) dimer reconstruction, and (c) missing row reconstruction for a cubic lattice with bulk lattice constant a.

characteristic of III–V compound semiconductor surfaces in general [6]. Analogous reconstructions occur for II–VI compounds as well.

For the III–V compounds, the driving force for such reconstructions is the minimization of surface energy that accompanies the lowering of dangling bonds on the Column V element and the raising of dangling bonds on the Column III element. Such surface relaxations are thus regular and characteristic features of semiconductor surfaces. This result also implies that surface bond energy minimization depends primarily on geometric structure rather than the detailed electronic structure of a specific structure.

13.6
Surface Lattices and Superstructures

Surface atomic layers can have different point group symmetries, Bravais lattices, and periodicities from those of the bulk crystal. Figure 13.6 shows a side view of three characteristic examples of atomic rearrangements for a simple cubic lattice of lattice constant a. In Figure 13.6a, the top row of atoms have relaxed back toward the bulk so that that $c_1 < c_2$. In principle, the second and lower rows may also relax toward smaller values. Figure 13.6b illustrates a dimerization of the top row atoms such that surface periodicity is now twice as large. In Figure 13.6c, every other atom in the top row is missing. This also gives rise to a surface periodicity that is twice as large. Similarly, the relaxation pictured in Figure 13.5 will also alter the periodicity. Indeed, this is evidenced in the LEED pattern of Figure 13.4. Thus, the LEED pattern alone is not sufficient to distinguish these different possibilities.

Surface adsorbates add another level of complexity. Figure 13.7 illustrates side views of adsorbates on a crystal lattice with different ratios of adsorbate versus substrate crystal lattice constants. Surface lattices with a different periodicity than that of the substrate are termed *superlattices*. The integer superlattice with $b/a = 2$ yields a LEED pattern with periodicity twice as large as that of the substrate, similar to cases (b) and (c) in Figure 13.6. The noninteger cases yield more complex patterns.

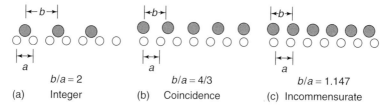

Figure 13.7 Adsorbate superstructures with different ratios of adsorbate-to-substrate lattice constant ratios.

The terminology to describe superlattice–substrate geometries is based on the two-dimensional translational vectors that determine the substrate unit mesh, namely,

$$r_m = m\mathbf{a}_1 + n\mathbf{a}_2 \tag{13.6}$$

where $\mathbf{m} = (m,n)$ signifies a pair of integers and the a_i are *unit mesh vectors*. The corresponding surface net of points denoting atomic positions in the topmost layer can then be described in terms of the substrate net vectors as

$$\mathbf{b}_1 = m_{11}\mathbf{a}_1 + m_{12}\mathbf{a}_2 \tag{13.7a}$$
$$\mathbf{b}_2 = m_{21}\mathbf{a}_1 + m_{22}\mathbf{a}_2 \tag{13.7b}$$

Equation 13.7 describe a 2×2 matrix with matrix elements m_{ij} that transforms substrate vectors into superlattice vectors. When the determinant of this matrix, $|\mathbf{b}_1 \times \mathbf{b}_2|/|\mathbf{a}_1 \times \mathbf{a}_2|$ is an integer, the superlattice is termed a *simple superlattice*. Figure 13.7a illustrates an example of such a lattice for $b/a = 2$. When the determinant is a rational number, for example, $b/a = 4/3$ as in Figure 13.7b, the superstructure is termed a *coincidence superlattice*. When the overlayer atoms are out of registry with the substrate net as in Figure 13.7c, the superstructure is termed *incommensurate*.

There exists a notation to describe superstructures by translational vector ratios and angles. According to this Wood notation [7], the surface superstructure of an element or compound M with surface Miller indices (hkl) can be described as

$$M(hkl) - \alpha \left[(a_s/a) \times (b_s/b) \right] \zeta - S \tag{13.8}$$

where $M(hkl)$ is the (hkl) crystal face of substrate M, α is either a p(primitive) or c(centered) unit mesh of composite system, $[(a_s/a) \times (b_s/b)]$ are the ratios of overlayer to substrate vectors, S is an overlayer element if present, and ζ is an angle between the overlayer and substrate mesh vectors $\mathbf{a}_s$ and $\mathbf{a}$.

Figure 13.8 illustrates several examples of adsorbate–substrate superlattices, their descriptions in Wood notation and their superlattices in reciprocal space [8]. Figure 13.8a–c represent adsorbed atoms on a (100) or (111) low index surfaces of, for example, a close-packed metal. In Figure 13.8a, the ratio of overlayer to substrate vectors $b/a = 2$ in both $\mathbf{a}_1$ and $\mathbf{a}_2$ directions and $\zeta = 0$. The reciprocal lattice vectors $\mathbf{a}_i^*$ that give rise to the diffraction spots are defined in terms of the real-space lattice vectors $\mathbf{a}_i$ according to $\mathbf{a}_1^* = 2\pi\,(\mathbf{a}_2 \times \mathbf{\check{n}})/|\mathbf{a}_1 \times \mathbf{a}_2|$, $\mathbf{a}_2^* = 2\pi\,(\mathbf{a}_1 \times \mathbf{\check{n}})/|\mathbf{a}_1 \times \mathbf{a}_2|$,

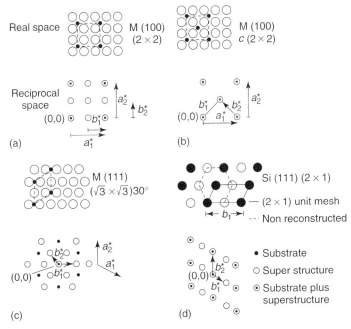

Figure 13.8 Representative superlattices of ordered adsorbate–substrate systems. Top-view real-space schematic illustrates (a) a square 2 × 2 array on a (100) closed packed metal substrate, (b) a square 2 × 2 array with center atom, (c) a trapezoidal √3 × √3 array rotated 30° with respect to a (111) substrate, and (d) a trapezoidal (2 × 1) array with respect to a (111) surface. After H. Lüth, Solid Surfaces, Interfaces, and Thin Films, 4th ed. (Springer, New York, 2001), p. 92.

where ň is a unit vector normal to the surface. Since the LEED beam can diffract from both the adsorbate superstructure and the outer monolayers of substrate, the overlayer contributes new diffraction spots whose symmetry, spacing, and registry reveal the arrangement of atoms in the outer monolayer. See, for example, [8].

This discussion pertains only to monolayers on surfaces and not to the cases where the adsorbate layer is (i) amorphous, (ii) interdiffused, or (iii) reacted to form an interface compound. Cases (ii) and (iii) for metals with semiconductors will be discussed in later chapters.

13.7
Silicon Reconstructions

One of the most important semiconductor reconstructions is that of the Si(111) surface. This surface provides a complex LEED pattern termed (7 × 7) according to the periodicity of the spot pattern. The spot pattern shown in Figure 13.9a [9] results from an atomic reconstruction illustrated in Figure 13.9b. After decades of study, the

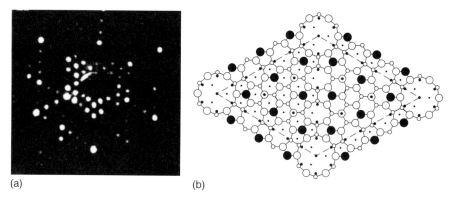

Figure 13.9 (a) Si (111) (7 × 7) LEED pattern; (b) Top view of dimer–adatom-stacking fault (DAS) model for the (7 × 7) surface reconstruction.

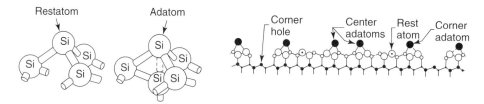

Figure 13.10 (a) Three-dimensional representation of rest and adatoms in Si(111) (7 × 7) surface and (b) cross-sectional view of corner hole, center, rest, and corner adatoms.

nature of this reconstruction was established by Takayanagi using a transmission electron microscopy technique with surface sensitivity [10]. STM measurements of surface morphology that are discussed in Chapter 14 are consistent with this model.

This complex surface reconstruction involves three double layers of atoms. The circles in Figure 13.9b indicate atoms whose decreasing diameter corresponds to positions lower down, away from the top surface. Figure 13.10 illustrates the configuration of the various surface atoms in cross section. The complexity of this *dimer adatoms and stacking fault* (DAS) model involving large numbers of surface atoms in an aggregate structure illustrates a major point. Extended, coherent features of the surface topology can help minimize strain and overall energy for structural stability.

Of more practical concern is the Si(100) surface, which is the wafer surface used in device technology. This face exhibits a (2 × 1) reconstruction that has also been the subject of extensive investigation. In this case, several alternative structures have been proposed to account for its periodicity. Figure 13.11 shows a top view of the ideal bulk-terminated surface geometry along with three alternative models with the same (2 × 1) periodicity. STM confirms the presence of surface dimers

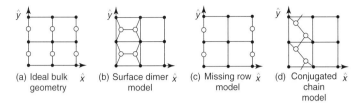

Figure 13.11 Top view of Si(100)(2 × 1) (a) ideal bulk geometry, (b) surface dimer model, (c) missing row model, and (d) conjugated chain model.

on this surface, supporting the surface dimer model shown in Figure 13.11b. As with the (7 × 7) reconstruction on the (111) surface, the Si(100)(2 × 1) surface requires an atom-scale probe technique to distinguish the actual surface atomic configuration.

13.8
III–V Compound Semiconductor Reconstructions

Compound semiconductors add another level of complexity since their stoichiometry can change with growth and processing as well as with crystal orientation. The surface reconstructions of III–V compound semiconductors are of particular importance since they are used in many device applications. In the case of GaAs, the (100) surface is used more widely than any other in crystal growth and heteroepitaxy. Researchers have discovered a wide array of surface reconstructions on this surface, depending on how they are grown and processed.

Figure 13.12 illustrates the various reconstructions reported for GaAs(100) surface as a function of As_4-to-Ga vapor flux ratio (also termed *beam equivalent pressure* (BEP) and substrate temperature during molecular beam epitaxy (MBE) growth. Figure 13.12 is a composition–temperature phase diagram, with surface phase plotted for a given composition and temperature. Reflection high-energy electron diffraction (RHEED) from the growth surface (to be discussed in the following section) enables measurement of these reconstructions inside the growth chamber and during the crystal growth.

The As_4-to-Ga vapor flux ratio and substrate temperature determine the reconstruction shown as well as surface faceting, droplet formation, and other surface degradation. This surface phase diagram emphasizes the importance of specific growth conditions in determining surface geometric structure. Figure 13.12 indicates that bulk growth conditions affect the surface reconstruction, and direct measurement of the surface elemental composition demonstrates that the surface reconstruction also depends on surface stoichiometry.

It is possible to obtain particular GaAs reconstructions by annealing in As vapor or in vacuum, then cooling to room temperature. However, while it is possible to change the surface reconstruction by changing the surface composition,

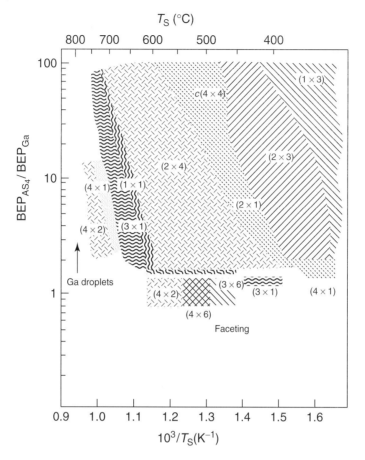

Figure 13.12 GaAs(100) reconstructions measured by RHEED as a function of As_4-to-Ga vapor flux (beam equivalent pressure) ratio and substrate temperature during MBE growth [11].

there is not a one-to-one correspondence between reconstruction and composition. Indeed, one can observe particular reconstructions over a range of surface stoichiometries. Figure 13.13 illustrates the range of surface compositions measured by AES and photoelectron spectroscopies for particular surface reconstructions. As the surface composition increases from As-deficient to As-rich, seven different reconstructions can be observed [12]. Notwithstanding the difference in reconstruction between these two limits, it is evident from Figure 13.13 that the same reconstruction can appear with a significant range of elemental composition [13–17].

This limited correlation between atomic structure and chemical composition may, in fact, be due to either (i) a composite of multiple reconstructions or (ii) a structure that is more complex than an extension of primitive unit cells.

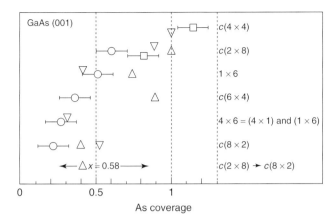

Figure 13.13 Fractional surface coverage of As on GaAs(100) surfaces measured by AES and photoelectron spectroscopy as a function of various LEED reconstructions. The data points (△ from van Bommel et al. [13], ○ from Drathen et al. [14], □ from Massies et al. [15], ▽ from Bachrach et al. [16], and Δx from Arthur [17], illustrate a correlation between atomic structure and chemical composition. (After [12].)

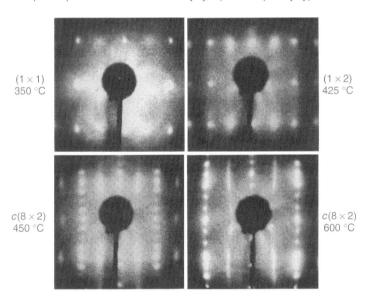

Figure 13.14 LEED patterns of the GaP(100) surface grown by MBE as a function of postgrowth annealing temperature in UHV [18].

Reconstructions can change after epitaxial growth is terminated. Figure 13.14 illustrates various reconstructions of the GaP(100) surface observed by LEED as a function of annealing temperature in UHV [18]. To avoid chemical contamination during air exposure between the termination of growth and the LEED measurements in a separate UHV chamber, the GaP surface was coated with As

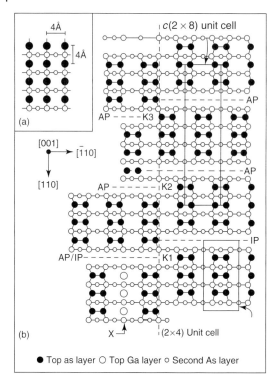

Figure 13.15 Geometric structure for the GaAs(001)(2 × 4)–c(2 × 8) reconstruction indicated by STM: (a) structure of the unreconstructed GaAs(100) and (b) the missing dimer model for the GaAs(100) (2 × 4) surface producing (2 × 4) or (2 × 8) structures, depending on the in-phase (IP) or antiphase (AP) nature of the missing dimer boundary, respectively. Different boundary kinks – k1, k2, and k3 – can arise, depending on the interaction of IP and AP domain boundaries. Disorder in the As pairing is denoted by X. (After [19].)

vapor after growth was completed. Such coatings are achieved by simply cooling the grown surface in the presence of As vapor. This procedure results in layers of As several hundred angstrom thick coating the surface. Such layers are effective in blocking oxidation and carbon contamination of the surface in air for limited periods of time. The layers can be removed by thermal desorption in UHV.

Figure 13.14 illustrates how sensitive the III–V compound semiconductor reconstruction can be with annealing in UHV below its growth temperature.

STM provides strong evidence that the surface geometric structure is determined by an extended array of primitive cells. Figure 13.15 illustrates a model for the GaAs(001)(2 × 4)–c(2 × 8) reconstruction based on STM measurements [19].

Figure 13.15a shows the simple 1 × 1 periodicity of the GaAs(100) bulk structure. Figure 13.15b illustrates the missing dimer model. Here, the fourfold periodicity consists of three top-layer dimer pairs of As (black) atoms and one missing dimer pair as shown in the upper right corner. These fourfold structures retain this periodicity and their (2 × 4) LEED pattern if they are *in-phase* (IP), that is, in line with their counterparts across the surface. If these structures have antiphase nature (AP), that is, the domains are shifted by one As atom, they have an eightfold periodicity and display a (2 × 8) LEED pattern. Different kinks due to the various domain boundaries appear in Figure 13.14 and are labeled K1, K2, and K3. The extended array of primitive cells provides a natural explanation for the range of stoichiometry measured for the same LEED reconstruction.

Other surface features that can contribute electron diffraction features include (i) adsorbed trimers and (ii) facets. The surface features described in this section lead to a wide range of reconstructions for specific orientations of different compound semiconductors. Table 13.1 provides a representative list of reconstructions reported for III–V and II–VI compound semiconductors with a variety of crystal structures [14,16, 20–50].

Inspection of this table shows that common reconstruction features occur for the same crystal orientation within the same family of III–V compounds.

13.9
Reflection High-Energy Electron Diffraction

RHEED has a surface sensitivity similar to LEED due to the grazing incidence angle of the primary beam. The RHEED technique is very useful as an *in situ* method for monitoring surface order during epitaxial growth. Figure 13.16 illustrates the geometry of the RHEED technique. The grazing incidence angle is sensitive to the surface morphology. The angular spread of the incident beam results in atomically ordered flat surfaces giving rise to streaks on a fluorescent screen illuminated by the diffracted beams. Phonons and lattice defects can add to such streak patterns. Figure 13.17a–c illustrate these streak patterns for three glancing incidence beam orientations of $SrTiO_3$ growth on Si(100). The spacing between streaks changes with the lattice periodicity along different crystal directions. On the other hand, crystalline islands or "droplets" on otherwise flat surfaces produce diffraction spots due to *transmission electron diffraction* (TED) as illustrated in Figure 13.17d.

The streak geometry and spacing can be used to identify particular reconstructions but also to study surface corrugation and film growth modes. On extremely flat and ideal surfaces with low angular beam spread and low thermal energy broadening ΔE of the incident beam energy, one can in principle observe sharp diffraction spots as well. For example, Figure 13.17d–f illustrates RHEED patterns taken from an atomically clean Si(111)(7 × 7) surface before and after deposition of silver adatoms [51]. Here, the sharp RHEED spots exhibit the periodicity corresponding to the Si(111)(7 × 7) pattern. The addition of Ag atoms blurs the streak pattern with the additional superstructure. Finally, a textured pattern arises due to the Ag layers. The sharp diffraction spots that appear require an ordered array within the third dimension, indicated the formation of Ag clusters at the surface.

Table 13.1 Preparation techniques and corresponding reconstructions of the polar faces of various binary compounds.

Crystal	Bulk structure	Surface	Preparation	ELEED pattern	Comments: structure	Selected refs.
ZnO	Wurzite	(0001)Zn	Cleaving	1×1	Highly stepped, facets	[20, 21]
			ArB + anneal <600 °C	1×1	Top Zn layer contracted by 0.2 ± 0.1 Å	[22–24]
			ArB + 600 °C anneal <850 °C	2×2; $\sqrt{3} \times \sqrt{3} - 30°$	Patches with different reconstruction	[23]
		(0001)O	Cleaving	1×1	Highly stepped, facets	[23]
			ArB + anneal <850 °C	1×1	Relaxation unknown	[23]
CdS	Wurzite	(0001)Cd	ArB + anneal ≤300 °C	1×1	Sixfold symmetry; relaxation unknown	[25]
		(0001)S	ArB + anneal ≤300 °C	1×1 + weak $\sqrt{3} \times \sqrt{3} - 30°$	Some evidence of facets	[25]
GaAs	Zinc blende	(100)	MBE, T = 550–660 °C	$c(8 \times 2)$	Ga dimers	[13], 16, 26
				4×6	Coexisting 2×6 and $c(8 \times 2)$ domains	[13], 16, 26
				2×6	Two As dimer/four missing As dimer cells	[13], 16, 26
				1×6	(2×6) Unit cell containing two As–As dimers	[26]
				2×4	Three As dimer/one missing As dimer cell; 2/3, 1/3 As dimer/missing As dimer	[19], 26, 46, 48
				$c(2 \times 8)$	Antiphase array of (2×4) cells	[19, 26]
				$c(4 \times 8)$	Triple As dimer adatoms	[26], 47, 49
				(1×1)	As saturated	[27, 28]
				(2×1)	1/2 monolayer of Ga vacancies	[19]

13.9 Reflection High-Energy Electron Diffraction

		(111)Ga	ArB + anneal 450 °C	2 × 2	–	[27–29]
			MBE, $T = 377$ °C	2 × 2	As stabilized, regular array of Ga vacancies	[29–31]
		(111)As	ArB + anneal 550 °C	2 × 2	1/4 missing Ga	[19]
			ArB + anneal 450 °C	3 × 3	(110) facets at $T \sim 600$ °C	[28]
			MBE, $T = 377$ °C	2 × 2	Trimers on ideal As plane	[30–32]
			MBE + anneal 300 °C	2 × 2	Unit cell: 7/4 As/Ga, adsorbed As trimer	[32]
			MBE + anneal 500 °C	$(\sqrt{3} \times \sqrt{3})R30°$	Intermediate between $\sqrt{19} \times$ and $\sqrt{3} \times$	[50]
			MBE, $T = 500$ °C	$\sqrt{19} \times \sqrt{19} - 23.4°$	Ga stabilized	[30]
			MBE + anneal 300 °C + anneal 500 °C	$\sqrt{19} \times \sqrt{19} - 23.4°$	As-capped hexagonal ring	[32, 33]
GaP	Zinc blende	(111)Ga	ArB + anneal 550 °C	2 × 2	1/4 missing Ga, nearly coplanar GaP bilayer	[33]
InP	Zinc blende	(100)	ArB + anneal 330 °C	4 × 1	Diffuse streaks	[34]
GaSb	Zinc blende		Prolonged annealing	4 × 2	–	
		(111)Ga	ArB + anneal 400 °C	2 × 2	–	[28]
InSb	Zinc blende	(111)Sb	ArB + anneal 400 °C	3 × 3	Facets for $T \sim 600$ °C	[28]
		(111)In	ArB + anneal	2 × 2	–	[27]
		(111)Sb	ArB + anneal	3 × 3	–	[27]
InAs	Zinc blende	(111)In	ArB + anneal	2 × 2	Possible contamination	[35]
			ArB + anneal	2 × 2	–	[36]
		(111)As	ArB + anneal	3 × 3	Facets form before superstructure	[36]

(continued overleaf)

Table 13.1 (continued)

Crystal	Bulk structure	Surface	Preparation	ELEED pattern	Comments: structure	Selected refs.
CdTe	Zinc blende	(111)Cd	ArB, no anneal	1 × 1	No comments on disorder	[37]
CoO	Cubic	(111)	Oxidation of Co(0001)	1 × 1	Top O layer contracted by 15%	[38]
MgO	Cubic	(111)	Chemical etch	1 × 1	(100) facets	[39]
NaO_2	CaF_2	(111)	ArB + anneal	1 × 1	Termination between two Na layers, bulk structure	[40]
TiO_2	Rutile	(100)	ArB + anneal + high T (1100 °C)	–	(110) facets with (2 × 1) reconstruction (114) facets	[41]
MoS_2	Molybdenite	(0001)	Cleaving	1 × 1	S–Mo layer spacing contracted 5% first van der Waals gap contracted 3%.	[42, 43]
$NbSe_2$	Molybdenite	(0001)	Cleaving	1 × 1	S–Nb layer spacing contracted 0.6% first van der Waals gap contracted 1.5%.	[42, 43]
TiS_2	CdI_2	(0001)	Cleaving	1 × 1	S–Ti layer spacing contracted 5%, first van der Waals gap contracted 5%	[44]
$TiSe_2$	CdI_2	(0001)	Cleaving	1 × 1	Se–Ti layer spacing contracted 5%, first van der Waals gap contracted 5%	[44]
LaB_6	CaB_6	(111)	ArB + anneal 1600 °C	1 × 1	Relaxed, La atoms move toward the surface	[45]
		(110)	ArB + anneal 1600 °C	c(2 × 2)	–	[45]

Adapted from Kahn [3] with permission. Kohl et al. [20], Van Hove and Leyson [21], Duke and Lubinsky [22], Chang and Mark [23], Lubinsky et al. [24], Chang and Mark [25], Bachrach et al. [16], Drathen et al. [14], Biegelsen et al. [26], Pashley et al. [19], Guichar et al. [27], MacRae [28], Haberern and Pashley [29], Cho [30], Tong et al. [31], Biegelsen et al. [32], Xu et al. [33], Moisson and Bensoussan [34], Haneman [35], Grant and Haas [36], Solzbach and Richter [37], Ignatiev et al. [38], Heinrich [39], Anderson et al. [40], Firment [41], Mrstik et al. [42], Van Hove et al. [43], Van Hove and Tong [44], Nishitani et al. [45], Larsen et al. [46], Biegelsen et al. [47], Larsen et al. [48], Sauvage-Simkin et al. [49], Jacobi et al. [50].

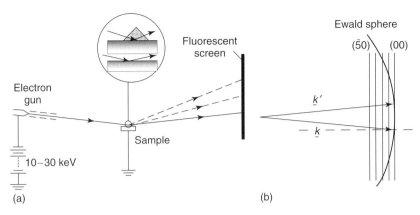

Figure 13.16 (a) Schematic of the RHEED experimental setup. The inset shows two different scattering situations on a highly enlarged surface area: surface scattering on a flat surface (below) and bulk scattering by a three-dimensional crystalline island on top of the surface (above). (b) The Ewald sphere and intersection points of the reciprocal lattice rods (hk) at specific $\mathbf{k}$ vectors. $\mathbf{k}$ and $\mathbf{k}'$ are primary and scattered wave vectors, respectively. The sphere radius $k = k'$ is much larger than the distance between the reciprocal lattice rods.

The energy broadening contributes to the angular spread 2β of the incident beam. This in turn establishes a *coherence length* over which surface atoms contribute IP to the diffraction pattern.

$$\Delta r_c \cong \lambda / \left\{ 2\beta \sqrt{\left[1 + \left(\frac{\Delta E}{2E}\right)^2\right]} \right\} \tag{13.9}$$

where λ is the wavelength of the incident electron. Hence, the larger the energy broadening ΔE is, the smaller is the coherence length of diffraction from the surface. Table 13.2 provides a comparison of LEED and RHEED coherence lengths. The higher energy and lower beam spread of the RHEED beam results in a lower angular spread and, for comparable thermal energy broadening, a coherence length that is twice as large as that for LEED.

Of course, another significant advantage for RHEED is that the grazing incident angle permits measurement of surface geometry during the MBE growth process.

Table 13.2 Comparison of energy E, energy broadening ΔE, angular spread 2β, and coherence length Δr_c for LEED versus RHEED.

	E (eV)	ΔE (eV)	2β (rad)	Δr_c (Å)
LEED	~100	~0.5	10^{-2}	~100
RHEED	1–5×10^4	~0.5	10^{-4}–10^{-5}	>~200

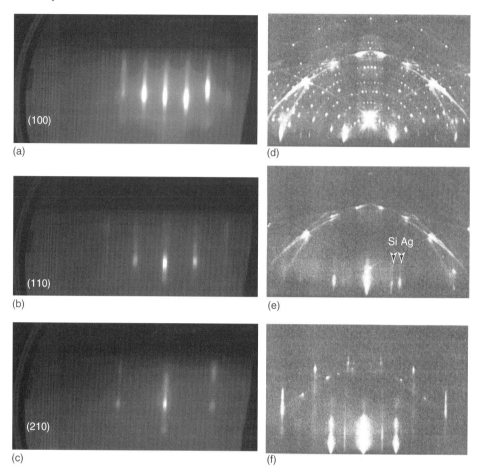

Figure 13.17 SrTiO₃ on Si(100) RHEED patterns taken with 15 keV along (a) (100), (b) (110), and (c) (210). RHEED patterns taken with a primary energy of E = 15 keV incident along the [112] direction on a Si(111) surface: (d) Clean Si(111) surface with a (7 × 7) superstructure. (e) After deposition of nominally 1.5 monolayers (MLs) of Ag streaks due to the Ag layers are seen on the blurred (7 × 7) structure. (f) After deposition of 3 ML of Ag [51]. The textured structure is due to the Ag layers that develop in place of the (7 × 7) structure.

LEED requires an electron gun and screen normal to the surface, which is typically precluded by the source flanges used for MBE growth.

13.9.1
RHEED Oscillations

The use of RHEED during growth has another advantage. By observing the intensity variations of the diffracted beam, one can obtain a direct measure of the growth rate and the crystal quality of the growth. The diffraction intensity oscillates between

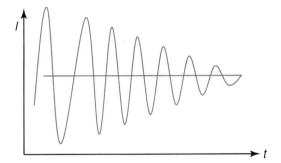

Figure 13.18 Schematic illustration of RHEED oscillations in intensity I as a function of elapsed growth time t.

high and low intensity when each layer of atoms is completed versus when each layer is only half complete. Figure 13.18 illustrates these RHEED oscillations in intensity as a function of elapsed time during growth. These oscillations allow the growth operator to "count" the number of atomic layers and the time interval between layers. As the epitaxial layer thickens, its surface can become less flat and the contrast between completed and half-completed layers becomes less pronounced.

13.10 Summary

This chapter presented a description of the diffraction techniques used to characterize atomic order at clean, adsorbed, or thermally annealed semiconductor surfaces.

Using LEED and RHEED techniques, researchers have found numerous atomic reconstructions that depend on surface orientation, surface stoichiometry, and thermal annealing.

The variety of different reconstructions involving large numbers of surface atoms in an aggregate structure demonstrates that extended, coherent features of the surface topology can help minimize strain and overall energy for structural stability.

These reconstructions can involve a variety of bond rotations, missing atoms, dimerization, and composites of simpler domain structures.

On the other hand, for the same crystal orientation, the reconstruction features are the same for the same family of III–V and II–VI compounds.

Within a class of III–V or II–VI compound semiconductors, the bond rotation model can be used to account for their similar reconstructions.

Different reconstructions with the same surface atom periodicity underscore the need for multiple techniques in order to distinguish between various models of reconstruction.

Problems

1. Derive Equation 13.1
2. For a hemispherical LEED screen located 5.0 cm from an ordered surface, 100 eV electrons produce a square (1 × 1) pattern of LEED spots with 1.05 cm spacing when projected as a flat pattern. What is the atomic lattice constant at the surface? Suppose the pattern becomes (2 × 1) with spots at twice the minimum spacing. What is the additional periodic spacing and the unit mesh? What are three possible reasons for this new periodicity?
3. Oxygen adsorption causes the (1 × 1) LEED pattern of a semiconductor to become (2 × 1) with half order spots. Describe two possible causes for this transition.
4. The transition in electron diffraction patterns is useful in calibrating the temperature of GaAs growth substrates. Identify a suitable reconstruction transition and calibration temperature.

References

1. Davisson, C. and Germer, L.H. (1927) *Phys. Rev.*, **30**, 705.
2. http://lasuam.fmc.uam.es/lasuam/img/glossary/leed1.gif. Surface Science Lab at Universidad Autónoma de Madrid (2009)
3. Kahn, A. (1983) *Surf. Sci. Rep.*, **3**, 193.
4. Duke, C.B. (1977) *J. Vac. Sci. Technol.*, **14**, 870.
5. Gatos, H.C., Moody, P.L., and Lavine, M.C. (1960) *J. Appl. Phys.*, **31**, 212.
6. Duke, C.B. (1992) *J. Vac. Sci. Technol. A*, **10**, 2032.
7. Wood, E.A. (1964) *J. Appl. Phys.*, **35**, 1306.
8. Moore, J.H. and Spencer, N.B. (2001) *Encyclopedia of Chemical Physics and Physical Chemistry*, Taylor and Francis, Abingdon, pp. 1551–1553.
9. Lander, J.J. and Morrison, J. (1964) *Surf. Sci.*, **2**, 553.
10. Takayanagi, K., Tanishiro, Y., Takahashi, M., and Takahashi, S. (1985) *J. Vac. Sci. Technol. A*, **3**, 1502.
11. Däweritz, L. and Hey, R. (1990) *Surf. Sci.*, **236**, 15.
12. Mönch, W. (1985) in *Molecular Beam Epitaxy and Heterostructures*, NATO ASI Series, vol. 87, Martinus Nijhoff, Dordrecht, p. 331.
13. van Bommel, A.J., Crombeen, J.E., and van Dirshot, T.G.J. (1978) *Surf. Sci.*, **72**, 95.
14. Drathen, P.W., Ranke, W., and Jacobi, K. (1978) *Surf. Sci.*, **77**, L162.
15. Massies, J., Etienne, P., Dezaly, F., and Linh, N.T. (1980) *Surf. Sci.*, **99**, 121.
16. Bachrach, R.Z., Bauer, R.S., Chiaradia, P., and Hansson, G.V. (1981) *J. Vac. Sci. Technol.*, **18**, 797.
17. Arthur, J.R. (1974) *Surf. Sci.*, **43**, 449.
18. Vitomirov, I.M., Raisanen, A.D., Brillson, L.J., Lin, C.L., McInturff, D.T., Kirchner, P.D., and Woodall, J.M. (1993) *J. Vac. Sci. Technol. A* 11, 841.
19. Pashley, M.D., Habererm, K.W., Friday, W., Woodall, J.M., and Kirchner, P.D. (1988) *Phys. Rev. Lett.*, **60**, 2176.
20. Kohl, D., Henzler, M., and Heilan, G. (1974) *Surf. Sci.*, **41**, 403.
21. Van Hove, H. and Leyson, R. (1972) *Phys. Status Solidi*, **9**, 361.
22. Duke, C.B. and Lubinsky, A.R. (1975) *Surf. Sci.*, **50**, 605.
23. Chang, S.C. and Mark, P. (1974) *Surf. Sci.*, **46**, 293.

24. Lubinsky, A.R., Duke, C.B., Chang, S.C., Lee, B.W., and Mark, P. (1976) *J. Vac. Sci. Technol.*, **13**, 189.
25. Chang, S.C. and Mark, P. (1975) *J. Vac. Sci. Technol.*, **12**, 629.
26. Biegelson, D.K., Bringans, R.D., Northrup, J.E., and Schwartz, L.-E. (1990) *Phys. Rev. Lett.*, **65**, 452.
27. Guichar, G.M., Sébenne, C.A., and Thuault, C.D. (1979) *J. Vac. Sci. Technol.*, **16**, 1212.
28. MacRae, A.U. (1966) *Surf. Sci.*, **4**, 247.
29. Haberern, K.W. and Pashley, M.D. (1990) *Phys. Rev. B*, **41**, 3226.
30. Cho, A.Y. (1970) *J. Appl. Phys.*, **41**, 2780.
31. Tong, S.Y., Mei, W.N., and Xu, G. (1984) *J. Vac. Sci. Technol. B*, **2**, 393.
32. Biegelson, D.K., Bringans, R.D., Northrup, J.E., and Schwartz, L.-E. (1990) *Phys. Rev. B*, **41**, 5701.
33. Xu, F., Hu, G.W.Y., Puga, M.W., Tong, S.Y., Yeh, J.L., Wang, S.R., and Lee, B.W. (1985) *Phys. Rev. B*, **32**, 8473.
34. Moisson, J.M. and Bensoussan, M. (1982) *J. Vac. Sci. Technol.*, **21**, 315.
35. Haneman, D. (1960) *J. Phys. Chem. Solids*, **14**, 162.
36. Grant, J.T. and Haas, T.W. (1971) *Surf. Sci.*, **26**, 669.
37. Solzbach, U. and Richter, H.J. (1980) *Surf. Sci.*, **97**, 191.
38. Ignatiev, A., Lee, B.W., and Van Hove, M.A. (1977) in *Proceedings of the 7th International Vacuum Congress and 3rd International Conference on Solid Surfaces, Vienna* (R.Dobrozemsky, F. Rüdenauer, F.B. Viehböck, and A. Breth), F. Berger und Söhne, Vienna, p. 1733.
39. Heinrich, V.E. (1976) *Surf. Sci.*, **57**, 385.
40. Anderson, S., Pendry, T.B., and Echenique, P.M. (1977) *Surf. Sci.*, **65**, 539.
41. Firment, L.E. (1982) *Surf. Sci.*, **116**, 205.
42. Mrstik, B.J., Kaplan, R., Reinecke, T.L., Van Hove, M.A., and Tong, S.Y. (1977) *Phys. Rev. B*, **15**, 897.
43. Van Hove, M.A., Tong, S.Y., and Elconin, M.H. (1977) *Surf. Sci.*, **64**, 85.
44. Van Hove, M.A. and Tong, S.Y. (1979) *Surface Crystallography by LEED*, Springer, Berlin.
45. Nishitani, R., Aono, M., Tanaka, T., Oshima, C., Kawai, S., Iwasaki, H., and Nakamura, S. (1980) *Surf. Sci.*, **93**, 535.
46. Larsen, P.K., van der Veen, J.F., Mazur, A., Pollman, J., Neave, J.H., and Joyce, B.A. (1982) *Phys. Rev. B*, **26**, 3222.
47. Biegelson D.K., Bringans R.D., and Schwartz L.-E. (1990) in *Surface and Interface Analysis of Microelectronic Processing and Growth*, Proceedings of SPIE, vol. 1186 (L.J.Brillson and F.H. Pollak), SPIE, Bellingham, WA, p. 136.
48. Larsen, P.K., Neave, J.H., van der Veen, J.F., Dobson, P.J., and Joyce, B.A. (1982) *Phys. Rev. B*, **27**, 4966.
49. Sauvage-Simkin, M., Pinchaux, R., Massies, J., Calverie, P., Jedrecy, N., Bonnet, J.E., and Robinson, I.L. (1989) *Phys. Rev. Lett.*, **62**, 563.
50. Jacobi, K., Mushwitz, C.B., and Ranke, W. (1979) *Surf. Sci.*, **82**, 270.
51. Hasagawa, S., Daimon, H., and Ino, S. (1987) *Surf. Sci.*, **186**, 138.

14
Scanning Tunneling Microscopy

14.1
Overview

The scanning tunneling microscopy (STM) technique has revolutionized our ability to image surfaces on an atomic scale. These images can reveal specifics of atomic arrangement, atomic orbital bonding, and electrostatic potential that were previously not possible. Combined with other experimental techniques and theory, STM provides a powerful tool for understanding electronic, chemical, and structural properties of surfaces.

STM is based on the principle that charge can tunnel across the vacuum between a metallic tip and a conducting surface. For extremely small tip–surface spacings (<10 Å), a small tunnel current, typically 10^{-12}–10^{-9} A, flows across the gap between them. If the metallic tip is sharpened to a single atom, it is possible to create surface maps that reflect the atomic morphology and orbital electronic states.

The basic STM hardware consists of a ultrahigh vacuum (UHV) chamber with an atomically sharp metallic tip that can be controlled both normal and transverse to a sample surface with a precision greater than 1 Å. Piezoelectric drivers are used to achieve this control. It is critical to eliminate mechanical vibration of the tip and sample. This basic STM configuration is displayed schematically in Figure 14.1 [1].

The tunnel current I_T depends sensitively on the tip–surface spacing according to the expression

$$I_T \propto (V/d) \exp\left(-kd\sqrt{\phi_{av}}\right) \qquad (14.1)$$

Here, V is the applied voltage and is much less that the work function of tip or sample, ϕ_{av} is the *average work function*, and $k \sim 1.025 \text{Å}^{-1} \, eV^{-\frac{1}{2}}$. To achieve sufficient height resolution, the distance d must be controlled to a precision of 0.05–0.1 Å. Lateral resolution must also be controlled to within 1–2 Å in order to obtain atomically resolved images.

There were two basic challenges to achieving atomic precision with STM: (i) reducing mechanical vibrations down to the sub-Å level and (ii) preparing tips with atomic sharpness. The first of these challenges was overcome only

Surfaces and Interfaces of Electronic Materials. Leonard J. Brillson
Copyright © 2010 WILEY-VCH Verlag GmbH & Co. KGaA, Weinheim
ISBN: 978-3-527-40915-0

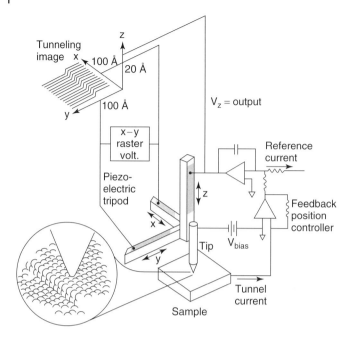

Figure 14.1 Schematic of the STM tip, sample surface, and electric circuit used to detect tunneling current. The current from tip to sample with an applied bias voltage V_{bias} is maintained constant by an electronic feedback system that controls the tip position normal to the sample by means of a piezoelectric tripod. The voltage applied to the z-leg of the tripod as the x–y tripod legs raster scan the tip laterally across the surface produce a tunneling image.

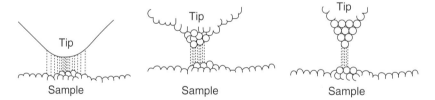

Figure 14.2 Visualization of the tunneling current (dashed lines) between tip and sample surface. A blunt tip consisting of large numbers of atoms (a) yields only low-resolution images, whereas tips with only a few atoms (b) or ideally one atom (c) achieve proportionally higher spatial resolution. (After [2].)

with ingenious design of piezoelectric positioners and vibration isolation. The second challenge has required painstaking preparation of tips using a variety of electrochemical and UHV techniques.

The STM tip should ideally consist of a single protruding atom in order to achieve surface images with atomic-scale resolution. Figure 14.2 represents the tunnel current between tip and surface for successively smaller numbers of atoms

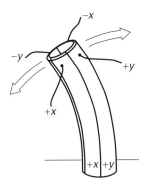

Figure 14.3 A piezoelectric scanner tube with electrodes attached to its side to drive changes in tube length along different directions.

in close proximity to a sample surface [2]. As the number of protruding atoms shrinks, the spatial resolution of the tip improves accordingly.

The STM images depend sensitively on the shape of the tip and its electronic properties. However, it is usually not possible to know what the tip's shape is prior to an experiment. Rather the images themselves can reveal whether or not the tip consists of a single atom. Contact between the tip and surface can often rearrange tip atoms, thereby changing subsequent images. The tip's electronic properties can also change with tip–surface contact as well as by adsorption of new atoms.

The tip itself is typically mounted on a piezoelectric *scanner tube* that moves by atomic-scale increments with applied voltage. Figure 14.3 illustrates such a tube fixed at its base with electrodes along its sides that cause the tube to change its length with applied bias. A combination of applied voltages drives the free end of the tube to move in two or three dimensions over distances that are controllable on an Å scale.

The reward for achieving atomic resolution can be seen from the remarkable images that are now widely available. Figure 14.4 illustrates such an image for the Si(111) (7 × 7) surface discussed in previous chapters. This image is notable in several respects. It illustrates the basic features of the (7 × 7) reconstruction discussed in Chapter 13 including the various superstructure atoms and the extended nature of its periodicity. The image also reveals changes in domain structure between the upper and lower halves of the topography. Finally, the image reveals wavelike features spreading across the lower half of the topography indicating a long-range coupling of the surface atoms. The high information content of such images illustrates the dramatic progress now possible in understanding surface atomic structure and its related chemical and electronic properties.

14.2
Tunneling Theory

STM is based on the tunneling of electrons from an atomically sharp, metallic tip to a conducting solid surface. In quantum mechanical terms, this involves the

Figure 14.4 STM image of the Si(111) (7 × 7) surface. (Courtesy, J. Pelz, Ohio State University.)

overlap of electron wavefunctions

$$\psi_T(z) = \psi_T^0 e^{-k(d-z)} \tag{14.2}$$

$$\psi_s(z) = \psi_s^0 \exp^{-kz} \tag{14.3}$$

of the tip and the surface, respectively. The wave vector

$$k = \left[2m(V-E)/\hbar^2\right]^{1/2} = [2m\phi]^{1/2}/\hbar \tag{14.4}$$

where ϕ represents the energy barrier for electrons with kinetic energy E relative to a local potential V. Figure 14.5 illustrates how wavefunctions from tip and solid surface extend away into vacuum. For metallic solids, electrons occupy a continuum of energies up to the Fermi level.

In this "bath" of electrons, those near the Fermi level are free to move into unoccupied states within the solid, so that they behave as traveling waves whose wavefunctions we can approximate by sinusoidal functions. Outside the tip and solid, however, these wavefunctions decay exponentially in amplitude over distances of only a few angstroms (Figure 14.5a). If the distance between the tip and solid is sufficiently small, then these "leaking" wavefunctions can overlap and there will be finite probability of tunneling between electrons in the two "baths" (Figure 14.5b). The shape of the barrier through which such tunneling takes place will depend on the distance between the two surfaces, their work functions, as well as the work function difference.

When the two solids are joined electrically, their Fermi levels align. A voltage applied between them shifts these Fermi levels by an amount as shown. Electrons

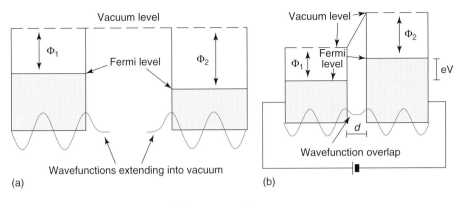

Figure 14.5 Simplified illustration of electrons near the Fermi level of two metallic solids with work functions Φ_1 and Φ_2.

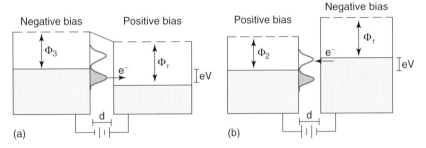

Figure 14.6 Tunneling from filled to empty densities of states at a given energy and position as a function of applied bias (electron volts).

in the solid sample can tunnel into unoccupied states in the tip so that a tunnel current I_T of electrons described by Equation 14.1 passes between solid and tip. Both the gap distance d and potential ϕ change locally, so that a record of I_T versus position produces a "map" as the tip is scanned laterally across the surface.

Figure 14.6a illustrates tunneling from a filled density of states (DOS) at a solid surface at negative bias into an empty DOS of the tip biased more positively. Similarly, Figure 14.6b shows tunneling into empty states of the same solid when the tip is biased more negatively than the solid. Therefore, by scanning the tip – solid bias between positive and negative voltages, one can probe both the filled and empty densities of states locally of a surface layer.

Figure 14.7 illustrates the difference in spatial distribution between filled and empty states. Theoretical state-density calculations indicate that the occupied state density is concentrated around the surface As atoms, while the unoccupied state density is situated around the Ga atoms [3]. Positive GaAs voltage induces tunneling

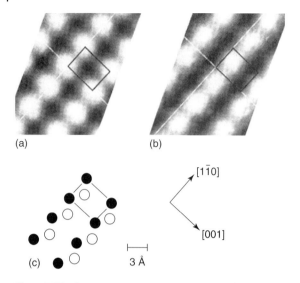

Figure 14.7 Constant-current STM images for the GaAs(110) surface simultaneously acquired at (a) +1.9 and (b) −1.9 V sample voltages. Surface height in grey scale ranges from 0 (black) to (a) 0.83 and (b) 0.65 Å (white). (c) A top view of the surface atoms represents As (open circles) and Ga (filled circles) atoms with the unit cell indicated in all three figures.

into empty state densities while negative GaAs bias produces tunneling from filled GaAs states into the tip.

From $(dI/dV)/(I/V)$ scans of the GaAs(110) surface versus applied bias, one obtains the ratio of differential to total conductivity that reach maxima at −1.9 eV due to tunneling into valence band states and +1.5 eV due to tunneling out of conduction band states. While both d and ϕ contribute to such I_T maps, it is possible to distinguish between these components based on their different spatial and gap dependencies. For example, one can extract values of ϕ since $\phi \propto (\partial \ln I / \partial d)^2$. See below. Contrast between images acquired with opposite bias voltages can reveal features due to filled versus empty states. Likewise, STM images can be compared with theoretical calculations of atomic positions to reveal deviations attributed to surface electronic variations [4].

Figures 14.7a,b show a shift of lateral position in tunneling intensity that is consistent with the unit cell structure indicated in Figure 14.7c. The spatial separation between occupied and unoccupied states is due to the charge transfer that occurs from Ga and As atoms at the reconstructed free (110) surface. This is consistent with the bond rotation model for the GaAs(110) reconstruction pictured in Figure 13.5. Thus, bias-dependent STM can reveal the spatial distribution of filled and empty bond orbitals for atoms at the surface of semiconductors.

For a tip positioned over a specific location, the bias dependence of I_T enables one to extract a DOS, defined as the number of states, filled or unfilled, at a given energy E. The *local density of states* (LDOS) at a given position r_0 is thus,

$\psi^*(r_0, E)\psi(r_0, E)$ and the tunneling conductance between filled and unfilled states at the same energy can be expressed as

$$\sigma(r_0, E) \propto \rho(r_0, E) \propto \psi_s^*(r_0, E)\psi_s(r_0, E) \tag{14.5}$$

where r_0 is the spatial location and E is the energy of the filled and unfilled states involved in the tunneling transition between the two solids. The tunneling conductance is also proportional to the LDOS of the tip at this same energy. Ideally, this parameter remains constant as the tip is scanned across a surface.

In general, the tunneling current can be expressed as

$$I = \frac{2e\pi}{\hbar} \sum_{TS} f(E_T)[1 - f(E_S + eV)] |M_{TS}|^2 \cdot \delta(E_T - E_S) \tag{14.6}$$

where the Fermi function $f(E) = 1/(1 + exp((E - E_F)/k_B T))$ and the tunneling matrix element between tip and surface is

$$M_{TS} = -\hbar^2/2m \int dS \bullet (\psi_T * \nabla \psi_S - \psi_S * \nabla \psi_T) \tag{14.7}$$

in which ψ_T and ψ_S are the tip and surface wavefunctions respectively and m is the free electron mass [5]. M_{TS} is integrated over the area S defined by the tip and sample surfaces.

Both wavefunctions decay into vacuum so that $\psi_S(z) = \psi_S^0 e^{-kz}$ and $\psi_T(z) = \psi_T^0 e^{-k(d-z)}$, where wave vector $k = [2m(V-E)/\hbar^2]^{1/2} = [2m\phi]^{1/2}/\hbar$ and ϕ = local barrier height or average work function. Thus, the tunnel current I is proportional to $|\psi_S^0|^2|\psi_T^0|^2 e^{-2kd}$, showing an exponentially decreasing current with increasing tip–surface separation.

For a spherical tip of radius R, the tip and surface wavefunctions can be expanded in a series of exponentially decaying plane waves [6], yielding

$$I = (32\pi^3 e^2 V/\hbar) D_T(E_F) R^2 k^{-4} e^{-2kr} \sum |\psi_s(r_0)|^2 \delta(E - E_F) \tag{14.8}$$

The tunneling conductance $\sigma(r_0, E)$ is obtained by dividing the voltage V and is therefore proportional to the LDOS

$$\sigma(r_0, E) \propto \rho(r_0, E) = \sum |\psi_s(r_0)|^2 e^{-2kr} \tag{14.9}$$

Taking the logarithm of the tunnel current yields

$$\ln I \propto -2kr = -2[2m\phi]^{1/2} r/\hbar \tag{14.10}$$

so that

$$\partial \ln I/\partial r \propto \phi^{1/2} \tag{14.11}$$

and

$$\phi \propto (\partial \ln I/\partial r)^2 \tag{14.12}$$

Thus, STM can determine local potential or barrier height from the variation in current with tip–sample separation. In turn, the variation of potential laterally on an atomic scale provides a means to distinguish between morphological and electronic effects on the tunnel current.

14.3
Surface Structure

The STM technique is capable of revealing the atomic configuration of surfaces in exquisite detail. See, for example, Figure 14.4 of the Si(111) (7 × 7) surface reveals an intricate unit cell pattern as well as wavelike features extending tens of unit cells across the surface. Examples of such extended features include step edges, dislocations, and atomic clusters. Figure 14.8(a) shows an STM image of a silicon surface intentionally miscut 0.5° off the (001) axis [7]. This vicinal surface exhibits a succession of (001) atomic terraces that step down from upper left to lower right. Also, evident on these terraces are alternating planes of Si atoms with dimers oriented either parallel to or orthogonal to the step edges. Depending on the orientation of the rows in these domains, the step edges are either rough or smooth.

Figure 14.8 also shows extended features corresponding to clusters of atoms. In some cases, these clusters exhibit surface facets that are crystallographically oriented, such as the STM image of a Ge "hut" (b) and its cross-sectional counterparts obtained by transmission electron microscopy (TEM) (c) and (d) [8].

STM can also image atomic-scale features of samples in cross section. This capability permits direct measurement of interface atomic and electronic features. For example, cross-sectional STM of AlAs/GaAs superlattices illustrates Al–Ga interdiffusion and As precipitate coarsening. The topographic image in Figure 14.9a reveals As precipitates that are typically spherical with diameters ranging from 1 to 5 nm [9]. Figure 14.9a shows a preferential precipitation on the GaAs side of the AlAs/GaAs interface. Annealing promotes nucleation away from the interface and affects the apparent superlattice period.

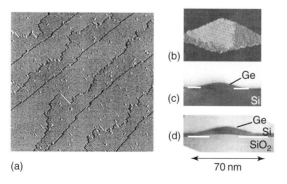

Figure 14.8 (a) STM image of a Si(001) surface misoriented by 0.5° [7]. A Ge quantum dot hut crystal on Si(001) and silicon-on-insulator (SOI) consisting of 1.5 nm of Ge deposited by MBE. (b) STM image, (c) TEM image on bulk Si, and (d) TEM image on SOI [8].

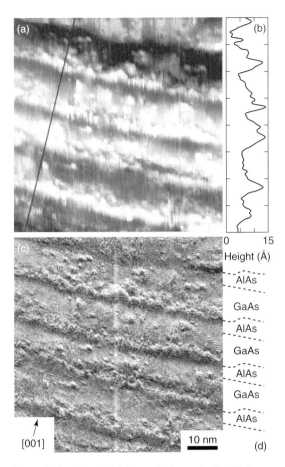

Figure 14.9 Figure 14.9 Nonstoichiometric AlAs/GaAs superlattice (a) Topographical map, (b) corresponding line profile along the [001]black line, (c) corresponding current image, and (d) identification of GaAs and AlAs layers [9].

On a yet larger scale, STM images reveal the effects of dislocations on surface atomic arrays. In Figure 14.10, a GaN(100) surface with a 6-nm overlayer of Pt displays features of the underlying semiconductor surface. The STM image exhibits steplike features as well as pits that distort or "pin" the steps into "wing"-like features [10]. These pits are known to form at the termination of threading or mixed threading/screw dislocations. The $\sim 10^8$ cm^{-2} density of these pits is in rough agreement with mixed dislocation densities measured by TEM.

Less common are spiral growth structures due to isolated screw dislocations. Here, also the steplike features extend over hundreds of nanometers. In this case, the ABAB packing of the GaN wurtzite lattice requires an even number of spiral arms.

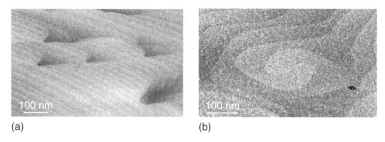

Figure 14.10 STM image of GaN(100) surface with a 6-nm thick Pt metal overlayer showing (a) monolayer steps and "wing" structures and (b) spiral growth structure. The granularity is due to nanometer-scale Pt clusters [10].

14.4
Atomic Force Microscopy

The scanning tunneling method of imaging surfaces is one of several scanning probe techniques. Other methods have been developed to measure surface morphology, as well as tribological, electrical, and magnetic forces. Figure 14.11 illustrates three detection modes of scanning probes. In Figure 14.11a, the tunneling current varies with morphology (dashed line) or DOS (dot-dashed line). In Figure 14.11b, forces between the tip and surface atoms drive deflections of a cantilever spring, causing changes in distance and thereby capacitance between the spring and an electrode. In Figure 14.11c, a laser beam reflected off a force-sensitive cantilever into a position-sensitive photodetector measures the cantilever deflection.

The cantilever is often fabricated from crystalline silicon by orientation-selective etching techniques developed for nanofabrication. The force exerted on the tip

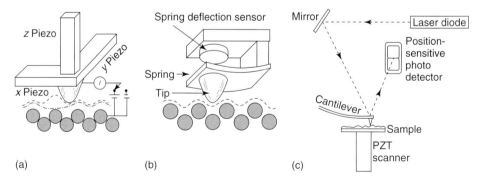

Figure 14.11 Detection modes of scanning probe techniques (a) scanning tunneling microscopy, (b) capacitance-sensitive force microscopy, and (c) optical-reflection-sensitive force microscopy.

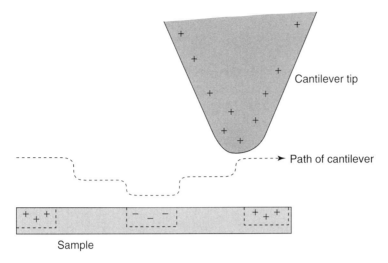

Figure 14.12 Electrostatic imaging by atomic force microscopy.

in proximity to a surface is given by Hooke's Law, that is, $\mathbf{F} = k \cdot \mathbf{d}$, where k is the *spring constant* of the cantilever and d is the cantilever's deflection. A major advantage of the force techniques is that they are sensitive to morphological and electrostatic features of all materials, regardless of their conductivity. Unlike STM, these atomic force microscopy (AFM) methods have proven useful in characterizing undoped or wide bandgap semiconductors and insulators including polymers and biological species. On the other hand, the spatial resolution of AFM is typically on a nanometer rather than an atomic scale.

Figure 14.11 illustrates methods to map morphology. AFM can also be adapted to measure electric potential or capacitance on a local scale. Figure 14.12 shows the path of a tip mounted on a cantilever as it passes over sample regions with different localized charge. Charge on the cantilever tip experiences an electrostatic force that is proportional to the charge on or below the surface. From the known tip–surface separation, this force can then be converted into maps of charge density, capacitance, or electric potential. Variations in these parameters across a surface can arise from composition, doping, morphological, and native point defects, as well as other charge trapping sites.

An alternative method of measuring surface electrostatic potential termed *Kelvin force probe microscopy* (KPFM) involves using the tip as a capacitor plate. Here, one measures the voltage required to null out the displacement current induced by the vibrating tip due to the voltage difference between tip and surface. This vibrating capacitance technique is also the basis for *surface photovoltage* (SPV) techniques to be discussed in Section 15.8. In the absence of excess charges, AFM is also sensitive to the much weaker force of van der Waals attraction.

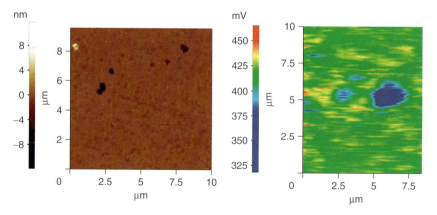

Figure 14.13 ZnO(000$\bar{1}$) maps of (a) morphology and (b) electric potential.

Surface potential and morphology measurements can be performed in tandem. For surface potential measurements, a constant average tip–surface separation is required. This separation is set either by (i) scanning an entire area and rescanning while maintaining constant separation or (ii) alternately adjusting the tip–surface spacing for morphology and capacitance in real time. For charged sites below a surface, the surface morphology and potential maps can differ significantly. Figure 14.13 illustrates morphology and electric potential maps for a ZnO(000$\bar{1}$) surface [11]. This surface exhibits a root-mean-square (rms) roughness of 0.63 nm and morphological features such as pits and scratches. The corresponding electric potential map displays variations in surface potential of hundreds of millielectronvolts. Significantly, these variations do not show a strong correlation, indicating that charges in states below the surface contribute to the changes in potential. The spatially localization of these features indicates their defect rather bulk lattice nature.

STM and AFM can also sense magnetic properties. In both cases, a magnetically coated and polarized tip scans across the surface. See Figure 14.14 [12]. In STM, spin polarized electrons tunnel into states of the same polarization in the sample. In AFM, the magnetic force between tip and surface determines the tip deflection as shown in Figure 14.14a. Figure 14.14b illustrates a *magnetic force microscopy* (MFM) image of the bits of a hard disk. Here, the image shows the micron scale of magnetic domains and their patterning. Such images are useful in understanding and controlling magnetic storage media.

Tribological measurements of friction are also possible with the scanning probe tip. By dragging the tip in contact with the surface, changes in the force required to deflect the cantilever provide a measure of surface friction or adhesion between an adsorbate and the surface. On the other hand, this *lateral force microscopy* (LFM) is not useful for measuring surface morphology or electronic features on atomic

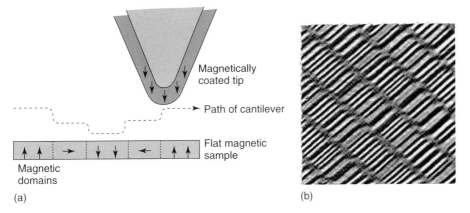

Figure 14.14 (a) MFM scanning of magnetic domains at a surface. (b) MFM image showing the pattern of bits of a hard magnetic storage disk. Field of view is 30 μm.

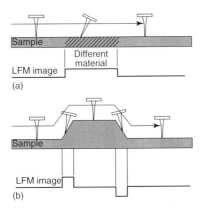

Figure 14.15 Lateral deflection of the cantilever from changes in surface friction (a) and changes in slope (b).

scale since both the surface and tip are significantly perturbed during the scanning process.

Figure 14.15 illustrates the LFM technique with changes in surface friction (a) or surface height (b) [13]. The LFM image scans exhibit positive or negative signal changes, depending on the increase or decrease in friction.

14.5
Ballistic Electron Emission Microscopy

Scanning probe microscopy can also reveal electronic features of semiconductor surfaces and metal–semiconductor interfaces. This is accomplished by introducing a third terminal in the tunneling process that controls the kinetic energy of the

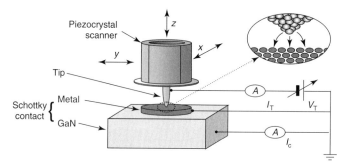

Figure 14.16 Schematic configuration of BEEM. (Courtesy, J. Pelz, Ohio State University.)

tunneling electrons. These ballistic electrons will pass into the sample if their kinetic energy exceeds the semiconductor's conduction band minimum. Figure 14.16 illustrates the *ballistic electron emission microscopy* (BEEM) experiment. A variable voltage V_T is applied between tip and metal overlayer, resulting in a current I_T of ballistic electrons. The collected current I_C exhibits an onset at a bias voltage that corresponds to the energy $E_C - E_F$ as shown in Figure 14.17a and represents the Schottky barrier $q\varnothing_B$. This onset in the BEEM spectrum is shown in Figure 14.17b. It represents a direct measure of the Schottky barrier at a particular location on a local scale. Additional onsets can be observed that correspond to tunneling into higher lying conduction bands [14].

The lateral resolution and signal intensity of BEEM is limited by scattering in the metal overlayer. On the basis of conservation of energy and momentum, the beam of electrons is sharply forward focused, exhibiting an intensity distribution of transmitted electrons

$$D(\theta) \propto \exp\left(-\frac{\sqrt{2m}}{\hbar} d \frac{E}{\sqrt{\Phi}} \sin^2 \theta\right) \qquad (14.13)$$

for angles θ around the forward direction. This transmission decreases with thickness and decreases with surface work function Φ. The more normal the angle of incidence θ, the narrower the cone of transmission and the higher the spatial resolution.

BEEM electron distributions typically spread by 1–2 nm after passing through a 5 nm thick metal film. This broadening is increased depending on the roughness of the metal film and/or the surface below. Hence, there is a tradeoff in terms of the minimum metal overlayer thickness required to achieve low scattering while maintaining uniform coverage and low resistivity compared with the rest of the BEEM bias circuit.

The BEEM images at energies near the threshold of collected current give information about the uniformity of the Schottky barrier between overlayer metal and the underlying semiconductor. Figure 14.18 illustrates a comparison between STM topography and BEEM collected current for a Au–GaAs interface prepared in UHV [15]. The STM spectra show a smoothly varying topography of

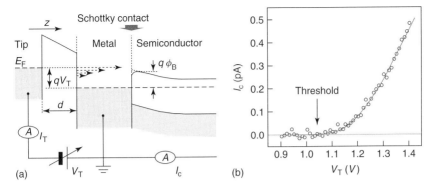

Figure 14.17 (a) Energy-level diagram of BEEM. (b) Representative BEEM spectrum and the fitting curve to extract local Schottky barrier height.

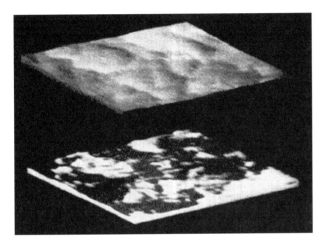

Figure 14.18 STM and BEEM images of a 10-nm Au film on a chemically prepared GaAs(100) Schottky barrier structure over a 510 × 390 Å^2 area. The light versus dark areas in the BEEM image correspond to a contrast of 14 pA with a 1.5-V tunnel bias voltage [15].

the MBE-grown GaAs surface, while the BEEM spectrum shows nanometer-scale domain structure attributed to outdiffusion of Ga and formation of As precipitates. A two-monolayer epitaxial AlAs layer deposited on the GaAs prior to Au deposition inhibits this outdiffusion, resulting in much more uniform BEEM spectra.

Besides Schottky barriers, BEEM can also be used to probe heterojunction band offsets [16], quantum wells [17], and morphological features such as dislocations [10]. Scanning probe tips can also be used to inject charge into semiconductors and induce luminescence. This technique is discussed in Chapter 16 as part of cathodoluminescence spectroscopy.

The advantages of the BEEM technique can be summarized as (i) nanometer-scale, real-space information about buried interfaces, (ii) energy resolution, typically 5 meV at room temperature, (iii) direct information about local conduction bands, (iv) direct information about local electric fields, and (v) the ability to study "hot" carriers in semiconductors.

The disadvantages of BEEM are (i) the technique requires a scanning tip that can and often change during experiments, (ii) BEEM requires nonleaky barriers to minimize noise, (iii) it requires a metal contact and a conducting substrate, thereby limiting its applicability, (iv) multiple interfaces are involved, potentially complicating analysis, and (v) the hot-carrier scattering at interfaces is not well known. Notwithstanding these complications, the BEEM technique represents a powerful method for determining local electronic band structure and barrier heights.

14.6
Atomic Positioning

The scanning probe tip can be used as a tool to position atoms on a surface. The resultant structures have produced marvelous imagery that exhibit quantum mechanical features of the surface electrons. A probe tip brought near an adsorbed atom on a surface induces a van der Waals force attraction that allows the tip to drag the atom across the surface. The atom remains in close proximity to the surface so that it is still adsorbed once the tip retracts.

Figure 14.19 illustrates the positioning technique for inert atoms such as Xe that are attracted only weakly to the surface [18]. This permits the tip to move the atom without making contact so that the atom remains on the surface as the tip is retracted. For adsorbed atoms that exchange charge with a surface, the tip can be biased so that the tip–atom attraction is stronger than the surface–atom bonding. Once the atom is repositioned, the tip polarity is reversed and the atom rebonds with the substrate. Needless to say, such methods require UHV conditions to prevent any adsorbate atoms from screening the forces among tip, adsorbate, and surface.

Atomic positioning provides visualizations of the wavefunction nature of atoms. Figure 14.20 shows two such examples. Figure 14.20a illustrates a quantum "corral" consisting of Fe atoms adsorbed on a Cu(111) surface and positioned in a circle using the tip of a low-temperature STM. Each peak corresponds to an individual Fe atom. The wave pattern is due to scattering of the Cu electrons in the surface off the Fe adatoms and point defects that produces an interference pattern [19].

Figure 14.20b shows the Fe atoms arrayed as a quantum "stadium" [19]. In this case, the quantum interference pattern exhibits two foci related to the two semicircles.

A practical application of atomic positioning may include the construction of electronic circuits that use orders of magnitude less energy than their macroscopic counterparts. Figure 14.20(c) shows a three-input sorter cascade of CO molecules arranged on a Cu substrate [17, 20]. An illustration of the equivalent electrical

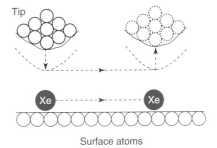

Figure 14.19 An STM tip brought close to an adsorbed atom can drag and reposition it on the surface before it is retracted. The atom remains in contact with the surface. ((a) and (b)) and from CO molecules on Cu (c).

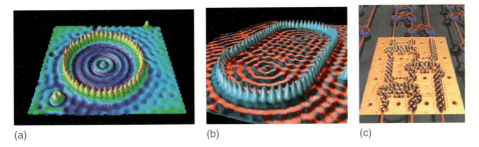

Figure 14.20 Quantum interference patterns from Fe atom arrays on Cu(111) surfaces ((a) and (b)) and from CO molecules on Cu (c).

schematic appears nearby. CO molecules positioned in registry with the substrate experience a mild repulsion on neighboring lattice sites. Their movement to more stable lattice sites can be triggered in sequence in analogy to a line of dominos. Suitably arranged at intersections, such arrays can perform logic functions. This example of nanometer-scale computer circuitry is estimated to use only 1 eV of energy, 100 000 less than the equivalent semiconductor circuitry [21].

14.7 Summary

Scanning probe techniques have opened a "window" into the structure and properties of atoms at surfaces. These advances have depended critically on the ability to isolate vibration and control the position of probe tips on a sub-Å scale.

The STM technique can measure (i) the arrangement of atoms within the surface unit cell, (ii) the extended distribution of atoms across tens or hundreds of nanometers and their relation to surface lattice strain and defects, (iii) the geometric positions of atoms adsorbed to surfaces, (iv) the formation of extended structures terrace and cluster structures driven by thermodynamics, (v) atomic diffusion and segregation at interfaces, and (vi) the densities of filled and empty states within individual unit cells. This information can be used to validate models of surface electronic and chemical structure.

AFM techniques have the capability to measure all surfaces, regardless of their insulating properties. On a nanometer-scale, AFM can measure (i) surface morphology, (ii) electric potential, (iii) capacitance, (iv) carrier concentration, (v) magnetic polarization, and (vi) friction. Even with lower spatial resolution than STM, AFM has found numerous applications in technology because of its versatility.

BEEM provides the means to determine (i) the alignment of local conduction bands at interfaces, (ii) Schottky barrier heights and their lateral inhomogeneity, (iii) local electric fields, (iv) heterojunction band offsets, and (v) "hot-carrier" effects, all with nanometer spatial resolution. While BEEM requires deposition of metal layers on the surfaces of interest, it is by nature ideal for determining Schottky barriers and the influence of surface electronic and chemical structure on barrier heights.

The ability to position atoms on surfaces with probe tips enables us to visualize the quantum mechanical nature of electrons near atoms. With this capability, researchers have demonstrated the construction of atomic arrays at the extreme limit of electronic circuitry, structures that represent orders of magnitude advances in packing density and energy consumption.

Problems

1. Calculate the wavefunction for an STM tip with 5.0-eV work function across a vacuum gap.
2. In BEEM, the beam of electrons is sharply forward focused. Calculate the angle at which the transmitted intensity decreases by $1/e$ for a metal thickness of 50 Å, a sample work function of 4.3 eV, and an electron kinetic energy of 1 V. What is the corresponding spatial spread?
3. In Figure 14.20, the pattern inside the quantum "corral" is due to interference between electrons in the Cu surface. Calculate the wavelength of these electrons. Is it consistent with the wave pattern shown? Hint: use the separation between Fe atoms to establish a length scale.

References

1. Golovchenko, J.A. (1986) *Science*, **232**, 48.
2. Quate, C. (1986) *Phys. Today*, **39**, 26.
3. Feenstra, R.M., Stroscio, J.A., Tersoff, J., and Fein, A.P. (1987) *Phys. Rev. Lett.*, **58**, 1192.
4. Tromp, R.M., Hamers, R.J., and Demuth, J.E. (1986) *Phys. Rev. B*, **15** (34), 1388.
5. Bardeen, J. (1961) *Phys. Rev. Lett.*, **6**, 57.
6. Hui, T.-W. at http://www.chembio.uoguelph.ca/educmat/

chm729/STMpage/ accessed 23 December 2009.
7. Swartzentruber, B.S., Kitamura, N., Lagally, M.G., and Webb, M.B. (1993) *Phys. Rev. B*, **47**, 13432.
8. Liu, F., Huang, M., Rugheimer, P.P., Savage, D.E., and Lagally, M.G. (2002) *Phys. Rev. Lett.*, **89**, 136101.
9. Lita, B., Ghaisis, S., Goldman, R.S., and Melloch, M.R. (1999) *Appl. Phys. Lett.*, **75**, 4082.
10. Im, H.-J., Ding, Y., Pelz, J.P., Heying, B., and Speck, J.S. (2001) *Phys. Rev. Lett.*, **87**, 106802.
11. Doutt, D.R., Zgrabik, C., Mosbacker, H.L., and Brillson, L.J. (2008) *J. Vac. Sci. Technol.*, **26**, 1477.
12. *http://veecolifesciences.com/images/modes/magnetic_force_microscopy_1733.jpg* accessed 21 December 2008.
13. *http://veecolifesciences.com/images/modes/latency_force_microscopy_1713.jpg* accessed 21 December 2008.
14. Bell, L.D. and Kaiser, W.J. (1988) *Phys. Rev. Lett.*, **61**, 2368.
15. Hecht, M.H., Bell, L.D., Kaiser, W.J., and Grunthaner, F.J.) (1989) *Appl. Phys. Lett.*, **55**, 780.
16. Dong, Y., Feenstra, R.M., Semtsiv, M.P., and Masserlink, W.T. (2004) *Appl. Phys. Lett.*, **84**, 227.
17. Ding, Y., Park, K.-B., Pelz, J.P., Palle, K.C., Mikhov, M.K., Skromme, B.J., Meidia, H., and Mahajan, S. (2004) *Phys. Rev. Lett.*, **69**, 041305.
18. Eigler, D.M. and Schweizer, E.K. (1990) *Nature*, **344**, 524.
19. Eigler, D. *http://www.almaden.ibm.com/vis/stm/corral.html* access 21 December 2008.
20. *http://domino.watson.ibm.com/comm/pr.nsf/pages/rsc.cascade.html* (Accessed 23 December 2009).
21. Heinrich, A.J., Lutz, C.P., Gupta, J.A., and Eigler, D.M. Sciencexpress, *www.sciencexpress.org/*. doi: 1/10.1126/science.1076768 (2002).

15
Optical Spectroscopies

15.1
Overview

Optical techniques using visible or near-visible infrared (IR) light can provide useful information on the properties of electronic materials' surfaces and interfaces. Purely optical techniques that involve photon excitation and detection include reflectance, ellipsometry, and inelastic light scattering spectroscopies such as Raman scattering and its variants. Optical techniques that induce electrical changes include photovoltage or photoconductivity spectroscopies. The advantages of optical techniques include their (i) excellent energy precision, (ii) nondestructive interaction with the sample, and (iii) capability for in situ or remote sensing. A significant disadvantage of most optical techniques is their inherently low surface sensitivity. This chapter provides an overview of these optical spectroscopies and the associated methods used to measure physical properties that are specific to surfaces and interfaces.

15.2
Optical Absorption

The propagation of electromagnetic radiation in a dielectric medium can be described by a wavefunction

$$\psi(r) = e^{i(\mathbf{k}\cdot\mathbf{r} - \omega t)} \tag{15.1}$$

where **k** is a complex wave vector expressed as

$$\mathbf{k}(\omega) = \frac{2\pi}{\lambda}\left[\eta(\omega) + i\kappa(\omega)\right] = \frac{2\pi n(\omega)}{\lambda} \tag{15.2}$$

ω is the radiation angular frequency, **r** is a position vector, λ is the optical wavelength, and $n(\omega)$ is a complex refractive index composed of *real* ($\eta(\omega)$) and imaginary ($\kappa(\omega)$) parts. Optical wavelength λ is related to the frequency ν according to $\lambda\nu = c$, where c is the speed of light. The refractive index components $\eta(\omega)$ and $\kappa(\omega)$ can be extracted from energy-dependent reflection measurements using a Kramers–Kronig analysis [1]. Figure 15.1 illustrates the energy dependence of

Surfaces and Interfaces of Electronic Materials. Leonard J. Brillson
Copyright © 2010 WILEY-VCH Verlag GmbH & Co. KGaA, Weinheim
ISBN: 978-3-527-40915-0

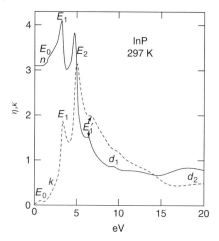

Figure 15.1 Refractive index n (i.e., $\eta(\omega)$) and absorption index k (i.e., $\kappa(\omega)$) of InP at room temperature.

$\eta(\omega)$ and $\kappa(\omega)$ for the typical III–V compound semiconductor InP from below its fundamental bandgap into the near-ultraviolet energy range [2].

The energy E_0 at which $\kappa(\omega)$ begins to increase corresponds to the fundamental highest valence to the lowest conduction band transition. Both $\eta(\omega)$ and $\kappa(\omega)$ exhibit strong peaks at energies labeled E_1, E_2, and E_1' that correspond to transitions involving higher energy valence to conduction band transitions.

From $\eta(\omega)$ and $\kappa(\omega)$, one obtains the *transverse dielectric constant*

$$\varepsilon_T(\omega) = |\eta(\omega) + i\kappa(\omega)|^2 \tag{15.3}$$

with real and imaginary components

$$\varepsilon_{T_1}(\omega) = \eta^2(\omega) - \kappa^2(\omega) \tag{15.4a}$$

$$\varepsilon_{T_2}(\omega) = 2\eta(\omega)\kappa(\omega) \tag{15.4b}$$

Since the *probability density function* for the radiation is proportional to $\psi(r)^*\psi(r) = |\psi(r)|^2$, the complex dielectric constant describes the absorption process in terms of

$$|e^{i(\mathbf{k}\cdot\mathbf{r}-\omega t)}|^2 = |e^{(i2\pi d/\lambda)(\eta(\omega)+i\kappa(\omega))}e^{-i\omega t}|^2$$

$$= |e^{(i2\pi d(\eta(\omega)/\lambda))}|^2 \cdot |e^{(-2\pi d\kappa(\omega)/\lambda)}|^2 \cdot |e^{-i\omega t}|^2 = e^{(-4\pi\kappa(\omega)d/\lambda)} \tag{15.5}$$

for a path length d of light penetrating into the dielectric medium.

In the energy range of optical absorption, the imaginary part of $n(\omega)$ dominates such that $n(\omega) \sim i\kappa(\omega)$. Then the intensity of an incident light beam with initial intensity I_0 is attenuated by absorption according to

$$I = I_0 e^{-\alpha d} \tag{15.6}$$

for which

$$\alpha = \frac{4\pi\kappa(\omega)}{\lambda} \tag{15.7}$$

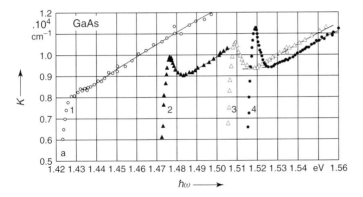

Figure 15.2 Absorption coefficient versus photon energy in the region of the excitonic absorption edge for four temperatures: (i) 294 K, (ii) 126 K, (iii) 90 K, and (iv) 21 K.

is the optical absorption coefficient. This absorption coefficient is defined such that, when $d = 1/\alpha$, incident light is attenuated by a factor $I/I_0 = e^{-4\pi\kappa(\omega)d/\lambda} = e^{-1}$.

The absorption coefficient becomes significant at energies above a semiconductor's bandgap and varies strongly at higher energies. Figure 15.2 illustrates the absorption coefficient for GaAs as a function of energy and temperature [3]. Absorption increases rapidly at energies above the absorption edge. The absorption edge itself increases with decreasing temperature. The peak feature at the absorption edge is due to exciton absorption a few tens of milli electronvolts below the conduction-to-valence-band energy separation. This excitonic feature becomes more pronounced with decreasing temperature. At the absorption edge, that is, the bandgap energy, α, is approximately 10^4 cm^{-1} at these four temperatures. According to Equation 15.6, $d = 1/\alpha = 10^{-4}$ cm = 10 000 Å, the probe depth at which the dielectric medium attenuates the light intensity by $1/e$. Above this energy, α continues to increase as higher lying energy transitions contribute to the absorption.

Consider a surface layer of thickness ~5 Å. The volume of the material contributing to the optical signal of this layer is then only ~5 Å/10 000 Å < 10^{-3} of the underlying bulk. This low ratio of surface to bulk signal represents the challenge for optical techniques in measuring the properties of the surface alone.

An alternative measure of optical absorption is *optical density* d_{opt}, defined as

$$d_{opt} = \log_{10}\left(\frac{I_I}{I_T}\right) \tag{15.8}$$

I_I and I_T are incident and transmitted light intensity, respectively. The energy dependence of optical density for compound semiconductors in the near band edge region exhibits pronounced onsets of absorption at the same critical point energies detected independently by reflection measurements. In general, these optical density measurements show that absorption increases logarithmically with the increase in photon energy for energies just above the fundamental bandgap.

Thus, near-surface optical sensitivity requires incident energies significantly above the bandgap so that light is absorbed strongly. For example, in the case of InP, the 4416 Å emission line of an He–Cd laser corresponds to a 2.80 eV excitation for which κ is ~ 2 in Figure 15.1, and $1/\alpha = \sim \lambda/8\pi$ for $\lambda = 4416$ Å is ~ 175 Å. Such depths are typically within the surface space charge regions of low-to-moderately doped semiconductors and thereby provide information related to surface properties. Shorter wavelength light yields proportionally shorter absorption depths, more so near critical points where κ exhibits local maxima.

15.3
Modulation Techniques

Besides using short wavelengths to maximize absorption, a number of optical techniques are available to enhance surface sensitivity. These include the following: (i) modulation spectroscopy to differentiate high background signals from the bulk, (ii) multiple surface scattering using total internal reflection or a multiple interface stack, (iii) comparative measurements with or without interface-specific features, and (iv) surface electrostatic measurements.

Figure 15.3 provides an example of a modulation spectroscopy measurement, in this case, *differential reflectance spectroscopy*. An IR light source is modulated in a square wave pattern and split in two. Half the light reflects off a specimen surface and into an optical detector while the other half passes directly into the detector. This technique can provide high precision measurements of $\Delta R/R$ since any variations in light intensity with photon energy or experimental artifacts are factored out by the split beam. Subtraction of the two beams permits precise measurements of small intensity differences due to reflection, even though the intensities of the two beams separately are high. The modulated beams permit cancellation of any

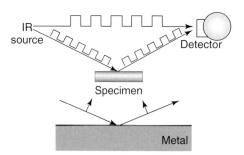

Figure 15.3 Split, modulated beam geometry for modulated reflectance spectroscopy. Polarization is in the plane of reflectance.

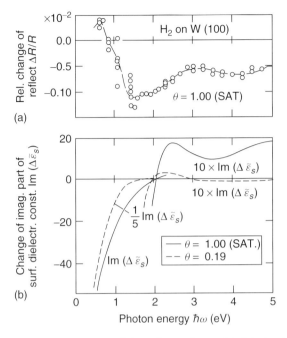

Figure 15.4 (a) Spectral dependence of relative reflectance change due to saturation coverage of hydrogen ($\theta = 1$) taken to be two H atoms per W atom. The theoretical (solid) curve is obtained from an oscillator fit of ε_s to reproduce the $\Delta R/R$ data points. (b) Calculated change of the imaginary part of the surface dielectric constant $\text{Im}\{\Delta \varepsilon_s\}$ as caused by hydrogen coverages $\theta = 0.19$ and 1. Assumed thickness for the surface layer (scale factor): 5 Å ($\theta = 1$) and 1 Å ($\theta = 0.19$), respectively [4].

background noise in both beams, further increasing signal-to-noise detectivity. The grazing incidence geometry permits light polarized in the plane of reflection to couple strongly to surface normal dipoles.

As an example, Figure 15.4a shows the $\Delta R/R$ response for H_2 adsorption on W(100). The IR absorption at the surface unbalances the two beams at the detector, resulting in a pronounced $\Delta R/R$ variation with photon energy [4]. An oscillator fit to the measured changes of $\Delta R/R$ permits one to calculate the change in the imaginary part of the surface dielectric constant $\text{Im}\{\Delta \varepsilon_s\}$. The pronounced increase $\text{Im}\{\Delta \varepsilon_s\}$ below 1.5 eV in Figure 15.4b corresponds to the reduced reflectance/increased absorbance, as shown in Figure 15.4a. Also note the sensitivity of $\text{Im}\{\Delta \varepsilon_s\}$ to the difference in surface coverage from $\theta = 0.19$ to $\theta = 1$.

Other modulation spectroscopies based on such comparative techniques include (i) electroreflectance spectroscopy (surface electric field modulation) [5, 6], (ii) thermoreflectance spectroscopy (temperature-dependent changes in band structure features) [7], (iii) wavelength modulation spectroscopy [8], and piezo-optical spectroscopies (pressure-dependent changes in band structure emissions) [9].

15.4
Multiple Surface Interaction Techniques

One can multiply the signal-to-noise ratio of surface measurements by allowing the probe beam to interact with a surface or interface multiple times or with many such interfaces. One method to promote multiple interactions is the *total internal reflection* technique. As shown in Figure 15.5, light passing between two surfaces is internally reflected multiple times. The smaller the angle θ between surface normal versus incident light and the longer the overall path length through the channel between the plates, the larger is the number of reflections. In this way, the photon–surface interactions can be magnified by orders of magnitude.

A striking example of such total internal reflection spectroscopy is the oxidation of Ge(111). As shown in Figure 15.6 lower inset, light is directed between the

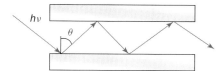

Figure 15.5 Parallel plate configuration for total internal reflection of light.

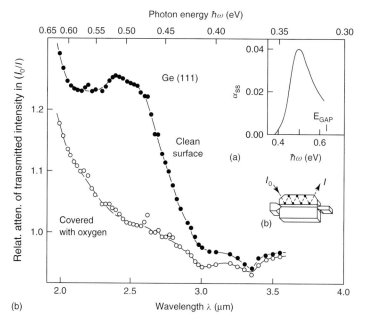

Figure 15.6 Natural logarithm of the incident versus transmitted light intensity ratio I_0/I versus wavelength for the clean cleaved and oxygen-covered Ge(111) surface (P(O$_2$) ~ 10^{-6} Torr). The surface absorption constant α_{ss} attributed to surface state transitions of the clean surface appears as the upper right inset. The experimental configuration is shown in the lower right inset [10].

opposing faces of a Ge crystal cleaved by wedge blades in ultrahigh vacuum (UHV) to expose (111) planes [10]. The Ge crystal has prism faces to produce multiple reflections at the cleaved surface (dashed lines).

Wavelength-dependent spectra are taken before and after exposure to an ambient O_2 pressure of 10^{-6} torr. The oxidation alters surface states present on the freshly cleaved surface, reducing the optical absorption by nearly an order of magnitude. This reduction occurs at $\hbar\omega \sim 0.5$ eV, corresponding to a surface state energy located ~ 0.5 eV above the valence band (upper inset). This experiment was among the first to demonstrate the existence of surface states within the fundamental bandgap.

In this experiment, the surface is not compared with a reference in real time. Rather, the comparison is between surfaces before and after a surface treatment. The difficulty with such a cleaved crystal approach is the demanding geometrical constraints imposed by UHV cleavage and the new crystal needed for each exposure.

Multiple reflection techniques have been used to enhance the detection of surface vibrational modes. Such IR vibrational mode spectroscopy has provided a great deal of useful information about gas molecule reactions and bonding with Si surfaces [11]. With such material, wet chemical treatments in air followed by cleaning in UHV provide a ready source of clean semiconductor surfaces.

15.5
Spectroscopic Ellipsometry

Spectroscopic ellipsometry (SE) is a powerful and nondestructive technique to measure optical properties of thin films and their surfaces. SE measures both the change in polarization and intensity as a function of wavelength and reflection angle. The surface sensitivity is due to the high precision of angular settings available. The ratio ρ of light intensity polarized parallel to versus perpendicular to the plane of reflection can be expressed as

$$\rho = \frac{R_\parallel}{R_\perp} = \tan\psi \, e^{i\Delta} \tag{15.9}$$

where Δ and ψ can be related to the real and imaginary parts of the dielectric constant, $\text{Re}\{\varepsilon\}$ and $\text{Im}\{\varepsilon\}$, respectively. The SE response can be interpreted using layer models with effective dielectric constants [12].

A distinct advantage of SE is the development of characteristic "fingerprints" of specific surface bond configuration. This has been used, for example, to adjust growth parameters during the molecular organic chemical vapor deposition (MOCVD) of GaAs(100) (2 × 4) surfaces [13].

A disadvantage of ellipsometry for surfaces and interfaces is the requirement to model the dielectric response of the layers at a surface or interface. In many cases, these surfaces and interfaces involve changes in elemental composition and bonding such that their dielectric properties are not well known.

15.6
Surface-Enhanced Raman Spectroscopy

Raman spectroscopy (RS) is a well-established technique to measure phonon frequencies in the solid state. A narrow-line source of excitation, typically a laser, incident on a specimen is absorbed and scattered by a number of excitations. Besides absorption per se, the incident photon can couple to the polarization fields of lattice vibrations, losing or gaining energy to optical and acoustic phonons. Figure 15.7 illustrates a Raman spectrum schematically for an incident photon of energy $E = h\nu_0$. The Raman process can be modeled as a three-step process that involves absorption, scattering, and re-emission. The absorption process creates a free electron–hole pair that interacts with the polarization fields of the lattice. Scattering that produces a phonon with energy E_{ph} results in re-emission at an energy $h\nu_0 - E_{ph}$. These are termed *Stokes emissions*. Similarly, scattering that annihilates a phonon results in reemission at an energy $h\nu_0 + E_{ph}$ and is termed an *anti-Stokes spectral line*. This scattering of photons by phonons and phonons coupled to other lattice excitation modes is termed *inelastic scattering*.

The RS cross section can be expressed as

$$S_j = \frac{dP_R}{P_0 d\Omega} = \frac{\omega_s^4}{c^4} VL \left| e_i \cdot \chi^j \cdot e_j \right|^2 \qquad (15.10)$$

where S_j is the scattering efficiency, P_0 is the incident power, $d\Omega = dA/R_2$ is the differential solid angle for a differential area A at a distance R^2, V is the volume equal to AL, L is the scattering length in the dielectric medium, ω_s is the scattered photon frequency, and χ^j is the dielectric susceptibility of the jth excitation. Equation 15.11 shows how S_j depends strongly on frequency ω_s, increasing as the fourth power. In general, *dielectric susceptibility* χ is defined for an ε according to

$$\varepsilon = 1 + 4\pi \chi V \qquad (15.11)$$

χ^j is a tensor whose components contain the symmetry and strength of Raman scattering.

Figure 15.8 illustrates a typical RS geometry in which the incident and specularly reflected light are outside the solid angle of a lens that collects and transmits

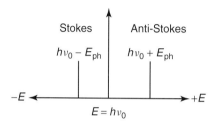

Figure 15.7 Schematic illustration of a Raman spectrum shows Stokes and anti-Stokes emission lines.

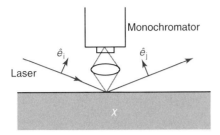

Figure 15.8 Schematic illustration of backscattering geometry for a Raman scattering experiment. Polarization of the incident and reflected light, e_i and e_j, respectively, is in the plane of reflection.

scattering light from the sample into a light monochromator. It is important to avoid collecting incident or reflected light since the RS scattering signals are typically four orders of magnitude below the incident light intensity.

The scattered light collected by the lens can be filtered so that only one polarization passes into the monochromator. The tensor components of χ^j are determined by the crystal symmetry. Thus polarization-dependent scattering measurements provide a method to determine the relative magnitudes of the tensor components as well as to confirm the effective crystal symmetry with extrinsic perturbations.

Since the RS process involves absorption of a photon, a method to increase the signal intensity is to use incident photons whose energy is tuned to an absorption edge as in Figures 15.1 and 15.2. This technique is termed *resonant Raman scattering*, which can enhance signals by more than an order of magnitude. A second method of increasing RS signals involves magnifying the electric fields induced at the surface by the incident light. A useful technique to achieve magnification is the deposition of metals such as Ag that form clusters on semiconductor surfaces that, in turn, couple strongly with the incident laser light. As a result, RS signals can be enhanced by orders of magnitude, making it possible to probe excitations within the top few monolayers of a semiconductor surface [14].

A third RS method of monitoring electrical properties near surfaces involves the so-called morphic effects. χ^j can be expanded in terms proportional to an externally applied force F_A such that, for small F_A,

$$\chi^j = \left(\frac{\partial \chi}{\partial \mathbf{Q}_j}\right)\mathbf{Q}_j + \left(\frac{\partial^2 \chi}{\partial \mathbf{Q}_j \partial F_A}\right)\mathbf{Q}_j F_A \text{ and higher order terms in } F_A \quad (15.12)$$

The parameter $\mathbf{Q}_j$ can be a lattice displacement $\mathbf{u}_j$ or an electric field $\mathbf{E}_j$ of a jth optical phonon mode. The first term in Equation 15.12 corresponds to the normal RS process. The second term represents the effect of the externally applied force $\mathbf{F}_A$. Morphic effects on Raman spectra have been demonstrated for mechanical forces that distort the lattice [15] and electric forces that result from changes in near-surface band bending [16]. Surface band bending is a

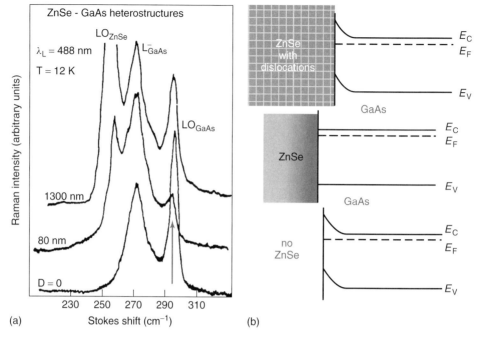

Figure 15.9 First-order Raman spectra of ZnSe–GaAs heterostructures for thickness $D = 0$, 80, and 1300 nm, the latter two below and above, respectively, the ZnSe critical thickness. For D below the critical thickness, the LO$_{GaAs}$ phonon mode decreases relative to either the bare surface or the thick heterojunction [19].

direct outcome of Schottky barrier formation so that metal deposition on a semiconductor can increase band bending and the electric fields below. In turn, these fields lower the crystal symmetry so that normally symmetry-forbidden phonon modes [17] or combinations of incident and scattered polarizations [18] become allowed.

Figure 15.9 illustrates the effect of band bending on normally symmetry-forbidden phonon modes. Raman spectra at a GaAs/ZnSe heterojunction show the presence of a strong longitudinal optical (LO) phonon signal for the clean GaAs(100) surface prior to overlayer coverage ($\theta = 0$) [19]. The corresponding energy bands versus depth are bent due to negative charges formed at the free surface. With the epitaxial deposition of 80 nm ZnSe over the GaAs surface, both surface band bending and the LO$_{GaAs}$ phonon mode decrease. For a thickness $D = 1300$ nm that exceeds the critical thickness, dislocations form in the ZnSe, causing band bending around the dislocation that reintroduces the field-induced LO$_{GaAs}$ phonon scattering.

The band bending within the surface space charge region is parabolic so that an exact analysis requires Airy functions [20]. The surface sensitivity is limited to the width of the space charge region, which can range from tens of nanometers to a micron, depending on the doping concentration.

15.7 Surface Photoconductivity

The surface photoconductivity (SPC)*surface photoconductivity* (SPC) technique provides a probe of surface states with energies within the semiconductor bandgap. The technique is based on photostimulated changes in band bending that alter the carrier concentration in the semiconductor depletion region. In Figure 15.10, the current between two contacts travels through both the semiconductor depletion region and the bulk. Surface illumination with light at subbandgap energies promotes carriers into and out of gap states, resulting in changes in surface charge. For states localized near the free surface, such changes alter the band bending and the carrier concentration in the depletion region. For a current density J and applied electric field $\mathcal{E}$ defined as

$$J = \sigma \mathcal{E} \tag{15.13}$$

where conductivity

$$\sigma = ne\mu \tag{15.14}$$

for carrier density n and mobility μ.

Figure 15.10 shows illumination by photons with energy $h\nu$ that promote electrons into gap states adding negative charge to the surface. In turn, this increases the n-type band bending, lowers the carrier density, and thereby lowers the current density within the space charge region. The change in band bending at well-defined photon energies with population or depopulation of surface energy levels has the advantage of identifying the position of the energy level with respect to the band edges. On the other hand, this technique has disadvantages (i) it requires contacts, which make UHV cleaning or other processing difficult and (ii) the surface sensitivity is limited by the bulk conduction that can occur in parallel.

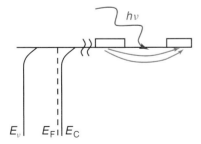

Figure 15.10 Schematic experimental geometry for surface conductivity and associated band bending within semiconductor depletion region.

15.8
Surface Photovoltage Spectroscopy

15.8.1
Theory of Surface Photovoltage Spectroscopy

The photostimulated changes in surface state occupancy enable another more powerful technique, *surface photovoltage spectroscopy* (SPS) [21]. Figure 15.11 illustrates photostimulated transitions that either fill (Figure 15.11a) or empty (Figure 15.11b) surface states. The SPS technique is based on measuring the surface work function as a continuous function of incident photon energy. This is accomplished by modulating a capacitor probe to induce a displacement current generated by the difference in work function between the sample and a reference probe. Figure 15.12a illustrates this potential difference and the dependence of semiconductor work function $q\Phi_S$ on the band bending qV_B in the surface depletion region. The work function is defined as $E_{vac} - E_F$ so that

$$E_{vac}^S - E_F^S = e\Phi^S \tag{15.15a}$$
$$E_{vac}^M - E_F^M = e\Phi^M \tag{15.15b}$$

With the Fermi levels aligned, the work function difference is defined as the *contact potential* U^{SM} such that

$$e(\Phi^S - \Phi^M) = E_{vac}^S - E_F^S - E_{vac}^M + E_F^M = E_{vac}^S - E_{vac}^M = eU^{SM} \tag{15.16}$$

represents the potential difference across the capacitor.

Taking into account the compensating voltage contribution U_{comp} to the circuit, the charge Q on each capacitor "plate" of the capacitance C is given by

$$Q = CV = C\left[-(\Phi_S - \Phi_M) + U_{comp}\right] \tag{15.17}$$

If the spacing of the reference probe from the semiconductor is modulated at a frequency ω such that spacing $d = d_0\sin(\omega t)$, the resultant vibrating capacitor generates a displacement current:

$$I = \frac{dQ}{dt} = \frac{dC}{dt}\left[-(\Phi_S - \Phi_M) + U_{comp}\right] \tag{15.18}$$

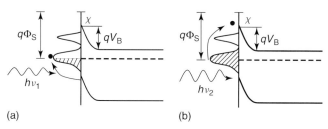

Figure 15.11 Surface state and/or bulk band transitions detectable by SPC and SPV spectroscopies for (a) filling and (b) unfilling optical transitions with work function $q\Phi_S$, semiconductor electron affinity χ, and band bending qV_B.

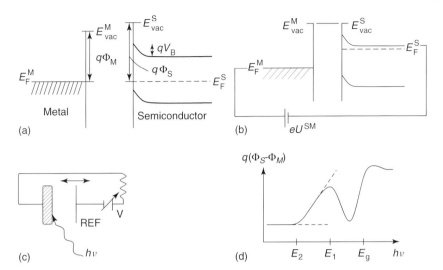

Figure 15.12 Metal and semiconductor with work functions $q\Phi_M$ and $q\Phi_S$, respectively, relative to vacuum levels E_{vac}^M and E_{vac}^S and (a) Fermi levels aligned and (b) with compensating voltage eU^{SM} that aligns E_{vac}^M and E_{vac}^S. (c) Vibrating Kelvin probe and capacitor with feedback circuit for measuring surface potential. (d) Schematic SPS spectrum showing $d\Phi^S/dh\nu$ slope changes at threshold energies E_1, E_2, and E_g corresponding to transitions in Figure 15.11.

This displacement current is typically small, for example, 10^{-10} A, requiring a high impedance preamplifier to generate significant voltages.

Figure 15.12c shows a sample and a vibrating metal reference probe that form a capacitor as part of a circuit with load resistor and variable voltage source. Equation 15.18 shows that $I = 0$ when the compensating voltage U_{comp} is set equal to $(\Phi_S - \Phi_M)$. Figure 15.12b shows the alignment of E_{vac}^S with E_{vac}^M with compensating voltage eU^{SM}.

SPS spectra are typically acquired by using a feedback circuit to null out the displacement current while scanning continuously in photon energy. Figure 15.12d shows a schematic illustration of the changes in $q(\Phi_S - \Phi_M)$ with photon energy at $h\nu_1$ that fills surface states or at energy $h\nu_2$ that empties surface states as shown in Figure 15.11. Hence, a transition that fills surface states makes the surface more negative, increasing the work function Φ_S. Conversely, surface state depopulation decreases negative charge on the surface, thereby decreasing Φ_S. Optical transitions that populate and depopulate the same level should be complementary, that is, $E_1 + E_2 = E_g$.

A particular advantage of the SPS technique is that small changes in surface state population cause large changes in band bending. Figure 15.13 illustrates the calculated changes in band bending as a function of surface state density and surface states per surface atom [22]. The insets at upper and lower left illustrate the energies E_{ss} separating the acceptor A_s and donor D_s surface states from the conduction and the valence bands, respectively. The band bending $|eV_s|$ calculated

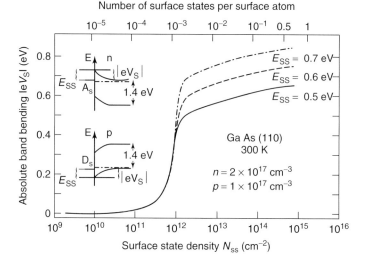

Figure 15.13 Left: Calculated absolute band bending $|V_s|$ due to an acceptor surface state level A_s and a donor level D_s for n- and p-type GaAs. $|V_s|$ is plotted versus the surface state density N_{ss} (lower scale) and related to the number of surface states per surface atom (upper scale). With the different definition of the energetic position E_{ss} for n- and p-type crystals (insets) the calculated curves for n- and p-type material are not distinguishable on the scale used [22].

for a given surface state density N_{ss} begins to increase rapidly for densities between 10^{11} and 10^{12} cm^{-2}. The absolute band bending exhibits relatively large changes that are comparable to the semiconductor bandgap, in this case, approximately 0.5–0.7 V, depending on E_{ss}, or nearly half the bandgap for GaAs. These changes can occur for surface state densities that are only 1/1000th of the surface atom density. Furthermore, with typical experimental precision of 1–10 meV, one can determine surface states with densities of $<10^9$ cm^{-2}. This detection limit is far below the typical 10^{12}–10^{13} cm^{-2} densities available with photoemission or other surface science techniques.

15.8.2
Surface Photovoltage Spectroscopy Equipment

A more detailed experimental configuration for SPS appears in Figure 15.14 [23]. In this arrangement, the Kelvin probe is a gold wire positioned within a fraction of a millimeters from a CdS surface. An oscillator and a solenoid, such as an audio amplifier, drive a speaker coil that deflects a rod that extends through a vacuum bellows into the vacuum chamber. The bellows serves to permit coarse probe positioning and fine vibrating motion. Alternatively, the Kelvin probe can be attached to a piezoelectric tuning fork, permitting greater control of probe-specimen separation.

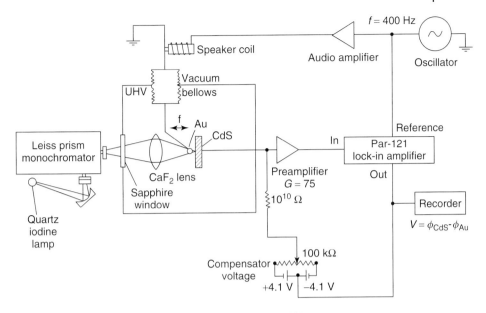

Figure 15.14 Surface photovoltage measurement apparatus [23].

The feedback circuit consists of a preamplifier with 10^{10} Ω input impedance and a gain of 75, a coarse compensator voltage set manually, and a lock-in amplifier coupled to the oscillator. The lock-in produces an output voltage equal and opposite to input voltage, resulting in a bias at the sample that nulls the contact potential difference. A monochromator and light source outside the UHV chamber direct light through a UV-IR sapphire lens that focuses on the sample. The capacitance can also be modulated with a stationary capacitor. Instead of varying the separation between Kelvin probe and specimen, one can modulate the intensity of incident light.

15.8.3
Surface Photovoltage Spectra and Photovoltage Transients

Figure 15.15 illustrates a set of SPS spectra that exhibit pronounced changes in $(\Phi_S - \Phi_M)$, also termed *contact potential difference* Δcpd, versus $h\nu$ [24]. Here, a CdSe crystal was cleaved in UHV to expose a clean, ordered $(11\bar{2}0)$ surface as verified by low-energy electron diffraction (LEED). The SPS spectra exhibit multiple slope changes Δcpd $= -d\Phi_S/dh\nu$ corresponding to numerous surface state energy positions in the CdSe bandgap. The large slope changes at 1.6–1.7 eV correspond to band to band transitions that supply both free electrons and holes and reduce the band bending. The sign of Δcpd at the bandgap energy is opposite to that of large changes at ∼1.1 eV and higher energies, which shows that molecular O exposure increases band bending. This is consistent with surface oxidation, which traps electrons at O bonding sites.

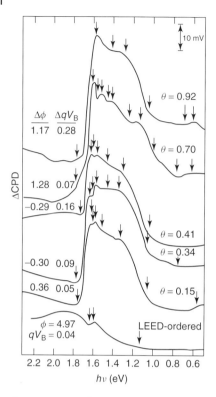

Figure 15.15 Surface photovoltage spectra of clean, LEED-ordered (1120) CdSe with increasing molecular oxygen surface coverage θ, as determined by XPS. Arrows indicate energies of maximum slope change corresponding to onsets of photovoltage transitions. The changes in surface work function $\Delta\phi^S$ and band bending $\Delta q v_{VB}$ are given for each curve [24].

The SPS technique can provide more than energy level and band bending information. SPS transient measurements at specific incident photon energies and densities yield information on surface state densities of states and photoionization cross sections. For a Schottky-type surface barrier, charge neutrality and Poisson's equation give

$$-Q_{ss} = Q_{sc} = (2\varepsilon k_B T n_b)^{1/2} |V_s|^{1/2} \tag{15.19}$$

with bulk dielectric permittivity ε, free electron density n_b in the bulk, surface barrier V_s in dimensionless units ($V_s \equiv qV_s/k_B T$) and $V_s^\circ = V_s$ with no light. Figure 15.16 illustrates the change in surface trap densities n_t and band bending V_s with photostimulated depopulation.

At the photon energy $h\nu = E_C - E_t$ required to depopulate the surface trap, the initial trap density n_t° decreases to $n_t^\circ - \delta n_t$, resulting in a band bending decrease $V_s^\circ - \delta V_s$. Thus, one obtains the variation in the number of electrons n_t trapped at E_t as a function of changes in barrier height δV_s (for $|\delta V_s/V_s^\circ| \ll 1$).

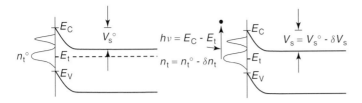

Figure 15.16 Photostimulated depopulation of surface trap with density n_t at energy E_t below the conduction band E_C and band bending V_s.

The rate of change in the density of occupied states can be expressed as the kinetic rate equation

$$\frac{dn_t}{dt} = -K_{ph}^d n_t I + K_n n_s p_t - K_n n_t n_1 \tag{15.20}$$

K_{ph}^d is the *capture cross section* of surface states for photon-stimulated depopulation. I is the *photon intensity*. $K_n = k_n v_T$ is the product of the capture cross section of surface states for electrons k_n multiplied by the *electron thermal velocity* $v_T \cdot n_s$ is the free electron density at the surface, n_t and p_t are the captured electron and hole densities, respectively, at the surface states, and n_1 is the emission constant, the rate of electron thermal emission from the surface state. The three terms contributing to the total change in surface state population dn_t/dt represent photoionization, recombination, and generation.

The rate of change in n_t can be approximated as [25]

$$\delta n_t \cong \frac{-(2\varepsilon k_B T n_b)^{1/2} \delta V_s}{(2q|V_s|^{1/2})} \tag{15.21}$$

The SPS transient measurement consists of monitoring Δcpd versus time for a sample initially in the dark as illumination switches on at time $t = 0$ and switches off at time $t = t_1$. In the dark, $I = 0$ so that $K_n n_s^\circ p_t^\circ - K_n n_t^\circ n_1 \cong 0$ at thermal equilibrium. Hence,

$$\frac{dn_t}{dt} \cong -K_{ph}^d n_t^\circ I \tag{15.22}$$

Then

$$\frac{dV_s}{dt} = \frac{\left(n_t K_{ph}^d 2q|V_s|^{1/2} I\right)}{(2\varepsilon k_B T n_b)^{1/2}} \tag{15.23}$$

Figure 15.17 shows a representative photovoltage transient with initial dimensionless voltage v_s° and voltage rate of change $v'_s{}^\circ$ with illumination [24]. The total Δcpd at time t_1 after the surface reaches a new equilibrium ($dn_t/dt = 0$) under illumination is v_s^1, the change due to illumination is $v_s^\circ - v_s^1 = \delta v_s^1$, and the rate of change with illumination off is $v'_s{}^1$.

For the oxygen on CdSe (1120) shown in Figures 15.15 and 15.17, total densities of states $N_t = n_t + p_t$ are in the mid-10^{10} cm^{-2} range with photon capture cross sections K_{ph} of $2-4 \times 10^{-17}$ cm^2, depending on the particular surface state.

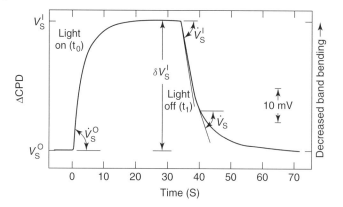

Figure 15.17 Surface photovoltage transient for $\theta = 0.15$ oxygen on CdSe and illumination $I = 3.8 \times 10^{15}$ photons cm^{-2} at $h\nu = 1.60$ eV.

SPS can also be used to stimulate transitions in the bulk that change surface potentials. Thus, transitions that depopulate states at the surface add a positive dipole extending into the bulk, whereas depopulation transitions below the surface can produce dipoles with the opposite signs, depending on the density of bulk gap states and their proximity to the surface. Thus, in systems with trap states in the bulk as well as at the surface, care needs to be taken in determining their net contribution. For an extensive review of SPS and its extension to measurements inside device structures, see [26]. A clear test of whether surface or bulk transitions are involved is the change in SPS spectra with only changes in surface conditions, for example, the surface oxidation discussed in this section or the deposition of metal on semiconductor surfaces to be discussed next.

15.8.4
Surface Photovoltage Spectroscopy of Metal-Induced Surface States

SPS permits the detection of electronic states induced by metals on semiconductors. Figure 15.18 illustrates SPS spectra that demonstrate that new electronic states form at metal–semiconductor interfaces. Experiments to detect such states rely on the ability to first create clean, low-defect surfaces in UHV. This can be accomplished by cleaving solid crystal blocks with an anvil-blade tool to expose clean flat surfaces. The cleaved surface spectrum in Figure 15.18a shows only a flat response at subbandgap energies as well as a relatively small response at the bandgap energy, indicating low surface state densities and band bending prior to depositing adsorbates.

With the initial deposition of 0.1 Å Au by thermal evaporation in UHV, new SPS features appear at subbandgap energies. With increasing Au coverage, the Δcpd slope change at 0.9 eV becomes quite pronounced, signifying a transition depopulating electrons from a state 0.9 eV below the conduction band. The Δcpd change at the GaAs bandgap energy (1.43 eV at room temperature) reflects the

15.8 Surface Photovoltage Spectroscopy | 275

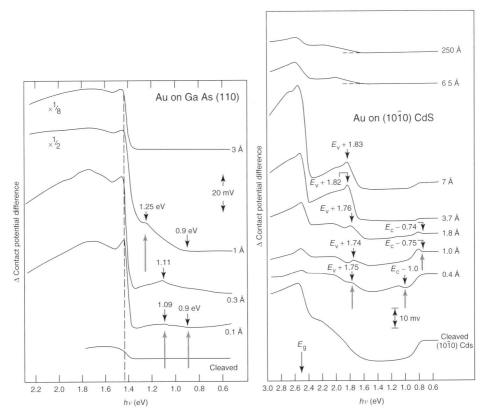

Figure 15.18 SPS spectra of metal-induced interface states for (a) Au on GaAs(110) [27] and (b) Au on CdS(10$\bar{1}$0) as a function of metal overlayer thickness [28]. Arrows indicate slope changes.

increased band bending due to the Au adsorbates [27]. Significantly, the states causing the 0.9 eV SPS onset lie at energies close to the Fermi level position of thick Au films on GaAs. In other words, metal-induced states that form at coverages below a single monolayer can account for the ultimate Schottky barrier height. Furthermore, there is no evidence for surface states prior to metal deposition. The significance of these results is discussed in later chapters.

Figure 15.18b illustrates analogous SPS spectra for incremental Au deposition on CdS(10$\bar{1}$0) obtained by UHV cleavage [28]. Again, the Δcpd response increases at the bandgap energy (2.5 eV at room temperature) with the increased band bending induced by Au deposition. However, the SPS spectrum of the cleaved surface already displays Δcpd transition onsets at energies below E_G. Such features could be due to states present at the clean surface or to trap states in the CdS bulk. Again, the deposition of Au induces new features at submonolayer coverages.

Above single monolayer Au coverage, a pronounced Δcpd feature at $\sim$1.8 eV signifies a transition populating electrons into a state 1.8 eV above the valence band, that is, 0.7 eV below the conduction band. As with Au on GaAs, this energy of this metal-induced state lies close to the final Fermi level position of thick Au films on CdS with a 0.78 eV Schottky barrier height [29].

GaAs and CdS are representative of two classes of binary semiconductors with covalent versus ionic bonding, respectively. The metal-induced features revealed by SPS in Figure 15.18 for these two compounds demonstrate that extrinsic phenomena at metal–semiconductor interfaces are common to a wide range of compounds.

15.9
Summary

A number of optical spectroscopies are available to measure the properties of semiconductor surfaces and interfaces. These techniques are able to overcome the inherent limitation of long absorption lengths with conventional photon sources.

Modulation techniques allow the separation of bulk and surface contributions to the optical response. A variety of modulation approaches have been developed including (i) pulsed optical stimulation in comparison with a reference surface, (ii) modulation of an applied electric field to the band bending region at the semiconductor surface, and (iii) modulation of the incident photon energy.

Multiple reflection techniques provide a method of magnifying single surface features even if absorption occurs well beyond the surface. Light is directed between parallel surfaces under investigation, obtained either by cleavage in UHV or by alternate cleaning methods. Spectra obtained before and after surface treatments reveal the effect of surface interactions.

Spectroscopy ellipsometry relies on the variation in surface reflectance intensity and phase as a function of energy, reflectance angle, and polarization to extract the real and imaginary parts of the dielectric function. This technique relies on knowledge of a given material's effective dielectric constants in order to interpret results using layer models. SE can provide characteristic "fingerprints" of specific bond configurations. However, it is limited for surface and interfaces that involve changes in elemental composition and bonding from those of the constituent materials.

Surface-enhanced Raman scattering provides information on band bending within a semiconductor's depletion region by monitoring optical phonon modes whose polarized intensities change as the crystal symmetry is lowered by morphic effects such as electric field.

Raman scattering can also provide information on surface vibrational modes when metal clusters at the surface enhance the electric fields of the incident radiation and thereby the Raman scattering efficiency.

SPC can provide information on surface and bulk states with energies in the semiconductor bandgap. Photostimulated population and depopulation of states at the surface alter the band bending and carrier density in the surface space charge region. However, unless conductive layers with thickness comparable to the depletion width are studied, the technique is limited by conduction through the semiconductor bulk.

SPS is highly sensitive to photostimulated population and depopulation transitions involving electronic states in the semiconductor bandgap. The onset of these transitions provides not only the energy but also the position of states relative to the band edges.

Transient measurements provide absolute densities and optical cross sections for specific states within the bandgap.

The SPS technique reveals that adsorbates such as gases and metals induce the formation of new electronic states in semiconductor bandgaps. In the case of metal overlayers, the energy position of induced states corresponds to the macroscopic Schottky barrier position.

Problems

1. A useful conversion between optical wavelength and energy is $E(eV) = 1.24$ eV/λ (μm). Thus, a photon with 1 eV energy has a wavelength of 1.24 μm. Derive this expression.
2. Calculate the excitation depth in InP for incident laser light of wavelength 6328 Å (He–Ne laser) versus 325 nm (He–Cd laser). Which of these laser energies would you choose to study: (i) a 50 nm epilayer of InP grown on lattice-matched InGaAs and (ii) the InP/InGaAs heterointerface? Explain your answers.
3. For a Si epilayer with doping $N_d = 10^{17}$ cm^{-3}, calculate film thickness needed to minimize competition from bulk conduction. (a) For an InP crystal with 10^{16} cm^{-3} doping initially with flat bands, how many electrons per square centimeter does a surface photovoltage excitation add to a surface state in InP if the measured work function increases by 20 meV? (b) If the semiconductor doping is increased to 10^{17} cm^{-3}, how much more incident light is needed to achieve the same band bending as in (a)?
4. A GaAs crystal with charged surface states has 10^{16} cm^{-3} doping and 0.8 eV equilibrium band bending in the dark. Photostimulated depopulation of these surface states with 10^{15} photons cm^{-2} s^{-1} intensity at $h\nu_d$ produces a 50 meV s^{-1} transient toward decreasing work function. The capture cross section for photostimulated depopulation is given as 5×10^{-17} cm^2. What is the trapped surface electron density at this energy?

References

1. Lucarini, V., Saarinen, J.J., Peiponen, K.-E., and Vartiainen, E.M. (2005) *Kramers-Kronig Relations in Optical Materials Research*, Springer, New York.
2. Cardona, M. (1965) *J. Appl. Phys.*, **36**, 2181.
3. Sturge, M.D. (1962) *Phys. Rev.*, **127**, 768.

4. Anderson, J., Rubloff, G.W., Passler, M., and Stiles, P.J. (1974) *Phys. Rev. B*, **10**, 2401.
5. Cardona, M. (1969) *Modulation Spectroscopy*, Academic Press, New York.
6. Pollak, F.H. (1994) in *Handbook on Semiconductors*, vol. 2 (ed. M.Balkanski), North-Holland, Amsterdam, p. 527 and reference therein.
7. Matatagui, E., Thompson, A.G., and Cardona, M. (1968) *Phys. Rev.*, **176**, 950.
8. Shaklee, K.L., Rowe, J.E., and Cardona, M. (1974) *Phys. Rev.*, **174**, 828.
9. Pollak, F.H. (1998) *Semiconductors and Semimetals*, vol. 55, Academic Press, New York.
10. Chiarotti, G., Nannarone, S., Pastore, R., and Chiaradia, P. (1971) *Phys. Rev. B*, **4**, 3398.
11. Chabal, Y.J. and Raghavachari, K. (2002) *Surf. Sci.*, **502–503**, 41.
12. Fujiwara, H. (2007) *Spectroscopic Ellipsometry: Principles and Applications*, John Wiley & Sons, Inc., New York.
13. Zettler, J.-T., Wethkamp, T., Zorn, M., Pristovsek, M., Meyne, C., Ploska, K., and Richter, W. (1995) *Appl. Phys. Lett.*, **67**, 3783.
14. Otto, A. (1983) in *Light Scattering in Solids*, vol. IV (eds M.Cardona and G. Güntherodt), Springer, Amsterdam.
15. Anastassakis, E., Iwasa, S., and Burstein, E. (1966) *Phys. Rev. Lett.*, **17**, 1051.
16. Maradudin, A.A., Ganesan, S., and Burstein, E. (1967) *Phys. Rev.*, **163**, 882.
17. Brillson, L.J. and Burstein, E. (1971) *Phys. Rev. Lett.*, **27**, 808.
18. Buchner, S. and Burstein, E. (1974) *Phys. Rev. Lett.*, **33**, 908.
19. Olego, D. (1987) *Appl. Phys. Lett.*, **51**, 1422.
20. Pinczuk, A. and Burstein, E. (1973) *Surf. Sci.*, **37**, 153.
21. Gatos, H.C. and Lagowski, J. (1973) *J. Vac. Sci. Technol.*, **10**, 130.
22. Lüth, H., Büchel, M., Dorn, R., Liehr, M., and Matz, R. (1977) *Phys. Rev. B*, **15**, 865.
23. Brillson, L.J. (1975) *Surf. Sci.*, **51**, 45.
24. Brillson, L.J. (1976) *J. Vac. Sci. Technol.*, **13**, 325.
25. Lagowski, J., Balestra, C.L., and Gatos, H.C. (1972) *Surf. Sci.*, **29**, 203.
26. Kronik, L. and Shapira, Y. (1999) *Surf. Sci. Rep.*, **37**, 1.
27. (a) Brillson, L.J. (1979 *Inst. Phys. Conf. Ser.* **43**, 765 (b) Brillson, L.J. (1979) *J. Vac. Sci. Technol.*, **16**, 137.
28. Brillson, L.J. (1978) *Phys. Rev.*, **18**, 2431.
29. Mead, C.A. (1966) *Solid-State Electron.*, **9**, 1023.

Further Reading

Kronik, L. and Shapira, Y. (1999) *Surf. Sci. Rep.*, **37**, 1–206.

16
Cathodoluminescence Spectroscopy

16.1
Overview

Cathodoluminescence spectroscopy (CLS) combines optical spectroscopy with electron excitation. CLS is similar to photoluminescence spectroscopy (PLS) except that electrons rather than photons produce the optical transitions. Both techniques involve the creation of free electron–hole pairs, their subsequent recombination and optical emission. The CLS technique possesses several advantages over PLS: (i) orders-of-magnitude higher free carrier generation rates; (ii) the ability to excite electron–hole pairs in wide bandgap semiconductors and insulators; (iii) an excitation depth that can be tuned from surface to bulk sensitivity; (iv) true surface sensitivity at low incident beam energies; (v) variable incident beam energies that permit nanoscale depth resolution; (vi) the ability to probe interfaces below the free surface; and (vii) spatial resolution on a scale of nanometers – achievable with focused electron beams.

The high free carrier generation rates available with electron beams provide more free carrier recombination and higher optical signal intensities than can be achieved with conventional light sources. The free electrons and holes produced by the incident electron beam have kinetic energies that exceed the bandgaps of all solid-state materials. Conventional laser sources have photon energies that extend from the infrared up to near 4 eV, limiting their use for wide bandgap semiconductors such as AlGaN (3.5–6.2 eV) and MgZnO (3.3–8 eV) or insulators such as SiO_2 (9.1 eV), HfO_2 (5.9 eV), diamond (5.5 eV), and Al_2O_3 (8.8 eV). Higher laser energies are available using frequency doubling techniques or excimer lasers. However, these require more elaborate experimental equipment.

Furthermore, photon excitation occurs at a single energy and a single absorption depth, whereas electron beam excitation can be varied from the free surface to microns within the bulk. At low incident beam energies – 1 keV and below, CLS excitation occurs primarily in the material's outer monolayer. The ability to vary the incident beam energy and thereby the excitation depth permits researchers to distinguish between features at different depths – for example, between the surface and the bulk or between a layer and its surrounding material. The same control of excitation depth enables CLS to probe "buried interfaces" below the free surface that are not accessible by conventional surface science techniques.

Surfaces and Interfaces of Electronic Materials. Leonard J. Brillson
Copyright © 2010 WILEY-VCH Verlag GmbH & Co. KGaA, Weinheim
ISBN: 978-3-527-40915-0

16 Cathodoluminescence Spectroscopy

The short excitation depths of low-energy electron beams and their variation with incident beam energy provide a technique to measure electronic properties with nanometer-scale depth resolution. Unlike photons, the incident electron beam can be focused down to nanometers or smaller, thereby permitting nanometer-scale lateral resolution as well.

Cathodoluminescence (CL) is one of several processes that occur with electron beam irradiation of solids. Figure 16.1 illustrates the various processes that occur within a pear-shaped volume of electron beam excitation. A primary electron beam focused to 10–100 Å on a solid surface produces (i) Auger electrons within the first few angstroms of the free surface (see Section 9.3);(ii) secondary electrons due to ionization of impacted atoms at nanometer or higher depths that depend on the incident beam energy; (iii) backscattered electrons at lower depths due to random collisions of electrons that have lost significant kinetic energy; (iv) X-rays characteristic of specific atomic transitions (see Section 9.3.2);(v) a continuum of X-ray energies resulting from secondary X-ray excitation; and (vi) fluorescent X-rays, that is, optical emission due to low-energy electrons initially excited by X-rays. The excitation volume pictured in Figure 16.1 can extend less than a few nanometers to tens of microns deep into the solid, depending on the incident beam energy. The pear-shaped width of the electron cascade has dimensions that are a significant fraction of this length.

While the Auger electrons generated by the electron beam preserve most of the incident beam's spatial resolution, the secondary and backscattered electrons generate electron–hole pairs over a larger volume. Nevertheless, at low incident beam energies, this volume can be only nanometers in scale so that the focused electron beam can provide site-specific information as well as mapping of extended areas.

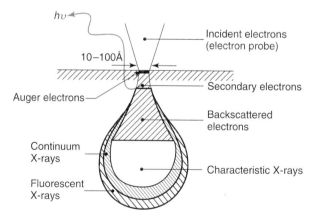

Figure 16.1 Schematic illustration of processes in a solid induced by electron beam irradiation. The cascade of secondary electrons creates free electron–hole pairs that recombine to produce cathodoluminescence that leaves the solid.

16.2 Theory

16.2.1 Scattering Cross Section

In order to understand the nanometer scale of excitation involved in CLS, let us first consider how electrons lose energy in a solid. Recall the particle – solid scattering formalism introduced in Chapter 9[1]. The change in particle momentum for a particle of charge Z_1 and impact parameter b is given by

$$\Delta p = \frac{(2Z_1 e^2)}{(bv)} \tag{9.9}$$

and the energy transfer T in a collision is

$$\frac{\Delta p^2}{(2m)} = \frac{(2Z_1^2 e^4)}{(b^2 m v^2)} = T \tag{9.10}$$

so that

$$T = \frac{(\Delta p^2)}{2m} = \frac{4e^4}{2b^2 m v^2} = \frac{e^4}{b^2 E} \tag{9.11}$$

for electron energy $E = 1/2\, mv^2$ and $Z_1 = 1$ for electron–electron collisions.

Again, the differential cross section for energy E between T and $T + dT$ is

$$d\sigma(t) = -2\pi b\, db = \frac{\pi e^4}{E} \cdot \frac{dT}{T^2} \tag{9.12}$$

The energy loss per unit path length can be defined as

$$\frac{-dE}{dx} = n_{T\min} \int^{T\max} T\, d\sigma \tag{16.1}$$

for n, the number of electrons per unit volume. Substituting for $d\sigma$ in Equation 9.12 yields

$$\frac{-dE}{dx} = n_{T\min} \int^{T\max} T \cdot 2\pi b\, db \tag{16.2}$$

Substituting for T in Equation 9.10 for a particle of charge Z_1 in general,

$$\frac{-dE}{dx} = 2\pi n \int \frac{(2Z_1^2 e^4 b)}{(b^2 m v^2)}\, db \tag{16.3}$$

$$= \frac{(4\pi n Z_1^2 e^4)}{(m v^2)} \int \frac{db}{b} \tag{16.4}$$

so that

$$\frac{-dE}{dx} = \frac{(4\pi n Z_1^2 e^4)}{(m v^2)} \ln\left(\frac{b_{\max}}{b_{\min}}\right) \tag{16.5}$$

We can now evaluate the limits of impact parameter b. The minimum impact parameter $b_{\min}$ corresponds to a head-on collision. In this case, the velocity v, changes to $2v$. Then the maximum energy transfer is given by

$$T_{max} = \frac{1}{2}m(2v)^2 = 2mv^2 \tag{16.6}$$

so that, using Equation 9.10 to replace T,

$$b_{min} = \frac{(Z_1 e^2)}{(mv^2)} \tag{16.7}$$

The maximum impact parameter b_{max} corresponds to a minimum energy T_{min} needed to raise an electron to an excited state. We can define an average electron excitation energy with a value I.

Again using Equation 9.10 with $T_{min} = I$,

$$b_{max} = \frac{(2Z_1 e^2)}{\sqrt{(2mv^2 I)}} \tag{16.8}$$

With these maximum and minimum values of impact parameter, Equation 16.5 now becomes

$$\frac{-dE}{dx} = \frac{(2\pi n Z_1^2 e^4)}{(mv^2)} \ln\left(\frac{2mv^2}{I}\right) \tag{16.9}$$

16.2.2
Stopping Power

Equation 16.9 is based solely on direct collisions with electrons in a solid. There is actually an additional energy loss per path length of comparable magnitude due to distant resonant transfer [2]. Considering both contributions, the total *"stopping power"* for an incoming particle of charge Z_1 is then

$$\frac{-dE}{dx} = \frac{(4\pi n Z_1^2 e^4)}{(mv^2)} \ln\left(\frac{2mv^2}{I}\right) \tag{16.9a}$$

One can generalize this expression to an incoming particle of mass M_1, charge Z_1, and energy $E = 1/2\, M_1\, v^2$ to yield

$$\frac{-dE}{dx} = \left(\frac{2\pi Z_1^2 e^4}{E}\right)(NZ_2)\left(\frac{M_1}{m}\right)\ln\left(\frac{2mv^2}{I}\right) \tag{16.10}$$

for atomic density N of the "stopping" medium and $n = NZ_2$. Note: Atomic density $N =$ density ρ/atomic weight A.

From this expression, one sees that stopping power depends on (i) the ratio E/I of particle energy to average excitation energy; (ii) the atomic density N; and (iii) the number of target electrons NZ_2. For nuclear particles, nearly all the target electrons participate.

The average excitation energy I for most elements is $\sim 10\, Z_2$ (in electronvolts) for $Z > 12$. For example, $I = 10\, Z_2 = 130$ for Al with an atomic number $Z_2 = 13$ equal to the number of "stopping" electrons.

16.2.3
Plasmon Energy Loss

Several mechanisms can contribute to the energy loss of electrons in solids. These include plasmons, X-ray generation, optical phonons, and impact ionization. Which of these loss mechanisms is dominant depends on the electron's energy, which is decreasing as the electron moves through the solid. Figure 16.2 illustrates the rate of energy loss per unit time of electrons within a solid as a function of incident electron energy [3].

This experimental graph shows that the dominant mode of energy loss for electrons in the 100–10,000-eV range of kinetic energies is the creation of plasmons. Plasmons are collective excitations of the electrons in a solid. For a metal or a conductive semiconductor, one may consider these electrons as a "gas" that displaces uniformly with respect to the positive ion background of the lattice. Figure 16.3 illustrates this concept, introduced first in Section 10.3[4].

A uniform displacement η of electrons vertically in Figure 16.3 produces an electric field $E = 4\pi \eta n e$ that acts as a restoring force. The equation of motion of a unit volume of electron gas is

$$\frac{nmd^2\eta}{dt^2} = -neE = \frac{-4\pi n^2 e^2 \eta}{\varepsilon} \tag{16.11}$$

or

$$\frac{d^2\eta}{dt^2} + \omega_p^2 \eta = 0 \tag{16.12}$$

where

$$\omega_p = \left(\frac{4\pi ne^2}{m\varepsilon}\right)^{1/2} \tag{16.13}$$

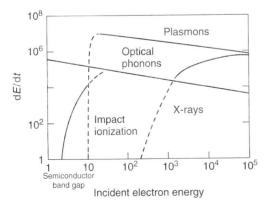

Figure 16.2 Rate of energy loss dE/dt of electrons in a solid due to X-rays, plasmons, optical phonons, and impact ionization [3].

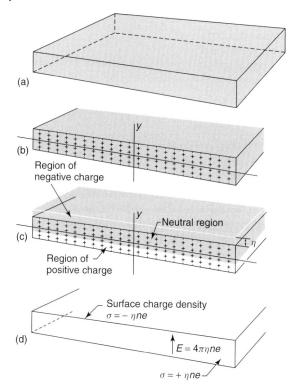

Figure 16.3 Uniform displacement of electrons in a conductive solid. (a) a conductive thin slab or film, (b) a cross-sectional view of the positive ion core (+ signs) and the electron "gas" density proportional to the shading, (c) uniform upward displacement of the electron gas by an amount η to produce regions of negative and positive charge, and (d) a resultant electric field $E = 4\pi\eta ne$ due to surface charge densities $\pm \eta ne$ [4].

The solution to Equation 16.12 is then

$$\eta = \eta_0 e^{\pm i\omega_p t} \quad (16.14)$$

In other words, the collective movement of electrons corresponds to an oscillation of frequency ω_p and finite wavelength whose energy $\hbar\omega_p$ depends on the electron density n, the electron mass m, and the dielectric constant ε of the medium.

Electron energy loss spectroscopy provides a direct method to determine these plasmon energies as already discussed in Section 10.3. For completeness, it is worth noting that the relative dielectric constant depends on both incident wave vector q and frequency ω according to

$$\varepsilon(q, \omega) = 1 + 4\pi\alpha(q, \omega) \quad (16.15)$$

where $\alpha(q, \omega)$ is the polarizability of interacting electron system. In the long wavelength limit, $\varepsilon(q, \omega) = 1 - \omega_p^2/\omega^2$, so that ε can be taken as ε_0 for incident electron energies above 100 eV.

The electron density n is $\sim 10^{22}$ cm^{-3} in metals or for the valence bands of semiconductors. For semiconductor conduction bands, $n \sim 10^{17}$ cm^{-3} or less, typically 5–6 orders of magnitude smaller. For comparison, ω_p values for metals range from 15 eV for Al, 5.8 eV for Na, and 7 eV for Li, whereas ω_p for the GaAs conduction band electrons is only 36.6 meV, 3 orders of magnitude smaller.

16.2.4
Electron Scattering Length

Consider now the stopping power of incident electrons in the 10–10 000 eV range. Equation 16.9 now becomes

$$\frac{-dE}{dx} = \frac{(\omega_p^2 e^2)}{(v^2)} \ln\left(\frac{2mv^2}{\hbar \omega_p}\right) \tag{16.16}$$

for $Z_1 = 1$, $I = \hbar\omega_p$, and $\omega_p^2 = 4\pi n e^2/m$ for $\varepsilon = \varepsilon_0 = 1$.

If plasmons dominate dE/dx, then the inverse scattering length $1/\lambda$ can be expressed as

$$1/\lambda = \frac{\left(\frac{-dE}{dx}\right)}{(\hbar \omega_p)}$$

$$= \left(\frac{\omega_p e^2}{\hbar v^2}\right) \ln\left[\frac{(2mv^2)}{(\hbar \omega_p)}\right] \tag{16.17}$$

Equation 16.17 shows that λ increases with kinetic energy $E = 1/2mv^2$. At very low energies, this expression breaks down since the loss mechanisms change rapidly in this regime. Nevertheless, at energies $E \approx \hbar\omega_p$, the $(1/v^2)$ factor in this equation indicates that λ also increases as E decreases. Physically, this is because there is steadily less energy and momentum to excite electronic transitions at very low energies. Therefore, λ passes through a minimum value between these high- and low-energy regimes. According to Equation 16.17, $\lambda = 5.45$ Å at 350 eV, $\cong 2$ Å at 92 eV, and >4 Å at 4 eV. This is consistent with a minimum at energies in the 50–100 eV range.

This energy dependence of the scattering length is the reason for the minimum in escape depth presented in Figure 6.1 and the basis on which surface science rests. Recall that the escape depth is the distance that electrons of well-defined energy can travel without losing energy. Only electrons within the escape depth of the surface can leave without energy loss.

Scattering lengths of electrons in solids can be measured as a function of energy by using thin film overlayers to gauge the energy-dependent escape depth phenomenon. The schematic cross section in Figure 16.4a illustrates the movement of electrons generated by incident excitation below the surface of a solid. Only electrons within the escape depth of the surface (shown as a dashed line in Figure 16.4a) can leave the solid without energy loss. This defines the depth of detection.

Consider a flux I_0 of electrons leaving the substrate shown in Figure 16.4a. Consider a solid with a density N' of scattering centers/cm^3, each of which has an

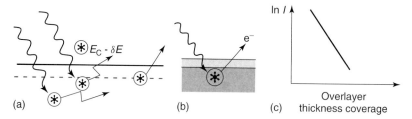

Figure 16.4 (a) Schematic cross section of excited electron trajectories below a solid surface. Only electrons within the escape depth (dashed line) leave the solid without inelastic scattering. (b) Thin film deposition attenuates escaping electrons. (c) The decrease in electrons that escape without energy loss decreases logarithmically with increasing overlayer thickness.

inelastic collision cross section σ. The number of electrons dI removed from the electron flux I is then σI per scattering center. Thus

$$-dI = \sigma I N' dx \tag{16.18}$$

per thickness increment dx. Integrating this expression yields

$$I = I_0 e^{-\sigma N' x} = I_0 e^{-x/\lambda} \tag{16.19}$$

where the mean free path between scattering events is defined as

$$\lambda = \frac{1}{N'\sigma} \tag{16.20}$$

and the cross section σ embodies the energy dependence expressed in Equation 16.10.

Figure 16.4b illustrates an electron escaping a solid through a uniform overlayer of an element or compound that differs from the substrate. With increasing thickness of this overlayer – obtained by careful deposition in vacuum, the intensity of escaping electrons from the substrate at a given kinetic energy decreases exponentially according to Equation 16.12. A plot of ln I versus overlayer thickness then yields the mean free path λ for a given material as represented in Figure 16.4c. Uniform coverage of the substrate by the overlayer is required to obtain this exponential decay; otherwise, the intensity of substrate electrons flux decays more slowly.

16.2.5
Semiconductor Ionization Energies

The increase in mean free path between scattering events at low energies can be related to the effective ionization energy, that is, the average energy E_{av} required for the generation of an electron–hole pair. This average energy to excite electron–hole pairs increases linearly with bandgap energy E_G, as shown in Figure 16.5 [5].

This average amount of energy given up by the incident radiation in the process of generating a single electron–hole pair is not a threshold energy [5]. The latter

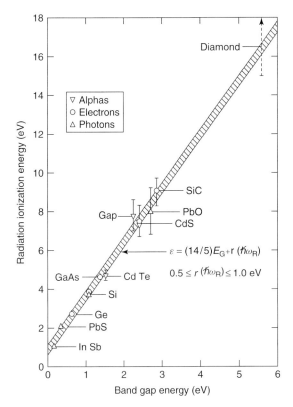

Figure 16.5 The effective ionization energy E_{av} to generate electron–hole pairs in semiconductors as a function of bandgap [5].

depends on the minimum energy required to conserve momentum. The energy dependence shown in Figure 16.5 is described by the relation

$$E_{av} = \left(\frac{14}{5}\right) E_G + r(\hbar\omega_R) \tag{16.21}$$

where $\hbar\omega_R$ represents the highest-frequency zero-wave-vector phonon detected by Raman scattering and r is the average number of phonons per pair generated during the initial stage of the impact ionization process. The latter can be viewed as the ratio of mean free path lengths for impact ionization (λ) versus phonon emission [5] so that $r(\hbar\omega_R)$ typically ranges from 0.5 to 1 eV, according to energy dependence of λ. Hence the bandgap dependence shown in Figure 16.5 reflects both the available states in momentum space at a given energy as well as the scattering by phonons.

The bandgapbandgap dependence shown in Figure 16.5 represents a semiempirical law for significant impact ionization in semiconductors as measured by a variety of experimental techniques. It shows that impact ionization by electrons or other particles requires energies equal to a significant multiple of the semiconductor bandgap in order to generate free electron–hole pairs efficiently. This requirement highlights a major advantage of CLS, namely, to provide high enough energies to produce efficient luminescence for spectroscopy. It also suggests that energy dissipation will limit the use of scanning probes for generating detectable CL since excitation in subnanometer-scale areas may introduce damage due to local heating.

16.2.6
Universal Range-Energy Relation

The energy dependence of the electron–electron scattering cross section (Equation 16.16) and the average ionization energy (Equation 16.21) indicates that an incident electron will produce impact ionization of secondary electrons that will continue to multiply until their kinetic energies fall below the average ionization energy. Thus the rate of electron–hole pair creation in this cascade will increase with increasing depth of electron penetration, then decrease as the electrons' kinetic energies decrease below the average ionization energy. Figure 16.6 illustrates this rate of electron–hole pair creation schematically. With increasing incident beam energy E_B, the cascade generates electron–hole pairs at successively deeper locations within the solid.

The number of *electron–hole pairs created per second per unit volume* for an electron current J_B penetrating a layer of material dx thick is

$$g(x) = \left(\frac{J_B}{eE_A}\right)\frac{dE}{dx} \tag{16.22}$$

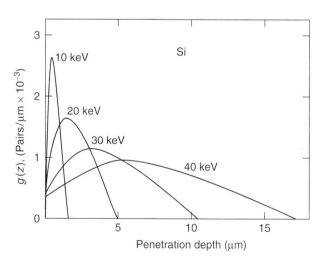

Figure 16.6 Rate of electron–hole pair creation g(z) versus penetration depth as a function of one incident electron's beam energy in Si. (After Everhart and Hoff. [6])

where E_A is the mean energy required for an energetic electron to create an electron–hole pair. For a *given incident beam energy* E_B, the rate of electron–hole pair creation reaches a maximum at intermediate depths that are a fraction of the total penetration depth.

A variant of Equation 16.10 expresses the number of scattering centers in terms of density ρ, atomic weight A, atomic number Z, and Avogadro's number N_A[3], 6:

$$\frac{-dE}{ds} = (2\pi N_A e^4) \left(\frac{Z\rho}{A}\right) \left[E^{-1} \ln\left(\frac{aE}{I}\right)\right] \quad (16.23)$$

where the *mean excitation energy loss* $I = (9.76 + 58.8 Z^{-1.19})Z$ (electronvolts) and $a = 1.1658$. This expression for the electron kinetic energy rate of change along path s contains an explicit dependence on material properties.

The maximum range over which the impact ionization occurs is given as the *Bohr–Bethe range* R_B. Here

$$R_B = \int_E^\infty \frac{dE}{\left(\frac{dE}{d(\rho s)}\right)} = \int_0^{\xi(E_B)} \frac{\xi\, d\xi}{\ln(\xi)} \quad (16.24)$$

where $\xi = 1.1658 E/I$.

R_B can be approximated by expressions of the form

$$R_B = C\xi^a \quad (16.25)$$

Equation 16.25 closely approximates Equation 16.24 over 3 orders of magnitude in energy [7]. The constant a is material independent and equals 1.29 for $\xi < 10$, the energy range around 1 keV [7]. The constant C in this expression is material dependent and equals $9.40 \times 10^{-12}\, I^2\, (A/Z)\, c'/\rho$. In turn, the constant c' is material independent and equals 1.48 for $\xi \leq 10$, the range of most interest for surface and interface studies. In its general form, Equation 16.24 provides a "universal" fit to experimental data.

The *maximum energy loss per unit depth* occurs at a depth U_0 whose energy dependence has been fitted to experimental measurements down to KeV energies. Here [7],

$$U_0(E > 1\,\text{keV}) = 0.069\, \xi^{1.71} \quad (16.26)$$

Similarly, Equation 16.25 has been fit to the Everhart–Hoff relation in Equation 16.23 over the same energy range:

$$R_B(E > 1\,\text{keV}) = 0.62\, \xi^{1.609} \quad (16.27)$$

whereas $R_B(E \sim 1\,\text{keV}) = 0.1.48\, \xi^{1.29}$. These R_B expressions apply to the same material in different energy ranges of E_B. The ratio of U_0/R_B varies between different materials but appears to be constant for a specific material at different energies [7]. From the U_0/R_B ratio at intermediate energies and $R_B(E \sim 1\,\text{keV})$, U_0 is then [8]

$$U_0(E \leq 1\,\text{keV}) = 0.1647\, \xi^{1.392} \quad (16.28)$$

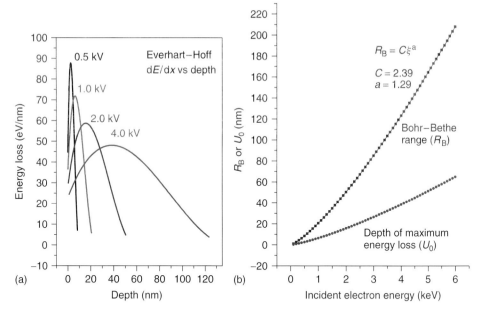

Figure 16.7 GaN depth dependence of electron-excited nanoscale luminescence spectroscopy. (a) Energy loss rate $-dE/dx$ versus depth and (b) Bohr–Bethe range R_B and depth of maximum energy loss rate U_0. (After Everhart and Hoff. [6])

Figure 16.7a illustrates energy loss rates based on Equation 16.23 versus depth in GaN for incident beam energies in the 0.5–4 keV range. For these energies, the depth of excitation extends only tens of nanometers into the semiconductor. For $E_B = 0.5$ kV, the peak excitation occurs at a depth only 1–3 nm. The depth range of electron–hole pair creation increases with increasing energy, extending only a few nanometers for $E_B \leq 1$ keV to > 100 nm for $E_B \geq 4$ keV.

Figure 16.7b illustrates the variation in R_B and U_0 as a function of incident electron beam energy using Equation 16.25 Over the 0.5–6 kV energy range, U_0 is approximately $R_B/3$ for a uniform material. Furthermore, these curves show that the depth of excitation varies over distances of only tens of nanometers or less for energies in the low kilovolt range.

Figure 16.7 illustrates why CLS at low keV incident beam energies qualifies as a surface science technique. The ability to excite electron–hole pair recombination within a few nanometers of the free surface and to vary the excitation depth on a nanometer scale permits measurements that distinguish between electronic properties of the surface versus the bulk. Furthermore, CLS at intermediate energies can maximize excitations at depths that correspond to interfaces between layers of materials. These surface-sensitive and depth-resolved capabilities have many applications for semiconductor surfaces and interfaces.

16.3
Monte Carlo Simulations

Monte Carlo simulations provide electron energy loss profiles analogous to those obtained by analytic expressions. Monte Carlo methods involve random number generators that determine the collision dynamics of incident electrons with electrons of the solid. The energies and momenta of the electrons after the collision then become the initial state of incident electrons that initiate the next stage of electron–electron collisions in the cascade process. The simulation calculates the energy loss per unit path length for an incident electron as its energy decreases from E_B to zero. This process is repeated for many electrons, typically several thousand, and summed in order to reveal an average electron cascade profile. Monte Carlo simulations have several major advantages over the energy loss calculations discussed in the previous section: (i) scattering cross sections tabulated versus particle energy in the cascade, (ii) experimentally determined stopping powers, (iii) capability to include backscattering contributions, and (iv) ability to simulate cascades in multiple-layer stacks with different stopping powers.

Figure 16.8 illustrates a set of Monte Carlo simulations using the CASINO program [9] for incident beam energies E_B in the 0.5–5 keV range without taking backscattering into account. The excitation profiles exhibit nearly the same features as those illustrated in Figure 16.7. For example, both figures show 1-keV profiles that extend to a depth of ∼20 nm and that reach a maximum at a depth of 6–7 nm. Figure 16.8 profiles are normalized to constant peak height. As in Figure 16.7, the U_0 and R_B values in Figure 16.8 vary on a nanometer scale for incident beam energies in the low keV range.

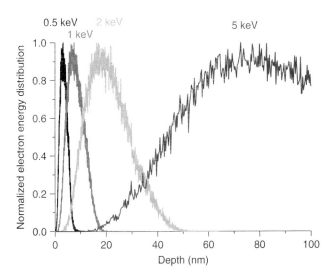

Figure 16.8 Monte Carlo simulation of electron excitation in GaN for an electron beam incident at 45° to the surface.

The excess carrier concentrations generated by the incident electron beam can be large enough to significantly change the free carrier concentration in the semiconductor. The excess carrier concentration N is given simply by the *generation rate* G multiplied by the *free carrier lifetime* τ.

$$N = G\tau \tag{16.29}$$

The generation rate for an incident beam of electrons can be approximated by

$$G = \left(\frac{I}{Ad}\right) \cdot \left(\frac{E_B}{E_0}\right) \tag{16.30}$$

for a beam current I, surface beam area A, effective penetration depth d, and average energy to create electron–hole pairs E_0. E_0 can be obtained from Figure 16.5, for example, 4.5 eV for InP or GaAs. Minority carrier lifetimes determine the recombination rate. These typically range from 10^{-5} to 10^{-10} seconds.

Example 16.1. As an example, consider a CLS measurement of wurtzite GaN ($E_G = 3.44$ eV) with $E_B = 1$ keV, $E_0 = 10$ eV, $I = 10$ µA, and $A = 10^{-3}$ cm². Assume a minority carrier lifetime $\tau = 10^{-10}$ seconds. Compare with photoluminescence with a 10 mW He–Cd laser with $h\nu = 3.82$ eV focused to the same area.

Cathodoluminescence: From Figure 16.7b, the Bohr–Bethe range is 20 nm. Then

$$N = \left[\frac{(10^{-5} A \cdot 6.24 \times 10^{18} \text{ electrons s}^{-1}/A)}{(10^{-3} \text{ cm}^2 \cdot 2 \times 10^{-6} \text{ cm})}\right]\left(\frac{10^3 \text{eV}}{10 \text{ eV}}\right)\tau(s)$$

$$= 3.12 \times 10^{24} \text{ electrons cm}^{-3} \text{ s}^{-1}\tau(s)$$

$$= 3.12 \times 10^{14} \text{ electrons cm}^{-3} \text{ within the top 20 nm.}$$

Photoluminescence: The absorption coefficient for wurtzite GaN at $h\nu = 3.82$ eV $= 1.2 \times 10^5$ cm^{-1} (Muth, J.F. et al. (1999) *MRS Internet J. Nitride Semicond. Res.*, **4S1**, G5.2). Using the conversion 1 W = 1 J s^{-1} = 6.242 × 10^{18} eV s^{-1},

$$10 \text{ mW} = \frac{6.242 \times 10^{16} \text{ eV/s}}{3.82 \text{ eV/photon}}$$

$$= 1.63 \times 10^{16} \text{ photons/s}$$

$$N = 1.63 \times 10^{16} \text{ photons/s}/(10^{-3}\text{cm}^2 \cdot 0.83 \times 10^{-5} \text{ cm})\tau(s)$$

and, assuming unity quantum efficiency,

$$N = 1.96 \times 10^{14} \text{ electrons cm}^{-3} \text{ within the top 830 Å}$$

Thus CL and PL generate comparable densities of carriers in this case. Continuous wave (CW) laser intensities can increase at least one order of magnitude without inducing thermal damage. (Incident photon energies in excess of the photon energies emitted at the bandgap are converted to phonons.) Electron beam intensities and densities can increase several orders of magnitude, particularly with a scanning electron microscope (SEM). While similar heating effects can occur, the band edge emission provides a sensitive indicator of temperature and displays CL temperature increases of only a few degrees for most semiconductors under the conditions given in this example.

Surface-sensitive CLS requires ultrahigh vacuum (UHV) conditions. At higher background pressures, residual gas molecules such as CO_2 in the vacuum chamber decompose under electron beam irradiation, building up a layer of carbon at the focal point. Such graphitized surfaces are efficient recombination centers and strongly reduce free carrier densities in their vicinity. As a result, researchers using conventional electron microscopes typically find that CL signals of semiconductors decrease strongly with $E_B \sim 5$ keV or less.

While the cascade profiles indicate that excitations can be spatially localized on a nanometer scale, carrier diffusion can degrade this depth resolution. Indeed, carrier diffusion can be gauged by using the incident electron beam to excite free carriers and monitor their decay as a function of distance from a collection electrode. This technique is termed *electron beam-induced current* (EBIC) and is used to measure free carrier lifetimes.

On the other hand, several factors tend to minimize this diffusion effect. First, the recombination volume is defined by the minority carrier diffusion, which is usually much lower than that of majority carriers. Second, diffusion length depends on recombination lifetime, which is typically high near surfaces and interfaces. Finally, many of the electronic materials currently being explored have minority carrier diffusion lengths that are also tens of nanometers or less. Indeed, the features of highest interest to researchers in understanding electronic material structures – native point defects, dislocations, and interfaces, are the very ones most likely to produce recombination and light emission. The next section provides examples that confirm the nanoscale control of CLS excitation at low beam energies.

16.4
Depth-Resolved Cathodoluminescence Spectroscopy

Depth-resolved cathodoluminescence spectroscopy (DRCLS) enables one to profile features of an electronic material from the surface to the bulk on a nanometer scale. This is particularly useful when studying the electronic properties of "buried" interfaces located more than a few nanometers below the free surface. This capability also permits investigations of films only a few nanometers thick as well as their interfaces with neighboring media. Figure 16.9 illustrates this capability for a 5-nm thick SiO_2 layer grown by plasma-assisted chemical vapor deposition on a Si surface treated with a remote plasma-assisted oxide (RPAO) and remote plasma-assisted nitride (RPAN) layer.

DRCLS spectra exhibit peaks at 2.0, 2.7, and 4.3 eV that can be associated with interface defects, SiO_2 E'_2 defects, and a bulk Si transition, respectively [10]. The top scale indicates E_B while the bottom scale displays the corresponding R_B and U_0 values. At $E_B = 0.5$ keV, only the outer surface of the SiO_2 layer is probed, compared with the interior of the SiO_2 at 0.6 keV ($R_B \sim 10$ nm $U_0 \sim 3$ nm), and the SiO_2/Si

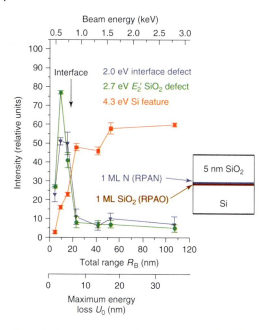

Figure 16.9 Intensities of DRCLS peaks at 2.0, 2.7, and 4.3 eV corresponding to interface defects, SiO_2 E'_2 defects, and a bulk Si transition, respectively.

interfaces at 0.75 keV ($R_B \sim 18$ nm $U_0 \sim 6$ nm). Thus the well-known E'_2 defect in SiO_2 reaches a maximum in the center of the 5-nm film, then decreases, while the interface state feature has maximum intensity slightly deeper. These intensity variations contrast with a bulk Si feature that increases steadily with increasing excitation depth. This example illustrates that DRCLS has nanoscale depth resolution comparable to that anticipated from Section 16.3.

A diagram of equipment for low-energy DRCLS appears in Figure 16.10 [11]. This equipment requires only simple modifications of conventional UHV surface science hardware. Shown here is a cross section of a UHV chamber with glancing incidence electron gun typically used for Auger electron spectrometry (AES). The sample is positioned at the focal point of the electron gun and in front of a UV–IR transmitting light lens with a manipulator. The light focuses through a sapphire viewport onto the entrance slit of a monochromator. A photomultiplier or photodiode at the exit slit of the monochromator provides an electrical signal proportional to the energy-resolved light intensity. The electron energy incident on the sample typically ranges from a few hundred volts to 5 kV with currents of a few microamperes and a focal spot diameter of $\sim 1/2$ mm. The ability to perform CLS under UHV conditions and at low electron beam energies permits studies of surface electronic properties as a function of chemical or structural changes. UHV conditions also minimize any beam decomposition of residual gases, which would complicate the interpretation of such results.

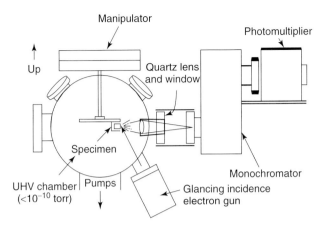

Figure 16.10 Cross-sectional schematic of low-energy CLS equipment [11].

16.4.1
Surface Electronic States

At incident beam energies of 1 keV, the surface sensitivity of CLS permits measurements of electronic states that are intrinsic to clean semiconductor surfaces. Figure 16.11 illustrates this capability to measure electronic states within the bandgap of clean semiconductor surfaces. Here, clean surfaces of GaAs grown by MBE were annealed to produce different surface reconstructions as measured by low-energy electron diffraction (LEED). (See Chapter 13.)

The deep-level emissions from these commonly observed reconstructions extend between photon energies of 0.8 and 1.3 eV and are relatively weak compared with the near band edge (NBE) peak intensity at 1.47 eV. Consistent with their surface origin, these features change with surface reconstruction and decrease in intensity with increasing E_B. Figure 16.11a shows a slight enhancement in the 0.8–1.1 eV range for the most As-rich (1 × 1) surface, which becomes weaker for the (2 × 4) and (4 × 2)–c(8 × 2) surface reconstructions (Figure 16.11b,c). At a temperature of 620 °C (Figure 16.11d), As desorption leads to the lowest As/Ga surface composition as measured concurrently by soft X-ray photoemission spectroscopy (SXPS), reducing the emission in the 1.1–1.2 eV range relative to the more As-rich surfaces [12]. Thus surface-sensitive CLS reveals surface states with emissions centered at ∼0.95 and 1.15 eV that change their relative intensities with As/Ga stoichiometry and reconstruction.

16.4.2
Interface Electronic States

The capability of DRCLS to probe interfaces below the free surface can provide localized electronic state information that crystal growers of electronic materials can use. As an example, the properties of the GaN/In$_x$Ga$_{1-x}$N/GaN quantum

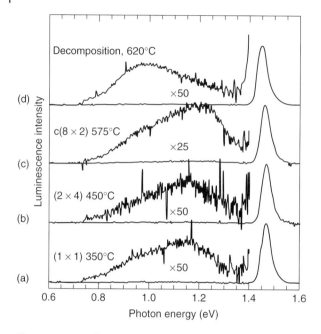

Figure 16.11 Surface-sensitive 1-keV CL spectra at $T \sim 180$ K from clean GaAs(100) surfaces versus surface reconstruction [12].

well structure used extensively for optoelectronics are sensitive to epitaxial growth conditions. This structure consists of 30 nm GaN on 1.5–2.5 nm $In_xGa_{1-x}N$ on a GaN buffer layer and substrate. Figure 16.12 shows this structure (a) and a series of CL spectra with E_B ranging from 0.75 to 4.5 keV (b and c) [13]. For $E_B = 0.75$ keV, both the outer GaN confinement layer and the InGaN quantum well are excited and emit light at 3.5 and 2.4 eV, respectively.

As E_B increases and excitation moves near the quantum well, the quantum well emission intensity grows while the GaN intensity decreases. As U_0 advances through the quantum well with increasing E_B, a new peak appears at 3.25 eV that corresponds to cubic GaN. This 3.25 eV feature reaches maximum intensity relative to the quantum well and the wurtzite GaN confinement layers at $E_B = 2.0$ eV. For higher E_B, both the quantum well and cubic GaN peaks decrease and the GaN peak increases. Measured relatively to the GaN peak intensity, the cubic GaN emission reaches its maximum intensity at $E_B = 2.0$ keV versus 1.0–1.25 keV for the quantum well. This difference in depth dependence indicates that the cubic GaN forms at the *lower* quantum well interface, that is, at the initial growth stage of the InGaN. This result is not surprising since the lattice constant of $In_{0.28}Ga_{0.72}N$ is sufficiently different from GaN and its growth temperature is significantly lower that the resultant strain produces stacking faults known to induce cubic GaN. The appearance of such interface states in optoelectronic structures is to be avoided since it introduces an alternative recombination pathway for free carriers that reduces the quantum well emission.

16.4 Depth-Resolved Cathodoluminescence Spectroscopy

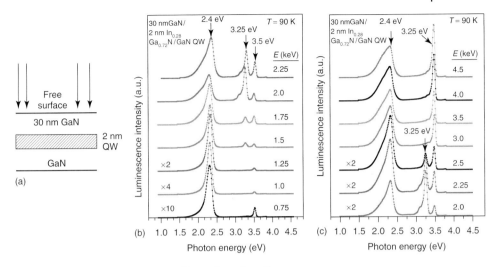

Figure 16.12 DRCL spectra at 90 K of 2-nm $In_{0.28}Ga_{0.72}N$ located 30 nm below free surface [13].

16.4.3
Localized CLS in Three Dimensions

The spatial extent of the electron beam permits localized CLS of electronic materials in three dimensions. Localized CLS on a nanoscale laterally is possible using a SEM. Figure 16.13 illustrates the equipment that can be used.

A focusing column with typical E_B ranging from <1 to 25 kV directs an electron beam with a spot diameter of a few nanometers or more onto a sample. The sample stage can be cooled cryogenically. The CL is collected by a parabolic mirror segment that directs in through a UV–IR window into a monochromator. A photomultiplier and IR diode detects the monochromatized light. In addition, electrons emitted from the sample are collected by a hemispherical energy analyzer and used for AES spectra or secondary electron threshold measurements. The pressure inside the chamber is less than 10^{-10} Torr, permitting measurements of clean and well-controlled surfaces.

The highly localized excitation volume of the electron beam enables CLS measurements of electronic structure at specific locations across a surface and at specific depths below those surface locations. Furthermore, the focused beam provides a tool for cross-sectional studies. For this application, the electron beam excites a cleaved surface of a specimen in a direction normal to the original surface before cleaving. In this way, CLS can probe individual layers and their interfaces directly.

The high energies available with an SEM permit excitation from deep within semiconductors and through various overlayers. One *caveat* in such studies involves the self-absorption of the CL out of the sample. At low E_B, the thicknesses involved are typically small enough such that self-absorption effects are small. For NBE emissions more than a few hundred nanometers below the free

16 Cathodoluminescence Spectroscopy

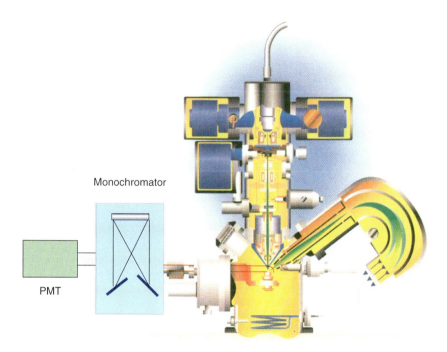

Figure 16.13 Schematic of CLS measurement apparatus using SEM and collection optics. (Courtesy, JEOL, Inc.)

surface, self-absorption can be significant, decreasing emissions near the absorption edge of the intermediate layers. For deep-level emissions at energies well below the bandgap energies, such absorption effects are negligible.

Similarly, for light transmission through metal overlayers, wavelength-dependent absorption can distort the CL spectrum. However, such absorption is quite uniform for most common metals at thicknesses of only a few tens of nanometers.

16.4.3.1 Wafer-Scale Analysis of 2-DEG Layers

The spatial resolution of the focused electron beam enables laterally localized measurements at specific depths. As an example, DRCLS can measure the electronic states with the two-dimensional electron gas (2-DEG) layers across the wafer surface of AlGaN grown epitaxially on GaN and capped with a thin GaN surface layer. These GaN/AlGaN/GaN wafers are commonly patterned to form high electron mobility transistors (HEMTs).

Figure 16.14a illustrates the experimental geometry of probe spots across a 2″ wafer radius (white area represents a section cut to fit into the microscope) while a cross-sectional view of the multilayer structure (b) shows the maximum excitation depth of 2-keV electrons located at the interface between the AlGaN and the GaN bulk layer. This excitation energy maximizes luminescence from

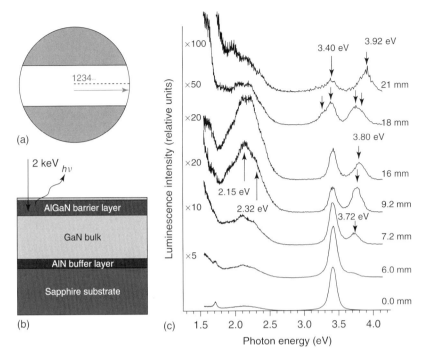

Figure 16.14 (a) wafer view of SEM spots for DRCLS measurements. (b) Cross-sectional view of GaN/AlGaN/GaN HEMT structure and excitation depth of 2-keV electrons. (c) DRCLS spectra (2 keV) across HEMT wafer showing variation in AlGaN composition and deep-level feature with radius [14].

the 2-DEG region, which contains the free carriers responsible for the transistor action. Spectra taken from a series of spots across the wafer radius as indicated in (a) appear in Figure 16.14c.

These spectra reveal three important features: (i) the electronic states in the vicinity of the 2-DEG layer vary considerably across the wafer; (ii) the bandgap emission of the AlGaN increases continuously with radial position, corresponding to increasing Al content at the wafer rim; and (iii) deep-level defects attributed to cation-vacancy-related native point defects are present across the wafer, reaching maximum concentration at intermediate radii. AES depth profiles at this intermediate radius reveal a substantial broadening of the interior AlGaN/GaN interface and an Al decrease that could introduce such cation vacancies [14]. These deep-level defects are resident in the AlGaN layer at concentrations estimated at $1-2 \times 10^{19}$ cm^{-3} from Poisson solver simulations. Their presence introduces acceptors at the AlGaN interface that alters the band profile, removing free carriers from the 2-DEG layer. This defect-driven change in band structure results in dramatic variations in HEMT current density and mobility.

16.4.3.2 Schottky Barriers

The ability to excite CL emission from interfaces is particularly useful for studying localized electronic states at metal–semiconductor interfaces. In turn, these states can be related to the Schottky barriers that form with charge transfer across the interfaces. Figure 16.15 illustrates CL spectra for ZnO($000\bar{1}$) surfaces with and without a metal overlayer [15]. The Figure 16.15a inset illustrates a metal overlayer on clean, ordered ZnO along with the incident electron beams and exciting photons. For the 30-nm thick metal overlayers shown, the probe depth of 2 keV into the bare ZnO is equivalent to the depth excited by 5 keV through the metal.

In the case of Au on ZnO($000\bar{1}$), both the metallized and bare surface ZnO show almost identical spectra with defect emission almost 3 orders of magnitude below the NBE emission. In the case of Al on ZnO($000\bar{1}$), Figure 16.15b, the metallized surface reveals an increase in the deep-level emission centered at 2.5 eV by nearly an order of magnitude [15]. This increase signifies the creation of new defect states within a depth of only a few nanometers from the metal interface. Notably, these new states occur at the same energy of native point defects that are intrinsic to the ZnO crystal.

In the case of the same Au–ZnO($000\bar{1}$) diode at different annealing temperatures, Figure 16.15c, little or no change takes place in the spectra for temperatures up to 550 °C. With a 650 °C anneal, however, a new emission appears centered at ~2.0 eV. This peak corresponds to emission frequently seen in bulk ZnO and has been attributed to Zn vacancies. Likewise, the 2.5-eV feature is commonly associated with oxygen vacancies [15]. Figure 16.15 leads to several conclusions : (i) metals on semiconductors can induce the formation of localized interface states, (ii) these new states appear at energies corresponding to point defects native to the bulk semiconductor, and (iii) the nature of the chemical interaction between metal and ZnO determines what states form and at what temperature. The appearance of new states in Figure 16.15c occurs at a temperature just above the Au–Zn eutectic temperature (642 °C), consistent with the extraction of Zn from the ZnO. Conversely, the formation of new states at Al–ZnO interfaces is consistent with the formation of a metal oxide that extracts oxygen from the ZnO. Overall, these

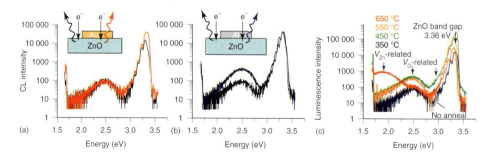

Figure 16.15 CL spectra of bare ZnO($000\bar{1}$) versus with and without an Au (a), Al (b), or annealed Au (c) overlayer on the semiconductor [15].

16.4 Depth-Resolved Cathodoluminescence Spectroscopy

interface-specific states indicate that native point defects can play an important role in Schottky barrier formation.

16.4.3.3 Electronic Devices

Spatially localized CL lends itself to probing arrays of electronic devices and their components. As an example, Figure 16.16 shows a blowup of an AlGaN/GaN transistor source-gate-drain layout within a quarter-wafer array of transistor die. The CLS spectrum obtained from the region indicated within the transistor appears above.

The CLS spectrum for a 2-keV incident beam exhibits NBE peaks due to the GaN and AlGaN layers within a few nanometers of the 2-DEG interface as in Figure 16.14. In addition, the spectrum exhibits deep-level emissions termed *yellow emission* (YL) and *donor–acceptor pairs* (DAPs). These defects can vary in density across the wafer section and within individual transistor dies. Variations in the AlGaN NBE energy and the defect intensities are related to the electrical properties of the transistors. Each die has a transmission line measurement (TLM) array for contact and sheet resistance measurements.

Figure 16.17a shows that the *contact resistance* ρ_C for transistors at different positions across the wafer increases and decreases with the ratio of these prominent

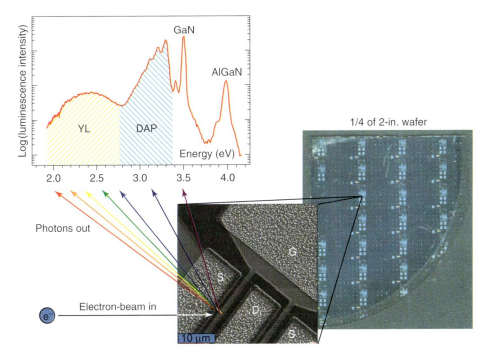

Figure 16.16 CL spectrum from the source-drain region of an AlGaN/GaN HEMT, the individual transistor emitting from the source-drain region, and the particular die within the quarter wafer of the device array.

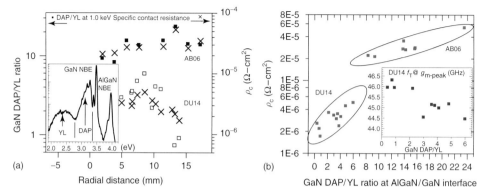

Figure 16.17 (a) Specific contact resistance for TLM in die across AlGaN/GaN wafer section versus position and DAP/YL defect emission intensity ratio at 1.0 keV. (b) Specific contact resistance ρ_C versus DAP/YL ratio for transistors grown on Al_2O_3 (AB06) and on SiC (DU14). In both cases, ρ_C increases with DAP/YL ratio. AlGaN/GaN HEMTs grown on SiC exhibit lower contact resistance. High-frequency cutoff f_T (inset) increases as DAP/YL ratio decreases [16].

defect emissions [16]. Independent of position, the ρ_C increases with increasing DAP/YL ratio. Furthermore, Figure 16.17b reveals that AlGaN/GaN layers grown on lattice-matched SiC display lower ρ_C and defect ratio compared to the poorly matched AlGaN/GaN on Al_2O_3 device structures. The inset in Figure 16.17b shows the impact of higher DAP/YL (and thereby ρ_C) on the *high-frequency cutoff* f_T. As DAP/YL intensity ratio increases, f_T decreases due in part to the increased RC time constant of the device.

These spatially localized measurements also show that (i) sheet resistance decreases as AlGaN NBE energy, and therefore the Al mole fraction, increases. This verifies the 2DEG concentration dependence on AlGaN/GaN lattice strain, namely, that sheet carrier concentration increases with lattice mismatch and Al content. (ii) For HEMTs in which YL-related defects dominate, ρ_C correlates directly with YL and specifically within the AlGaN layers [17].

These results highlight the value of such nondestructive diagnostic techniques in predicting device performance – and thereby avoiding the time and expense of producing devices on lower quality material, before devices are fabricated on the semiconductor wafer.

16.5
Summary

CLS provides a wide range of electronic information about surfaces, interfaces, and bulk properties of electronic materials. These include (i) luminescence center concentrations and their spatial distributions; (ii) the concentration and distribution of extended defects such as dislocations; (iii) the composition of materials from their band edge emission; (iv) carrier diffusion lengths; and (v) new interface compound formation.

The advantages of CLS include (i) nanoscale depth resolution; (ii) nanoscale lateral localization; (iii) high excitation intensity; (iv) sensitivity to defects, impurities, and compound formation at the 10^{11} cm^{-2} and $<10^{15}$ cm^{-3} level; and (v) the ability to probe near and below surfaces.

The limitations of CLS are (i) the difficulty of quantifying CL intensities in terms of defect densities; (ii) the dependence of relative peak heights on injection levels and thereby saturation effects; and (iii) thermal beam damage. All three limitations are common to both CL and PL. However, the CL electron beam is often more likely to break bonds of organic solids.

It is possible to calibrate CL defect densities using ancillary techniques such as deep-level transient spectroscopy (DLTS) or deep-level optical spectroscopy (DLOS). In turn, these techniques require rectifying contacts to gauge capacitance. Both CL and DLTS can be compared at a common depth. In turn, CL using this calibration can then describe defect densities at all depths within range of the incident electron beam.

Using incident beam energies of a few kilo electronvolts or less, it is possible to probe surface and subsurface electronic states of electronic materials in bulk or as ultrathin films.

By varying incident beam energy, one can profile electronic states through a multilayer structure and identify features localized at specific interfaces.

By using an SEM, one can gauge the properties of semiconductors laterally to correlate electronic and morphological features. In semiconductor devices, this lateral resolution provides a method to correlate the performance of individual devices distributed across a wafer scale with their local electronic properties.

Problems

1. For an incoming electron ($Z_1 = 1$) scattered by an atom with Z_2 electrons, the distance d of closest approach is given by equating the kinetic energy E with the potential energy $Z_1 Z_2 e^2/d$. Calculate the distance of closest approach for a 2-keV electron with Si. Compare this to the Bohr radius.
2. Calculate the stopping power of a 2-keV incident electron beam in Ge.
3. The incident electron beam used to excite CL can produce high electron–hole densities. Calculate the free electron density for an $n = 10^{17}$ cm^{-3} ZnO semiconductor with minority carrier lifetime $\tau_P = 10^{-10}$ s, incident beam current $I_B = 10^{-9}$ A, and beam diameter 20 nm with an incident beam energy (a) $E_B = 1$ keV and (b) 5 keV. (c) Which is more likely to produce significant flattening of the semiconductor bands? (Hint: for large cascade distances, approximate the pear-shape as a sphere.)
4. Given an experimentally determined plasmon frequency ω_p corresponding to 15 eV, calculate the effective electron density within the metal.

5. Calculate the Bohr–Bethe range for a 2-keV electron beam cascade in GaAs.
6. Calculate the Bohr–Bethe range for a 25-keV electron beam cascade in ZnO. Use $a = 1.62$ and $c' = 0.68$ for $10 < \xi < 100$.

References

1. Feldman, L.C. and Mayer, J.W. (1986) *Fundamentals of Surface and Thin Film Analysis*, Prentice Hall, p. 20.
2. Feldman, L.C. and Mayer, J.W. (1986) *Fundamentals of Surface and Thin Film Analysis*, Prentice Hall, p. 44.
3. Rose, A. (1966). *RCA Rev*, **27**, 600.
4. Kittel, C. (1986) *Introduction to Solid State Physics*, 6th edn, John Wiley & Sons, Inc., New York, p. 261.
5. Klein, C.A. (1968) *Phys. Rev.*, **39**, 2029.
6. Everhart, T.E. and Hoff, P.H. (1971) *J. Appl. Phys.*, **42**, 5837.
7. Shea, S.P. (1984) *Scan. Electron Microsc.*, **1**, 145–151.
8. Brillson, L.J. and Viturro, R.E. (1988) *Scanning Microsc.*, **2**, 789–799.
9. Hovington, P., Drouin, D., and Gauvin, R. (1997) *Scanning*, **19**, 1–14. The source code is available at Web site *http://www.geology.wisc.edu/courses/g777/Scanning/CASINO_Scanning.pdf* (2009).
10. Brillson, L.J., Young, A.P., White, B.D., Schafer, J., Niimi, H., Lee, Y.M., and Lucovsky, G. (2000) *J. Vac. Sci. Technol. B*, **18**, 1737.
11. Brillson, L.J. (2001) *J.Vac. Sci. Technol. B*, **19**, 1762.
12. Vitomirov, I.M., Raisanen, A.D., Finnefrock, A.C., Viturro, R.E., Brillson, L.J., Kirchner, P.D., Pettit, G.D., and Woodall, J.M. (1992) *J. Vac. Sci. Technol. B*, **10**, 1898.
13. Brillson, L.J., Levin, T.M., Jessen, G.H., and Ponce, F.A. (1999) *Appl. Phys. Lett.*, **75**, 3835.
14. Bradley, S.T., Young, A.P., Brillson, L.J., Murphy, M.J., Schaff, W.J., and Eastman, L.F. (2001) *IEEE Trans. Electron. Dev.*, **48**, 412.
15. Brillson, L.J., Mosbacker, H.L., Hetzer, M.J., Strzhemechny, Y.M., Jessen, G.H., Look, D.C., Cantwell, G., Zhang, J., and Song, J.J. (2007) *Appl. Phys. Lett.*, **90**, 102116.
16. Jessen, G.H., Fitch, R.C., Gillespie, J.K., Via, G.D., White, B.D., Bradley, S.T., Walker, D.E.Jr. , and Brillson, L.J. (2003) *Appl. Phys. Lett.*, **83**, 485.
17. Jessen, G., White, B., Bradley, S., Smith, P., and Brillson, L.J. (2002) *Solid-State Electron.*, **46**, 1427.

Further Reading

Yacobi, B.G. and Holt, D.B. (1986) *J. Appl. Phys.*, **59**, R1–R24.

17
Electronic Materials' Surfaces

17.1
Overview

This chapter examines the geometric, chemical, and electronic structure of electronic materials' surfaces. Chapter 4 highlighted the importance of surface electronic features with respect to the charge transfer by charge carriers across interfaces. It also showed how local variations in surface atomic geometry can have major effects on electronic properties. This chapter extends the concepts introduced in Chapter 4 for a more comprehensive description of electron materials' surfaces.

17.2
Geometric Structure

17.2.1
Surface Relaxation and Reconstruction

The geometric structure of semiconductor surfaces offers a natural introduction to the different properties of the semiconductor–vacuum interface compared with the bulk properties. Such structure encompasses not only the surface relaxation and reconstruction within the individual unit cell described in Chapters 4 and 13 but also the two- and three-dimensional structures that can extend across the surfaces on a larger scale.

Atoms at ordered surfaces change their positions in two ways to reflect the discontinuity of the crystal lattice. First, smoothing of the surface electron distribution alters the electrostatic equilibrium of the ion cores, resulting in a surface relaxation that modifies the bond angles and lengths of the surface atoms to lower the overall free energy. Generally, the in-plane size and shape of the unit cell structure remains the same as the bulk structure. The ionic character of the chemical bond is a major factor in determining the bulk crystalline structure of a solid [1]. For semiconductors, the directional nature of the chemical bonds between atoms favors the tetrahedral coordination of the zinc-blende and wurzite structures [2]. Prepared under different conditions, the surfaces of such lattice structures can assume a variety of complex atomic bonding arrangements, which modify the size

Surfaces and Interfaces of Electronic Materials. Leonard J. Brillson
Copyright © 2010 WILEY-VCH Verlag GmbH & Co. KGaA, Weinheim
ISBN: 978-3-527-40915-0

and shape of the unit cell. Such reconstructions depend sensitively on the surface stoichiometry, orientation, and thermal/ambient atmosphere processing history. Furthermore, surface reconstructions can involve multilayer atom displacements that serve to minimize the overall elastic and electronic energy.

17.2.2
Extended Geometric Structures

In addition to the bonding changes associated with surface relaxation and reconstruction, extrinsic features of the semiconductor surface introduce new geometric structural features. For clean surfaces, these extrinsic features include multiple domains, steps, facets, and point defects. Besides their effects on the electronic structure, such morphological features play a dominant role in crystal growth, epitaxy, and etching.

17.2.2.1 Domains
Domains with different geometric types of order can coexist on a scale of only tens of nanometers or less, as already illustrated for the GaAs(001) $(2 \times 4) - c(2 \times 8)$ in Figure 13.14. Depending on the domain size relative to the coherence length of an incident probe beam, typically several hundred angstroms Å, low-energy electron diffraction (LEED) can provide evidence of such domains from spot broadening/splitting and a rise in background noise [3]. Scanning tunneling microscopy (STM) can provide images of these domains and the unit cell structures within them. For examples, see Figures 14.7 and 14.8.

Low energy electron microscopy (LEEM) provides a direct method of observing the growth of domains, particularly at high temperatures, and underscores the importance of impurities, elastic strain, and dislocations [4]. LEEM is especially useful in revealing the kinetics of semiconductor growth, that is, the temperature-dependent movement of atomic layers formed as atoms are deposited and move across a surface.

17.2.2.2 Steps
Steps also break the translational invariance of the surface. Controlled misorientation of a crystal surface away from a low-index plane can provide a regular array of steps with an average step spacing defined by the misorientation angle and the step height. Indeed, regular step arrays even provide a way of introducing a grating structure with periodic chemical and electronic properties. Figure 17.1 illustrates steps formed at vicinal surface whose average orientation differs slightly from the normal of a low-index plane.

Steps due to slight misorientation are desirable in crystal growth since they serve as nucleation sites that increase the rate of film growth. Thus in Figure 17.2, atoms deposited on a crystal plane diffuse across a surface until they bond to a partially completed step with higher chemical activity on the same plane. The rate at which such *step flow* occurs determines the rate of crystal growth. Without such steps and

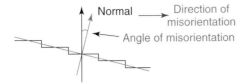

Figure 17.1 Schematic illustration of a vicinal surface. The misorientation angle, the misorientation direction, and step height determine the average step spacing.

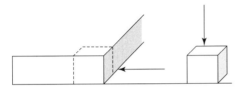

Figure 17.2 Block illustration of deposited atom (right) migrating toward step and adding to atomic layer on the same plane.

their *nucleation sites* to promote layer growth, rates of crystal growth are lower and the layers themselves are less homogeneous.

Steps also expose new crystal faces and different chemical bonds from those of the unstepped surface. Depending on the direction of misorientation, steps in different directions have different bond terminations at their surface. An example is the low-index GaAs(100) surface oriented 2° toward the [111] A versus B face. Figure 17.3 illustrates the surface atomic geometry of the vicinal GaAs(100) 2° → [111]B surface.

The [111] oriented planes exposed by the misoriented (100) plane have As atoms with sp^3 bonds that extend out of the plane. As discussed in Chapter 4, such *dangling bonds* give rise to localized states that affect charge transfer. The density of such localized states scales in nearly one-to-one proportion with the density of exposed As sites [5]. As discussed later, surfaces with such dangling bonds are also chemically more active than their low-index counterparts [6].

Figure 17.4 illustrates step arrays for a vicinal GaAs(001) surface cut in two, perpendicular, vicinal directions [7]. For these GaAs(100) 2° → [111]A and B surfaces, STM images exhibit step spacings consistent with the vicinal angle. Each step has a single-layer height and terrace widths that vary from 50 to 120 Å. In addition to disruption of the lateral invariance and introduction of dangling bonds, the steps introduce kink sites whose density depends on the misorientation direction.

Figure 17.4 illustrates steps along the dimer direction (a), which are relatively smooth, and steps perpendicular to the dimer direction (b), which exhibit a high kink density. As shown, these kinks occur in units of 16 Å, that is, in dimensions of the (2 × 4) unit cells. Higher spatial resolution STM images confirm the identity of these features. Since the (2 × 4) unit cell contains the correct number of electrons to fill the As dangling bonds and empty the Ga dangling bonds, this structure and extensions of such island structures along the 2x direction are stable. Such structures grow out from steps in the [110] misorientation and result in the ragged B-type step edges pictured in (b). In contrast, A-type edges are relatively straight, a result attributed to island structures growing out from A-type step edges having an insufficient number of electrons to fill all the As dangling bonds and thus being thermodynamically less favorable to grow out from the steps [8]. This example shows that local bonding associated with the surface reconstruction is responsible for the contrast in the extended vicinal surface morphology.

Analogous effects occur for vicinal Si surfaces. For the Si(001) plane, surface reconstruction also results in the formation of dimers. Owing to the symmetry of the underlying diamond lattice, the dimer rows are perpendicular to each other on terraces separated by an odd number of monoatomic steps [9]. For surfaces cut toward the [110] direction, each terrace has a dimer reconstruction running perpendicular to those of the previous step. Furthermore, adjacent steps are inequivalent, for example, smooth (S_A) versus rough (S_B), analogous to the GaAs(001) case as in Figure 17.4.

The energies associated with step atoms and long-range strain both influence step height distributions and spacings. In the vicinal Si(001) case, a step of type S_A connects a higher 2×1 terrace to a lower 1×2 terrace with dimer rows parallel to S_A and energy ε_{SA}. A step of type S_B connects these terraces in reverse order with dimer rows perpendicular to S_B and energy ε_{SB}. Kinks in one kind of step are composed of dimer segments of the other type of step. For higher misorientation angle, terraces exhibit the same (2×1) or (1×2) periodicity but with a double-layer step height. The transition between the single- and double-layer step morphologies

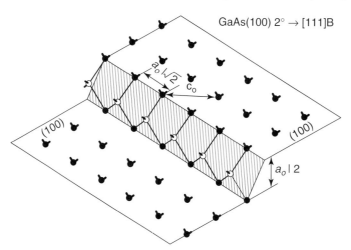

Figure 17.3 Vicinal GaAs surface atomic geometry. Ga atoms (dark circles) terminate the (100) surface. The [111] surface (shaded) exhibits As (light circles) dangling bonds extending out of the plane.

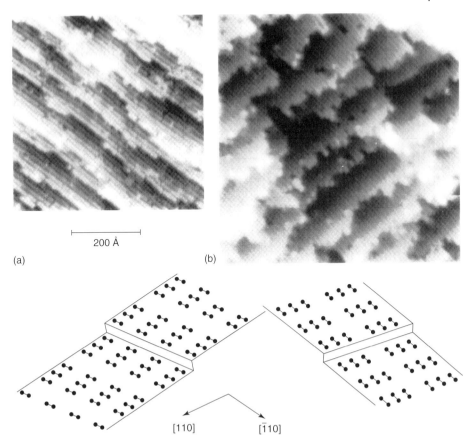

Figure 17.4 STM filled-state images● of steps on GaAs(001) misoriented 2° toward (a) the (111)A (Ga) surface and (b) the (111)B (As) surface. The corresponding schematic diagrams of the step edges with respect to the dimer direction appear below. Two misorientations exhibit different in-step raggedness and kink densities. For the GaAs(100) 2° → [111]A orientation, steps run parallel to the dimer rows and are relatively straight. For the GaAs(100) 2° → [111]B orientation, steps run perpendicular to the dimer rows and exhibit high kink densities [7].

is believed to occur at $\theta > \sim 4°$. The transition between these two surfaces has been predicted theoretically to be $\theta_C \sim 1.2 - 2.5°$ based on minimization of step plus strain relaxation energies [10] as well as the entropy associated with the single/double step distributions at thermal equilibrium [11].

Thermodynamic analysis of the distributions of the kink separations and kink lengths can provide step and kink energies. For a simple lattice with energy $E(n) = n\varepsilon + C$ and independent kinks of length n atoms and a Boltzmann distribution

$$N(n) \propto e^{-E(n)/k_B T} \tag{17.1}$$

one obtains $\varepsilon_{SA} = 0.028$ eV per atom for the smooth step edge, $\varepsilon_{SB} = 0.09$ eV per atom for the rough step edge, and $C = 0.08$ eV per atom [9]. Hence these energies are comparable to $k_B T$ at room temperature. For a comprehensive review of the thermodynamics of surface morphology, see [12].

Long-range strain fields associated with step wandering also contribute to the free energy of steps – effectively, a repulsive potential due to a reduction in entropy. For example, such strain fields are believed to influence the Si(111)(7 × 7) reconstruction pictured in Figure 13.9 [13]. Therefore, the details of surface preparation, specifically, the sample cleaning history, can determine subsequent metastable structures. The short-range kink energetics govern the step fluctuations, while the long-range strain fields control the extended average positions of the step segments on a scale comparable to the terrace widths. These long-range strain fields determine the distribution of the terrace widths. Furthermore, the (7 × 7) reconstruction correlates strongly across the single steps of slowly cooled surfaces so that the widths of terraces come in multiples of the distance between adjacent rows of corner holes [14]. As with the III–V compound surfaces in Figure 17.4, these vicinal Si features show that reconstruction plays a role in forming the step configuration.

17.2.2.3 Defects

Various defects can contribute to surface geometric structure. In addition to steps, the surface may be faceted. That is, it may incorporate macroscopic, nonuniform morphologies with new crystallographic planes exposed. Impurities or adsorbates can alter the step/facet structure by either the kinetic or thermodynamic mechanisms. Examples include carbon contamination and As adsorption, both of which cause macroscopic faceting with very low coverages, for example, 0.3% C or 0.2 monolayer (ML) As.

Point defects are also found at semiconductor surfaces. For example, *advacancies* can be viewed as the inverse analog of surface adsorbates [15]. For the surface pictured in Figure 17.5, the terraces contain both adatoms and advacancies that diffuse toward step edges, causing the terraces to advance or retreat.

STM observations of such surface defects include anion and cation vacancies for GaAs (16, 17), anion vacancies for InSb, and adjacent anion and cation vacancies (Schottky defects) [18]. Ga vacancy point defects appear to be negatively charged. The quality of crystal cleave and cleavage direction can determine terraces with different step directions [16]. Edge atoms at such monoatomic steps displayed high-enough densities ($>10^{13}$ cm^{-2}) to have significant electrostatic effects, as noted in Section 4.4.

STM and high-resolution photoluminescence from surfaces of the ternary compound CuInSe$_2$ has yielded evidence for several different defect types – Cu vacancies, Cu at In sites, and In at Cu sites [19] as well as specific energies associated with these defect sites. Theoretical studies of CuInSe$_2$ reveal a complex of defects, $(2\,V_{Cu}^{-} + In_{Cu}^{++})^0$, to be neutral and more stable than isolated defects, producing a Cu-vacancy reconstruction at (112) surfaces and a carrier-type inversion from p- to n-type [20].

Overall, these domains, steps, and defects reduce the lattice geometric order, adding features to the semiconductor surface, which generally exhibit electrical activity at the interface, which are discussed in the following sections.

17.3
Chemical Structure

The chemical structure of the clean semiconductor surface can involve either the addition to, movement on, or removal from the outer layers of atoms of the same constituents as the bulk crystal. This section deals with these topics in terms of crystal growth, the dynamics of growth, diffusion, and evaporation, as well as the role of surface structure and bonding in etching phenomena.

17.3.1
Crystal Growth

17.3.1.1 Bulk Crystal Growth

Numerous techniques are available for bulk crystal growth. These include (i) Czochralski (CZ) growth from the melt, (ii) horizontal Bridgman growth from the melt, (iii) float-zone crystal growth, and (iv) liquid phase-techniques.

Figure 17.6 illustrates the CZ growth method. In Figure 17.6a, a crystalline seed of Si provides a template for molten Si in a crucible to enlarge the crystal as it slowly rotates and pulls away from the 2000 °C melt. Figure 17.6b shows the crystal boule as it emerges from the molten bath (21, 22). Figure 17.6c shows a cassette of wafers obtained by wire sawing of a Si crystal boule.

The Horizontal Bridgman method shown in Figure 17.7 also involves growth from a seed, in this case moving through a temperature gradient within a tube furnace. Here, the molten material in contact with the seed within an elongated

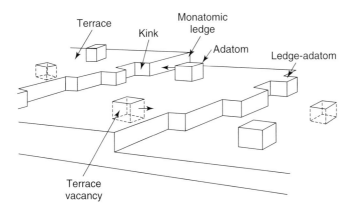

Figure 17.5 Schematic illustration of adatoms, advacancies, and kinks on a stepped surface.

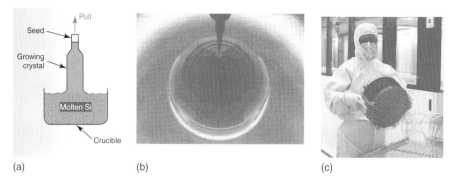

Figure 17.6 Pulling of a Si crystal from the melt (Czochralski method): (a) schematic diagram of the crystal growth process; (b) an 8-in. diameter, <100> oriented Si crystal being pulled from the melt [21]; and (c) A cassette of 300 mm diameter Si wafers [21]. Courtesy of MEMC Electronics Intl.

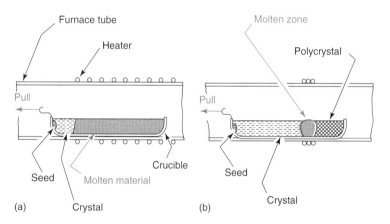

Figure 17.7 Horizontal Bridgman crystal growth from the melt in a crucible: (a) solidification from one end of the melt and (b) melting and solidification in a moving zone [21].

crucible crystallizes and enlarges the seed as both move out from the molten zone to lower temperatures.

An alternative to CZ growth is the float-zone (FZ) method. In Figure 17.8, polycrystalline material typically in rod shape is crystallized by local melting, starting at the interface between the single crystal seed and the polycrystalline material [23]. The melt is not in contact with a crucible or active gases so that no impurities dissolve into the growing crystal. FZ crystals have particularly low oxygen concentrations. The FZ method results in very high-purity single-crystal Si but is limited in size relative to the CZ method.

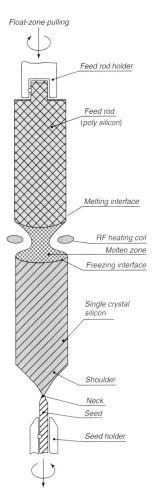

Figure 17.8 Float-zone pulling method in which a seed and polycrystalline rod moves through a melt zone heated by an RF coil [23].

Liquid-phase epitaxy (LPE) is a growth technique intermediate between the bulk- and thin-film epitaxy approaches. Here, a lattice-matched bulk substrate provides the template or seed for crystallization. For example, Figure 17.9a shows a GaAs substrate covered with a molten solution of Ga, Al, and As. The pull rod in Figure 17.9b slides this substrate into various pockets containing different elemental compositions. Here, two pockets are provided, containing melts for AlGaAs and GaAs growth. To grow an AlGaAs/GaAs/AlGaAs quantum well structure for a light-emitting diode or laser, the GaAs substrate on the slider moves first into the pocket containing Al, Ga, and As. After an AlGaAs epitaxial layer grows on this substrate (a), the slider assembly wipes the excess molten solution off as the substrate is pulled into another pocket containing just GaAs. The process is repeated for the next pocket again containing Al, Ga, and As. Other pockets can be used to add intermediate compositions and dopants. Growth rates are sufficiently high that this method is employed to manufacture GaAs-based optoelectronics.

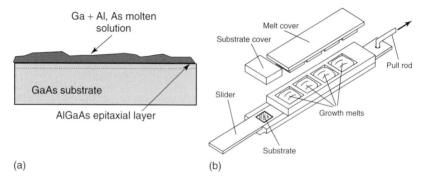

Figure 17.9 Liquid-phase epitaxial growth of AlGaAs and GaAs layers on a GaAs substrate: (a) cross section of the sample in contact with a Ga-rich melt containing Al and As and (b) carbon slider used to move the GaAs substrate between various melts.

17.3.1.2 Epitaxial Layer Crystal Growth

Layer-by-atomic-layer deposition to form crystalline thin films is termed *epitaxy* and is the most controllable form of crystal growth. Epitaxial crystal growth techniques include (i) vapor-phase epitaxy (VPE) of solid elemental or compound sources, (ii) LPE described in Figure 17.9, and (iii) molecular beam epitaxy (MBE) which involves sequential deposition of molecules from individual sources. A variant of MBE is chemical beam epitaxy (CBE) or molecular organic chemical vapor deposition (MOCVD) which employ molecules that decompose at the growth surface, leaving behind constituents of the growing film.

Epitaxial growth techniques performed under vacuum conditions provide the greatest opportunity for analyzing the structural and energetic parameters that govern atom addition to and incorporation within the crystal lattice. The vacuum environment enables one to use many of the surface science techniques described in previous chapters.

In the case of MBE, neutral atom beams evaporated from crucibles impinge on a heated crystalline substrate and interact to form ordered overlayers [24, 25]. The most studied of such MBE-grown systems has been the GaAs(001) surface. Typically, monoatomic Ga and As tetramers or dimers evaporate from Knudsen cells and are deposited on a clean GaAs substrate. The latter is prepared by etching to eliminate surface contaminants and to produce a protective, volatile oxide, which is desorbed prior to growth in vacuum. Additional Knudsen sources supply dopant elements, which can be controlled to produce free carrier concentrations in the $10^{14}-10^{19}$ cm^{-3} range. The ultrahigh vacuum (UHV) environment during growth is essential to obtain the low contamination levels characteristic of high-quality semiconductor material.

The adsorption, dissociation, and desorption of the impinging species can be summarized in a mass balance equation for the surface concentration of GaAs atoms (Θ) during growth [26]

$$\left(\frac{d\Theta}{dt}\right) = J_{Ga} - nJ_{As_n}S_{As_n} + 2R_{As_2} + D_{Ga} - D_{As} \tag{17.2}$$

where Θ is the surface concentration of Ga atoms during growth, J_{Ga} and J_{As_n} ($n=2$ or 4) are the incident fluxes of Ga and As_n, respectively, S_{As_n} is the As_n sticking coefficient, R_{As_2} is the dissociation/desorption rate of As_2 molecules from the substrate, and D_{Ga} and D_{As} are the diffusion rates of Ga and As atoms, respectively, from the bulk to the surface. Over a range of substrate temperatures (e.g., 450–650°C for GaAs), all of the Group-III element adsorbs onto the surface [24]. The resident lifetime of Ga is much longer than that of As, which desorbs rapidly unless adsorbed Ga is present (e.g., R_{As_2} is composition- and temperature-dependent). Upon bonding to Ga, S_{As_n} increases and then decreases again when the excess Ga is consumed. Thus a 1 : 1 ratio of Ga to As can be maintained provided that $J_{As_n} > J_{Ga}$ [27]. Depending on the annealing temperature, duration, quench rate, crystal constituents may diffuse out from the bulk to the surface, producing surface accumulation and subsurface depletion of species [28]. As already noted in Figures 13.12 and 13.13, reconstructions are based on stoichiometry, but particular surface compositions do not necessarily specify a particular structure.

Figure 17.10 provides a schematic illustration of a UHV MBE chamber. A set of effusion cells and gas sources are aimed at a crystal substrate mounted on a heated manipulator that rotates on an axis normal to the growth plane. An electron beam from a reflection high-energy electron diffraction (RHEED) gun glances off the growing film and illuminates a fluorescent screen with a diffraction pattern. Liquid nitrogen cools the shrouds surrounding the effusion cell ports. Mass flow controllers and leak valves admit gases at controlled pressures. Heated filaments serve to "crack" the gas molecules to increase their chemical reactivity. Transfer rods enable the epilayers to be transferred between such growth chambers and analysis chambers without exposure to atmosphere [29].

Observation of the RHEED diffraction pattern in two orthogonal directions or from a single backscattered LEED pattern permits one to determine the symmetry of the growing crystal surface. The RHEED analysis procedure involves monitoring the change in diffraction streaks due to surface disorder over the course of crystal growth [30]. Deviations from perfect lattice ordering restrict the average size of the ordered regions, thereby broadening the reciprocal lattice rod and increasing its area of intersection with the Ewald sphere. The resultant broadened or elongated rods are observable in the diffraction pattern. From the integral or half-integral position of the streaks, it is also possible to extract the periodicity of the surface disorder [26]. Other *in situ* techniques to monitor the crystal surface during growth include ellipsometry and mass spectrometry.

Ellipsometric measurement of the surface dielectric properties [31] is particularly valuable for higher pressure methods such as MOCVD. Modulated beam mass spectrometry of the kinetic features of adsorption and desorption provides

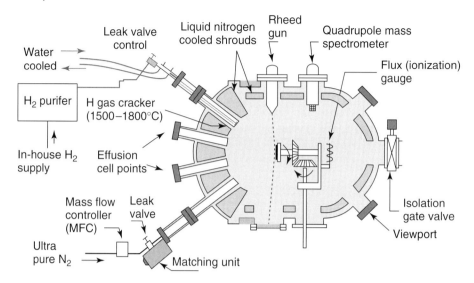

Figure 17.10 Top-side view of a solid-source UHV MBE chamber. (Courtesy Cliff Fonstad, MIT, Yoon Soon Fatt, used with permission.) [29].

the surface lifetimes, the energies of the binding states, the sticking coefficients, the thermal accommodation coefficient of molecules interacting with the surface, and the orders of reaction. From such measurements, it is possible to derive detailed models that account for the features of the geometrical structure observed [26].

The variation of RHEED intensities during crystal growth also provides information about the nature of crystalline layer formation [32, 33]. Figure 17.11 schematically illustrates the oscillatory behavior of the specular RHEED spot intensity usually associated with layer-by-layer two-dimensional growth. As depicted in the thin-film growth model, the onset of growth results in the formation of two-dimensional centers, distributed randomly across the surface. For a single-scattering process with the electron wavelength much shorter than the crystal step height, that is, ~0.1 Å versus 2.83 Å for GaAs, diffuse scattering increases and specular scattering decreases correspondingly.

Such diffuse scattering reaches a maximum at half-layer coverage and a minimum at full monolayer coverage. Hence the oscillations have a period corresponding to monolayer-by-monolayer growth. The monotonic decrease of peak specular amplitude relates to the completeness of each layer prior to the growth of the next layer. Kinetic models of growth use such RHEED behavior to distinguish between alternative processes and rates of atom adsorption, diffusion, and lattice incorporation [34, 35]. Notwithstanding multiple-scattering effects, *in situ* diffraction analysis of MBE growth affords considerable information about the atomic layer features associated with crystal growth.

17.3 Chemical Structure

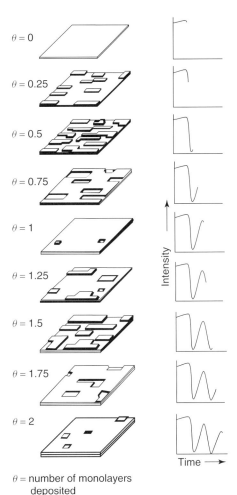

Figure 17.11 Schematic representation of the RHEED intensity variations associated with the laminar growth of epitaxial monolayers. Diffuse scattering is a maximum at half-layer coverage and a minimum at full monolayer coverage, leading to the oscillations shown [26].

θ = number of monolayers deposited

17.3.2
Kinetics of Growth, Diffusion, and Evaporation

Regular RHEED oscillations provide a monitor of both growth and evaporation, provided they occur in a layer-by-layer fashion. For evaporation rates measured as a function of temperature, one can derive activation energies of sublimations [33]. RHEED oscillations also probe the surface diffusion parameters inherent in the growth process. Here vicinal surfaces are used whose steps act as sinks for the growth of deposited atoms. For a terrace width W and diffusion length L of the rate-controlling species – for example, Ga for GaAs, no growth occurs on the terrace as long as $L > W$. See Figure 17.5. For a high-enough incident flux or low-enough temperatures such that $L < W$, growth occurs on the terraces, and the RHEED oscillations from beams directed along terrace edges begin to show disorder. For example, Ga on GaAs(001) (2 × 4) exhibits an activation energy $E_D \approx 1.3$ eV and a diffusion coefficient $D_0 \approx 10^{-5}$ cm^2 s^{-1} where $D = D_0 e^{-E_D/K_B T}$. Such parameters determine the terrace widths under actual growth conditions and are critical in establishing the atomic-scale roughness of epitaxial interfaces.

RHEED also provides rates of surface smoothing from the increased specular spot intensities evident with interrupted growth. The RHEED intensities recover with both fast and slow exponential time constants. Values of the fast time constant as a function of temperature yield an apparent activation energy for the smoothing process of ~2.3 eV for GaAs, considerably higher than the activation energy for surface migration already discussed and somewhat higher than the cohesive energy of GaAs (~1.7 eV) [26]. These dynamics suggest that a bond-breaking mechanism is involved. Thus the energies associated with dissociation and surface diffusion are of central importance in determining the smoothness of multiple quantum wells, superlattices, as well as simple heterojunctions under a given set of growth conditions. Furthermore, inhomogeneities on this scale lead to geometric structure features that can give rise to a variety of electronic features.

17.4
Etching

17.4.1
Etch Processes

The removal of atoms by etching provides another aspect of the roles surface structure and bonding play in the macroscopic features of the semiconductor surface. Etch processes are diverse and include (i) wet chemical, (ii) reactive ion, (iii) plasma, (iv) thermal, (v) sputter, (vi) electrolytic, (vii) ion bombardment, and (viii) melt etching. Plasma etching, in particular, affords submicron and nanoscale structures with high aspect ratios (vertical depth versus lateral precision). Examples of such structures include gratings, mesas, highly anisotropic holes and grooves, and selectively patterned heterostructures. This feature of dry (i.e., plasma-assisted or reactive ion) etching lends itself to patterning on the ever-decreasing scale of microelectronics as well as microfabrication on a nanoscale [36]. Dry etching performed in a low-pressure gaseous discharge provides increased control, selectivity, and definition for semiconductor etching [37]. Both reactive ion and plasma etching techniques rely on chemical reactions to produce volatile compounds. However, the activated-ion process is sufficiently energetic to disrupt the local bonding geometry and create subsurface defects.

17.4.2
Wet Chemical Etching

This section is restricted to wet chemical etching only since it illustrates the role of atomic-scale properties in forming macroscopic surface features. Wet etching is useful not only for preparing (e.g., cleaning, functionalizing) surfaces but also as a probe for (i) surface defects, (ii) the distribution of impurities, as well as (iii) forming specific geometric structures for devices. The two primary aspects of wet etching are the macroscopic rate of etching and the resultant microstructure of the surface. The geometric structure features that affect wet etching are (i) the surface orientation, (ii) surface or near-surface defects, and (iii) impurities at the surface.

17.4.3
Orientation Effects on Wet Chemical Etching

The difference in chemical reactivity between surfaces with different orientations are not usually observed on a macroscopic scale, due to the small differences in their surface free energies and the lack of atomically flat faces in "real" macroscopic surfaces. Such differences are evident, however, in relatively slow reactions, in which the geometric structure on an atomic scale introduces kinetic limitations to the process. Such orientation effects may be amplified by impurity adsorption, which may be specific to the particular surface [38].

This difference in chemical reactivity with crystal orientation manifests itself clearly in III–V compound semiconductors. Surface termination with Group-V atoms (B{111} surfaces) exhibit more reactivity and higher etch rates than those terminating with Group-III atoms (A{111} surfaces) [39]. Figure 17.12 illustrates such orientation effects on wet chemical etching for GaAs(111) versus GaAs($\bar{1}\bar{1}\bar{1}$) surfaces [40]. The etching characteristics of the B surface are much more pronounced, consistent with more aggressive chemical attack. Similar behavior is manifest for A versus B faces of InSb, GaSb, AlSb, InAs, and InP.

The ability to donate unshared electrons at dangling bond sites provides a basis for understanding the etch rate differences between different crystallographic planes, as well as the differences at edge dislocations and other lattice nonuniformities [41]. Figure 17.13 illustrates the zinc-blende crystal structure of the III–V compounds [42].

This figure shows the difference in bond density along the <111> direction between layers. The bond density is highest between the A (Group-III atoms, dark) and B' (Group-V atoms, light) planes indicated where three times as many bonds must be broken as between the A and B planes marked accordingly.

Figure 17.14 shows this difference in more detail. In two dimensions (a), the low bond density between the AA and BB planes in Figure 17.13 favors separation of the zinc-blende crystal lattice such that there is only one type of atom at the surface. In Figure 17.14b, the fourfold-coordinated A and B atoms are triply bonded to the lattice and share a pair of electrons (I). After a cut of the bonds between AA and BB planes, it is energetically unfavorable for each separated A and B atom to have one electron each and thereby charged A^- and B^+ surfaces (II) since the ionization potential associated with formation of the resultant B^+ is much greater than the electron affinity associated with formation of A^-. The electron configuration of the neutral Group-V atoms with an unshared pair of

17 Electronic Materials' Surfaces

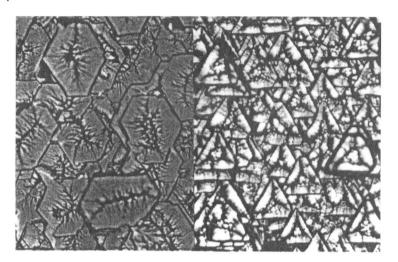

Figure 17.12 Micrograph of etched A and B parallel surfaces of a GaAs(111) wafer. (0.2 N Fe^{3+} in 6 N HCl, 10 minutes) The B surface exhibits much more pronounced etching features [40].

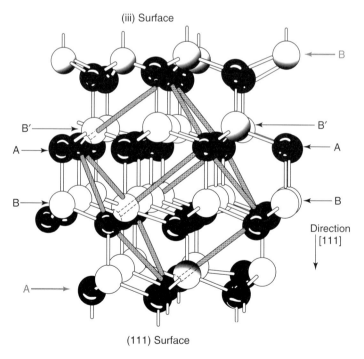

Figure 17.13 The zinc-blende III–V compound crystal structure showing the termination of the A and B faces based on the lowest density of broken bonds [42].

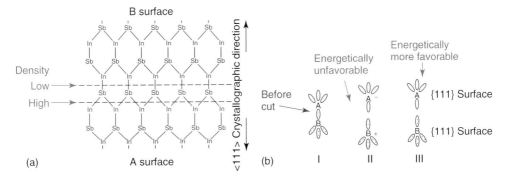

Figure 17.14 (a) Two-dimensional structure of the zinc-blende structure along the polar direction. (b) Formation of an A and B surface atom by a cut of a bond between planes AA and BB in Figure 17.13. The formation of an A and B surface atom before (I) and after (II and III) a cut of the bonds between AA and BB planes [42].

electrons leads to sp³ bonding and dangling bonds in (III), whereas the Group-III surface atoms have only six electrons, leading to sp² hybrid bonding and a planar configuration.

The difference between these two surface bonding configurations manifests itself in the etch rate variations across other crystallographic faces with mixed composition. The measured etch rates for III–V compound semiconductors decrease in the sequence B{111} ≥ {100} > {110} > A{111}, consistent with the B{111} dangling bonds, the average of two electrons from {100} A and B atoms, the equal numbers of triply bonded {110} A and B atoms, and the A{111} sp² planar configuration.

In addition, the planar A{111} bonding is distorted from the bulk sp³ lattice, leading to more strain in the surface plane. As a result, mechanical damage is observed higher at A{111} versus B{111} surfaces [43].

Similarly, the increased reactivity observed for edge dislocations and the creation and propagation of etch pits is related to the chemical bonding and structure of the atoms along the dislocation line. Here the reactivity of the dislocation relative to the surface plane determines the etch pit growth. Molecular kinetic theories of etching describe the etch kinetics in terms of the rate of step formation and movement [44]. Tables of chemical etch treatments for III–V, II–VI, and several other families of semiconductors appear in [31, 41, 45].

17.4.4
Impurities, Doping, and Light

Impurities can alter the rate of chemical etching, either by passivating or creating chemically active atomic sites. This is particularly evident for crystal dislocations, where impurity segregation to dislocation sites can change etch patterns, depending on the reactivity of the dislocation relative to the impurity and the adsorption site of the impurity.

Semiconductor crystal doping [46] and light [47, 48] also affect etch rates. In both cases, the addition of free charge carriers accelerates the surface chemical processes. Photoelectrochemical etching provides a remarkable illustration of the effect that free carriers have on etching. The photograph in Figure 17.15a shows free-standing nanowires of GaN of diameter 10–50 nm up to a micron in length on a GaN substrate [49]. These nanowires are formed when a GaN wafer was etched under bandgap illumination.

The spacing of these nanowires corresponds to the density and orientation of dislocations in the GaN(100)-oriented bulk wafer. These nanowires form because the etch rate is much slower at these locations than in the surrounding crystal. This difference in etch rate is due to the much lower density of free carriers generated by light in the vicinity of dislocations. Figure 17.15b,c provide schematics to illustrate this difference. In (b), an incident photon with $h\nu > E_G$ produces free electron–hole pairs that contribute to the surface etch reaction. In (c), the photogenerated free carriers recombine via a localized electronic state within the bandgap. Electronic states near the middle of the GaN bandgap are often associated with native point defects. The strain associated with such morphological defects can attract such defects, which then form a "Cottrell cloud" around the dislocation [50]. These defect states plus states intrinsic to the dislocation itself represent efficient recombination centers that decrease free carrier concentrations by several orders of magnitude. Hence, the free carriers are no longer available to participate in the etch reactions, leaving unetched GaN in

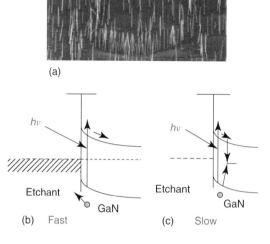

Figure 17.15 (a) Scanning electron micrograph of free-standing GaN on a GaN(100) substrate formed by 0.02 M KOH under bandgap illumination. (b) Schematic of electronic processes corresponding to free electron–hole pairelectron–hole pair creation and correspondingly fast etching versus recombination and slow etch rates [49].

the vicinity of these dislocations as the photochemical etching of the bulk GaN proceeds.

17.5
Electronic Implications

The bond termination differences that account for etching characteristics have electronic implications. Figure 17.14 shows that the A face is deficient in electrons, while the B face has an electron excess. Intuitively, this suggests that acceptors form at the A face while donors form at the B face. Elegant yet simple measurements of carrier concentration confirm this: edge dislocations terminate along the dislocation line in either triply bonded A (α dislocation) or B (β dislocation) atoms. Plastic deformation, in this case, bending around specific crystallographic axes, can introduce excess α versus β dislocations preferentially [51]. Hall coefficient and resistivity measurements confirm that formation of excess α dislocations decreases electron concentrations, whereas excess β dislocations correlate with increased electron concentration.

17.6
Summary

Semiconductor surfaces with domains, steps, and defects show that morphological features can dominate growth, epitaxy, and etching processes. The growth of domains is influenced strongly by the presence of impurities, elastic strain, and dislocations.

Steps are desirable in crystal growth since nucleation sites increase the rate of growth. Steps expose new crystal faces and different chemical bonds. Steps in different directions have different defects that depend on surface reconstruction. STM measurements measure activation energies for kinks at steps and the distribution of step heights. They reveal that short-range kink energetics govern step fluctuations and that long-range strain fields control the spacing of extended average step spacings.

Depending on their preparation, surfaces can have facets and point defects with densities high enough to influence charge transfer at interfaces significantly.

Of the numerous crystal growth techniques that have been developed, epitaxial growth methods provide control of crystal structure at the atomic level. Epitaxial techniques currently enable the design and fabrication of state-of-the-art electronic devices.

The removal of atoms by chemical etching exhibits a dependence on crystal orientation that reflects the importance of electronic bonds at the free semiconductor surface.

Overall, each of the growth, diffusion, and etching phenomena described in this chapter provide evidence for the role of the atomic bonding geometry on macroscopic chemical and electronic properties.

Problems

1. A GaAs(100) unreconstructed surface is misoriented 2° toward [111]B with As dangling bonds perpendicular to the steps. See Figure 17.3. (a) Calculate the area density of As dangling bonds. (b) For GaAs(100)2°-[110] surfaces, steps are parallel to the [100] direction. What is the corresponding area density of As sites?
2. Consider the diffusion of atoms on MBE growth surfaces. (a) At 600 °C, how far can a surface As or Ga atom diffuse in 1 second if its activation energy $E_A = 1.3$ eV and its diffusion coefficient prefactor $D_0 = 10^{-5}$ cm^2 s^{-1}? (b) If this diffusion length matches the terrace width of an off-cut GaAs (100) crystal, calculate the vicinal angle. (Assume a step height of $1/2$ lattice constant.)
3. Chemical etching can be used to distinguish between polar crystal surfaces of compound semiconductors, as pictured in Figure 17.13. At 4 °C, the A surface of a III–V compound semiconductor is observed to etch with a 25-kcal mol^{-1} activation energy and etches 10 times slower than the B surface. Calculate the activation energy of the B surface.
4. A Si wafer is to be diced into device chips. What is the upper limit of electrically active defect density on the wafer surface that will permit a 70% yield of chips with area 200 μm^2? (b) A GaAs wafer is to be diced into chips for lasers and has a surface etch pit density corresponding to dislocations of 10^3 cm^{-2}. Calculate the square chip dimension needed to achieve the same yield.

References

1. Phillips, J.C. (1970) *Rev. Mod. Phys.*, **42**, 317.
2. Zangwill, A. (1988) *Physics at Surfaces*, Cambridge University Press, Cambridge.
3. Henzler, M. (1985) in *The Structure of Surfaces*, Vol. 2 (eds M.Van Hove and R. Howe), Springer, New York, p. 351.
4. Bauer, E. (1990) in *Chemistry and Physics of Solid Surfaces*, Vol. 8 (eds R.Vanselow and R. Howe), Springer, Berlin, p. 267.
5. Chang, S., Vitomirov, I.M., Brillson, L.J., Rioux, D.F., Kirchner, P.D., Pettit, G.D., and Woodall, J.M. (1991) *Phys. Rev. B*, **44**, 1391.
6. Chang, S., Brillson, L.J., Kime, Y.J., Rioux, D.S., Kirchner, P.D., Pettit, G.D., and Woodall, J.M. (1990) *Phys. Rev. Lett.*, **64**, 2551.
7. Pashley, M.D., Haberern, K.W., and Gaines, J.W. (1991) *Appl. Phys. Lett.*, **58**, 406.
8. Pashley, M.D. (1989) *Phys. Rev.*, **40**, 10481.
9. Swartzentruber, B.S., Mo, Y.-M., Kariotis, R., Lagally, M.G., and Webb, M.B. (1990) *Phys. Rev. Lett.*, **65**, 1913.

10. Alerhand, O.L., Berker, A.N., Joannopoulos, J.D., Vanderbilit, D., Hamers, R.J., and Demuth, J.E. (1990) *Phys. Rev. Lett.*, **64**, 2406.
11. de Miguel, J.J., Aumann, C.E., Kariotis, R., and Lagally, M.G. (1991) *Phys. Rev. Lett.*, **67**, 2830.
12. Williams, E.D. and Bartelt, N.C. (1991) *Science*, **251**, 393.
13. Alerhand, O.L., Vanderbilt, D., Meade, R.D., and Joannopoulos, J.D. (1988) *Phys. Rev. Lett.*, **61**, 1973.
14. Wang, X.S., Goldberg, J.L., Bartelt, N.C., Einstein, T.L., and Williams, E.D. (1990) *Phys. Rev. Lett.*, **65**, 2430.
15. Lapoujoulade, J. (1990) in *Interactions of Atoms and Molecules with Solid Surfaces* (eds V.Bortolani, N.H. March, and M.P. Tosi), Plenum, New York, pp. 381–405.
16. Cox, G., Graf, K.H., Szynka, U., and Urban, K. (1990) *Vacuum*, **41**, 591.
17. Stroscio, J.A., Feenstra, R.M., Newns, D.M., and Fein, A.P. (1988) *J. Vac. Sci. Technol. A*, **6**, 499.
18. Whitman, L.J., Stroscio, J.A., Dragoset, R.A., and Celotta, R.J. (1991) *J. Vac. Sci. Technol. B*, **9**, 770.
19. Aboie-Elfotouh, F.A., Kazmerski, L.L., Moutinho, H.R., Wissel, J.M., Dhere, R.G., Nelson, A.J., and Bakry, A.M. (1991) *J. Vac. Sci. Technol. A*, **9**, 554.
20. Persson, C. and Zunger, A. (2003) *Phys. Rev. Lett.*, **91**, 266401.
21. Streetman, B.G. and Banerjee, S. (1995) *Solid State Electronic Devices*, 4th edn, Prentice Hall, Upper Saddle River, p. 14.
22. EMC Electronics, Intl., Sherman, Texas. Reprinted with general permission.
23. Foll, H. (2009) University of Kiel -<hf@www.tf.uni-kiel.de <mailto:hf@www.tf.uni-kiel.de>> http://www.tf.uni-kiel.de/matwis/amat/elmat_en/kap_6/illustr/fz_color.gif .
24. Arthur, J.R. (1968) *J. Appl. Phys.*, **39**, 4032.
25. Cho, A.Y. and Arthur, J.R. (1975) *Progr. Solid State Chem.*, **10**, 157.
26. Joyce, B.A., Dobson, P.J., and Larsen, P.K. (1988) in *Surface Properties of Electronic Materials*, The Chemical Physics of Solid Surfaces and Heterogeneous Catalysis, Vol. **5** (eds D.A.King and D.P. Woodruff), Elsevier, Amsterdam, pp. 271–307.
27. Panish, J.B. (1980) *Science*, **208**, 916.
28. Massies, J., Etienne, P., Dezaly, F., and Linh, N.T. (1980) *Surf. Sci.*, **99**, 121.
29. Prof. Yoon Soon Fatt, National Technical University, Singapore, with permission granted 7 September 2009. http://ocw.mit.edu/NR/rdonlyres/Electrical-Engineering-and-Computer-Science/6-772Spring2003/B5D923F5-9B4C-4436-A1F1-0343B35E1928/0/lect8_part1.pdf.
30. Joyce, B.A., Neave, J.H., Dobson, Pl.J., and Larsen, P.K. (1984) *Phys. Rev. B*, **29**, 814.
31. Aspnes, D.E. and Studna, A.A. (1981) *Appl. Phys. Lett.*, **39**, 316.
32. Neave, J.H., Joyce, B.A., Dobson, P.J., and Norton, N. (1983) *Appl. Phys. A*, **31**, 1.
33. Van Hove, J.M., Lent, C.S., Pukite, P.R., and Cohen, P.I. (1983) *J. Vac. Sci. Technol. B*, **1**, 741.
34. Madhukar, A. and Ghaisas, S.V. (1988) *CRC Crit. Rev. Solid State Mater. Sci.*, **14**, 2.
35. Cohen, P.I., Dabiran, A., and Pukite, P.R. (1990) in *Kinetics of Ordering and Growth at Surfaces* (ed. M.G.Lagally), Plenum, New York, p. 225.
36. Madou, M. (2002) *Fundamentals of Microfabrication*, 2nd edn, CRC Press, New York.
37. Mogab, C.J. (1983) in *VLSI Technology*, Chapter 8 (ed. S.M.Sze), McGraw-Hill, New York.
38. Gatos, H.C. and Lavine, M.C. (1960) *J. Appl. Phys.*, **31**, 743.
39. Gatos, H.C., Moody, P.L., and Lavine, M.C. (1960) *J. Appl. Phys.*, **31**, 212.
40. Gatos, H.C. and Lavine, M.C. (1960) *J. Electrochem. Soc.*, **107**, 427.
41. Gatos, H.C. and Lavine, M.C. (1965) in *Progress in Semiconductors*, Vol. **9** (eds A.F.Gibson and R.F. Burgess), Temple, London, pp. 1–45.
42. Gatos, H.C. (1994) *Surf. Sci.*, **299/300**, 1 and references therein.
43. Warekois, E.P., Lavine, M.C., and Gatos, H.C. (1960) *J. Appl. Phys.*, **31**, 1302.
44. Heimann, R.B. (1982) in *Crystals*, Vol. **8** (ed. H.C.Freyhardt), Springer, Berlin, pp. 175–224.

45. Kern, W. and Deckert, C.A. (1978) in *Thin Film Processes* (eds J.L.Vossen and W. Kern), Academic Press, New York, pp. 432–498.
46. Winters, H.F. and Haarer, D. (1987) *Phys. Rev. B*, **36**, 6613.
47. Morrison, S.R. (1977) *The Chemical Physics of Surfaces*, Plenum, New York.
48. Houle, F.A. (1989) *Phys. Rev. B*, **39**, 10120.
49. (a) Youtsey, C., Romano, L.T., and Adesida, I. (1998) *Appl. Phys. Lett.*, **73**, 797; (b) Youtsey, C., Romano, L.T., and Adesida, I. (1999) *Appl. Phys. Lett.*, **75**, 3537.
50. Hirth, J.P. and Lothe, J. (1968) *Theory of Dislocations*, McGraw-Hill, New York.
51. Gatos, H.C., Finn, M.C., and Lavine, M.C. (1961) *J. Appl. Phys.*, **32**, 1174.

Additional Reading

Brillson, L.J. (1992) Surfaces and interfaces: atomic-scale structure, band bending and band offsets, in *Handbook on Semiconductors* (ed. P.T. Landsberg), Elsevier Science Publishers B.V., pp. 281–417.

Vanselow, R. and Howe, R. (eds) (1990) *Chemistry and Physics of Solid Surfaces*, Vol. **8**, Springer, Berlin, p. 267.

18
Adsorbates on Electronic Materials' Surfaces

18.1
Overview

The previous chapter examined the geometric, chemical, growth, and etching properties of semiconductor surfaces. Adsorbates on semiconductors and insulators involve many of the same phenomena and physical principles, but the addition of overlayer atoms and films introduces a new level of complexity. This chapter treats the geometric, chemical, and electronic properties of both adsorbed atoms as well as multilayer films since both comprise a bridge between electronic materials' surfaces and their interfaces with macroscopic overlayers. As in the previous chapter, a unifying theme is the sensitivity of the geometric, chemical, and electronic properties to atomic-scale processes. Understanding these processes can enable new ways to control their macroscopic electronic properties.

18.2
Geometric Structure

Adsorption and layer growth of atoms on semiconductors and insulators is, in general, polycrystalline or amorphous; however, a considerable body of information exists for ordered adlayers and overlayers. These geometric features offer an additional insight into the chemical and electronic properties of semiconductor surface sites as well as the kinetics and thermodynamics of heteroepitaxial growth. This section provides an overview of such adlayers and multilayer structures with an emphasis on (i) site specificity; (ii) the influence of adsorbates on surface reconstruction; and (iii) the properties of epitaxical overlayer growth.

The advent of scanning tunneling microscopy (STM) has enabled significant advances in our knowledge of chemisorbed species on semiconductor surfaces and their effects on geometric structure. As with clean surfaces, adsorbate-bonding studies have centered on the semiconductors Si and GaAs.

Surfaces and Interfaces of Electronic Materials. Leonard J. Brillson
Copyright © 2010 WILEY-VCH Verlag GmbH & Co. KGaA, Weinheim
ISBN: 978-3-527-40915-0

18.2.1
Site Specificity

Atoms adsorbed on Si exhibit several characteristic features relative to the clean-surface geometry. For Si on Si(111), Si atoms adsorb with a marked preference for orientations that nearly align the (7 × 7) unit cells of the substrate with the growing islands [1]. These island growths form domains with nucleation sites that promote further growth at their boundaries. See, for example, the motion of adatoms on terraces to step edges in Figure 17.5. Such domains exhibit a marked anisotropy due to the higher energy of the type B (e.g., Figure 17.4) versus type A step edges, with evidence for both activated kinetics [2] and equilibrium thermodynamic [3] contributions.

Within the Si(111) (7 × 7) unit cell shown in Figure 18.1a, adsorbates display a preference for particular chemisorption sites. Figure 18.1b illustrates the unit cell in cross section, with inequivalent unit-cell sites – corner holes, center adatoms, rest atoms, and corner adatoms of Si labeled. Figure 18.1c illustrates the local bonding and geometry for rest and adatoms. Si and Ge are believed to adsorb onto such T_4 "top" sites of the Si (111) (7 × 7) reconstruction as well as for both Si in a $c(2 \times 2)$ reconstruction, and Ge in a $c(2 \times 8)$ reconstruction, respectively.

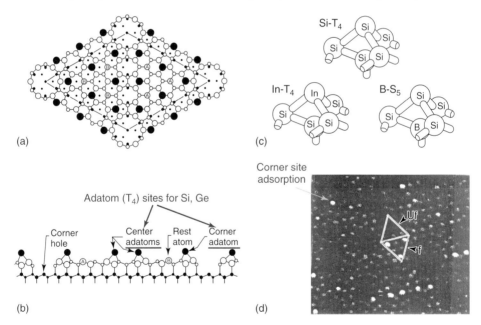

Figure 18.1 (a) Si(111) (7 × 7) unit cell structure. (b) Corresponding cross section with various surface atom types. (c) Ball-and-stick representation of local bonding structure for adatoms on Si (Si–T_4), In on Si (In–T_4), and B in Si (B–S_5) [4]. (d) Corner site adsorption of oxygen on Si(111) (7 × 7) [5].

Oxidation of Si(111) (7 × 7) reveals an asymmetry of O adsorption within the unit cell, with adsorption preferred 3 : 1 on the "faulted" versus "unfaulted" half and corner atom sites preferred by a 2 : 1 margin, as illustrated in Figure 18.1d. Si surface reactions with H_2O, NH_3, phosphine (PH_3), and disilane (Si_2H_6) indicate that rest atoms are more reactive than adatoms and, among adatoms, center sites appear to be more reactive than corner sites [4, 5]. These results emphasize the role of local geometry and site-specific bond energies in overlayer growth.

18.2.2
Metal Adsorbates on Si

Metal adsorbates within the Si(111) (7 × 7) unit cell have particular significance in the context of Schottky barrier formation. Numerous studies find that the Column-III metals Al, Ga, and In on Si(111) adsorb substitutionally on T_4 adatom sites. In contrast, B is believed to adsorb by substitution directly below an Si adatom as shown in Figure 17.1. This "subsurface" S_5 site exhibits a total energy lower than that of a T_4 site [4]. Furthermore, such chemisorbed B drastically alters the surface chemistry of NH_3 on Si (7 × 7) as well as the charge transfer between the alkali metal K and the Si substrate [6].

The alkali metal Cs forms periodic, one-dimensional chainlike structures on Si and GaAs [7], and other alkali metals can exhibit coverage-dependent morphologies [8, 9].

Relatively unreactive metals such as Au can form one-dimensional structures with striking quantum-mechanical features. Thus angle-resolved soft X-ray photoemission spectroscopy (SXPS) and STM of Au on Si(111) reveal metallic bands and Fermi surfaces that depend on the atom spacing, orientation, and spacing between atom lines [10].

The transition metals (TMs) Fe, Co, Ni, and Cu form silicides that exhibit two (1 × 1) variants rotated by 180° plus Si adatoms and trimers [11]. Pd, Ag, and Li appear to chemisorb preferentially on the "faulted" half of the Si(111) (7 × 7) unit cell. In all these adsorbate-Si systems, preferential adsorption appears related to the relative bond energies associated with specific adsorption sites.

18.2.3
Metal Adsorbates on GaAs

Adsorbate studies on other semiconductors are more limited. Most such STM studies have focused on GaAs. Table 18.1 summarizes these results [12–23].

Al on GaAs(110) provides an example of staged adsorption, reaction, and adsorbate incorporation into the semiconductor lattice. At very low coverages, Al exhibits mobile chemisorption, presumably in twofold sites [17, 18]. Figure 18.2 shows how clustering occurs at higher coverages [19, 20], which supplies the energy needed to overcome the activation barrier for bond breaking and Al–Ga place exchange. With further coverage and annealing, Al replaces Ga in second- and third-layer sites, which finally forms a diffusion barrier to further interdiffusion [22, 23].

Table 18.1 Adsorbates on GaAs.

Adsorbate	GaAs orientation	Observation	References
Au	(110)	Preferential bonds to Ga sites	[12]
Si	(100)	Multiple inequivalent sites	[13]
Cl	(110)	Anion (As) sites	[14]
Sb	(110)	Zigzag chains in place of top-layer anions and cations	[15]
Sm	(110)	Zigzag chains, chemically disruptive at $\theta > 0.01$ ML	[16]
Al	(110)	Twofold adsorption sites at very low coverages	[17, 18]
Al	(110)	Clustering at intermediate coverages	[19, 20]
Al	(110)	Staged Al–Ga place exchange reaction	[21–23]
Al	(110)	Al place exchange in second- and third-layer sites with annealing	[22, 23]

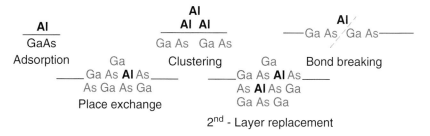

Figure 18.2 Schematic illustration of Al adsorption on GaAs(110), Al clustering, bonding breaking, Al–Ga place exchange, and second layer replacement of Ga atoms with annealing.

The condensation energy associated with Al–Al bonding provides energy to break bonds in the GaAs surface, thereby driving the exchange reaction. The reaction Al + GaAs → AlAs + Ga is energetically favorable since the GaAs and AlAs heats of formation are −17 and −27 kcal mol^{-1}, respectively. Once Al atoms replace Ga in the second layer of GaAs, each second-layer Al is then bonded to four As atoms. Since there is no further energy gain for Al diffusing further into the GaAs, this AlAs layers acts as a diffusion barrier against further Al exchange.

The Al–GaAs interface provides an example of interface bonding that alters the top monolayers of semiconductor. Thus the junction is no longer an atomically abrupt interface between a metal and a semiconductor. The exchange reaction at this interface illustrates the relevance of local chemical reactions and thermodynamics to the atomic and thereby the electronic structure of metal–semiconductor interfaces.

18.2.4
Epitaxical Overlayers

The vast majority of metal and other overlayer interfaces with semiconductors are polycrystalline; yet there exists a subset of epitaxically ordered metal–semiconductor interfaces. Such interfaces can, in principle, display the influence of interface geometric structure on electronic properties without complications due to lattice disorder, grain boundaries, or other inhomogeneities.

18.2.4.1 Elemental Metal Overlayers

For elemental metals, abrupt epilayers on semiconductors are extremely rare. A notable example is Pb on Si(111), which grows in parallel epitaxy with (111) $Pb_{||}(111)Si$ and $[1\bar{1}0]Pb_{||}[1\bar{1}0]Si$. Prepared at room temperature, this interface displays a commensurate Pb/Si(111)(7 × 7) structure in parallel orientation with the substrate [24–26]. See Figure 18.3.

After high-temperature annealing or growth, an incommensurate $Si(111)(\sqrt{3} \times \sqrt{3})R30°$–Pb structure appears, which differs only in the structure of the first layer of Pb and Si atoms at the interface with otherwise identical bulk structures in both cases. Significantly, the Si(111) (7 × 7)–Pb structure yields a Schottky barrier of 0.70 eV, whereas the $Si(111)(\sqrt{3} \times \sqrt{3})R30°$–Pb structure has a barrier height of 0.93 eV. This large Schottky barrier difference highlights the importance of local geometric and thereby local electronic structure rather than bulk properties alone in forming Schottky barriers at ordered interfaces [26].

18.2.4.2 Metal Silicides on Si

In contrast to elemental metals, metal silicides on Si provide numerous examples of epitaxically ordered metal–semiconductor interfaces. Epitaxical silicides on Si include Ni, Ti, V, Cr, Fe, Co, Zr, Nb, Mo, Pd, Ta, W, and Pt. See Chen *et al.* for a tabulation of their properties [27].

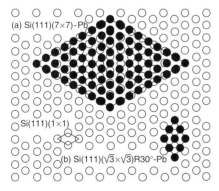

Figure 18.3 Atomic arrangement of a Pb monolayer on Si(111): (a) the Si(111) (7 × 7)–Pb structure; (b) the $Si(111)(\sqrt{3} \times \sqrt{3})R30°$–Pb structure. A (111) plane of the Si substrate appears as reference. (After Heslinga *et al.* [26]).

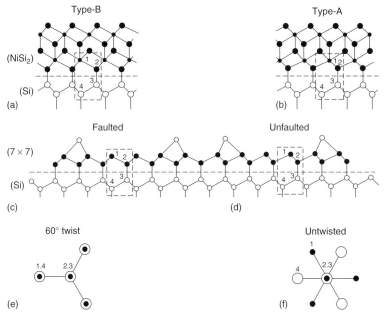

Figure 18.4 Ball-and-stick models for (a) type-A and (b) type-B epitaxial NiSi$_2$/Si interfaces and, for comparison, the (c) unfaulted, and (d) faulted sides of a Si(111) (7 × 7) unit cell, all viewed in the <110> direction. For type-A NiSi$_2$/Si(111), the silicide retains the same orientation as the substrate, whereas the type-B silicide is rotated 180° about the surface normal. The dashed lines separate surface/interface atoms (1,2) from the substrate atoms (3,4). Dashed boxes indicate interface units for both systems. Top views of these interface units appear in (e) and (f). There is a one-to-one correspondence in mapping silicide interface units to the 7 × 7 interface units. In both cases, the two variants are related by a simple 60° twist along the <110> direction between atoms 2 and 3. (After Yeh. [28])

One of the most important of these epitaxically ordered metal–Si interfaces is the NiSi$_2$/Si(111) junction. NiSi$_2$ has a cubic CaF$_2$-type structure with a lattice constant only 0.44% smaller than that of Si ($a_{0,Si}$ = 5.4307 Å at room temperature). Furthermore, NiSi$_2$ on Si(111) has two sevenfold coordinated epitaxial variants, rotated by 180°, which exhibit significant Schottky barrier differences. Figure 18.4 illustrates the two epitaxial variants of the NiSi$_2$/Si(111) interfaces [28]. For "type A" NiSi$_2$/Si(111), the silicide keeps the same orientation as the substrate, whereas the "type B" silicide is rotated.

Exclusive growth of untwinned type-A epitaxy or twinned type-B epitaxy is possible, depending upon the recipe for surface preparation, template structure, and annealing conditions [29]. Figure 18.4 also shows for comparison ball-and-stick models for the unfaulted (c) and faulted (d) sides of an Si(111)(7 × 7) unit cell viewed along the same <110> direction [28]. One can relate the interface units (shown as the dashed boxes) of the A- and B-type variants in the same way as the faulted and unfaulted parts of the (7 × 7) unit cell, that is, by a simple 60° twist along the <111> direction between atoms 2 and 3 as illustrated in Figure 18.4e,f. As with the epitaxially ordered Pb–Si interface, there is a significant difference in Schottky barrier heights, again underscoring the role of local interface bonding on macroscopic electronic properties.

18.2.4.3 Metal Epitaxy on Compound Semiconductors

Several metals exhibit epitaxical growth on compound semiconductors. Among the most notable are Al [30] and Ag [31, 32] on GaAs(001). Both exhibit multiple variants, depending on surface termination and growth temperature [33]. Epitaxical Fe on GaAs(110) and (100) [34] is distinguished by its ferromagnetic properties and good lattice match ($a_{0,Fe} = 0.286$ nm vs $\frac{1}{2}a_{0,GaAs} < 1.2\%$). Table 18.2 lists epitaxial metals on GaAs [31,34–60]. In addition to a process-dependent epitaxy,

Table 18.2 Epitaxical elemental metal films grown on UHV-prepared GaAs surfaces.

Metal	Melting point (°C)	Epitaxy on GaAs	Stability and morphology	References
Al	660	{100}Al ||{100}GaAs <010> Al || < 011 > GaAs < 1.8% mismatch, multiple {100} variants	Stable on GaAs, Al–Ga exchange on GaAs, reactive in As vapor, 3D growth, prone to agglomeration	[31, 37–43]
Ag	961	Single unrotated variant possible for growth above 200 °C on {100}	Stable on GaAs in closed system, can "etch" GaAs in an open system, 3D Volmer–Weber growth, prone to agglomeration	[31, 32, 44–47]
Fe	1536	Unrotated body centered cubic (b.c.c.) Fe on {100} and {110} GaAs	Not stable on GaAs, forms arsenides or Fe–Ga–As phases during deposition, annealing and exposure to As vapor, smooth films on {100} GaAs after coalescence of 3D nuclei	[34, 48–52]
Au	1064	Single variant on {100}, {110}, and {111} possible, often twinned	Similar to Ag	[53–55]
Co	1495	Unrotated b.c.c. Co on {100} and {110} GaAs, similar to Fe	Similar to Fe, Co–Ga–As phase formed during annealing	[56–58]
In	156	Axes of pseudocubic In unit cell parallel to cube axes of GaAs on {100} GaAs	Stable on InAs but dissolution and regrowth of (In,Ga)As on GaAs substrates, 3D growth	[59, 60]

Adapted from Sands. [35]

there is evidence of chemical interdiffusion and/or reaction for most systems, as is the case for most, if not all, metals on GaAs at elevated temperatures. This table shows that metal epitaxy on GaAs is highly-process dependent. Furthermore, chemical reaction and diffusion are evident for most systems at elevated temperatures.

A given elemental metal film may meet one or more of the multiple criteria for growth of stable epitaxial metal/III–V semiconductor heterostructures. However, no single elemental film meets all the criteria of (i) single-variant epitaxy, (ii) smooth morphology, and (iii) phase stability.

Metallic alloys provide greater flexibility than elemental metals in achieving lattice-matched interfaces to III–V compound semiconductors. Furthermore, films of such metallic alloys can achieve all three criteria for stable epitaxial interfaces. Many TM–Group III-metal intermetallic phases crystallize in the CsCl structure, an ordered derivative of the body-centered cubic structure. Figure 18.5a shows that the TM–III intermetallic phases have lattice parameters that are approximately one-half those of many III–V compound semiconductors. In terms of growth, their advantages include a wide range of compositional homogeneity and abrupt junctions – the latter due to the presence of the Group-III constituent in both metal and semiconductor.

As with the silicides, epitaxial variants are possible for TM–III/III–V interfaces that are associated with the rotational orientation of metal to semiconductor. Similarly, their preparation is highly dependent on the growth temperature, initial

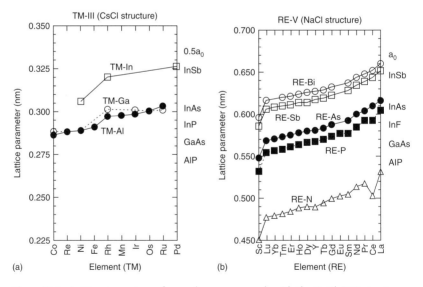

Figure 18.5 Lattice parameters of several III–V compound semiconductors in comparison with the lattice parameters of (a) transition-metal TM–III phases with the CsCl (B2) structure and (b) the rare-earth RE–V compounds with the NaCl (B1) structure. Depending on the TM or RE element, it is possible to achieve lattice matching between the overlayer phase and the semiconductor substrate [35].

surface reconstruction, and the composition of the III–V surfaces [35]. Rare earth (RE) lanthanides, actinides, plus Sc and Y form monopnictides and monochalcogenides with a NaCl structure that exhibits a large range of lattice parameters as well. Figure 18.5b shows that, for the monopnictides of Sc, Y, and the RE elements (all termed $RE-V$ here), the lattice parameter range and chemistry are quite suitable for epitaxical and stable growth of III–V heterostructures as well with lattice constants that span the range of many III–V compounds. Figure 18.5 illustrates the range of lattice parameters for TM–III and RE–V compounds in comparison with those of several III–V compounds [35].

Besides the flexibility to match lattice constants with common III–V compound semiconductors, these intermetallic alloys provide a high thermal stability due to their phase equilibria [61] as well as morphological stability (low film agglomeration) due to their high melting points. By careful deposition of template structures and control of growth temperatures, it is possible to achieve buried III–V/metal/III–V epitaxy [33, 62]. Furthermore, these epitaxial metal–semiconductor systems provide excellent structures for probing the properties of electronic material junctions since they represent ideal, ordered interfaces.

18.2.4.4 Epitaxical Metal–Semiconductor Applications

Epitaxical metal–semiconductor interfaces offer the possibility to create a number of useful electronic device applications. For example, surface-emitting laser diodes rely on Bragg reflection of light within the laser cavity in order to achieve amplification. In Figure 18.6, this reflection is due to destructive interference of the cavity radiation by a periodic variation in the refractive index in a plane parallel to the emitting surface. In turn, this requires compositions that vary periodically over thicknesses of optical wavelengths. Since compositions must vary significantly to achieve the refractive index changes needed, this presents a trade-off among increased reflectivity, thickness, and lattice strain. If instead, metal layers substitute for such index variations, one can increase absorption in the active semiconductor layer without increasing the semiconductor volume as well as any defects within this volume.

Another device enabled by these epitaxial interfaces is the permeable base transistor. As shown in Figure 18.6b, an ultrathin metal layer acts as a gate with openings to permit charge to flow between a source and drain. This internal gate

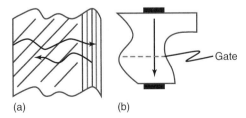

Figure 18.6 Applications of epitaxical metal–semiconductor interfaces. (a) Buried III–V/metal/III–V Bragg reflectors and (b) permeable base transistors.

permits a vertical architecture with possible size and power reductions. Because the metal is in registry with the semiconductor, the entire structure is grown with the same lattice constant.

Schottky barriers for spintronics represent a third application of epitaxial metal–semiconductor interfaces. Here, the Schottky barrier performs a function of injecting spins from a ferromagnetic metal into a semiconductor. Bonding disorder at the interface can introduce lattice scattering that reduces the number of spin-polarized electrons crossing the interface. Efforts to reduce such scattering have focused on metals such as Fe, shown in Table 18.2.

These interfaces may also find use in solar cells or indeed any other device enabled by embedded conductive structures that control the movement of light and charge carriers.

18.3
Chemical Properties

Ultrahigh vacuum surface science techniques show that the deposition of metals on clean semiconductor surfaces leads to interface-specific chemical effects [63]. These phenomena can alter composition and bonding near the interface and lead to new chemical phases with properties that differ from those of the bulk semiconductor and overlayer constituents. Depending on the nature of the local bonding interactions, overlayer growth can proceed in several ways – each controlled by the specific kinetics and thermodynamics. The local bond strength and interface processing will also determine the nature and extent of the chemical reaction and diffusion. Indeed, these exhibit systematic behavior on a microscopic scale. Conversely, this dependence on interface parameters has led to several new approaches to control the interface chemistry via atomic-scale techniques. Both the interface chemical phenomena and their atomic-scale control are significant since they can lead to new electronic properties.

18.3.1
Metal Overlayers on Semiconductors

The growth of overlayers can occur in several ways, depending on the specifics of the chemical bonding, atomic structure, and processing at the interface. On a monolayer scale, gas-phase adsorption on semiconductors has been the subject of considerable research, focused primarily on the role of local chemical bonding in catalytic processes and in the formation of electronic device structures.

On a micron scale, the area of thin-film condensation, nucleation, and growth is well developed, particularly for microelectronics [64]. Between the adsorption and micron regimes is the domain of adatom growth into thin films. Here growth depends on (i) surface mobility; (ii) adatom–adatom and adatom–substrate bonding energies; (iii) substrate surface defects and their density; as well as (iv) the temperature and method of deposition.

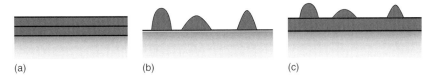

Figure 18.7 Overlayer growth modes: (a) Frank Van der Merwe, (b) Volmer–Weber, and (c) Stranski–Krastanov.

18.3.1.1 Overlayer Growth Modes

When atoms are incident on a surface, they form bonds with the substrate atoms by chemisorption or physisorption. If no strong chemical reaction occurs to form an intermediate layer, overlayer growth may proceed in several modes:

1) layer by layer, with surface kinetics already described in Section 17.2.1;
2) islanding, that is, with three-dimensional growth commencing immediately following nucleation;
3) a combination of (1) and (2) in which a full monolayer grows before islands begin to form.

Figure 18.7 shows these three modes of thin-film growth.

The layer-by-layer formation in Figure 18.7a is termed *Frank Van der Merwe growth*. Island formation is termed *Volmer–Weber growth*. Island formation over an initial uniform layer is termed *Stranski–Krastanov growth*.

Laterally uniform growth depends on the surface mobilities of adsorbates, which in turn depend on both kinetic and thermodynamic factors. Thermodynamic bond strengths between surface and adatoms directly affect adatom surface mobilities, whereas adatom–adatom bond strengths are important only in the limit of high adatom mobility. Thus growth uniformity involves the strength of adatom–substrate versus adatom–adatom bonding, as well as the epitaxical relationship between adsorbate and substrate. Epitaxial growth promotes lateral versus island growth and may be commensurate or incommensurate. See Figure 13.7.

Figure 18.8 illustrates the full set of pathways for interfaces to evolve, taking into account both subsurface diffusion and reaction [65, 66].

18.3.1.2 Thermodynamic Factors

Figure 18.8 illustrates the various noninteracting configurations for (a) isolated adatoms and (b) groupings of adatoms, respectively, depending on their epitaxical nature. Low adatom mobility due to strong adatom–substrate bonding or a high density of adatom-defect bonding sites favors more uniform lateral growth. Strong, preferential, and lattice-matched adatom–substrate bonding promotes ordered overgrowth, whereas bonding dominated by randomly distributed defects or incommensurate lattice matching leads to more disordered overlayers. Strong adatom–adatom bonding associated with high heats of fusion or condensation favors cluster formation, shown in Figure 18.8c, which can provide sufficient energy to break substrate bonds and initiate chemical reactions, diffusion, or other defect

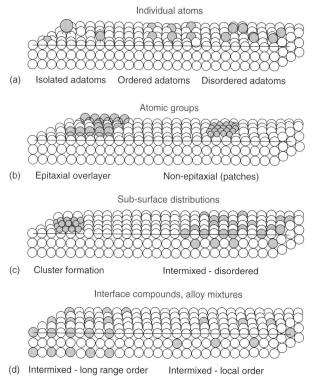

Figure 18.8 Alternative pathways of interface evolution for atoms deposited on a surface. The panels illustrate atomic distribution and bonding with increasing deposition and/or time elapsed, starting with (a) individual atoms, (b) atomic groupings, (c) subsurface distributions, and (d) interface compounds or solutions [65, 66].

formation [19]. Figure 18.8d illustrates the penetration of surface adatoms into the substrate either in an ordered or a disordered array.

A disordered solid solution is more likely to form when the chemical interactions between constituent and like atoms are nearly equivalent. In this case, entropy drives the diffusion process, that is, entropy is the primary difference between the free energies of the separate versus interdiffused systems. Conversely, chemical phases with more long-range order and well-defined boundaries form when the energy gained by bonding between dissimilar atoms dominates the interaction [66]. Segregation of subsurface constituents to the growing film surface can also occur in this regime.

18.3.1.3 Nonequilibrium Energy Processing

A number of techniques can provide energy to produce nonequilibrium conditions that alter the interface chemistry and modify the growth process. These techniques include the following:

1) Heat (e.g., thermal and rapid thermal annealing)
2) Light (e.g., laser annealing; high-energy bond-breaking excitation, i.e., photolysis)
3) Energetic particles (e.g., ion bombardment).

Surface science research on metal–semiconductor interfaces has focused on the bonding and interface atomic distributions pictured in Figure 18.8. In general, the fully developed interface structure is not atomically abrupt for most metal–semiconductor (and other) interfaces even at room temperature. Instead, such junctions can involve subsurface regions of interdiffusion and chemical reaction, as well as overlayer phases with dielectric properties unlike those of the chemical constituents [63]. The systematic relation of these chemical properties to Schottky barrier formation is the subject of Chapter 20.

18.3.2
Macroscopic Interface Reaction Kinetics

18.3.2.1 Silicide Phase Formation

The microelectronics industry provided the earliest driving force for understanding metal–semiconductor interfaces. The fabrication of microelectronics involves multilayer metal coverages and elevated temperatures such that reactions, diffusion, and new phase formation occur on a scale of microns. Among the leading concerns in silicon-based microelectronics has been the formation of highly conductive, ohmic electrical contacts that meet the requirements of (i) smooth morphology, (ii) controlled thickness, (iii) strong adhesion, and (iv) thermodynamic stability.

Metal silicides formed by thermally driven diffusion and reaction at the metal–silicon interface can satisfy these requirements. Rutherford backscattering spectroscopy (RBS) and transmission electron microscopy (TEM) have provided much of our knowledge of the buried interfaces at thick metal–semiconductor junctions for both Si and III–V compounds [67]. RBS measurements of metal–Si junctions show that interface formation occurs well below the lowest liquid phase eutectic temperature and, in most cases, at temperatures of only one-half to one-third of the Kelvin melting point [68]. In general, a sequence of phases can evolve that depends on the (i) metal, (ii) the metal thickness, (iii) temperature, (iv) crystal structure, and (v) impurities present.

Depending on which element is consumed first, the interface system moves toward equilibrium by formation of compounds richer in the remaining element. In other words, phase formation in these submicron thin films is dictated both by kinetics as well as by thermodynamics.

To illustrate this, Figure 18.9 shows schematically the sequence of phases formed at the Si–Ni interface for Ni-rich versus Si-rich cases [69]. With increasing temperature, a variety of phases form, depending on both kinetics and thermodynamics. As one of the elements is consumed, the system is driven to equilibrium by the formation of compounds richer in the remaining element. Hence the end phase in the Ni-rich system is Ni_3Si, whereas $NiSi_2$ is the end phase for the Si-rich case. The

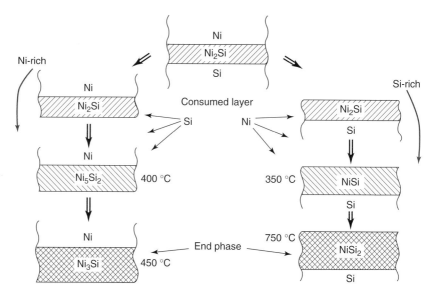

Figure 18.9 Schematic diagrams showing the phase formation in Ni-rich (left) versus Si-rich Si–Ni interface systems. The indicated temperatures are rough guides for the formation of phases for films a few thousand angstrom thick and an annealing time of 1 hour. (After Ottaviani. [69])

Table 18.3 Silicides observed following thermal anneals of a thin-film metal on a Si substrate, listed in order of their appearance, the first formed listed uppermost.

	Ti	V	Cr	Mn	Fe	Co	Ni
	Ti$_5$Si$_3$(?)	V$\underline{Si}_2$	Cr$\underline{Si}_2$	MnSi	Fe$\underline{Si}$	$\underline{Co}_2$Si	Ni$_2$Si
	TiSi(?)			MnSi$_2$	FeSi$_{1.75}$	$\underline{Co}_2$Si	Ni$\underline{Si}$
	Ti$\underline{Si}_2$					Co$\underline{Si}_2^a$	Ni$\underline{Si}_2^a$
Y	**Zr**	**Nb**	**Mo**			**Rh**	**Pd**
YSi$_{1.7}^a$	ZrSi$_2$	Nb$\underline{Si}_2$	Mo$\underline{Si}_2$			Rh$\underline{Si}$	Pd$_2$Si
						Rh$_4\underline{Si}_5^a$	Pd$\underline{Si}^a$
						Rh$_3\underline{Si}_4^a$	
Tb/Er	**Hf**	**Ta**	**W**			**Ir**	**Pt**
TbSi$_{1.7}^a$	Hf$\underline{Si}$	Ta$\underline{Si}_2$	W$\underline{Si}_2$			Ir$\underline{Si}$	Pt$_2$Si
ErSi$_{1.7}$	Hf$\underline{Si}_2$					Ir$\underline{Si}_{1.75}$	Pt$\underline{Si}$

? Tentative assignment
 The diffusing element is underlined.
a A laterally nonuniform compound.
 (After Mayer and Lau [64].)

Ni–Si equilibrium phase diagram describes each of the phases pictured at their given composition [70].

Other silicide-forming metals exhibit analogous behavior. Table 18.3 lists the experimentally observed sequence of silicide phases formed by thermal annealing of thin metal films on a silicon substrate. This table indicates two categories of solid-state reactions: (i) laterally uniform growth with well-defined kinetics and temperature dependence and (ii) nonuniform growth that depends critically on anneal temperature. The nucleation stage controls the initial silicide composition and uniformity for thin-film reactions.

In general, the growth of metal–Si interfaces can be viewed in three stages [71]:

1) an incubation stage in which the deposited metal reduces barriers to reaction and/or intermixing via rebonding, dielectric screening [72], or other mechanisms;
2) an initial interaction involving lateral diffusion, indiffusion, lattice defect formation, clustering, and/or inhomogeneous adlayer growth; and
3) a thicker film regime in which the specifics of these atomic-scale features influence the formation of various combination of reacted/intermixed, homogeneous/inhomogeneous, and abrupt/extended features.

18.3.2.2 Thin Film versus Bulk Diffusion

The intermediate phases that form at metal–Si interfaces require more than thermodynamics or the rate of compound growth to predict. The identity of the moving species and its concentration at the reaction interface are essential ingredients of the process.

Consider a thin film versus a bulk diffusion couple [64]. If only one phase forms, then the reaction proceeds until one of the elements of the binary system is completely consumed. If two phases start to form, one overtakes the other until a critical (diffusion) thickness is reached. Figure 18.10 illustrates this competitive phase growth case [73]. Here, the driving force for a phase to grow is

$$\frac{-\partial \bar{u}}{\partial x} \cong \frac{-\Delta H_R}{x} \tag{18.1}$$

where $\bar{u}$ is the chemical potential and x the layer thickness of the growing phase. ΔH_R is the heat of reaction, approximately the change in Gibbs free energy ΔG.

In analogy to Einstein's relation for the diffusion of electric charges q,

$$\frac{D_q}{\mu_q} = \frac{k_B T}{q} \tag{18.2}$$

where D_q is the diffusion coefficient and μ_q the mobility for electrons or holes, the diffusion of atoms can be given by

$$\frac{D}{\mu_a} = k_B T \text{ per diffusing species} \tag{18.3}$$

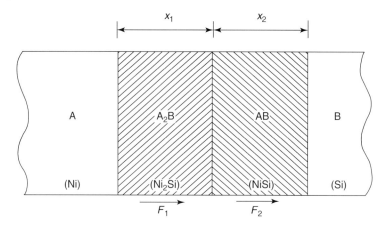

Figure 18.10 Competitive growth of two phases A_2B (e.g., Ni_2Si) and AB (e.g., NiSi), assuming A atoms (Ni) are mobile in A_2B and AB. F_1 and F_2 are the fluxes of atom A in A_2B and AB, respectively [73].

Here D is the diffusion coefficient and μ_a is an effective species mobility under a chemical driving force such that diffusion velocity v is

$$v = \mu_a \left(\frac{-\Delta H_R}{x} \right) \tag{18.4}$$

The diffusion flux F of atoms for a given concentration C is then given by

$$F = v \cdot C = \left(\frac{DC}{k_B T} \right) \left(\frac{-\Delta H_R}{x} \right) \tag{18.5}$$

Figure 18.10 shows the phase front of A_2B moving to the right as A atoms diffuse into A_2B and AB. The rate at which A_2B grows is

$$\frac{dx_1}{dt} = F_1 - F_2 \tag{18.6}$$

whereas phase AB grows at a rate

$$\frac{dx_2}{dt} = F_2 - dx_1/dt = F_2 - (F_1 - F_2) = 2F_2 - F_1 \tag{18.7}$$

For diffusion-controlled growth, x is proportional to $\sqrt{t}$, the diffusion time, and both A_2B and AB grow together. For interface reaction-controlled growth, x is linearly proportional to t and AB will not grow until A_2B phase grows to a critical thickness. For this growth mode, the reaction rate is slower than the transport of atomic species to the interface. This growth mode persists until the phase reaches a critical thickness x_c, above which the diffusion rate decreases below the interface reaction rate. Here, $F_2 = \gamma$ is a constant while $F_1 = \alpha/x_1$ with α a constant related to diffusivity, temperature, and heat of reaction. The critical thickness is then defined as

$$x_1^C \geq \frac{\alpha}{2\gamma} \tag{18.8}$$

above which multiple phases begin to form. As an example, NiSi and $NiSi_2$ begin to form only after Ni_2Si reaches 25–30μm thickness. Similar critical thickness parameters account for the intermediate phases shown in Table 18.3.

18.3.2.3 Mechanical and Morphological Effects

Besides metal reactions with Si or other semiconductors, metal–metal interactions can produce serious microelectronic fabrication problems. Multiple metals are often used to meet the requirements stated in Section 18.3.2.1 as well as to bond wires on chips to external circuitry. Al on Si was an early candidate for low-resistivity interconnects, while the ductile metal Au is commonly used to wire bond to the conductor pads on chips. However, at clean interfaces, reactions take place between Al and Au at room temperature, indicated by the Au–Al eutectic phase diagram. The initial Au_5Al_2 and other Au–Al compounds that form (termed *"purple plague"* from their color) are brittle and break under the stress of thermal compression bonding. Morphological effects at metal–Si junctions include the formation of alloy spikes that degrade Al contacts to Si and SiO_2 at the Al–Si eutectic temperature. Similar interactions occur at $Al–Pd_2S–Si$ junctions. See [64] for a comprehensive overview of metal alloys with Si. The interactions between multiple metals and semiconductors remain an ongoing concern as microelectronics shrinks ever smaller, and the higher electrical and thermal conductivity requirements demand new materials. These effects further emphasize the importance of interface chemistry in fabricating electronic devices.

18.3.2.4 Diffusion Barriers

New materials that prevent the interdiffusion between two metals or between a metal and a semiconductor are termed *diffusion barriers*. Between two materials A and B, a layer X that is an effective diffusion barrier must satisfy the following requirements: (i) transport of A and B across X must be small; (ii) the loss rate of

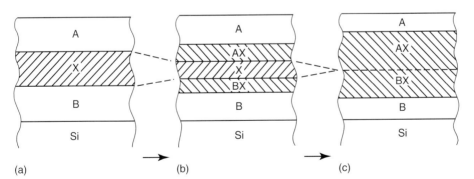

Figure 18.11 Sacrificial diffusion barrier: (a) barrier layer X between layers A and B, (b) intermediate compounds AX and BX formed at the A–X and B–X interfaces, and (c) full consumption of X into compounds AX and BX. (After Nicolet. [74])

X into A and B must be small; (iii) X must be thermodynamically stable against A and B; (iv) X must have strong adhesion to both A and B; (v) the specific electrical resistance of X to A and B must be small; (vi) X must be laterally uniform in thickness and in structure; (vii) X must be resistant to mechanical and thermal stress; and (viii) X must be highly conductive both electrically and thermally. See Figure 18.11a. It is usually quite difficult to meet all these requirements with a single diffusion barrier.

One useful approach has involved sacrificial barrier layers that achieve chemical stability by reaction to new, thermodynamically stable phases. Figure 18.11 pictures such a sacrificial diffusion barrier schematically [74]. In (b), reactions between X with A and B produces compounds AX and BX that diminish the layer thickness of X. In (c), reactions have completely consumed layer X.

Once the reactions of the barrier layer X with A and B are known and the layers are laterally uniform, the time and temperature required to achieve reacted layers AX + BX can be predicted. As long as layer X is not completely consumed, A and B remain separated. Thus there is a minimum thickness of X that will maintain separation for a given time–temperature reaction cycle, and this thickness can be predicted.

"Stuffed" barriers are another approach to prevent interdiffusion. This method is based on the elimination of short-circuit pathways in polycrystalline materials. Figure 18.12 schematically illustrates the grain boundaries between two metals (Mo and Ti), an intermediate layer X, and impurities (closed circles) from this layer that prevent metal atoms that diffuse across X from diffusing rapidly through the second metal's grain boundaries versus through the bulk film grains themselves. In Figure 18.12, the metals are Mo and Ti, which intermix when in intimate contact. However, with an oxide layer between them, metal interdiffusion is removed, even at elevated temperatures [74].

Amorphous layers represent a third form of diffusion barrier. In this case, the intermediate reaction layer is designed to eliminate grain boundaries. In general,

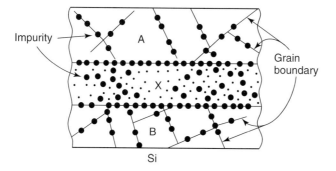

Figure 18.12 "Stuffed" polycrystalline grain boundaries that prevent interdiffusion of Mo and Ti metal layers separated by intermediate layer X with impurities (closed circles). (After Nicolet. [74])

amorphous layers are metastable. They tend to crystallize at elevated temperatures. Furthermore, single-element films do not form amorphous phases at room temperature.

The design of amorphous alloys is based on the following principles: the larger the difference in (i) atomic size, (ii) crystal structure, or (iii) electronegativity, the easier it is to form metallic amorphous alloys. An example is the combination of a near-noble metal and a refractory metal, for example, Ni_xW_{1-x} or Cu_xTa_{1-x} with $x \sim 0.5$.

Rapid quenching from the melt is another method to achieve amorphous alloys, for example, Ni–B or Sm_2Fe. However, such quenching techniques may be impractical for metals on most semiconductors.

In general, diffusion barriers are important for reducing the silicon consumed in silicide formation, especially as contact dimensions reduce into the nanometer range in large-scale integration.

18.3.2.5 Atomic-Scale Metal-Si Reactions

In addition to the phase changes observed on a micron scale, secondary phases may be present at thicknesses and volumes undetectable by RBS. In the Ni-rich Ni_2Si–Si case shown in Figure 18.9, X-ray photoemission spectroscopy (XPS) reveals a Ni-rich silicide transition region, while the Si-rich Ni_2Si–Si system has a corresponding Si-rich silicide region. Such transition regions are only 10–20 Å in thickness. As with Ni monolayers on Si, such transition regions involve Si atoms displaced from lattice sites and Ni atoms diffusing to tetrahedral interstitial sites in the Si lattice [75]. Analogous interdiffusion occurs for other metals. For epitaxial silicides such as Pd_2Si on Si(111), TEM lattice images reveal structurally sharp interfaces with a Si-rich silicide, misfit dislocations, and atomic steps within several angstroms of the interface [76]. Overall, phase formation at metal–Si junctions demonstrates the influence of the local bond strength and of the atomic structure on the macroscopic composition and morphology. Furthermore, since these microscopic phases are in intimate contact with the semiconductor, they can directly affect charge transfer and Schottky barrier formation.

18.3.3
Compound Semiconductor Reactions

Thick metal films on III–V compounds exhibit a rich variety of chemical phenomena, analogous to the phase formation of their elemental counterparts. In addition, the stoichiometry of the semiconductor near the metal interface can change, adding further complexity. For metals on binary semiconductors, the three-component system will evolve during annealing to a combination of phases predicted by the appropriate isothermal section of the equilibrium phase diagram for the constituents [77, 78]. Such phases include not only compounds but also elemental clusters of these constituents – for example, arsenic clusters at GaAs interfaces [79]. This assumes that species are not added or lost, a boundary condition that is not always satisfied in real processes. Single-phase regions are separated from one another by two-phase regions consisting of "bundles" of tie lines [35]. For system compositions within a two-phase region, the composition of the two phases is given by the end points of the appropriate tie line, with the relative amounts of each phase being determined by the *lever rule* [68].

The application of equilibrium bulk thermodynamics to metal–compound semiconductor reactions has led to the design of thermally stable, as-deposited metallizations. Deposition and prolonged annealing of a reactive elemental metal film on a compound semiconductor leads to the formation of new, thermally stable phases. In contrast, codeposition of the thermally stable phase results in a thermally stable contact that does not react with the semiconductor [35]. Such thermally stable phases, along with the epitaxial lattice properties and the high metal melting points required to prevent agglomeration, are basic requirements for the growth of abrupt, multilayer metal–semiconductor heterostructures [80].

Compound overlayers may also display metastability against chemical reaction as long as the compound phases formed by the initial overlayer–substrate reaction contain no low-melting phases to facilitate further interactions [81].

Microscopically abrupt semiconductor junctions with compound overlayers are also achievable via solid-phase regrowth [82]. While these concepts have been demonstrated for bulklike metal films on III–V compounds, the presence of surfaces, interfaces, and the associated strain are expected to alter the phase equilibria. Nevertheless, metal films as thin as ~100 Å have exhibited reactions consistent with the phase diagrams determined from bulk diffusion couples.

In the chapters to follow, we examine the systematics of chemical reaction and diffusion at microscopic metal interfaces with elemental and compound semiconductors. Overall, these results show that the complexities of reaction, diffusion, and phase formation exhibit systematic features that are defined both by the kinetic and thermodynamic parameters of the macroscopic metal–semiconductor couples as well as the detailed bonding at their microscopic interfaces.

18.4
Electronic Properties

The electronic structure of adsorbates on semiconductors derives from the local bonding and the associated charge transfer that takes place. In turn, this charge transfer alters the semiconductor electron affinity, the surface work function, and the semiconductor band bending within the space charge region. These changes depend on the specific arrangement of atoms or molecules on or within the outer surface layers.

18.4.1
Physisorption

Physisorption represents the weakest form of interaction between an adsorbate and a surface. For an adsorbed molecule, its electronic structure is only slightly

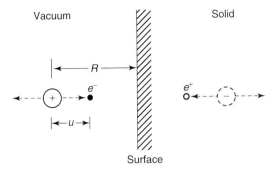

Figure 18.13 Schematic illustration of physisorbed atom consisting of a positive ion and valence electron. Oscillations normal to the solid surface follow classical dynamics with attractive forces due to image charges.

perturbed upon adsorption. In analogy to the van der Waals bonding between molecules, the bonding between the molecule and a surface decreases with distance r as r^{-3} instead of r^{-6}. The attractive force is due to correlated charge fluctuations between adsorbate and surface, that is, mutually induced dipole moments. Figure 18.13 illustrates this mutually induced dipole effect. Here the physisorbed atom consists of a positive ion and a valence electron. The potential is given as

$$V(r) \cong -\frac{(q^2\mu^2)}{(4r^3)} \qquad (18.9)$$

for dipole moment at a distance r from the surface. Since $V(r)$ increases because of repulsion of the outer adsorbate and surface atom electrons, the overall potential exhibits a minimum at an equilibrium value r_0 pictured in Figure 18.14 [83]. This minimum value is approximately $\frac{1}{2}a_0$, the lattice constant. At this potential minimum, classical mechanics describes the motion of the valence electron in a potential well:

$$V(z) \propto -(r - r_o)^{-3} \qquad (18.10)$$

The binding energy of the adsorbed atom or molecule to the surface is weak, typically ~10–100 meV, as opposed to $k_BT = 25.9$ meV at room temperature. The separation from the surface is relatively large, typically 3–10 Å, as shown in Figure 18.14.

18.4.2
Chemisorption

In contrast to physisorption, chemisorption involves strong chemical bonding and charge transfer between adsorbate and surface. As an example, consider a molecule with a partially filled molecular orbital and a TM with a partially filled d-band. As shown in Figure 18.15, one expects covalent bonding between the

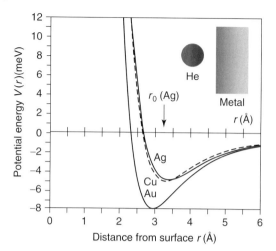

Figure 18.14 Physisorption potentials for He on the noble metals. Here the metal is described by a "jellium" model (sea of electrons in a mean density of positive ionic charge) [83].

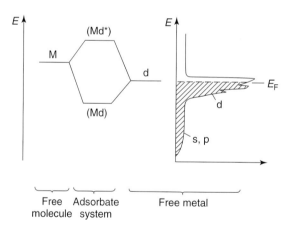

Figure 18.15 Hybridized bonding (Md) and antibonding (Md*) electronic levels due to bonding between a molecule (M) and a d-band metal [84].

partially filled orbitals of the chemisorbed molecule and the metal surface [84]. In this simple picture, the orbital overlap between molecule and metal leads to hybridized orbitals with bonding and antibonding states, termed Md and Md*, respectively. Interactions with s- and p-state orbitals are neglected. One may describe this adsorbate–metal system by an approximate wavefunction

$$\psi = a\psi_1 + b\psi_2 = a\psi_1(M^-, d^+) + b\psi_2(M^+, d^-) \tag{18.11}$$

Wavefunction ψ_1 describes a state in which an electron transfers from the metal d-state to the molecule, whereas ψ_2 describes the reverse charge transfer. This wavefunction permits calculation of the new chemisorption energy levels by minimizing an energy functional $\tilde{E}$ defined (in *bra–ket* notation) by

$$\tilde{E} = \frac{\langle \psi | H | \psi \rangle}{\langle \psi | \psi \rangle} \tag{18.12}$$

H is the total Hamiltonian of the molecule and metal substrate and ψ is a trial wavefunction that describes the charge transfer required to minimize $\tilde{E}$. One can define $H_1 = \langle \psi_1 | H | \psi_1 \rangle$ for the electron transfer to the molecule, $H_2 = \langle \psi_2 | H | \psi_2 \rangle$ for the electron transfer to the metal, $H_{12} = H_{21} = \langle \psi_1 | H | \psi_2 \rangle$ for the interaction between the two charge transfer states, and $S_{12} = \langle \psi_1 | \psi_2 \rangle$ as the overlap integral between the two states. $\tilde{E}$ is minimized by setting $\partial \tilde{E} / \partial a = 0$ and $\partial \tilde{E} / \partial b = 0$. Assuming weak overlap between ψ_1 and ψ_2 and neglecting second-order terms in S and H_{12}, the linear approximation yields

$$\tilde{E}_{\pm} = \frac{1}{2}(H_1 + H_2) \pm \left[\left(\frac{(H_1 - H_2)}{2} \right)^2 + H_{12}^2 \right]^{1/2} \tag{18.13}$$

The two solutions correspond to the antibonding (+) and bonding (−) states pictured in Figure 18.15. The total wavefunction ψ used here neglects the contributions of other metal states. More sophisticated wavefunctions and approaches are available, for example, cluster models with finite numbers of substrate atoms. In general, one applies the methods of quantum chemistry to the chemical bonding. Examples of high interest for electronics include the oxidation of III–V and II–VI compounds.

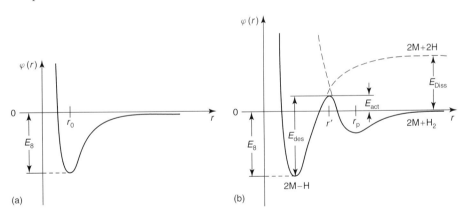

Figure 18.16 (a) Qualitative chemisorption potential for an atom or molecule with binding energy E_B and equilibrium distance r_0 on a solid surface. (b) Superposition of chemisorption and physisorption potentials for dissociative hydrogen bonding on a metal (M) surface. Distance r_p is the equilibrium distance for physisorption [85].

The rearrangement of electronics orbitals changes the shape of the adsorbate due to the new chemical bonds with the substrate. This can lead to new chemical species, including dissociation of the adsorbed molecule into atoms. Perhaps the most notable example of such dissociative adsorption is *hydrogen catalysis* – the dissociation of H_2 molecules on TMs such as Pd into H atoms. In this case, H atoms have a much higher binding energy E_B and shorter equilibrium distance r_0 than H_2 molecules. Figure 18.16a provides a schematic illustration of the chemisorption potential versus distance for an atom or molecule, whereas Figure 18.16b shows a qualitative superposition of a chemisorption and a physisorption potential for dissociative hydrogen molecule bonding on a metal surface [85]. The equilibrium distance r_0 is ~1–3 Å and binding energy E_B is $\approx$ 1 eV. E_{diss} is the *dissociation energy* of H_2 into 2H in the gas phase, E_B is the binding energy in the chemisorption state that includes H_2 and two metal atoms $2M-H_2$, E_{act} is the activation energy for adsorption of H_2, and E_{des} is the activation energy for desorption of 2H.

The higher binding energy E_B of chemisorbed atoms leads to dissociation of physisorbed H_2 molecules at distance r_p into H atoms at distance r_0. The two potential curves for H_2 and H intersect at a position r' such that a hydrogen molecule with enough kinetic energy to exceed the activation barrier E_{act} is chemisorbed as two H atoms that form two metal–H bonds. Since $E_{act} < E_{diss}$, dissociation by atomic chemisorption on the metal is favored over molecular dissociation in the gas phase. This decrease of activation barrier for molecular dissociation is characteristic of catalytic decomposition. Desorption of the isolated H atoms requires *desorption energy*

$$E_{des} = E_B + E_{act} \qquad (18.14)$$

Such desorption is typically promoted by substrate heating.

18.4.3
Work Function Effects

The bonding of adsorbates to metal or semiconductor surfaces involves charge transfer and dipole formation that induce changes in surface work function and, for semiconductors, band bending within the space charge region. These charge transfer and potential changes depend on the specific adsorption geometry of the adsorbate to the substrate atoms. Depending on the specific processes used to adsorb these particles, one can prepare surfaces with desirable sensor or transducer properties.

18.4.3.1 Charge Transfer
Charge transfer between adsorbate and surface can be understood in terms of work functions, the energies required to remove electrons from inside the bulk to a distance far enough from the surface that image forces can be neglected. In general, this distance is ~1 μm. Removing an electron from the Fermi level into

vacuum at infinite separation requires an energy

$$e\Phi = E_{vac} - E_F \tag{18.15}$$

The work function can also be defined thermodynamically. Thus the change in energy E_N of a system with N electrons as one electron is removed to the vacuum is given by

$$e\Phi = E_{N-1} + E_{vac} - E_N = E_{vac} - (E_N - E_{N-1}) \tag{18.16}$$

$$= E_{vac} - \left(\frac{\partial F}{\partial N}\right)_{T,V} = E_{vac} - \mu \tag{18.17}$$

where the free energy $F = \mu N$ and μ is the electrochemical potential of an electron in the bulk. Comparing Equations (18.15) and (18.17), the electrochemical potential of an electron is equivalent to the Fermi energy.

From this work function definition, let us now consider its various components near a semiconductor surface. For a charge q, Figure 18.17a illustrates clean a semiconductor surface with electron affinity χ_{SC}, band bending qV_B, and Fermi level positioned $E_C - E_F$ below the conduction band E_C. Here the work function consists of three contributions:

$$q\Phi_{SC} = \chi_{SC} + qV_B + (E_C - E_F)_{bulk} \tag{18.18}$$

Adsorbed atoms, molecules, or overlayers will induce charge transfer to or from the semiconductor. This charge transfer changes both the band bending and the electron affinity as illustrated in Figure 18.17b. Beyond the surface band-bending region, the Fermi level position relative to the conduction and valence bands does not change.

Photoemission spectroscopy can detect changes in both band bending and electron affinity. Figure 18.18 illustrates the spectral changes that occur before and after change transfer at a semiconductor surface. Here photons of energy $h\nu$ excite electrons from the valence band edge E_V and filled states below into empty states

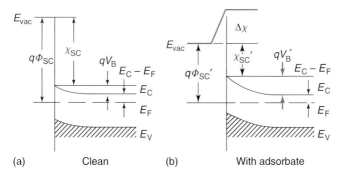

Figure 18.17 Schematic energy band diagram of (a) clean and (b) adsorbed semiconductor surface. With adsorption, both band bending and effective electron affinity change.

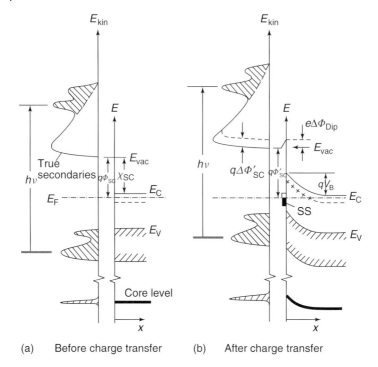

(a) Before charge transfer   (b) After charge transfer

Figure 18.18 Adsorbate-induced changes in the photoemission spectra (a) before and (b) after adsorption. Electrons promoted to states above the Fermi level exhibit core-level and valence band energies that are $h\nu$ higher than their equilibrium values and a low energy cutoff at the vacuum level.

above the vacuum level, from which they can leave the crystal with a kinetic energy E_{KIN} above the vacuum level E_{vac}. An electron detector measures these energies relative to a common Fermi level E_F if the sample and analyzer are connected electrically. See also Figure 7.9. As discussed in Chapter 7, these photoelectrons are superimposed upon a background of secondary electrons, photoelectrons that have lost energy due to scattering as they left the solid. In Figure 18.18a, E_F is positioned just above donor levels (dashed line) near E_C with no band bending. The highest kinetic energy electrons correspond to those at energy E_V. The low energy cutoff of secondary electrons defines E_{vac}.

With adsorption, Figure 18.18b shows the formation of n-type (upward) band bending qV_B within a *surface depletion layer* due to electron transfer from the interior of the semiconductor to states at the surface. Here donor levels above E_F are now "depleted" of charge. In this case, the outer surface becomes negatively charged and is balanced by the positive charge within the depletion layer. This represents a dipole with a spatial extent equal to the width of the depletion region.

If there are no states localized at the surface, then the surface work function change, $\Delta\Phi = q\Phi - q\Phi'$, is just equal to the new qV_B. However, if some of the charge transfer involves filling or unfilling of surface states (SSs), an additional local dipole contributes to $\Delta\phi$. Then from Equation (18.18),

$$q\Delta\Phi = \Delta\chi + q\Delta V_B \tag{18.19}$$

With or without surface states, the highest energy electrons at the valence band edge after charge transfer are higher by an amount qV_B since $E_F - E_V$ decreases by this amount. Thus, following Equation (18.22),

$$\Delta qV_B = \Delta(E_F - E_V) \tag{18.20}$$

On the other hand, electrons at the low energy cutoff define the work function. Thus the change in width ΔEDC of the energy distribution curve (EDC) equals $q\Delta\Phi$, which includes both components. The difference between $\Delta EDC = q\Delta\Phi$ and ΔqV_B yields $\Delta\chi$, the surface dipole.

18.4.3.2 Dipole Formation

In general, a *dipole* can be defined simply as a parallel plate capacitor.

$$q\Delta\Phi_{dipole} = -q\mathcal{E}d \tag{18.21}$$

where, according to Gauss' equation, electric field $\mathcal{E} = n_{dipole}/\varepsilon_o$ for a number n_{dipole} of dipoles, each with charge q, and d is the separation between opposite charges. Then

$$q\Delta\Phi = -qn_{dipole}P/\varepsilon_o \tag{18.22}$$

where dipole moment $P = qd$.

Note that the dipole moments at surfaces are not, in general, well known since the dipole separations occur on an atomic scale. Furthermore, the presence of dipoles on neighboring lattice sites as in Figure 18.19 produces a depolarization due to van der Waal interactions, where the field of one dipole affects the charges on its neighbors. Depolarization reduces the effective electric field $\mathcal{E}_{eff}$ according to

$$\mathcal{E}_{eff} = \mathcal{E} - f_{depol}\mathcal{E}_{eff} \tag{18.23}$$

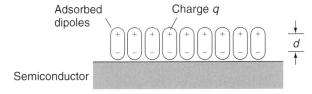

Figure 18.19 Adsorbed dipoles on a surface, each with charges q and $-q$ separated by distance d.

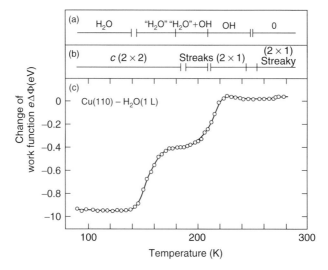

Figure 18.20 Work function changes and LEED ordering of H_2O on Cu(110) versus annealing temperature [87].

which takes into account the presence of fluctuating dipole fields in their vicinity. For an array of dipoles with polarizability α in a square pattern, termed a *Topping array* [86],

$$f_{depol} \cong \frac{9\alpha n_{dipole}^{3/2}}{4\pi \varepsilon_o} \tag{18.24}$$

such that

$$e\Delta\Phi = -qP\, n_{dipole}/\varepsilon_o \left[1 + (9\alpha\, n_{dipole})/(4\pi \varepsilon_o)\right]^{-1} \tag{18.25}$$

Adsorbate polarizabilities are also not well known due to their atomic scale and the effect of atomic-scale probes.

18.4.3.3 Ordered Adsorption

Surface science techniques enable measurements of work function changes with different ordering of adsorbates on a surface. For example, combined ultraviolet photoemission spectroscopy (UPS) and low-energy electron diffraction (LEED) measurements reveal pronounced differences in the work function of Cu(110) surfaces with changes in ordering of adsorbed water molecules [87]. In Figure 18.20, the Cu(110) surface with an initial exposure of 1 L H_2O exhibits a $c(2 \times 2)$ LEED pattern and valence band features characteristic of H_2O chemical composition. As annealing temperature increases above ~150 K, Φ increases by ~0.6 eV and the valence band features change from physisorbed to chemisorbed H_2O. Above ~200 K, photoemission indicates an OH composition, LEED streaks appear, and the work function increases by an additional ~0.4 eV. Further annealing results

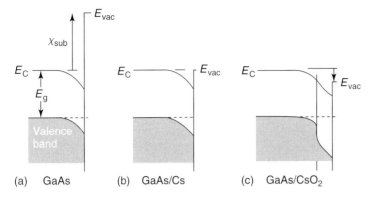

Figure 18.21 Schematic energy band diagram of p-type GaAs before and after Cs and O adsorption.

in a streaky (2 × 1) LEED pattern, an atomic O-covered surface, but no further changes in Φ.

18.4.3.4 Negative Electron Affinity

Ordered adsorption normal to the semiconductor surface can form dipoles that enable important electronic applications. For example, surface dipoles that reduce electron affinity sufficiently can produce *negative electron affinity* (NEA), where $E_{vac} < E_C$. In such cases, there is no barrier for surface electrons in the semiconductor conduction band to escape the solid.

Figure 18.21 illustrates this effect for Cs on GaAs. With one monolayer of Cs on GaAs, the p-type (downward) band bending at the clean surface (a) increases strongly since Cs is highly electropositive, donating electrons to the semiconductor and lowering the vacuum level relative to the near-surface conduction band (b). Subsequent oxidation produces a dipole (c) that reduces the substrate electron affinity χ_{sub} further, lowering E_{vac} below E_C. The resultant surface is a high flux source of electrons useful for, among other applications, electron gun filaments, photocathodes, and electron injectors in high-energy storage rings, particularly when generated by pulsed excitation sources.

Such NEA surfaces can also provide a source of highly spin-polarized electrons. Valence electrons photoexcited from the spin–orbit–split valence band to the conduction band minimum in GaAs can be strongly polarized [88], providing a major source of spins for spintronic device structures [89].

18.4.3.5 Reconstruction Changes

Surface reconstruction changes vary the positions of atoms normal to a semiconductor surface, resulting in sizable changes in work function. Figure 18.22 illustrates these changes in work function for the GaAs(100) surface with different reconstructions. Here a Kelvin probe provides work function measurements of GaAs surfaces prepared by molecular beam epitaxy (MBE) and capped with As to

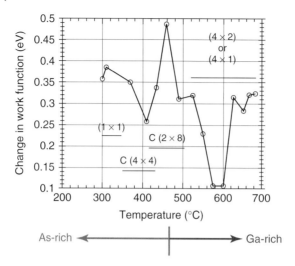

Figure 18.22 Variation of the work function measured by contact potential difference as a function of decapping temperature [91].

preserve their cleanliness during transfer from growth to analysis chambers. Clean surfaces are recovered once the As cap layer is removed by annealing in ultrahigh vacuum (UHV) [90]. As desorption at different temperatures results in different reconstructions as measured by LEED: (1×1) and $c(4 \times 4)$ for lower temperature, As-rich reconstructions, $c(2 \times 8)$ at near stoichiometry, and (4×2) or (4×1) for Ga-rich conditions. The work function varies with reconstruction from one to another over a range of ~0.4 eV.

The variations can be understood by considering the local bonding and charge transfer for each reconstruction. Figure 18.23 illustrates these differences for three such reconstructions. The alternating positive and negative charges of the As-rich $c(4 \times 4)$ surface correspond to a neutral surface layer, whereas the $c(2 \times 4)$ – $c(2 \times 8)$ reconstruction results in a more negative outer layer of As atoms. The electron affinity and work function therefore increase, consistent with Figure 18.22. Similarly, the Ga-rich (4×2) – $c(8 \times 2)$ reconstruction leads to a more positive outer layer of Ga atoms. Now χ and Φ both decrease, again consistent with Figure 18.22 and the charge transfer between the top two Ga and As layers of GaAs. These reconstruction-dependent measurements demonstrate the importance of surface atomic bonding on macroscopic work functions.

18.5
Summary

Adsorbates on semiconductors and insulators involve many of the same phenomena exhibited by semiconductors alone. This chapter serves as a

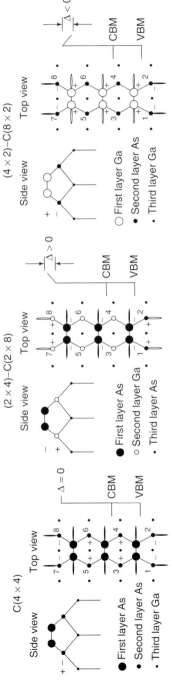

Figure 18.23 Side and top views of atomic bonding geometries for (a) $c(4\times4)$, (b) $(2\times4) - c(2\times8)$, and (c) $(4\times2) - c(8\times2)$ reconstructions. The relative changes in electron affinity with electron distribution appear to the right of each top view. After [91].

bridge between the properties of clean surfaces and their interfaces with macroscopic overlayers. The geometric, chemical, and electronic properties of adsorbates on semiconductors further emphasize their sensitivity to atomic-scale properties.

The geometric structure of adsorbates on common semiconductors such as Si and GaAs shows that (i) adsorption can be site selective, (ii) adsorbates can alter surface reconstruction, and (iii) adsorbate ordering on surfaces can enable epitaxical overlayer growth.

Metal adsorbates on clean semiconductor surfaces can produce interface chemical reactions, leading to new chemical phases with properties different from either constituent. Kinetics and thermodynamics determine the subsequent overlayer growth at such interfaces, leading to systematic behavior on a microscopic scale and the potential to control such interface chemistry by surface science techniques.

Overlayer growth depends on the specifics of chemical bonding, atomic structure, and processing. Local chemical bonding of adsorbates on semiconductors controls catalytic processes and the formation of electronic devices.

Macroscopic thin-film growth depends on (i) surface mobility, (ii) adatom–adatom versus adatom–substrate bonding, (iii) substrate surface defects, and (iv) the temperature and method of deposition. Growth modes may be (i) uniform layer-by-layer (Frank Van der Merwe), (ii) three-dimensional or islanding (Volmer–Weber), and (iii) a combination of both islands and layers (Stranski–Krastanov).

Local bonding and charge transfer strongly affects the electronic structure of adsorbates on semiconductors. These determine the semiconductor electron affinity, the surface work function, and the band bending within the surface space charge region. All depend on the specific arrangement of atoms or molecules at the adsorbate–semiconductor interface.

Physisorbed atoms or molecules are only weakly bonded, producing only a small electronic perturbation.

Chemisorbed species have strong chemical bonds and involve significant charge transfer between adsorbate and semiconductor. These lead to charge transfer and work function changes that manifest themselves macroscopically. The dipoles formed by such charge transfer and atomic rearrangement can produce large changes in electron affinity and band bending that provide important electronic applications for electron injection and sensing, the latter to be discussed in the following chapter.

Overall, these adsorbate–semiconductor features reveal that local atomic structure at the monolayer scale can have a major impact on macroscopic electronic devices.

Problems

1. Calculate the driving force for an Al–In exchange reaction (a) at the InP(100) surface and (b) one monolayer below. Explain why this reaction produces a diffusion barrier to semiconductor outdiffusion from the bulk semiconductor and specifically to which element(s)?
2. Which metallic epitaxical layer would be a potential candidate for growth on a GaAs(100) substrate if it is to be stable at high temperature and form either an arsenide or gallide during growth? Explain your reasoning.
3. Consider the diffusion of Si atoms into a 100-μ;m-thick Al overlayer. At 327 °C, the diffusion length is 16.33 μm after 5 minutes. (a) What is the value of the diffusion coefficient D? (b) After annealing After annealing for 5 minutes at 527 °C, the diffusion length is 1.67 times longer. What is the activation energy of diffusion?
4. If 1000 Å of Pt is deposited on a Si(100) surface and annealed to form a Pt_2Si low-resistance contact, (a) what thickness of Si is consumed by this much Pt in the formation of a low-resistance Pt_2Si contact? (Hint: atomic density $N = N_A \rho / A$ where N_A is Avogadro's number and A is atomic weight.) (b) What thickness of Pt is required to form one monolayer of Pt_2Si?
5. Consider a metal–semiconductor interface with a heat of reaction $\Delta H_R = -200$ kJ mol^{-1}. Assuming an activation energy for diffusion of 1.5 eV and an activated hopping prefactor for diffusion $D_0 = 10^{-1}$ cm^2 s^{-1}, what is the initial velocity of diffusing interface atoms across the first 10 Å at room temperature? (Note: 1 eV per molecule = 96.48 kJ mol^{-1}.)
6. Ti is used between Al and Si to form a laterally uniform $TiAl_3$ interfacial layer that delays the formation of Al "spikes" into Si. The effective diffusion constant D for the formation of $TiAl_3$ is $0.15 \exp(-1.85$ eV $k_B T^{-1})$. What is the thickness of the interfacial layer required to prevent the onset of spiking at 500 °C after 1 hour?
7. How thick does an Au overlayer on a semiconductor have to be to prevent more than 3% interface oxidation at room temperature after 24 hours? How thick would an analogous film of In have to be? Assume a diffusion prefactor of 6.1×10^{-16} cm^2 s^{-1}, an activation energy $E_A = b \cdot |\Delta H_0|$ where $b = 1.3 \times 10^{-5}$ (calories per mole), and no diffusion through the edges of the film. (Hint: calculate the diffusion length and assume 100% oxidation of the free metal surface.) For In oxide, $\Delta H_0 = -220$ kcal mol^{-1}.
8. Calculate the energy required to desorb a physisorbed molecule from a metal, where the molecule's equilibrium distance from the metal is 3.5 Å and the molecule has 0.5 electron separated by 1 Å.

9. A UV photoemission spectroscopy experiment with $h\nu = 21.2$ eV of clean, cleaved n-type ZnS in UHV yields an EDC with $E_{max} = 13.78$ eV and $E_{min} = 0$ relative to E_{VAC} of the electron analyzer. Comparison with a clean Au foil in electrical contact shows that $E_F - E_C = 0.1$ eV at the surface. *In situ* adsorption of one-half monolayer of oxygen rigidly shifts the EDC such that $E_{max} = 15.76$ eV. However, $E_{min} = 0.3$ eV. (a) What is the change in band bending ΔqV_B, work function $\Delta\Phi_{SC}$, and dipole $\Delta\chi$? (b) Neglecting depolarization and assuming a dipole length of 10 Å, calculate the adsorbed oxygen's dipole moment?

References

1. Kohler, V., Demuth, J.E., and Hamers, R.J. (1989) *J. Vac. Sci. Technol. A*, **7**, 2860.
2. Mo, Y.-M., Swartzentruber, B.S., Kariotis, Y., Webb, M.B., and Lagally, M.G. (1989) *Phys. Rev. Lett.*, **63**, 2393.
3. Alerhand, O.L., Berker, A.N., Joannopoulos, J.D., Vanderbilt, D., Hamers, R.J., and Demuth, J.E. (1990) *Phys. Rev. Lett.*, **64**, 2406.
4. Avouris, P. and Lyo, I.-W. (1990) in *Chemistry and Physics of Solid Surfaces*, Vol. 8 (eds R. Vanselow and R. Howe), Springer, Berlin, p. 371.
5. Avouris, P. and Lyo, I.-W. (1991) *Surf. Sci.*, **242**, 1.
6. Ma, Y., Row, J.E., Chaban, E.E., Chen, C.T., Headrick, R.L., Meigs, G.M., Modesti, S., and Sette, F. (1990) *Phys. Rev. Lett.*, **65**, 2173.
7. Stroscio, J.A., Feenstra, R.M., Newns, D.M., and Fein, A.P. (1988) *J. Vac. Sci. Technol. A*, **6**, 499.
8. Hashizume, T., Hasegawa, Y., Kamiya, I., Ide, T., Sumita, I., Hyodo, S., Sakurai, T., Tochihara, H., Kubota, M., and Murata, Y. (1990) *J. Vac. Sci. Technol. A*, **8**, 233.
9. Badt, D., Brodde, A., Tosch, St., and Neddermeyer, H. (1990) *J. Vac. Sci. Technnol. A*, **8**, 251.
10. Crain, J.N., McChesney, J.L., Zheng, F., Gallagher, M.C., Snjders, P.C., Bissen, M., Gundelach, C., Erwin, S.C., and Himpsel, F.J. (2004) *Phys. Rev.*, **69**, 125401.
11. Kubby, J., Wang, Y.R., and Greene, W.J. (1991) *Phys. Rev. Lett.*, **68**, 329.
12. Feenstra, R.M. (1989) *Phys. Rev. Lett.*, **63**, 1412.
13. Bringans, R.D. and Bachrach, R.Z. (1990) in *Synchrotron Radiation Research: Advances in Surface Science* (ed. R.Z. Bachrach), Plenum, New York.
14. Margaritondo, G., Rowe, J.E., Bertoni, C.M., Calandra, C., and Manghi, G. (1979) *Phys. Rev. B*, **20**, 1538.
15. Mårtensson, P. and Feenstra, R.M. (1989) *Phys. Rev. B*, **39**, 7744.
16. Trafas, B.M., Hill, D.M., Siefert, R.L., and Weaver, J.H. (1990) *Phys. Rev. B*, **42**, 3231.
17. Ihm, J. and Joannopoulos, J.D. (1982) *Phys. Rev. B*, **26**, 4429.
18. Daniels, R.R., Katnani, A.D., Zhao, T.-X., Margaritondo, G., and Zunger, A. (1982) *Phys. Rev. Lett.*, **49**, 895.
19. Zunger, A. (1983) *Thin Solid Films*, **104**, 301.
20. Skeath, P., Lindau, I., Su, C.Y., and Spicer, W.E. (1983) *Phys. Rev. B*, **28**, 7051.
21. Brillson, L.J., Bachrach, R.Z., and Bauer, R.S. (1979) *Phys. Rev. Lett.*, **42**, 397.
22. Duke, C.B., Paton, A., Meyer, R.J., Brillson, L.J., Kahn, A., Kanani, D., Carelli, T., Yeh, J.L., Margaritondo, G.,

and Katnani, A.D. (1981) *Phys. Rev. Lett.*, **46**, 440.
23. Kahn, A., Kanani, D., Carelli, T., Yeh, J.L., Duke, C.B., Meyer, R.J., Paton, A., and Brillson, L.J. (1981) *J. Vac. Sci. Technol.*, **18**, 792.
24. Estrup, P.J. and Morrison, J. (1964) *Surf. Sci.*, **2**, 465.
25. Le Lay, G., Peretti, J., and Hanbücken, M. (1988) *Surf. Sci.*, **204**, 57.
26. Heslinga, D.R., Weitering, H.H., van der Werf, D.P., Klapwijk, T.M., and Hibma, T. (1990) *Phys. Rev. Lett.*, **64**, 1589.
27. Chen, L.J., Cheng, H.C., and Lin, W.T. (1986) in *Thin Films – Interfaces and Phenomena*, MRS Symposia Proceedings, Vol. 54 (eds R.J. Nemanich, P.S. Ho, and S.S. Lau), *Materials Research Society* Pittsburgh, PA p. 245.
28. Yeh, J.-J. (1989) *Appl. Phys. Lett.*, **55**, 1241.
29. Tung, R.T., Gibson, J.M., and Poate, J.M. (1983) *Phys. Rev. Lett.*, **50**, 429.
30. Ludeke, R., Chang, L.L., and Esaki, L. (1973) *Appl. Phys. Lett.*, **23**, 201.
31. Ludeke, R. (1984) *J. Vac. Sci. Technol. B*, **2**, 400 and references therein.
32. Massies, J., Delescluse, P., Etienne, P., and Linh, N.T. (1982) *Thin Solid Films*, **90**, 112.
33. Sands, T., Harbison, J.P., Tabatabaie, N., Chan, W.K., Gilchrist, H.L., Cheeks, T.L., Flores, L.T., and Keramidas, V.G. (1990) *Surf. Sci.*, **228**, 1.
34. Prinz, G.A. and Krebs, J.J. (1981) *Appl. Phys. Lett.*, **39**, 397.
35. Sands, T., Palmstrøm, C.J., Harbison, J.P., Keramidas, V.G., Tabatabaie, N., Cheeks, T.L., Ramesh, R., and Silberberg, Y. (1990) *Mat. Sci. Rep.*, **5**, 99.
36. Palmstrøm, C.J. and Morgan, D.V. (1985) in *Gallium Arsenide: Materials, Devices and Circuits* (eds M.G. Howes and D.V. Morgan), John Wiley & Sons, Inc., New York, p. 195.
37. Cho, A.Y. and Dernier, P.D. (1978) *J. Appl. Phys.*, **49**, 3328.
38. Missous, M., Rhoderick, E.H., and Singer, K.E. (1986) *J. Appl. Phys.*, **59**, 3189.
39. Svensson, S.P., Landgren, G., and Andersson, T.G. (1983) *J. Appl. Phys.*, **54**, 4474.
40. Tadayon, B., Tadayon, S., Spencer, M.G., Harris, G.L., Rathbun, L., Bradshw, J.T., Schaffer, W.J., Tasker, P.W., and Eastman, L.F. (1988) *Appl. Phys. Lett.*, **53**, 2664.
41. Okamoto, K., Wood, C.E.C., Rathbun, L., and Eastman, L.F. (1982) *J. Appl. Phys.*, **53**, 1532.
42. Chambers, S.A. (1989) *Phys. Rev. B*, **39**, 12664.
43. Oh, J.E., Bhattacharya, P.K., Singh, J., Dos Passos, W., Clarke, R., Mestres, N., Merlin, R., Chang, K.H., and Gibala, R. (1990) *Surf. Sci.*, **228**, 16.
44. Farrow, R.F.C., Speriosu, V.S., Parkin, S.S.P., Chien, C., Bravman, J.C., Marks, R.N., Kirchner, P.D., Prinz, G.A., and Zonker, B.T. (1989) in *Thin Films – Stresses and Mechanical Properties*, Materials Research Society Symposium Proceedings, Vol. 130 (eds J.C. Bravman, W.D. Nix, D.M. Barnett, and D.A. Smith), Materials Research Society, Pittsburgh, p. 281.
45. Panish, M.B. (1967) *J. Electrochem. Soc.*, **114**, 516.
46. Pugh, J.H. and Williams, R.S. (1986) *J. Mater. Res.*, **1**, 343.
47. Lilienthal-Weber, Z., Miret-Goutier, A., Newman, N., Jou, C., Spicer, W.E., Washburn, J., and Weber, E.R. (1988) in *Epitaxy of Semiconductor Layered Structures*, Materials Research Society Symposium Proceedings, Vol. 102 (eds R.T. Tung, L.R. Dawson, and R.L. Gunshor), Materials Research Society, Pittsburgh, p. 241.
48. Ruckman, M.W., Joyce, J.J., and Weaver, J.H. (1987) *Phys. Rev. B*, **33**, 7029.
49. Waldrop, J.R. and Grant, R.W. (1979) *Appl. Phys. Lett.*, **34**, 630.
50. Krebs, J.J., Jonker, B.T., and Prinz, G.A. (1987) *J. Appl. Phys.*, **61**, 2596.
51. Chambers, S.A., Xu, F., Chen, H.-W., Vitomirov, I.M., Anderson, S.B., and Weaver, J.H. (1986) *Phys. Rev. B*, **34**, 6605.
52. Harris, I.R., Smith, N.A., Cockayne, B., and MacEwan, W.R. (1987) *J. Cryst. Growth*, **82**, 450.

53. Takeda, K., Hanawa, T., and Shimojo, T. *Jpn* (1974). *J. Appl. Phys.*, **2**, (Suppl. 2, Part 2), 589.
54. Snyman, L.W., Vermaak, J.S., and Auret, F.D. (1977) *J. Cryst. Growth*, **42**, 132.
55. Leung, S., Milnes, A.G., and Chung, D.D. (1983) in *Interfaces and Contacts*, Materials Research Society Symposium Proceedings, Vol. 18 (eds R. Ludeke and K. Rose), Materials Research Society, Pittsburgh, p. 109.
56. Prinz, G.A. (1985) *Phys. Rev. Lett.*, **54**, 1051.
57. Xu, F., Aldao, C.M., Vitomirov, I.M., and Weaver, J.H. (1987) *Appl. Phys. Lett.*, **51**, 1946.
58. Lind, D.M., Idzerda, Y.U., Prinz, G.A., Jonker, B.T., and Krebs, J.J. (1988) *J. Vac. Sci. Technol. A*, **6**, 819.
59. Ding, J., Washburn, J., Sands, T., and Keramidas, V.G. (1986) *Appl. Phys. Lett.*, **49**, 818.
60. Savage, D.E. and Lagally, M.G. (1986) *J. Vac. Sci. Technol. B*, **4**, 943.
61. Palmstrøm, C.J., Cheeks, T.L., Gilchrist, H.L., Zhu, J.G., Carter, C.B., and Nahory, R.E. (1998) in *Electronic, Optical and Device Properties of Layered Structures* (eds J.R. Hayes, M.S. Hyberston, and E.R. Weber), Materials Research Society, Pittsburgh, p. 63.
62. Palmstrøm, C.J., Harbison, J.B., Sands, T., Ramesh, R., Finstad, T.G., Mounier, S., Zhu, J.G., Carter, C.B., Florez, L.T., and Keramidas, V.G. (1990) *Mater. Res. Soc. Symp. Proc.*, **198**, 153.
63. Brillson, L.J. (1982) *Surf. Sci. Repts.*, **2**, 123.
64. Mayer, J.W. and Lau, S.S. (1990) *Electronic Materials Science: for Integrated Circuits in Si and GaAs*, MacMillan, New York.
65. Weaver, J.H. (1988) Analytical techniques for thin films, in *Treatise on Materials Science and Technology* (eds K.N. Tu and R. Rosenberg), Academic Press, New York, pp. 15–53.
66. Weaver, J.H. (1986) *Phys. Today*, **39**, 24.
67. Poate, J.M., Tu, K.N., and Mayer, J.W. (1978) *Thin Films – Interdiffusion and Reactions*, Wiley-Interscience, New York.
68. Mayer, J.W. and Tu, K.N. (1971) *J. Vac. Sci. Technol.*, **11**, 86.
69. Ottaviani, G. (1979) *J. Vac. Sci. Technol.*, **16**, 1112.
70. Hansen, M. and Elliot, R.P. (1986) *Constitution of Binary Alloys*, Genium, Schenectady.
71. Braicovitch, L. (1988) Surface properties of electronic materials, in *The Chemical Physics and Heterogeneous Catalysis of Semiconductor Surfaces* (eds D.A. King and D.P. Woodruff), Elsevier, Amsterdam, pp. 235–269.
72. Hiraki, A. (1983) *Surf. Sci. Rep.*, **3**, 355.
73. d'Heurle, F.M. and Gas, P. (1986) *J. Mat. Res.*, **1**, 205.
74. Nicolet, M.-A. (1978) *Thin Solid Films*, **52**, 415.
75. Cheung, N.W., Grunthaner, P.J., and Grunthaner, F.J. (1981) *J. Vac. Sci. Technol.*, **18**, 917.
76. Schmid, P.E., Ho, P.S., Föll, H., and Rubloff, G.W. (1981) *J. Vac. Sci. Technol.*, **18**, 937.
77. Beyers, R.K., Kim, B., and Sinclair, R. (1987) *J. Appl. Phys.*, **61**, 2195.
78. Sands, T. (1989) *Mater. Sci. Eng. B*, **1**, 289.
79. Woodall, J.M. and Freeeouf, J.L. (1981) *J. Vac. Sci. Technol.*, **19**, 794.
80. Brillson, L.J. (1992) Surfaces and interfaces: atomic-scale structure, band bending, and band offsets, *Handbook on Semiconductors*, Basic Properties of Semiconductors, Vol. 1, Chapter 7, Elsevier, New York.
81. Murakami, M., Kim, H.J., Shih, Y.-C., Price, W.H., and Parks, C.C. (1989) *Appl. Surf. Sci.*, **41/42**, 195.
82. Marschall, E.D., Lau, S.S., Palmstrøm, C.J., Sands, T., Schwartz, C.L., Schwarz, S.A., Harbison, J.P., and Florez, L.T. (1989) Materials Chemistry and Defects in Semiconductor Heterostructures Research Society Proceedings, Vol. **148**, Materials Research Society, Pittsburgh, p. 163.
83. Zaremba, E. and Kohn, W. (1977) *Phys. Rev. B*, **15**, 1769.
84. Lüth, H. (2001) *Solid Surfaces, Interfaces, and Thin Films*, Springer-Verlag, Berlin, p. 503.

85. Lüth, H. (2001) *Solid Surfaces, Interfaces, and Thin Films*, Springer-Verlag, Berlin, p. 506.
86. Topping, J. (1927) *Proc. R. Soc. London A*, **114**, 67.
87. Spitzer, A., Ritz, A., and Lüth, H. (1982) *Surf. Sci.*, **120**, 376.
88. Pierce, D.T., Celotta, R.J., Wang, G.-C., Unertl, W.N., Galejs, A., Kuyatt, C.E., and Mielczarek, S.R. (1980) *Rev. Sci. Instrum.*, **51**, 478.
89. Wolf, S.A., Awshalom, D.D., Buhrman, R.A., Daughton, J.M., von Molnár, S., Roukes, M.L., Chtchelkanova, A.Y., and Trager, D.M. (2001) *Science*, **294**, 1488.
90. Kowalczyk, S.P., Schaffer, W.J., Kraut, E.A., and Grant, R.W. (1982) *J. Vac. Sci. Technol.*, **20**, 705.
91. Chen, W., Dumas, Mao, M.D., and Kahn, A. (1992) *J. Vac. Sci. Technol. B*, **10**, 1886.

Additional Reading

Brillson, L.J. (1992) Surfaces and interfaces: atomic-scale structure, band bending, and band offsets, *Handbook on Semiconductors, Basic Properties of Semiconductors*, Vol. 1, Chapter 7, Elsevier, New York.

Mayer, J.W. and Lau, S.S. (1990) *Electronic Materials Science: for Integrated Circuits in Si and GaAs*, MacMillan, New York.

19
Adsorbate–Semiconductor Sensors

19.1
Adsorbate–Surface Charge Transfer

The charge transfer between adsorbates and semiconductor surfaces described in Section 18.1 leads to many important electronic applications. Historically, the most common among these are applications for gas sensors. More recently, adsorbate–surface charge transfer has become an enabling feature of chemical and biosensors. This chapter examines the underlying physical mechanisms and materials that determine the properties of these surface device structures.

As already discussed, charge transfer can induce band bending that alters the carrier concentration within the surface space charge region. Band bending that induces charge accumulation near the surface changes the surface conductivity in a manner similar to transistor action. For adsorbate-induced band bending, however, charge transfer between the surface and the adsorbate replaces the gate bias action of the transistor. This effect is especially pronounced for wide bandgap semiconductors such as GaN, ZnO, and other metal oxides whose intrinsic carrier concentrations are very low.

19.1.1
Band-Bending Effects

The surface conductivity of metal oxide semiconductors is strongly dependent on specific gas adsorption. Figure 19.1 illustrates the effect of ZnO exposure to oxygen or atomic hydrogen on the semiconductor's charge distribution, band bending, and electron concentration [1]. Oxygen adsorption on clean ZnO surfaces results in charge transfer of electrons to the O atoms, producing a negatively charged surface and positively charged donors within the surface space charge region. This n-type (upward) band bending results in a surface barrier qV_B and a carrier concentration $n(x)$ that decreases toward the surface.

Conversely, adsorption of atomic hydrogen on ZnO results in charge transfer from the adsorbate to the semiconductor. This produces a positively charged surface and negatively charged acceptors within the surface space charge region. This p-type (downward) band bending moves the conduction band edge below the Fermi level, inducing an accumulation of electrons near the surface. In both cases,

Surfaces and Interfaces of Electronic Materials. Leonard J. Brillson
Copyright © 2010 WILEY-VCH Verlag GmbH & Co. KGaA, Weinheim
ISBN: 978-3-527-40915-0

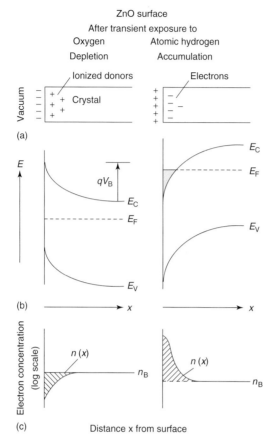

Figure 19.1 ZnO surface (a) charge distribution, (b) band bending, and (c) electron concentration before and after transient exposure to oxygen and atomic hydrogen. Oxygen depletes the surface of electrons, while atomic hydrogen induces electron accumulation [1].

the carrier concentration $n(x)$ near the surface differs by orders of magnitude from the bulk carrier concentration n_B.

19.1.2
Surface Conductance

The adsorbate-induced changes in $n(x)$ affect the sheet conductance of the semiconductor surface. Sheet conductance $g_\square$ is defined as

$$g_\square = q \int_0^l \mu \cdot n \, dx \tag{19.1}$$

integrated over the total layer thickness l.

This integral can be separated into three integrals that reflect the bulk and surface contributions, as well as the difference between surface and bulk mobilities.

$$g_\Box = q\int_0^l \mu_B n_B dx + q\int_0^d \mu_B[n(x) - n_B]dx - q\int_0^d (\mu_B - \mu_S)n(x)dx \quad (19.2)$$

where, given that $\mu_e > \mu_h$ for ZnO, μ_B and μ_S are the bulk and surface electron mobilities, respectively. l is the total semiconductor layer thickness, and d is the thickness of the surface space charge region where $d < l \cdot n_B$ is the bulk electron concentration and $n_s = n(x)$ is the surface concentration where $x < d$, neglecting the third term in Equation 19.2,

$$g_\Box = \sigma_B l + \Delta \sigma \quad (19.3)$$

Here, σ_B is the bulk conductivity, $\Delta\sigma = \Delta N e\mu_S$ is the surface conductivity, and

$$\Delta N = \int_0^l [n(x) - n_B]dx \quad (19.4)$$

is the excess electron concentration in the surface space charge region. Since $n(z)$ can exceed n_B by orders of magnitude in ZnO and many other wide gap materials, $\Delta\sigma$ can dominate sheet conductance $g_\Box$. This principle underlies the function of many semiconductor sensors.

19.1.3
Self-limiting Charge Transfer

Chemisorption of atoms or molecules at the semiconductor interface causes a charge transfer. The resultant change in surface conductance with carrier concentration reaches an equilibrium value that depends on the surface band bending. Figure 19.2 illustrates this principle. Here, a chemisorbed oxygen atom creates a localized state in the semiconductor bandgap that traps electrons near the surface. As a result, the surface becomes more negative and the semiconductor bands bend up. Electron transfer from below the surface is now retarded by the energy barrier in the surface depletion layer. As the number of trapped electrons at the surface increases, the band bending qV_B increases until further electron transfer from the bulk to the surface is pinched off. This is, in fact, analogous to the balance between carrier drift and diffusion that leads to the Einstein equation $D/\mu = k_B T/q$.

For band bending qV_B, the probability of finding a conduction electron from the bulk at the surface is then

$$P_S = \frac{n_S(T)}{n_B(T)} = e^{-qV_B/k_B T} \quad (19.5)$$

For a band bending $qV_B = 0.3\,\text{eV}$ at room temperature, Equation 19.5 yields $P_S = 9.3 \times 10^{-6}$. Assuming that the surface trap state lies below E_F, and approximating the attempt frequency of electrons as $\nu = k_B T/h = 6.25 \times 10^{12}\,\text{s}^{-1}$ at room temperature, the rate of charge flow toward the surface

$$r = P_s \cdot n_B \cdot \nu \quad (19.6)$$

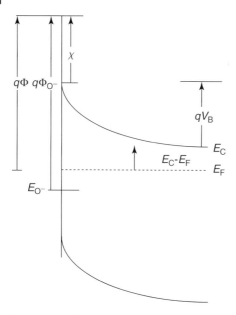

Figure 19.2 Schematic conduction band diagram of surface barrier model for self-limiting band bending.

At room temperature and $n_B = 10^{17}$ cm^{-3}, $r = 1.25 \times 10^{19}$ cm^{-2} s^{-1}, so that oxygen traps with a density $N_T = 10^{14}$ cm^{-2} would fill within $t = N_T/r < 10^{-5}$ s. For band bending $qV_B = 0.9$ eV, the same trap density would require almost 10^5 seconds. The final band bending will be determined by the density of trap states at the surface and their energy within the bandgap. Once this gap state saturates or fills to $E = E_F$, charge transfer is no longer energetically favorable. Thus the energy and density of states determine the final band bending and sheet carrier concentration, but kinetics determines the rate of approach toward these values.

This self-limiting process is strongly temperature-dependent. The attempt frequency, the probability for conduction band electrons at the surface, and the conduction band electron concentration all increase with temperature. As a result, the rate of charge transfer increases, making more electrons available for additional adsorption and/or saturating the available surface traps more rapidly.

For ZnO, the intrinsic n-type carrier concentration is associated primarily with oxygen vacancies or their complexes.[1] Oxygen adsorption creates surface levels in the bandgap that localize electrons extracted from the lattice. In addition, this oxidation can also reduce the oxygen vacancy concentration within the outer few ZnO atomic layers, thereby depleting n_S within the space charge region further. Conversely, hydrogen adsorption donates charge to the surface, forming a positive

---

1) Isolated oxygen vacancies are calculated to form deep levels in ZnO and therefore are not considered effective electron donors. However, complexes of oxygen vacancies with other lattice defects or impurities may form shallow donor levels.

surface layer. Hydrogen also binds with oxygen in the lattice, thereby increasing oxygen vacancy concentration. Both effects act to increase σ_S near the surface. Thus oxygen vacancy concentration, hydrogen diffusion, and the passivation of defects are all important factors in the response of oxide semiconductors such as ZnO to adsorbed species.

19.1.4
Transient Effects

The strong temperature dependence of charge transfer indicated in Equations 19.5 and 19.6 is evident in the time dependence of conductivity changes with gas adsorption. Figure 19.3 illustrates the change in current along the surface of ZnO needles exposed to molecular hydrogen at different temperatures [2]. At time $t = 0$ and initial exposure to 16 At H_2, current increases from an initial value i_0. The normalized rate of change $(i - i_0)/(i_s - i_0)$ decreases with time until the current reaches a saturation value i_s. With the removal of H_2, current decreases gradually returning to its initial starting value before gas exposure. *Both the magnitude and this reversibility are important parameters for sensor applications.*

Another significant parameter is the rate of response. Figure 19.3 shows this response rate increasing with increasing temperature, as expected. At 505 °C, $(i - i_0)/(i_s - i_0)$ requires more than 2000 seconds to reach its saturation value and ~1000 seconds to recover its starting value. These times decrease with increasing temperature so that at 795 °C, saturation and recovery both occur in less than 15 seconds. This pronounced increase in response rates highlights the value of elevated temperature for gas sensors that are based on the charge transfer across a depletion region.

19.1.5
Orientational Dependence

Surface composition and bonding can significantly affect gas adsorption and charge transfer. As an example, conductivity changes of ZnO surfaces with gas exposure exhibit an orientation dependence. Hydrogen exposure increases σ faster on oxygen-terminated surfaces while oxygen exposure increases σ faster on Zn-terminated surfaces. Figure 19.4 illustrates the change in sheet conductance for atomic hydrogen on a Zn- versus O-terminated ZnO surface.

With initial H exposure, $g_\square$ increases to values over an order-of-magnitude higher on the oxygen face versus the zinc face. Ultimately, $g_\square$ values for the two surfaces converge since the final band bending and sheet conductance depend on the energy and density of the states, regardless of the rate of charge transfer. Nevertheless, Figure 19.4 shows that the chemical composition and bonding of the outer surface layer affects the rate of charge transfer at adsorbate–semiconductor interfaces.

Since hydrogen donates electrons to the ZnO for both polar surfaces, one expects the same band bending and charge transfer once the adsorbate–semiconductor

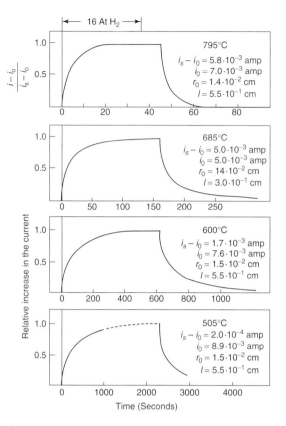

Figure 19.3 Variation in current versus exposure time to 16 At H_2 at different temperatures [2].

bonding forms the same localized states. However, the surface atom termination affects the rate at which chemical bonding and impurity centers form and at which charge transfer takes place.

19.2
Sensors

19.2.1
Sensor Operating Principles

Sensors are an important feature of modern life. We use them to regulate the environment in the buildings we inhabit, to enable our optical and electronic communications networks, and to protect our water and food supplies, to name but a few. Sensors operate on a wide variety of physical principles. For example, photosensors produce an electronic signal proportional to photogenerated electron–hole pairs. Pressure sensors generate a voltage due to the displacement of atoms and

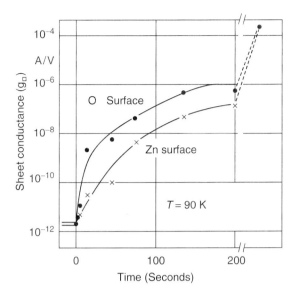

Figure 19.4 Sheet conductance versus time for atomic hydrogen exposure to Zn-terminated versus oxygen-terminated ZnO surfaces. The O-terminated surface exhibits over an order-of-magnitude larger response [1].

the resultant electric polarization with the material's unit cell. Material growth monitors change their resonant frequency with the addition of deposited atoms on thin-film crystal oscillators. Magnetic field sensors generate inductive currents in the presence of changing magnetic fields.

This section is restricted to sensors that depend on charge transfer at semiconductor interfaces and specifically to sensors based on the field effect principle. Here, the transfer of charge to a semiconductor surface produces electric fields that alter the band bending within the semiconductor's surface space charge region. The exponential effect of charge transfer on the carrier concentration and thereby the sheet conductance provides such sensors with very high sensitivity. This field effect principle alone is useful for a wide range of applications.

19.2.2
Oxide Gas Sensors

Oxide semiconductors have proven to be excellent candidates for gas sensors since (i) their bandgaps are typically large, resulting in very low intrinsic carrier concentrations; (ii) they form native point defects at metal contacts that produce ohmic contacts; (iii) they can be manufactured economically either in single crystal or polycrystalline form; and (iv) their high thermodynamic stability permits their use in high temperature or chemically corrosive environments.

Within the family of oxide sensors, electronic conductance sensors that detect molecules in the gas phase can involve both surface and bulk reactions.

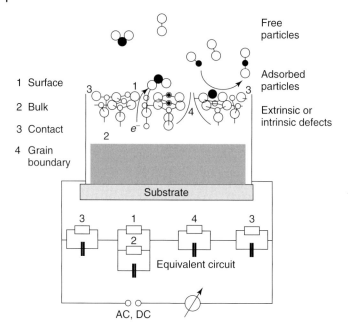

Figure 19.5 Elementary steps during the detection of molecules (free particles) in the gas phase with electronic conductance sensors. The equivalent circuit diagram illustrates the contribution from surface, bulk, contact, and grain boundary. (After Schierbaum et al. [3])

Figure 19.5 illustrates specific stages of such gas-phase detection [3]. These include the surface space charge layer changes (i) already discussed, which can be either frequency-independent or have a characteristic frequency response.

The semiconductor bulk (ii) contributes to the overall conductance and capacitance response. Electrical contacts (iii) represent a three-phase boundary involving the semiconductor, a metal, and the adsorbate. Two ohmic contacts to the semiconductor are typically required. Many oxide sensors incorporate polycrystalline grains that (i) increase surface area and (ii) provide channels for diffusion through the grains to underlying layers. In general, detection of specific molecules requires design of particular three-dimensional architecture, substrate doping, contact geometry, operating temperature, and electrical frequency.

In the case of a Schottky barrier sensor, a semiconductor such as TiO_2 is overlaid with a metal such as Pt. Here, the surface conductivity changes due to band bending as adsorbates such as H diffuse through the metal. However, the metal itself may also diffuse into the oxide surface. Overall, band bending can take place in several regions.

The channels for diffusion can also alter the chemical composition of species that reach layers beneath. An example is the potentiometric NO_x sensor developed for monitoring automotive emissions. Automotive sensors present challenges in

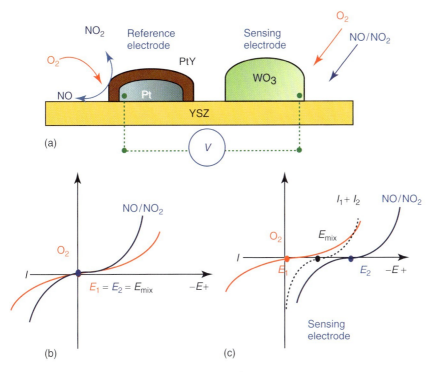

Figure 19.6 Potentiometric NO$_x$ gas sensor (a) with WO$_3$ sensing electrode and PtY-covered Pt reference electrode mounted on a zirconia substrate. O$_2$ and NO/NO$_x$ are in equilibrium in (b) but not in (c) such that NO/NO$_x$ in the presence of O$_2$ is detected as a voltage between them [4].

terms of their harsh chemical and thermal environment as well as the interference between the various gases present. Among promising approaches are sensors that compare potential differences between electrodes with different chemical and electrochemical reactivity [4]. As an example, Figure 19.6a shows one electrode composed of Pt is covered with Pt-containing zeolite Y (yittrium zirconium oxide or zirconate), termed *PtY*. A second electrode is composed of WO$_3$. The PtY electrode is able to equilibrate a mixture of NO and O$_2$ at elevated temperatures, whereas the WO$_3$ electrode does not. Figure 19.6b shows gas content versus electromotive force for O$_2$ and NO/NO$_2$ at the PtY electrode. Here, the NO$_2$/NO diffusion through PtY results in an NO$_x$ equilibrium with O$_2$ so there is no thermodynamic driving force, that is, $E_1 = E_2$, between them. Figure 19.6c shows mixed potentials between O$_2$ and NO/NO$_2$, that is, $E_1 \neq E_2$, permitting detection of NO$_x$ in the presence of O$_2$. The PtY filter in Figure 19.6 can, in fact, remove interferences to NO$_x$ detection from high concentrations of CO, propane, ammonia, as well as minimize effects of O$_2$, CO$_2$, and H$_2$O. In general, the sensitivity of gas sensors increases with increased concentration of gas molecules. Connecting several such sensors in series can further improve their sensitivity.

A major goal within the sensor community is the development of an electronic "nose" that can discriminate between different gas molecules in a manner similar to our own noses. Such sensors involve an array of sensors with varying sensitivity to multiple gases combined with electronic logic circuitry that computes the absolute concentration of each gas.

19.2.3
Granular Gas Sensors

Polycrystalline semiconductors are common in gas sensors since they provide large surface areas for gas adsorption and they are much more economical to manufacture than the same semiconductor in single crystal form. Figure 19.7 illustrates a granular film along a conduction pathway and the electronic bands in one dimension [5].

Here, the adjoining crystallites have grain boundaries that can be viewed as back-to-back Schottky barriers. The conductivity $\sigma = ne\mu$ can be expressed as

$$\sigma = N_C e\mu f(a)g(d)\exp\left\{\frac{-(E_{C,S} - E_F)}{kT}\right\} \tag{19.7}$$

N_C is the conduction band density of states, $f(a)$ is a contact area factor, $g(d)$ is a diameter of grains factor, and $E_{C,S}$ is the conduction band minimum at the surface. The intergranular conductivity therefore depends on the carrier concentration, which in turn depends on N_C and the Schottky barrier height $E_{C,S} - E_F$. The conductance $G = \sigma A/L$ depends on the average contact area A at the grain boundary and the average path length L through the film. The conductivity changes with $E_{C,S}$ as adsorbates oxidize or reduce the surface.

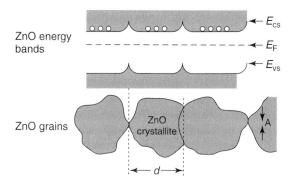

Figure 19.7 Schematic representation of model for calculating powder resistance in ZnO crystallites. (After Morrison. [5])

19.2.4
Chemical and Biosensors

The same field effect principle used in gas sensors enables the detection of chemicals and biological species. Instead of a gas, these so-called ChemFET and BioFET sensors detect molecules in solution that adsorb on semiconductor surfaces. Figure 19.8 illustrates a design for a sensor that detects chemical or biological species in solution. Again, the sensor has ohmic source and drain contacts. The semiconductor channel region has a thin surface oxide that is exposed to a micro-reservoir that contains the liquid. This particular design is based on an AlGaN/GaN high electron mobility transistor (HEMT) structure. A thin AlGaN epitaxial layer grown on GaN induces band bending such that an accumulation layer forms at their interface. The two-dimensional electron gas (2DEG) that forms has high carrier density as well as high mobility, undiminished by scattering with dopant atoms, all of which reside outside this region (see also Section 2.2). Such 2DEG layers have enabled high power transistors for radiofrequency (RF) communications at high gigahertz frequencies.

The oxidized AlGaN surface is chemically inert to most ions in solution, thereby providing an additional advantage for sensors in contact with liquids. Chemical inertness overcomes a serious concern for ChemFETs and BioFETs – the diffusion of ionic species into or through the semiconductor surface oxide. Ions diffused into field oxides can strongly alter the dielectric properties, creating electronic traps and irreversible capacitance changes that dominate the charge transfer signals due to molecular adsorption. This is a particular concern for Si and SiO_2-based transistors exposed to counterions such as Na in solution [3].

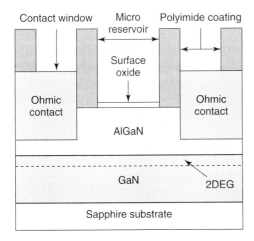

Figure 19.8 Sensor design for detecting molecules in solution. Molecules that adsorb and transfer charge to a semiconductor alter conduction through a two-dimensional gas channel [6].

Another important consideration for sensors with liquids is the screening of charge transfer by counterions that normally provide charge balance for the molecules of interest (termed "*target molecules*") in solution. It was widely believed that such counterions would effectively screen out charge transfer between molecule and semiconductor substrate. However, recent studies of semiconductors exposed to biomolecules in solution demonstrate that such screening can be avoided, so that systematic conductivity changes with adsorption are achievable [6].

19.2.4.1 Sensor Selectivity

Sensor selectivity is a particularly important parameter for biosensors. Sensors are now used to detect air- or liquid-borne contaminants, impurities, or pathogens. Unless one can distinguish the desired adsorbate's signal from others, the environment containing such species can induce electrical responses that yield false positives or that obscure the response from target species of interest. False positives in sensors used to detect pathogens are especially undesirable since they require rapid human response.

Methods to achieve selectivity involve engineering the semiconductor surface to produce a charge transfer response only to the desired adsorbate. Among ways to accomplish this are (i) attach "receptor" molecules to the surface that bond selectively only with the "target" molecule. A prime biological example is the binding of streptavidin molecules to biotin molecules attached to surfaces [7]. (ii) Restrict the available space for molecules to bind with the substrate. Figure 19.9 illustrates binding sites on a substrate that restrict adsorbates to those below a

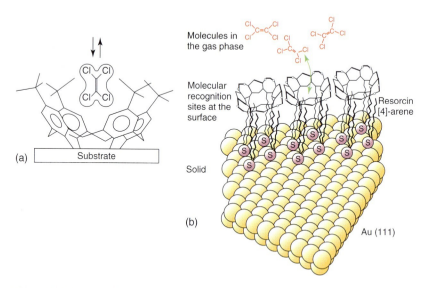

Figure 19.9 (a) Organic cage molecule and Cl-bearing adsorbate molecule. The cage molecule restricts the size of the adsorbate [3] Cage molecule array on Au(111) substrate. (Adapted from. [8])

maximum size [8]. (iii) Use capillaries or other restrictions to filter the molecules available for adsorption. The use of micron-scale channels to feed solutions into a sensor chamber is a common technique.

19.2.4.2 Sensor Sensitivity

Sensitivity to trace amounts of a specific molecule is the primary metric for chemical and biosensors. The field effect transistor provides high sensitivity since charge transfer induces an exponential change in conductivity. Indeed the FET provides a useful tool to gauge the number of charges deposited on a semiconductor surface in terms of the channel current at a given voltage bias.

In the case of an n-channel metal–oxide–semiconductor field effect transistor (MOSFET), the source-drain current I_D below saturation is given as

$$I_D = \left(\frac{\mu Z C_i}{L}\right)\left[(V_G - V_T)V_D - \frac{1}{2}V_D^2\right] \tag{19.8}$$

where μ is the surface electron mobility, Z is the channel width, C_i is the surface oxide capacitance, L is the channel length, V_G is the gate voltage, V_T is the threshold voltage, and V_D is the drain to source voltage [9].

Since the biosensor has no applied gate voltage, $V_G = 0$, using Equation 19.8, one can measure the equivalent threshold voltage set up by adsorption of the biomolecules on the semiconductor surface. One can then compare I_D before and after adsorption. Since V_T depends on the surface charge $Q_S = -Q_d$ according to [9]

$$V_T = V_{FB} - \left(\frac{Q_d}{C_i}\right) + 2\phi_F \tag{19.9}$$

where V_{FB} is the flat-band voltage, Q_d is the depletion charge, and $\phi_F = E_i - E_F$ reflects the doping concentration, then the change in surface charge on the insulator is obtained from the relation [10]

$$-\Delta Q_d = C_i \Delta V_T = \Delta Q_{biomaterial} \tag{19.10}$$

The charge transfer per molecule follows by dividing ΔQ_S by the number of adsorbed molecules, which can be determined by atomic force microscopy, fluorescence, or photoemission spectroscopy.

In order to increase sensitivity to low concentrations of a target species, air or liquid containing these species can be pumped past the channel surface to increase the absolute number of adsorbates brought in contact with the semiconductor surface.

19.2.5 Other Transducers

Besides field effect transducers, many other devices are capable of converting chemical to electrical information. For example, adsorption can alter the surface conductivity or capacitance between metal electrodes. Figure 19.10 shows a device consisting of a comb-like pair of metal electrodes mounted on an insulating

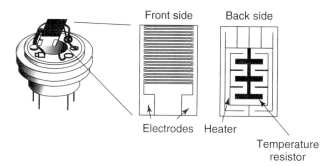

Figure 19.10 Comb structure and heater for surface conductivity or capacitance measurement mounted on electrical pin plug.

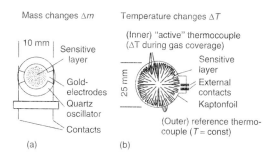

Figure 19.11 (a) Mass-sensitive crystal oscillator consisting of a vibrating crystal and metal electrodes and (b) temperature-sensitive "active" and reference thermocouples. Both generate voltage outputs.

substrate. Gas molecules adsorbed on the front surface alter the capacitance by changing the effective dielectric constant between electrodes. The comb pattern (left) increases the linear dimension for surface conductivity or capacitance or capacitance measurements. The back side contains a resistive heater to maintain elevated temperatures during operation. Single crystalline sapphire substrate and platinum electrodes are typical materials and several hundred degrees (°C) are typical operating temperatures.

Mass-sensitive devices are another method of detecting adsorption. The increase in mass due to adsorption on a surface alters the resonant frequency of a vibrating film. The change in frequency is related to the mass change according to

$$\frac{\Delta f}{f} = \frac{-\Delta m * f}{(\rho_Q F N)} \qquad (19.11)$$

where F is the unit area, ρ_Q is the vibrating film density, and m_Q is the starting film mass as described in Section 5.6. The vibrating film is piezoelectric, generating a small voltage at the vibration frequency. Figure 19.11a illustrates such a deposit thickness monitor. Quartz is one of the most commonly used oscillator materials. Au electrodes are sufficiently inert to minimize contamination or chemical reaction with adsorbates, which could otherwise complicate the film's mass change and resonant frequency. Si cantilevers micromachined by chemical etching on a nanometer scale provide another method to gauge adsorption. The changes in their resonant frequency generate capacitance signals and are also useful in scanning probe devices.

Adsorbed species can change the temperature upon condensation or reevaporation. Figure 19.11b shows a thermocouple with densely folded surface area, consisting of an active thermocouple separated by a Kapton foil from a reference thermocouple. External contacts supply the difference in thermocouple voltages [3]. All three adsorption sensors have found application in vacuum systems, the crystal oscillator to monitor deposition, the capacitance and temperature sensors to measure pressure.

19.2.6
Electronic Materials for Sensors

A wide variety of electronic materials are available for sensor applications. Metals such as Pt, Pd, Au, and Ag are relatively inert, permitting use in aggressive chemical environments. Pd and Pt, in particular, are used for hydrogen sensors since these metals permit rapid hydrogen diffusion through their bulk to the metal–semiconductor interface.

Semiconductors typically used for sensing adsorption include wide bandgap semiconductors such as CdS, ZnS, GaN as well as micromachined semiconductors such as Si.

Other ionic compounds used for gas adsorption include (i) *electronic conductors* such as ZnO, SnO_2, TiO_2, Ta_2O_5, and InO_x and $In_xSn_{1-x}O$; (ii) *mixed conductors* (i.e., surface and bulk conductor) such as $SrTiO_3$, $La_{1-x}Sr_xCo_{1-y}Ni_yO_3$, and Ga_2O_3; and (iii) *ionic conductors* such as ZrO_2, LaF_3, CeO_2, CaF_2, Na_2O_3, and β-alumina (Al_2O_3).

Organic compounds for gas sensors include (i) *molecular crystals*, for example, *pthalocyanines* such as PbPc, $LuPc_2$, and LiPc; (ii) Langmuir–Blodgett films such as pthalocyanines and polyacetylene; (iii) conducting polymers including polyacetylene, polysiloxanes, polypyrroles, and polythiophenes; and (iv) bimolecular functional systems such as phospholipids and enzymes.

In addition to gauging adsorption, semiconductors serve as sensors for a wide range of other stimuli. For example, Si, Ge, InSb, GaAs, and In_xGa_{1-x}As are commonly used in photomultipliers and photodetectors for measuring light intensity.

19.3
Summary

The interface charge transfer between atoms or molecules on semiconductors produces band bending and carrier concentration changes that strongly affect surface conductivity.

This charge transfer is self-limiting. Charge trapped at the surface produces a retarding field that repels further charge transfer from the bulk.

The charge transfer is a time-dependent process. The rate of charge transfer and ultimate saturation increases with increasing temperature. The magnitude of charge transfer and thereby the saturated change in surface conductivity depends on the energy and density of surface states within the bandgap at the surface. The reversibility of surface conductivity changes with atom or molecule desorption is an important feature of sensors that monitor adsorption changes continuously.

Surface atomic composition and bonding can affect the rate of charge transfer between adsorbed species and the semiconductor. The crystal orientation dependence of ZnO exposed to hydrogen and oxygen illustrates a two-step process involving (i) the adsorption and creation of localized states and (ii) the change of electron concentration due to charge transfer between surface and bulk.

Many adsorption sensors are available; oxides are commonly used as gas sensors, making use of the field effect principle. Changes in band bending and diffusion occur with gas adsorption that can affect not only the semiconductor film but also its electrical contacts and morphological features such as grain boundaries.

Comparative measurements between electrodes on the same substrate permit potentiometric measurements that can distinguish between different gases in the same environment.

Granular gas sensors can magnify the electrical effects of gas adsorption due to their higher surface area and back-to-back Schottky barrier interfaces.

Sensors based on the field effect principle have high sensitivity based on their exponential response to charge transfer. One can also enhance sensitivity by mechanically increasing the flow of target molecules past semiconductor surface receptors.

Sensors for detecting molecular species in solution require semiconductors that do not interact chemically with the adsorbate's solution. Chemical and biological sensors require receptor species to bind preferentially to target molecules. Restricted adsorption sites and filter structures are additional structures to increase sensor selectivity.

Besides the field effect principle, sensors that the monitor surface conductivity, capacitance, mass, and temperature change can convert chemical to electrical information.

A wide range of metals, semiconductors, ionic compounds, organic, and biological compounds are available as electronic materials for sensors.

Problems

1. Calculate the final rate of charge transfer between bulk and surface for O on ZnO with a carrier concentration of 10^{17} cm^{-3} at 200 °C. What is the final band bending for $E_C - E_T = 0.5$ eV. Assuming that E_F stabilizes at

this energy, how long does it take to reach equilibrium for an O trap density $N_T = 5 \times 10^{14}$ cm^{-2} at this rate? (Hint: intrinsic carrier concentration $n_i = 1.1 \times 10^{19}$ $(T/300)_{3/2}$ exp$(-E_G/2k_B T)$ where $E_G = 3.34$ eV at 200 °C.)

2. In Problem 1, how much charge does the calculated band bending correspond to for a ZnO(0001) surface? (Hint: use the depletion approximation.) Is the assumption of nearly full band bending throughout the charge transfer process reasonable?

3. Calculate the room temperature change in surface conductance of ZnO with $n = 10^{18}$ cm^{-3} assuming flat bands initially and that adsorption results in charge transfer such that $E_C - E_F = 0.5$ eV at the surface. $\mu_e = 50$ cm^2 (V s)$^{-1}$ for polycrystalline ZnO (Hint: Use $n_0 = n_i$ exp$((E_F - E_i)/k_B T)$ at the surface to obtain the band bending.) Is this type of conductivity sensor more sensitive as a free standing crystal or a thin film?

4. GaN nanowires are to be grown for liquid sensors based on the field effect mechanism. Assuming a typical Fermi-level stabilization with liquid contact of 0.9 eV below E_C, what is the optimal wire diameter for GaN doped $N = 5 \times 10^{17}$ cm^{-3}?

5. Design a photosensor patterned on a surface similar to the serpentine pattern in Figure 19.10. The semiconductor film is 5-μm thick with electron mobility $\mu_n = 50$ cm^2(V-s)$^{-1} \gg \mu_p$, a donor density $N_d = 10^{14}$ cm^{-3}, electron lifetime $\tau_n = 2 \times 10^{-6}$ s, and a dark resistance of 10 MΩ. The photosensor must fit inside a 0.6 cm^2 area. Hint: $R = \rho L/A$ where L is length and A is cross-sectional area. (b) For bandgap illumination that produces 10^{21} electron–hole pairs /cm^3-s, what is the change in resistance?

6. An n-channel biosensor is made with source and drain on a p-type Si substrate with $N_a = 5 \times 10^{15}$ cm^{-3}, $\mu_e = 1250$ cm^2(V s)$^{-1}$, and $Z/L = 10$. The SiO$_2$ thickness is 1 nm in the bare channel region. $\varepsilon_i(SiO_2) = 3.9\varepsilon_0$. The effective interface charge is due to charge transfer between adsorbed molecules and the SiO$_2$. (a) Calculate the insulator capacitance C_i. (b) Given a threshold voltage $V_T = -0.284$ V, calculate the source-drain current I_D for $V_D = 1$ V. (c) If I_D increases by 5% after adsorption of 10^{10} molecules /cm^2, calculate the charge transfer per molecule.

References

1. Heiland, G. (1969) *Surf. Sci.*, **13**, 72.
2. Heiland, G., Mollwo, E., and Stockmann, F. (1959) in *Solid State Physics*, Vol. **8** (eds F.Seitz and D. Turnbull), Academic, New York.
3. Schierbaum, K.D., Wi-Xing, X., Fischer, S., and Göpel, W. (1993) in *Adsorption on Ordered Surfaces of Ionic Solids and Thin Films*, Springer Series in Surface Sciences, Vol. **33** (eds E.Umbach and H.J. Freund), Spring-Verlag, Berlin, pp. 268–278.
4. Yang, J.-C. and Dutta, P. (2007) *Sens. Actuators*, **125**, 30.

5. Roy Morrison, S. (1971) *Surf. Sci.*, **27**, 586.
6. Gupta, S., Elias, M., Wen, X., Shapiro, J., Brillson, L.J., Lu, W., and Lee, S. (2008) *Biosens. Bioelectron.*, **24**, 505.
7. Lee, S.C., Keener, M.T., Tokachichu, D.R., Bhushan, B., Barnes, P.D., Cipriany, B.R., Gao, M., and Brillson, L.J. (2005) *J. Vac. Sci. Technol. B*, **23**, 1856.
8. Schierbaum, K.D., Weiss, T., Thoden van Velzen, E.U., Engersen, J.F.J., Reinhoudt, D.N., and Göpel, W. (1994) *Science*, **265**, 1413.
9. Streetman, B.G. and Banerjee, S. (2000) *Solid State Electronic Devices*, 5th edn, Prentice Hall, Upper Saddle River, pp. 286–288.
10. Cipriany, B. (2003) Development of a hybridized field effect transistor for biosensing, University Honors Thesis, The Ohio State University, Columbus.

20
Semiconductor Heterojunctions

20.1
Overview

Semiconductor heterojunctions offer unique electrical and optical properties not otherwise found in nature. It is possible to select combinations of semiconductors with bandgaps and band offsets for desirable physical properties such as light absorption and emission, photon and/or charge carrier confinement, carrier injection, and electron tunneling. Figure 20.1 illustrates the three types of energy band alignments possible for two semiconductors with different bandgaps. The simplest energy band lineup is the *straddling* or *nested* hetererojunction as shown in Figure 20.1a. When either the conduction or the valence band of one semiconductor lies outside the bandgap of the other, the heterojunction is said to be *staggered*. When both valence and conduction bands do not overlap, the heterojunction is termed *broken-gap*.

The enormous range of physical properties and applications available by combining different semiconductors is restricted by a number of geometrical, chemical, and electronic constraints. For semiconductor heterojunctions to provide the physical properties expected in principle, their interfaces must have high crystalline perfection, high thermodynamic stability, and correspondingly low densities of electronic defects. This chapter examines the geometrical, chemical, and electronic properties of semiconductor heterojunctions. As will be seen, the electronic structure of semiconductor–semiconductor interfaces depends sensitively on atomic-scale control of the local bonding and chemical structure.

20.2
Geometric Structure

20.2.1
Epitaxial Growth

High-quality heterojunctions require a nearly exact match between the two semiconductor lattices, both in terms of the lattice constants as well as their surface geometric net. A near-lattice match near room temperature allows for (i) a

Surfaces and Interfaces of Electronic Materials. Leonard J. Brillson
Copyright © 2010 WILEY-VCH Verlag GmbH & Co. KGaA, Weinheim
ISBN: 978-3-527-40915-0

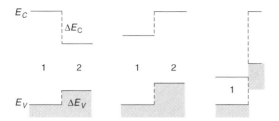

Figure 20.1 Heterojunction classes of energy band lineups.

defect-free continuation of the lattice overlayer growth and (ii) minimum lattice strain with no subsequent generation of imperfections. The interface structure and any defects at or near the junction will ultimately control the device performance. This control becomes even more important as epilayer dimensions shrink from macroscopic or micron thicknesses into the nanoscale regime.

20.2.2
Lattice Matching

The design of electronic heterojunctions involves choosing combinations of semiconductors that satisfy both the electronic properties desired as well as the close lattice match. One of the most important of such electronic properties is the bandgap of one or both of the semiconductor constituents. The bandgap will determine the energies of optical absorption, emission, and quantum confinement.

Another important property is carrier mobility. The movement of charge within epitaxial layers will determine the speed at which the semiconductor layer can conduct and its maximum frequency of operation in an electronic circuit.

Yet another significant parameter is the *thermal coefficient of expansion* and its difference between the two semiconductors. Epitaxial growth takes place at elevated temperatures. As the completed heterostructure returns to room temperature, differences in the contraction of the two lattices can lead to high strains and possible cracking or delamination.

A practical aspect to epitaxial growth on a bulk substrate is the substrate cost, quality, and availability. This consideration along with the speed and the cost of the epitaxial growth itself factor in to the manufacturability of a semiconductor heterojunction device.

20.2.2.1 Lattice Match and Alloy Composition

Many semiconductors and their alloys offer a considerable array of choices to achieve both desired energy gaps and good lattice match. Table 20.1 lists alloys of Column III–V compound semiconductors and their direct energy gaps. For just Column III elements Al, Ga, and In combined with Column V elements P, As, and Sb, the range of bandgap energies extends from 0.18 to 2.45 eV.

Figure 20.2 illustrates the variation of lattice constant for a given III–V ternary crystalline solid solutions as a function of mole fraction x according to *Vegard's law*

Table 20.1 Energy bandgap versus alloy composition.

Compound	Direct energy gap E_G (eV)
$Al_xIn_{1-x}P$	$1.351 + 2.23x$
$Al_xGa_{1-x}As$	$1.424 + 1.247x^a$
	$1.424 + 1.455x^b$
$Al_xIn_{1-x}As$	$0.360 + 2.012x + 0.698x^2$
$Al_xGa_{1-x}Sb$	$0.726 + 1.129x + 0.368x^2$
$Al_xIn_{1-x}Sb$	$0.172 + 1.621x + 0.43x^2$
$Ga_xIn_{1-x}P$	$1.351 + 0.643x + 0.786x^2$
$Ga_xIn_{1-x}As$	$0.36 + 1.064x$
$Ga_xIn_{1-x}Sb$	$0.172 + 0.139x + 0.415x^2$
GaP_xAs_{1-x}	$1.424 + 1.150x + 0.176x^2$
$GaAs_xSb_{1-x}$	$0.726 - 0.502x + 1.2x^2$
InP_xAs_{1-x}	$0.360 + 0.891x + 0.101x^2$
$InAs_xSb_{1-x}$	$0.18 - 0.41x + 0.58x^2$

Source: After Casey and Panish [1].
Reference [1] with permission.
[a] (0 < x < 0.45).
[b] (0 < x < 0.37); Kuech et al. [2].

[1]. Here, the end components for each solid solution differ by less than 0.05 nm. The dashed lines indicate alloy regions in which miscibility gaps can occur [1]. To use this figure for lattice matching, one finds the intersection of a given alloy series with the lattice constant of a desired substrate. For example, the lattice constant of both AlAs and GaAs is 0.565 nm. A horizontal line at this lattice constant intersects the $Ga_xIn_{1-x}P$ and $Al_xIn_{1-x}P$ alloy series line at approximately $x = 0.49$. Similarly, InP provides a lattice-matched substrate for the growth of $Ga_xIn_{1-x}As$ at $x = 0.48$.

Good lattice match is nominally defined as a lattice constant difference of less than 0.5% of the substrate lattice constant.

A plot of lattice constant versus energy gap can serve as a guide to lattice-matched heterojunction growth. Figure 20.3 relates Si, Ge, and III–V compound semiconductors with a given lattice constant to their energy bandgap [3]. The solid lines between binary compounds denote direct bandgaps for ternary alloys, while the dashed lines denote indirect bandgap compounds. Lines within the boundaries set by multiple compounds indicate the lattice-matching conditions for quaternary alloys [3]. Thus, $Ga_xIn_{1-x}As_yP_{1-y}$ compounds are possible with compositions and energies within the boundary lines between GaAs, InAs, and InP.

Note the close match between lattice constants of Ge with GaAs and AlAs, indicating good epitaxial growth of Ge/GaAs, Ge/AlAs, and GaAs/AlAs epilayers. On the other hand, with the exception of GaP, there are no semiconductors that are lattice matched to Si.

Figures 20.4 and 20.5 illustrate two additional bandgap energy versus lattice parameter plots for a wider range of semiconductors. Figure 20.3 shows that close

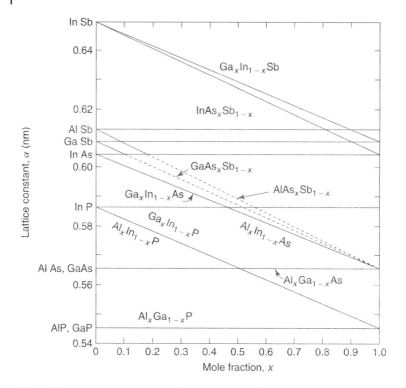

Figure 20.2 Lattice match versus alloy composition [1].

lattice matching is possible between ZnS and Si, ZnSe with GaAs and Ge, and CdTe and InSb. However, ZnSe and ZnSe have wurzite crystal structures, while Si, Ge, and the III–V compounds are diamond zincblende in structure. Note the wider range of energies available with the inclusion of ZnS and HgTe.

Figure 20.5 shows a similar bandgap energy versus lattice constant plot that includes wide bandgap (WBG) oxides, nitrides, SiC, in addition to the rocksalt structure compounds MgS and MgSe. The close lattice match between AlN and SiC is useful for growing AlGaN compounds on wurzite 6H–SiC. Similarly, the close lattice constants of GaN and ZnO indicate the usefulness of large area ZnO for epitaxical growth of nitride alloys. Note the wide range of bandgap energies possible with the Al, Ga, and In nitrides. This wide range is useful in designing optoelectronic structures with emission energies ranging from the deep ultraviolet to the infrared.

To design optoelectronic devices with desired optical and electronic properties, one must select semiconductor compositions with appropriate energy bandgaps while minimizing the lattice constant mismatch and strain effects.

20.2.2.2 Lattice-Mismatched Interfaces

The epitaxial growth of crystalline materials with dissimilar lattice constants leads to a distortion of the crystal lattice of the growing film. As the film thickness

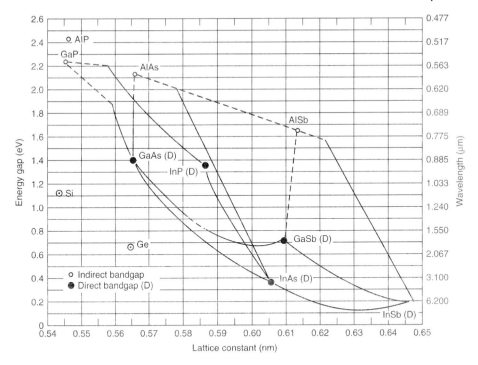

Figure 20.3 Lattice constants and corresponding energy gaps for the common III–V compounds plus Ge and Si [3]. The left axis is in units of energy, while the right axis denotes the corresponding optical wavelengths.

increases, so does this strain until, at a *critical thickness*, dislocations form. The schematic illustration of crystal lattice spacing shown in Figure 20.6 illustrates the formation of misfit dislocations. In Figure 20.6a, the lattice constant a_f of the growing film is less than a_s of the substrate. As the film grows, the epilayer unit cell distorts to accommodate this mismatch, expanding in the interface plane and contracting in the growth direction. In Figure 20.6b, the epilayer is under *tensile* strain. This *pseudomorphic* growth continues to a critical thickness h_c, beyond which the crystal "relaxes" by forming misfit dislocations. In Figure 20.6c, this relaxation consists of additional lattice planes of the smaller lattice constant epilayer that intersects the interface plane (dashed line) with no continuation into the substrate.

These morphological defects are two-dimensional *edge dislocations*, consisting of lines of atoms at the edges of planes. In turn, the atomic-scale disruption alters the lattice potential locally, leading to electrically active sites such as recombination centers and traps. Such defects strongly affect electrical device properties, for example, lowering carrier lifetimes, diffusion lengths, current amplification, and light emission.

20 Semiconductor Heterojunctions

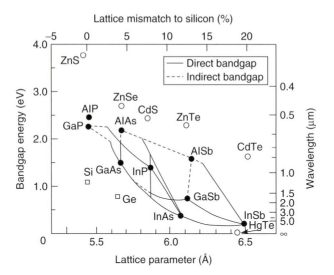

Figure 20.4 Direct and indirect bandgaps versus lattice parameter for elemental, II–V, and III–V compound semiconductors. (J. Schetzina, with permission.)

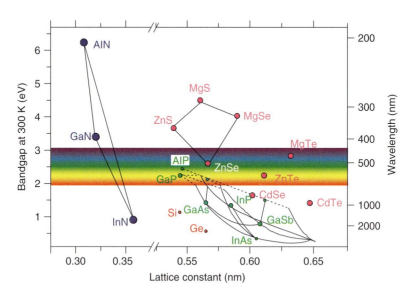

Figure 20.5 Direct bandgaps versus lattice parameter for oxides, nitrides, SiC, and Mg chalcogenides, as well as the elemental, II–V, and III–V compound semiconductors. (J. Schetzina, with permission.)

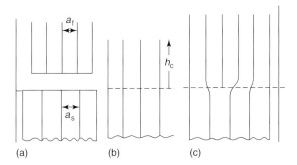

Figure 20.6 (a) Epitaxical growth of epilayer on a substrate with dissimilar lattice constants. (b) Pseudomorphic growth below critical thickness. (c) Formation of misfit dislocations above critical thickness. (After Mayer and Lau. [4])

20.2.2.3 Dislocations and Strain

Dislocation formation can be understood in terms of the buildup in strain energy, which depends on the *lattice mismatch f*, defined as

$$f = \frac{\Delta a}{a_{\text{avg}}} \tag{20.1}$$

where $\Delta a = a_f - a_s$ and $a_{\text{avg}} = (a_f + a_s)/2$. The strain energy will depend not only on f and the film thickness h but also on the specific crystal structure, elastic constants, and strain direction.

Assuming that the strain energy is taken up entirely in the epilayer, the elastic strain energy per unit volume is $(Y/2)f^2$, where Y is Young's modulus expressed in dyne per square centimeter or newton per square centimeter, the ratio of stress to strain for a given material. The strain energy σ_s per unit area is then

$$\sigma_s = \left(\frac{Y}{2}\right) hf^2 \tag{20.2}$$

for film thickness h. Hence, strain energy builds up quadratically with lattice mismatch [4].

On the other hand, the energy of a grid of edge dislocations Γ_e can be expressed as

$$\Gamma_e \simeq \alpha G b^2 \simeq \alpha Y b^2 \tag{20.3}$$

with the approximation $Y \simeq G$, which is the *shear modulus*. Here, α is a geometrical factor $\simeq 1$ and b is a *Burgers vector*, which expresses the magnitude and the direction of the lattice distortion associated with a dislocation. The Burgers vector for an edge dislocation is perpendicular to the dislocation line and is in the slip (or glide) plane of the dislocation. When a dislocation slips on its slip plane, the atoms move in the direction of the Burgers vector, approximately the same distance as a_f and a_s.

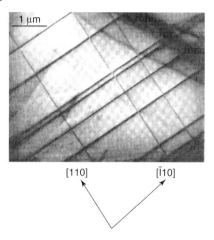

[110]   [1̄10]

Figure 20.7 Transmission electron micrograph of partially relaxed $In_{0.08}Ga_{0.92}As$ on InP(100) showing array of misfit dislocations in the interface plane [5].

Figure 20.7 illustrates a transmission electron microscopy (TEM) micrograph of a dislocation grid for 100 nm $In_{0.08}Ga_{0.92}As$ on InP(100) [5]. This film is partially relaxed with more relaxation in the [110] versus the [1̄10] direction.

Following Mayer and Lau [4], we assume a square dislocation grid with length D, such that the grid area is D^2 and the length of the dislocation associated with a square grid is $2D$ since each dislocation is shared by two grids. Then the dislocation energy σ_D per unit area is

$$\sigma_D = \frac{2D\Gamma_e}{D^2} = \frac{2\alpha Y b^2}{D} \tag{20.4}$$

Assuming that the dislocations take up the entire mismatch, the grid spacing D is then

$$D = \frac{a_{avg}}{f} \tag{20.5}$$

$$\sigma_D = \frac{2\alpha Y b^2 f}{a_{avg}} \simeq 2\alpha Y b f \tag{20.6}$$

where Burgers vector $b \simeq a_{avg}$. Hence, the dislocation energy builds up linearly with lattice mismatch f. Figure 20.8 shows a schematic plot of Equations 20.2 and 20.6 The strain energy is initially below the energy required to form dislocations. The dislocation energy increases linearly with b and f. Above the intersection of these curves, it is energetically favorable to form dislocations rather than to increase strain. At the intersection point, $\sigma_s = \sigma_D$, so that

$$h_c = \frac{4\alpha b}{f} \tag{20.7}$$

Equation 20.7 shows that the critical thickness increases as the lattice mismatch decreases. A derivation of critical thickness at a heterojunction that takes into

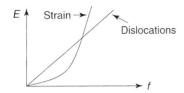

Figure 20.8 Schematic energy versus strain plot of energies due to strain or dislocation formation. Above the critical thickness h_c, the formation of dislocations is energetically favorable.

account the orientation of slip plane and slip direction, as well as the *Poisson ratio* $\mu = -\varepsilon_X/\varepsilon_Z = -\varepsilon_Y/\varepsilon_Z$ of lateral-to-longitudinal strain under uniaxial elastic deformation, is [6]

$$h_c = \frac{\left[b\left(1 - \mu \cos^2 \Theta\right) \ln\left(\left(\frac{h_c}{b}\right) + 1\right)\right]}{\left[2\pi f (1 + \mu) \cos \Theta\right]} \quad (20.8)$$

where Θ is the angle between dislocation line and b, Φ is the angle between slip direction and the direction in the film plane that is perpendicular to the line of intersection of the slip plane and the interface. For growth in the <100> direction, both Θ and Φ are at 45°. Note the similar b/f dependences of Equations 20.8 and 20.7.

The critical thickness h_c versus lattice mismatch based on Equation 20.8 is plotted in Figure 20.9 [7]. The solid line indicates h_c, the boundary between pseudomorphic, two-dimensional growth and the relaxed epilayer regime with buried two-dimensional arrays of glide misfit dislocations. For $f > \sim2\%$ and $h < h_c$, pseudomorphic growth at a monolayer scale becomes three-dimensional, that is, Stranski–Krastanov. For $f > \sim2\%$ and $h > h_c$, the relaxed layers have buried arrays of edge dislocations in addition to high densities of threading dislocations/stacking faults.

For $f < 1$–2%, this curve can be approximated as

$$h_c = \frac{a}{2f} \quad (20.9)$$

For example, Equation 20.9 yields $h_c = 294$ Å for $a_0(\text{InP}) = 5.87$ Å and $f = 1\%$, similar to the $h_c = 300$ Å value shown in Figure 20.9.

Figure 20.10 illustrates the generation of misfit dislocations from threading dislocations. In Figure 20.10a, a threading dislocation extends through an interface. In Figure 20.10b, shear forces at the interface between mismatched layers can displace or "bow" the dislocation in a direction within the interface plane. In Figure 20.10c, the shear displacement extends or "glides" the dislocation along the interface, generating a misfit dislocation in the plane of the interface. The Burgers vector shown is oriented out of the interface plane. A network of misfit dislocations appears in Figure 20.7.

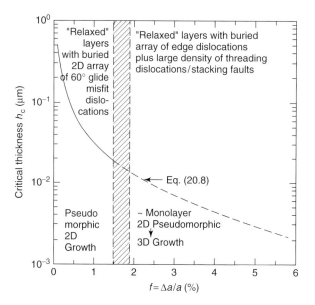

Figure 20.9 Critical thickness h_c versus lattice mismatch $f = \Delta a/a$ [6]. For $f \sim 1.5\%$ (dashed line), pseudomorphic two-dimensional growth continues up to a few monolayers, followed by three-dimensional growth. For $h > h_c$, dislocations and stacking faults form to relax the epilayer strain. (After Woodall et al. [7])

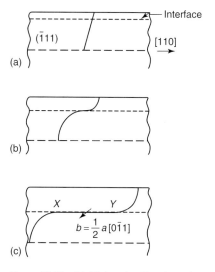

Figure 20.10 (a) Dislocation line through interface. (b) Shear-induced dislocation "bowing". (c) Dislocation "glide" along interface [6].

20.2.3
Two-Dimensional Electron Gas Heterojunctions

Crystal strain at unrelaxed heterojunctions can introduce desirable electronic effects. Among these are two-dimensional electron gas (2DEG) layers – sheets that combine high carrier densities with high carrier mobility. Figure 20.11 illustrates this 2DEG concept. Figure 20.11a shows the band lineup for the two semiconductors without strain. Dashed lines indicate the Fermi levels aligned within the semiconductors.

Lattice mismatch introduces strain within the outer semiconductor layer. If this material is piezoelectric, for example, AlGaN, the strain produces an electric field that bends the bands within the lattice-strained layer. See Figure 20.11b. In turn, this field induces charge at the free surface and at the heterojunction such that an accumulation layer forms inside the narrower gap semiconductor. Growth of these lattice-mismatched structures requires ultrathin film dimensions since critical thicknesses required to achieve maximum electric fields and 2DEG carrier densities are only a few tens of nanometers. Beyond this limit, dislocation formation degrades the 2DEG layer.

The high density of electrons within this layer is achieved without a correspondingly high density of dopant atoms, which would otherwise scatter carriers and reduce their mobility. The high mobility and current densities of these 2DEG layers in AlGaN/GaN heterostructures are particularly useful for high-power, high-frequency transistors in RF communications. The high mobility in nanoscale confinement structures also enables the observation of striking quantum phenomena such as the fractional quantum Hall effect.

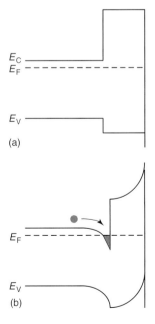

Figure 20.11 Heterojunction interface between a wide gap insulating semiconductor and a narrower gap n-type semiconductor. Piezoelectric strain introduces electric fields within the wide gap material that bend the bands in both materials such that charge accumulates at the heterojunction.

20.2.4
Strained Layer Superlattices

Strained layer superlattices (SLSs) offer advantages of higher critical film thicknesses and novel electronic structures. The superlattice consists of alternating epitaxial layer of two mismatched materials, each with a thickness $h < h_c$. Figure 20.12 illustrates such an alternating structure schematically [8]. Here, the two unstrained films grown separately have different lattice constants, but when joined, the smaller lattice constant material is under biaxial tension, whereas the large lattice constant film is under biaxial compression. The resulting superlattice has an average lattice constant that is intermediate between the two separate films. The alternating compression and tension within the film relaxes the lattice match condition such that the critical thickness can increase by greater than an order of magnitude.

Figure 20.13 demonstrates the increasing critical thickness on InGaAs grown on GaAs across a wide range of In composition [9]. At $x \sim 0.5$, h_c of the superlattice is $>10\times$ that of a uniform $In_{0.5}Ga_{0.5}As$ film.

Superlattices can be uniform in their layer composition and thickness or they can be graded. The graded layer shown in Figure 20.12 illustrates how it is possible to create a new growth template with a lattice constant significantly different from that of the substrate. Such graded layers can be used to "bridge" otherwise large lattice mismatches between a desired semiconductor and a required substrate.

20.2.4.1 Superlattice Energy Bands
The superlattices pictured in Figure 20.12 enable the creation of semiconductors with unique electronic properties. The electronic properties of single heterojunctions appear in a later section of this chapter. The electronic properties of superlattices depend on their layer widths as well as their individual layer

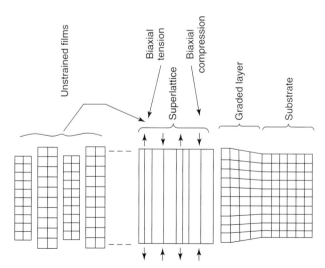

Figure 20.12 Alternating epitaxial layers of two mismatched materials, each with $h < h_c$ [8].

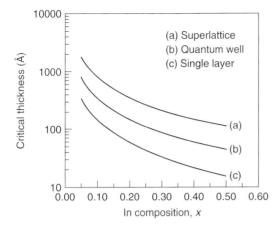

Figure 20.13 Critical thickness of InGaAs grown on GaAs for In composition $0.5 < x < 0.5$ as a (a) superlattice, (b) quantum well, and (c) single layer [9].

properties. Figure 20.14 shows schematic representations of the conduction and the valence-band structures for superlattices with "thick" versus "thin" layer widths.

The isolated films with nanometer-scale thicknesses have narrow energy levels corresponding to quantum confinement in the direction normal to the film growth direction. When joined as a superlattice, these quantum levels broaden into *minibands* and extend throughout the multilayer film. The miniband positions define new conduction and valence bands of the composite film with new effective bandgaps. For layer thickness of more than a few lattice constants, these minibands are relatively narrow. As the layer thickness decreases, the minibands broaden and the energy gap increases. Furthermore, the effective masses m_e^* and m_h^* decrease with decreasing layer width. These decreases in effective mass provide an avenue to increase carrier mobilities by the way of these man-made layer structures.

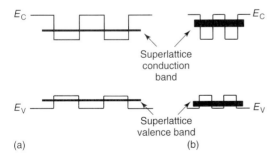

Figure 20.14 Individual layer conduction and valence-band edges versus energy bands of (a) a thick versus *a* (b) thin superlattice.

20.2.4.2 Strain-Induced Polarization Fields

The polarization fields induced by strain can be used to control the optical bandgap of superlattices. Figure 20.15 exhibits the conduction and the valence miniband energies of a $Ga_{0.47}In_{0.53}As–Al_{0.70}In_{0.30}As$ strained-layer superlattice with equal numbers of molecular layers N_a and N_b within each layer [10]. Here, the *polarization electric field* $F_i = -(\varepsilon_{14}\varepsilon_{jk})/(\varepsilon_0\varepsilon)$, where ε_{jk} is the symmetrized strain component, ε_{14} the piezoelectric constant, ε the low frequency dielectric constant, and ε_0 the free space permittivity. The inset in Figure 20.15 shows the tilted conduction and the valence bands of the constituent layers ($E_G(Ga_{0.47}In_{0.53}As) < E_G(Al_{0.70}In_{0.30}As)$) versus distance in the presence of a polarization field. As in Figure 20.14, the superlattice bandgap decreases with increasing layer thickness.

Optical illumination with $h\nu > E_G(Ga_{0.47}In_{0.53}As–Al_{0.70}In_{0.30}As)$ creates free electron–hole pairs that screen the electric field. Without these free carriers,

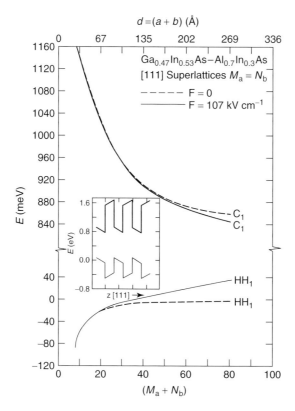

Figure 20.15 The calculated energies of conduction (C_1) and heavy hole valence-band (HH_1) levels as a function of superlattice layer thickness. $M_a = N_b$ are the numbers of molecular layers of materials *a* and *b* in a superlattice cycle for $Ga_{0.47}In_{0.53}As–Al_{0.70}In_{0.30}As$ strained-layer superlattice. Inset shows conduction and valence bands for the individual constituents in real space. Band flattening increases the bandgap as shown by the dashed lines for C_1 and HH_1. (After Mailhiot and Smith. [10])

$F_i \sim 10^5 \, \text{V cm}^{-1}$ for $f = 1\text{--}2\%$ in III–V compound semiconductors in general. Such high fields produce an effective narrowing of the superlattice bandgap by the *Franz–Keldysh effect*.

The optically induced free carriers act to screen the polarization fields in each layer. As a result, the bandgap increases with reduced polarization field as shown by the dashed lines for C_1 and HH_1 in Figure 20.15. This effect is extremely useful in optoelectronic devices such as nonlinear optoelectronic switches. Thus, bandgap illumination prepares the state of a polarized superlattice to either absorb or pass a second light beam. Such structures can serve as logic elements in optoelectronic circuits. The bandgap employed in these optoelectronic devices can be tailored by (i) layer composition, (ii) layer width, and (iii) strain, providing significant latitude in their design.

20.3
Chemical Structure

The chemical structures of semiconductor–semiconductor interfaces can be distinct from those of their bulk constituents because new chemical phases can form and/or because the epitaxial growth can vary. Interdiffusion and chemical reaction as well as interfacial template structures highlight the key role of local bonding and interface preparation on the resultant interface properties.

20.3.1
Interdiffusion

Analogs to the chemical structure of adsorbates on semiconductors described in Chapter 18, several chemical phenomena can act to render the semiconductor heterojunction less than atomically abrupt. In principle, lattice-matched heterojunctions provide atomically abrupt and ordered interfaces and physical properties derived solely from their bulk constituents. Nevertheless, numerous semiconductor classes of heterostructures exhibit interdiffusion and chemical reaction, particularly at temperatures elevated above those required for epitaxial growth.

20.3.1.1 IV–IV Interfaces
The elemental Ge–Si interface exhibits chemical features that illustrate the thickness and the temperature dependences of interdiffusion. Although monolayer adsorption leads to ordered overlayers (11–13), thicker interfaces exhibit significant interdiffusion for temperatures above $\sim 200\,°\text{C}$, as well as islanding and surface roughening at even higher temperatures [14].

For Ge on Si, a Ge–Si alloy layer shown in Figure 20.16a forms at their interface at a temperature of $350\,°\text{C}$ [15]. For Ge/Si superlattices, the activation energy for this interdiffusion is $3.1 + 0.2\,\text{eV}$ in the range $640\text{--}780\,°\text{C}$ temperature range. For Ge-rich GeSi alloy films on Si, Ge segregation occurs into Si.

20.3.1.2 III–V Compound Heterojunctions

Diffusion is also common at III–V compound semiconductor heterojunctions. Ge growth on GaAs at temperatures of 320–360 °C yields atomically abrupt interfaces [16]. At higher temperatures, diffusion begins to occur. For example, at 650 °C, Ge diffuses hundreds of nanometer into GaAs, reaching concentrations of 10^{17}–10^{18} cm^{-3} [17]. Figure 20.16b illustrates this schematically. Ge diffusion into GaAs is important since Ge acts as a p-type dopant. Even at low impurity concentrations of only parts per million, this diffused impurity can easily exceed bulk crystal doping levels. Its diffusion length at elevated temperatures may even exceed the dimensions of micron-scale electronic devices. Conversely, by controlling the depth of diffusion with anneal time and temperature, one can employ solid-phase regrowth to create degenerately-doped GaAs surface layers and shallow depth ohmic contacts to GaAs and other III–V compound semiconductors [18].

Interdiffusion at heterojunctions is dependent on composition. For example, AlAs–GaAs, AlGaAs, and GaAsP–GaAs superlattices interfaces grown by organometallic chemical vapor deposition (OMCVD) at 700–750 °C are atomically abrupt as measured by secondary ion mass spectroscopy (SIMS). On the other hand, Auger electron spectroscopy (AES) measures interdiffusion of GaAs–Al$_x$Ga$_{1-x}$As–GaAs heterojunctions grown by molecular beam epitaxy (MBE) at temperatures of 850–1100 °C [19]. This diffusion decreases with increasing Al content.

Furthermore, intermixing of semiconductor superlattices increases significantly in the presence of dopant impurities, such as Zn [20]. This impurity alters the defect statistics of the bulk crystal, which in turn dominate atomic diffusion through the semiconductor lattice. Zn implantation can be used to create alloyed patterns across superlattice films. This *impurity-induced disorder* creates regions with different bandgaps that are used for carrier and optical confinement in vertical cavity surface emitting lasers (VCSELs).

The presence of misfit dislocations at heterojunctions can also enhance interdiffusion since the lattice disruptions at dislocations provide pathways for atoms to diffuse with lower activation energy than in the perfect crystal.

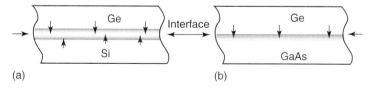

Figure 20.16 (a) Interdiffusion at Ge–Si heterointerface. A Ge–Si alloy layer forms at their junction. Arrows indicate the movement of both Ge and Si. (b) Ge diffusion into GaAs.

20.3.2
Chemical Reactions

Chemical reactions can occur at semiconductor–semiconductor interfaces, producing monolayers of new compounds at the intimate heterojunction. For example, TEM and in situ X-ray photoemission spectroscopy (XPS) studies reveal the formation of a zinc blende Ga_2Se_3 interfacial layer at the ZnSe–GaAs interface [21, 22]. Analogous reactions occur at the CdTe/InP heterojunction, where XPS measurements show the formation of In–Te layers whose thickness depends on substrate's temperature [23]. A high Cd overpressure suppresses this interfacial reaction during growth [24]. The degree to which such interface compounds form depends primarily on the availability of constituents at the intimate heterojunction as well as on the rate of atomic diffusion from the substrate and within the growing film.

Interlayers of compounds may be introduced at heterojunctions to retard interdiffusion and undesirable chemical reaction. For example, a monolayer-thick AlAs layer grown on GaAs before ZnSe overgrowth lowers strain, reduces Zn and Ga interdiffusion, and improves the epilayer crystal quality.

The growth of epitaxial semiconductor layers entails a trade-off: high-quality crystal growth requires elevated substrate temperatures while at the same time minimizing thermally activated diffusion and chemical reaction.

20.3.3
Template Structures

Interlayers have a variety of uses at semiconductor–semiconductor heterojunctions including the following: (i) "bridge" layers, (ii) monolayer films for substrate bond passivation, (iii) specific crystal orientations, (iv) monolayer surfactants for improved morphology, and (v) specific atomic structure and stoichiometry to control dipole fields.

20.3.3.1 Bridge Layers

"Bridge" or *buffer* layers of changing composition and lattice constant are used to create gradual transitions in epitaxial growth between semiconductors of significant differences in lattice constant. These bridge layers are typically hundreds of nanometers thick. They may involve *pseudobinary* alloys of the two semiconductors (changing composition with thickness). An example is $Ga_xIn_{1-x}As$ with compositions varying from $0 \leq x \leq 1$ to bridge the 7% lattice mismatch between GaAs and InAs [25]. Bridge layers also include combinations of new materials with intermediate lattice properties analogs to superlattices. For example, layers of fluoride structure BaF_2 on CaF_2 minimize the 9% lattice mismatch between the chalcogenide PbTe with cubic rocksalt structure versus elemental Si with diamond zincblende structure [26]. Another multilayer example is the growth of barium hexaferrite or BaM, $BaFe_{12}O_{19}$ on SiC using interwoven layers of BaM and rocksalt structure MgO on wurtzite SiC [27].

20.3.3.2 Monolayer Passivation

Monolayers can promote the growth of different epitaxical variants. For example, the initial coverage of Ni on Si(111) controls which epitaxial variant of $NiSi_2$ forms [28]. See Figure 18.4. Monolayer adsorbates are also effective in promoting heteroepitaxial growth between two semiconductors. This is evident from the dependence of epilayer crystal quality on substrate stoichiometry and reconstruction.

Overcoming the lattice mismatch of GaAs growth on Si has been an important goal for integrating III–V optoelectronics on Si microelectronics. However, atomic-scale studies reveal the complexity of this interface. Thus, TEM and SXPS measurements of GaAs grown on Si show that the lattice mismatch, the transition between covalent and partially ionic bonding, and the sublattice selection all influence the mode of GaAs overlayer growth [29]. In this case, SXPS core-level measurements reveal the predominance of Si–As bonding at the epitaxial GaAs on Si(111) interface, suggesting a Ga–As double layer atop the As-terminated Si(111) surface. However, the polar bonding required is energetically unstable [30], a consideration to be discussed later in this chapter. Indeed, total energy calculations reveal that an interface consisting of three-fourths layer Si above one Al layer and one Ga layer yields a stable structure and ARPES results indicative of a Ga–Si exchange reaction support this atomic geometry [31]. For GaAs on the Si(100) surface, there is additional complexity involving mixed compositions in islands and interfacial layers [29].

20.3.3.3 Crystal Orientations

Substrate crystal orientation can strongly affect the quality and nature of heterojunction growth. It is believed that the superior growth morphology of GaAs(110) on Ge is due to the absence of surface reconstructions at (110) interfaces due to (i) the energetic instability of interfaces other than (110) interfaces that must reconstruct [30] and (ii) the free-surface reconstruction of the diamond zincblende (110) surface favors subsequent growth of zincblende structures without antiphase domain boundaries [32]. The high fields associated with interface dipoles at non-(110) surfaces can disrupt heteroepitaxial growth and lead to rough interfaces with poor electrical properties.

Surface misorientation provides additional possibilities for the growth of quantum structures. By controlling the nucleation and growth kinetics with epitaxy on intentionally misoriented surfaces, it is possible to create laterally periodic changes in a surface atomic composition and so-called tilted superlattices. Figure 20.17 illustrates two types of quantum-scale structures obtained by growth on vicinal surfaces. In Figure 20.17a, a surface tilted only a few degrees θ away from a low-index plane provides a series of wide terraces and steps with monolayer height. Overlayers grown on these terraces produce lattices with periodicities that vary in a direction normal to the tilt plane [33].

The step edges associated with such vicinal surfaces can provide nucleation sites for quantum wires. Figure 20.17b shows such a quantum semiconductor structure in which each terrace supports only single monolayers of the compound semiconductor. Although quantum wires such as carbon nanotubes or ZnO

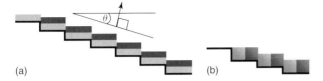

Figure 20.17 (a) Tilted superlattice structure. (b) Quantum wire structure. The two colors signify cation and anion lattice planes.

nanorods can also be grown freestanding by vapor deposition techniques, epitaxial growth enables the growth of such wires at single monolayer dimensions with controlled composition and orientation.

20.3.3.4 Monolayer Surfactants

Monolayer adsorbates that "float" on the growing epitaxial film can passivate the bonding not only at the intimate semiconductor–semiconductor interface but also within the film itself. Figure 20.18 shows an As monolayer deposited on Si(001) under an As overpressure that facilitates dislocation-free growth of epitaxial Ge films [34]. The As layer suppresses the tendency of Ge on Si to form islands, the edges of which can promote misfit dislocations reaching down to the substrate as they coalesce [7]. In its place, the As film promotes a uniformly distributed, unconnected set of novel defects that relieve the strain in the growing film. The surfactant makes it possible to grow epitaxial films free of misfit dislocations above the critical thickness shown in Figure 20.9.

20.3.3.5 Dipole Control Structures

Specific atomic structures and layer compositions can control the otherwise large electrostatic fields that can form with the epitaxial growth of heterovalent semiconductor-semiconductor interfaces. The growth of II–VI semiconductors on III–V compound substrates is of particular electronic interest. For example, n-type ZnSe growth on p-type GaAs enables p–n junctions with bandgaps in the blue and ultraviolet spectral range. As with the choice of crystal orientation, the dipole fields can be avoided by (i) graded or mixed interfaces with equal numbers of III–VI and II–V bonds (i.e., Ga–Se and Zn–As bonds for ZnSe on GaAs) or (ii) specific planar reconstructions that achieve the same mixed bonding, for example, by equating the number of Ga- and As-based interface bonds in the case of GaAs substrates [35]. Such reconstructions require careful control of substrate temperature and the ratio of constituent atom fluxes during deposition.

Figure 20.18 As monolayer surfactant "floating" on Ge film as it grows on Si(100).

20.4
Electronic Structure

20.4.1
Heterojunction Band Offsets

The electronic properties of semiconductor–semiconductor interfaces have enabled a wide range of fundamental physical phenomena and state-of-the-art electronic device technology. With the advent of epitaxical growth techniques, these engineered crystal structures have led to novel band structures with features such as quantum confinement, ballistic transport, and record-breaking carrier mobilities. These features can be applied to high-speed, high-power transistors, ultrasensitive photodetectors, and lasers extending in wavelengths from the far infrared to the ultraviolet spectral range.

A central feature responsible for these physical phenomena and device applications is the band offset between the semiconductors. In turn, the band offset along with the band bending within each surface space charge region strongly depends on the interface dipoles and localized states already discussed in previous chapters. Finally, these dipoles and localized states involve the specific geometric and chemical properties of the junction on an atomic scale.

Here, we describe the following: (i) the band structure at the semiconductor–semiconductor interface and the contributions of interface dipoles and band bending, (ii) the techniques employed to measure heterojunction band offsets, (iii) the various interface dipole models developed to account for these measurements, and (iv) the atomic-scale technique now available to modify or control heterojunction band offsets.

Figure 20.19 shows a schematic energy band diagram of a straddling semiconductor energy band lineup. Band bending contributes to the variation of E_C and E_V across the interface. Here, semiconductor A has an n-type (upward) band bending due to ionized impurities within the depletion region, while semiconductor B has a p-type (downward) band bending due to charge accumulated in an inversion region.

The single-headed arrows denote the sign convention for the ΔE_C and ΔE_V discontinuities. In this figure, ΔE_C is positive when the conduction-band edge of semiconductor A is above that of B. ΔE_V is positive when the valence band of A is below that of B. Hence, ΔE_C and ΔE_V shown in Figure 20.19 are both positive. The convention used here is that $E_G(A) > E_G(B)$ so that

$$\Delta E_G = E_G(A) - E_G(B) \tag{20.10}$$

In general, there are a wide range of template structures available to modify the epitaxical semiconductor–semiconductor growth. These template structures are emblematic of the role that atomic-scale interface features have on the macroscopic semiconductor properties.

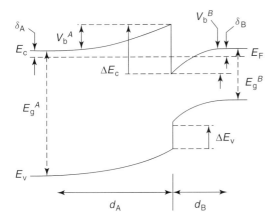

Figure 20.19 Schematic energy diagram of a semiconductor heterojunction, showing conduction-band minimum E_C and valence-band maximum E_V as functions of position across the interface between two n-type semiconductors with different energy bandgaps $E_G(A)$ and $E_G(B)$. δ_A and δ_B are the corresponding bulk $E_C - E_F$ separations from the majority carrier band edges. Single-headed arrow conventions shown are positive from the larger (A) to the smaller (B) gap semiconductor.

is > 0 and

$$\Delta E_G = \Delta E_C + \Delta E_V \tag{20.11}$$

The vacuum levels of the two semiconductors are, in general, not aligned after they join. Instead, they are shifted by the built-in potential V_{bi}, which depends on the difference in semiconductor surfaces and interface dipoles. This built-in potential is equal to the sum of the individual band bendings V_b^A and V_b^B, analogs to the built-in potential of a p–n homojunction:

$$V_{bi} = V_b^A + V_b^B \tag{20.12}$$

The band edge energies as a function of position, as shown in Figure 20.19, follow from the individual band bendings V_b^A and V_b^B and their depletion widths as calculated from Poisson's equation (Equation 3.4). A complete calculation of heterojunction band diagrams involves ΔE_C and ΔE_V as well as V_b^A and V_b^B, δ_A and δ_B from the semiconductor doping levels, and the dielectric constants ε_A and ε_B. See, for example, [1]. Approximate band diagrams can be drawn based on values of the band offsets, the semiconductor doping levels, and the dielectric constants [36].

The optimal cases for calculating or estimating the band discontinuities are based on lattice-matched junctions with low interfacial strain. Lattice-mismatched heterojunctions induce interfacial strain and crystal imperfections that complicate these calculations. Nonabrupt changes in composition at the interface produce graded band structure features whose spatial variation can be calculated by superimposing band offsets for the spatially dependent composition onto the potential across the band bending regions and the built-in potential.

The band alignment between the two semiconductors involves the following: (i) the dipoles V_S at the surfaces of the individual semiconductors, (ii) the voltage drops V_b across each semiconductor's band-bending region, and (iii) an interface dipole V_i between the two semiconductors after contact. To account for the surface dipoles, we introduce an internal potential S_S for each semiconductor such that $\chi = V_S + S_S$ and

$$S_S^A = \chi_A - V_S^A \tag{20.13a}$$
$$S_S^B = \chi_B - V_S^B \tag{20.13b}$$

The internal potential S_S allows us to express the surface dipole in terms of the electron affinity χ, a macroscopic measurable parameter. Equation 20.13 are represented schematically in Figure 20.20a,b. S_S is characteristic of the bulk, whereas V_S depends on the surface atomic composition and bonding.

With E_F as a common reference level, one can use the internal potentials to equate the voltage drops on either side of the interface [37]. Thus,

$$S_S^A + V_b^A + E_C^A - E_F = S_S^B + V_b^B + E_C^B - E_F + V_i \tag{20.14}$$

Rearranging terms and using Equation 20.13,

$$\Delta E_C^S = E_C^A - E_C^B = (\chi_B - \chi_A) + (V_i + V_b^B - V_b^A + V_S^A - V_S^B) \tag{20.15}$$

And finally,

$$\Delta E_C^S = E_C^A + V_b^A - (E_C^B + V_b^B) \tag{20.16}$$
$$= (\chi_B - \chi_A) + (V_i + V_S^A - V_S^B) \tag{20.17}$$

Equation 20.16 expresses the conduction-band offset in terms of the difference between the two semiconductors of their bulk conduction band in addition to their surface band bending. This difference has the two components shown in Equation 20.17 – the difference in electron affinities plus a term that is just the difference

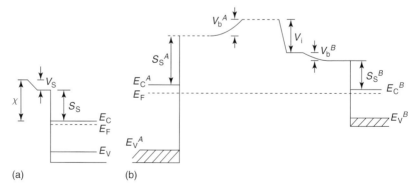

Figure 20.20 Schematic heterojunction band diagram illustrating the voltage drops across (a) an individual semiconductor surface and (b) the entire interface. (After Mailhiot and Duke [37], with permission.)

between the interface dipole and the two surface dipoles of the separated semiconductors. The electron affinities and their difference are measurable quantities. However, the dipoles are not. One extracts their values only from their effects on the measurable quantities χ_B, χ_A, V_b^A, and V_b^B.

For the case in which the dipole term is 0, Equation 20.17 reduces to just the difference in electron affinities. This relation is known as the *electron affinity rule* [38] and is commonly used to approximate band alignments. However, experimental measurements of band offsets show that this classical relation does not predict band offsets accurately. Therefore, we can conclude that the dipole term in Equation 20.17 makes a significant contribution to the heterojunction band offset.

20.4.2
Band Offset Characterization

The measurement of heterojunction band offsets involves many of the same macroscopic and microscopic techniques used to characterize the metal–semiconductor interface. Table 20.2 provides a comparison among the various techniques used to measure heterojunction band offsets.

20.4.2.1 Macroscopic Electrical and Optical Methods

Macroscopic measurements involve charge transport, optical excitation, or both. In the case of transport methods, C–V measurements can extract the built-in potential V_{bi}, assuming a square-root dependence for C on $V_{bi} - V$, as with Equation 3.16 One can obtain ΔE_C and ΔE_V by combining V_{bi} with δ_A and δ_B.

Thus, equating energies on either side of Figure 20.19, one obtains

$$\Delta E_C - V_b^A - \delta_A = V_b^B - \delta_B \qquad (20.18)$$

and since the built-in potential $V_{bi} = V_b^A - V_b^B$,

$$\Delta E_C = V_{bi} + \delta_A - \delta_B \qquad (20.19)$$

Table 20.2 Macroscopic and atomic-scale methods for measuring band offsets along with their limitations.

Technique	Measured parameter	Limitations/requirements
C–V	V_{bi}	Impurity gradients, near-interface traps, nonohmic contacts
J–V	V_{bi}	Same as C–V plus recombination, tunneling, shunt currents
IPS	$J_{Threshold}$	Intense, monochromatic near-IR light
PL	ΔE_{QW}	Self-consistent fit to model
Raman	V_b, n, E_{QW}	Self-consistent fit to model
STM	$J_{Threshold}$	Cross-sectional geometry in UHV
SXPS	ΔE_{core}, ΔE_V	XPS at UHV growth interface, line shape fits

Within each semiconductor, the free carrier concentration $n = |N_d - N_a|$, where N_D and N_A are donor and acceptor densities, and N_C is the conduction-band density of states (DOS). Then,

$$\delta_A = kT \ln\left(\frac{n_A}{N_C^A}\right) \tag{20.20a}$$

$$\delta_B = kT \ln\left(\frac{n_B}{N_C^B}\right) \tag{20.20b}$$

and

$$\delta_A - \delta_B = kT\left[\ln\left(\frac{n_A}{n_B}\right) + 1.5\ln\left(\frac{m_B^*}{m_A^*}\right)\right] \tag{20.21}$$

where m_B^* and m_A^* are the effective electron masses in the two semiconductors. Since the free carrier concentrations and effective masses are known, ΔE_C follows in a straightforward way from the measurement of V_{bi}. For a heterojunction with B a p-type semiconductor, Equation 20.19 becomes

$$\Delta E_C = V_{bi} - E_G^B + \delta_A + \delta_B \tag{20.22}$$

The built-in voltage V_{bi} follows from the capacitance, which is obtained from Poisson's equation in one dimension. Requiring that the electric displacement be continuous across the heterojunction boundary, one obtains

$$\varepsilon_A \varepsilon_A = \varepsilon_B \varepsilon_B \tag{20.23}$$

for dielectric permittivity and electric field ε. Considering the heterojunction as two capacitors in series and with the depletion approximation, the capacitance per unit area is [39]

$$C^2 = \frac{[qN_A N_B \varepsilon_A \varepsilon_B]}{[2(\varepsilon_A N_A + \varepsilon_B N_B)(V_{bi} - V)]} \tag{20.24}$$

for an applied bias V. ΔE_C follows from Equation 20.19 using the intercept value of V_{bi} in a plot of $1/C^2$ versus V.

This analysis is complicated by any impurity gradients or near-interface charge [39]. Among such capacitance techniques, the most reliable are measurements of interface charge at the heterojunction with known doping profiles [32]. Likewise, carrier recombination, tunneling, and shunt currents can seriously affect the interpretation of I–V measurements.

Various trap spectroscopies are available to gauge the localized states near the heterojunction that contributes to the capacitance. A capacitance method coupled to thermal excitation such as *deep-level transient spectroscopy* (DLTS) gauges the activation of carriers out of trap states and the resultant capacitance changes [40]. An optical variant of this technique, DLOS, enables carrier promotion from traps at higher energies from the band edges of WBG semiconductors [41].

Photoexcitation combined with transport is the basis for internal photoemission spectroscopy (IPS) of heterojunctions. Analogous to the IPS technique for measuring Schottky barriers described in Chapter 14 , the incident photons excite electrons from a band populated with carriers in one semiconductor to an empty band in the other, which produces measurable photocurrent. The threshold energy for conduction marks the minimum transition energy across the junction. Figure 20.21a illustrates the photoexcitation of electrons from the valence band of one semiconductor to the conduction band of the other. In this case, photon energies exceeding the smaller bandgap excite the transitions.

Internal photoemission from the conduction band of an accumulated to an empty conduction band also measures ΔE_C. In Figure 20.21b, the excitation of electrons from the accumulation layer in the smaller bandgap to the conduction band of the larger gap semiconductor requires intense, monochromatic infrared photons with energies $h \sim 0.1$–0.5 eV. This requirement can be satisfied with, for example, free electron lasers [42]. The IPS methods typically require special material structures, for example, superlattices, so that optical excitation across the interface is the dominant contribution to photoconduction. However, it is the most direct of the transport techniques.

Optical techniques involving photons both in and out of the heterojunction include absorption, luminescence, and light scattering spectroscopy. Optical absorption measures transitions between energy subbands within heterojunction quantum wells. The depth of these quantum wells is defined by the conduction-band discontinuities between larger gap semiconductors on either side of a smaller gap material [43]. Optical emission between these subbands provides luminescence spectra that yield energy-level positions from a self-consistent fit to a model of the quantum well based on its depth and width. Luminescence features versus hydrostatic pressure are sensitive to states on either side of the heterojunction [44], as are indirect transitions versus electric field for staggered heterostructures [45]. The polarization and intensity dependence of intersubband transitions in light scattering can also gauge band offsets [46]. However, all these optical techniques are indirect methods since they require a self-consistent calculation of the quantum-confined states and, by induction, the well depths and hence band offsets. As with the IPS techniques, these light scattering techniques must be

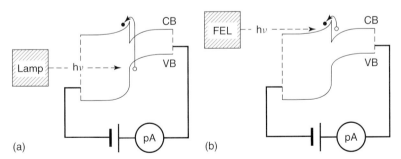

Figure 20.21 Internal photoemission at semiconductor heterojunctions involving (a) valence-to-conduction band transitions from the smaller to the larger bandgap semiconductor and (b) conduction-to-conduction band transitions from the accumulation layer of the smaller gap material to the larger gap semiconductor. In (b), free electron laser (FEL) provides intense infrared light to promote the interconduction band transitions [42].

used with multilayer structures whose features depend on thickness and which are geometrically abrupt, chemically uniform, and free of electronic defects.

Therefore, each of these macroscopic transport and optical techniques has drawbacks that can complicate the measurement of band offsets, particularly if chemical interactions and localized electronic states are present.

20.4.2.2 Scanned Probe Techniques

Scanning tunneling microscopy (STM) techniques can also gauge heterojunction band offsets. Analogous to BEEM measurements described in Chapter 14, a tunneling tip scans across a cross-sectioned specimen that exposes the heterointerface. As shown in Figure 20.22a, electrons from a metal probe tip are injected into the conduction band of a semiconductor. The current–voltage threshold V_{bias} for injection identifies $E_C - E_F$ in a semiconductor. Comparison of these threshold biases between the two sides of the junction yields ΔE_C as shown in Figure 20.22c. See, for example, the GaAs/AlAs band offset measurement by Albrektsen et al. [47]. Abraham et al. found that the onset of injection can also induce light emission as illustrated in Figure 20.22b [48], providing an inverse photoemission method to image the difference in conduction band at the heterojunction.

Similar to cross-sectional STM techniques, the focused electron beam of an ultrahigh vacuum (UHV) scanning electron microscope (SEM) can scan across a UHV-cleaved heterojunction to yield changes in secondary electron threshold (SET) as the vacuum level E_{VAC} changes relative to E_F. The onset of secondary electron emission from a solid, referenced to the Fermi level of an electron energy analyzer E_F^A, is a measure of the work function. In turn, the difference in E_{VAC} across the interface, corrected for the difference in electron affinities, is a measure of ΔE_C. Figure 20.23 illustrates the measurement of $E_{VAC} - E_F$. As with the secondary electron cut-off energy in photoemission (Chapter 7), only electrons with energies above E_{VAC} can escape the solid. An applied bias voltage V_{bias} insures that $E_{VAC}^{SC} > E_{VAC}^A$. Note that $E_F^{SC} = E_F^A$ since the analyzer and semiconductor are in electrical contact. Hence, the kinetic energy ΔE measured by the analyzer plus its work function Φ^A is equal to $E_{VAC} - E_F$ of the semiconductor.

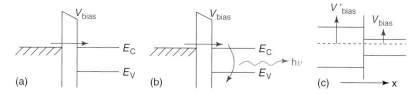

Figure 20.22 STM methods for measuring heterojunction band offsets. (a) Electron injection versus bias voltage measures $E_C - E_F$. (b) Electron injection induces light emission. (c) A scan across the interface yields the difference in thresholds for light emission and hence the relative E_C positions.

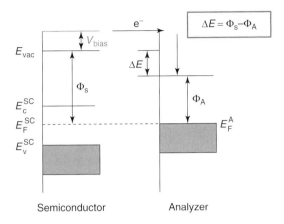

Figure 20.23 Principle of the SET work function measurement.

Recent studies of cleaved III–V compound heterojunctions have employed SEM beams to measure the work function across the heterojunction with nanometer-scale resolution [49]. These studies show that MBE growth of InGaAs on InP depends on specific growth conditions at the transition between InP termination and InGaAs growth. This transition requires an overlap of P and As beams in order to maintain a smooth interface. Relatively short overlap periods (20–40 s) yield abrupt interfaces as measured by SIMS, whereas extended overlap periods (180 s) result in broadened interfaces.

Figure 20.24 shows that ΔE_C is 0.2 eV larger for the abrupt interface, consistent with the 0.23 eV value measured by C–V [50]. This broadening has a large effect on ΔE_C, lowering the 0.2 eV offset in 20.23b to nearly zero in (a). The experimental

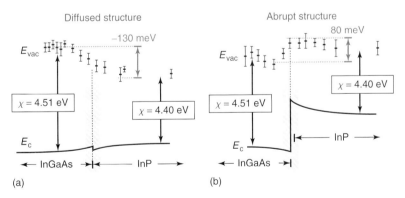

Figure 20.24 SET measurements of E_{VAC} of an InGaAs/InP junction with (a) and without (b) an extended anion soak at the onset of InGaAs growth. Without the extended anion soak, the interface is abrupt and has much larger ΔE_C (0.2 eV). SIMS depth profiles reveal interfacial broadening with increasing soak time that decreases the effective InGaAs–InP ΔE_C [49].

values of E_{VAC} versus position normal to the interface also follow the expected band bending within the two materials.

Both the STM and SEM techniques represent powerful tools to measure band offsets at fully formed heterojunctions directly, notwithstanding the difficulties of cleaved interface preparation and band bending normal to the cleaved surface.

20.4.2.3 Photoemission Spectroscopy Techniques

Photoemission spectroscopy is capable of measuring heterojunction band offsets during the initial stages of interface formation. This technique combines measurement of both core levels and valence band to monitor the relative energy displacement in the densities of states as the first few monolayers of semiconductor overlayer form on the substrate [51]. Figure 20.25 illustrates how the difference in known core-level energies for two semiconductors is translated into the difference of their valence-band edges.

First, XPS spectra provide the energy separation $E_{CL} - E_V$ of a given core level from that semiconductor's valence band for A and B separately. Second, one measures the same core levels for semiconductor A on B (or vice versa) to obtain ΔE_{CL}. Finally, the relation

$$\Delta E_V = (E_{CL}^B - E_V^B) - (E_{CL}^A - E_V^A) + \Delta E_{CL} \qquad (20.25)$$

yields the valence-band offset. The conduction-band offset follows from Equation 20.10

This technique requires a well-defined determination of ΔE_V from a lineshape fit of the XPS features to a theoretical valence-band DOS [52] and a deconvolution of core-level lineshapes to distinguish bulk from chemically shifted components.

Another photoemission approach to measure valence-band offsets involves measuring the composite valence-band spectra of the overlayer plus the substrate.

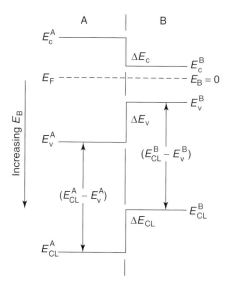

Figure 20.25 Schematic energy band diagram for a semiconductor heterojunction. XPS spectra provide the valence-band offset from core-level binding energies and valence-band edge energies.

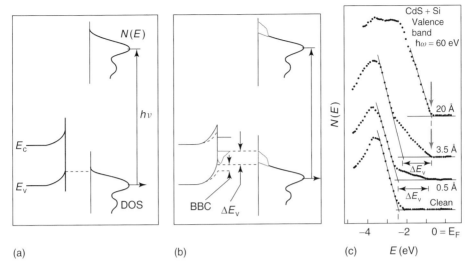

Figure 20.26 Schematic energy band and density-of-states diagram for photoemission measurement of ΔE_V (a) before and (b) after depositing a second semiconductor over the clean surface of the initial semiconductor. (c) A double-edged valence-band spectrum for amorphous Si on CdS(10$\bar{1}$0) [54].

In this method, the band offset follows from the separation of valence-band edges in the double-edge valence-band spectrum [53]. Figure 20.26 shows the energy band diagram of the clean semiconductor substrate as well as the initial valence-band DOS. Photoexcitation with energy $h\nu$ produces the energy distribution curve $N(E)$ shown schematically in Figure 20.26a.

Figure 20.26b illustrates the change in the schematic energy band diagram and valence-band DOS with several monolayers or less of a second semiconductor. This overlayer has a valence-band edge that lies within the bandgap of the substrate. BBC denotes the band-bending component of the valence-band spectrum. The difference in valence-band onsets in the energy distribution curve is equal to ΔE_V.

Figure 20.26c shows an example of measured double-edge spectra. Here, amorphous Si is deposited on a UHV-cleaved CdS(10$\bar{1}$0) surface. The Si valence-band edge is visible with the first 0.5 Å of coverage. This feature increase in magnitude with thickness and dominates the full spectrum at $\Theta = 20$ Å [54].

This method provides a direct measurement of ΔE_V for clear double-edged spectra. If such features are not distinct, only indirect measurements are possible using shifts of the composite features' leading edge and taking band bending into account. Photovoltaic effects that flatten the bands can simplify such analyses of band bending.

The precision of these photoemission techniques is limited by the XPS energy resolution of 0.1–0.2 eV versus $\Delta h\nu = 0.01$ eV of the optical techniques. Nevertheless, photoemission energy shifts can provide significantly better accuracies than the nominal XPS resolution.

20.4.2.4 Band Offset Results

The various techniques described here have provided heterojunction band offset values for many combinations of elemental, III–V, and II–VI compound semiconductors. These results are compiled in Table 20.3. This table lists average values and the techniques used for band offsets from multiple techniques applied to the same heterojunction. Prefixes *a-* and *c-* denote amorphous and crystalline materials, respectively. These values form a database on which various theoretical models of heterojunction band offsets and the contribution of interface dipoles can be evaluated.

Measurements of heterojunction values for WBG semiconductors such as SiC, ZnO, GaN, AlN, InN, and their nitride alloys are only now becoming available. Figure 20.27 illustrates these approximate band lineups from an estimate of their band lineups based simply on electron affinities. This figure reflects the latest E_G value of 0.7 eV for InN. Note the extremely large ΔE_C and ΔE_V values between WBG semiconductors with Si and GaAs. In situ photoemission measurements of band offsets for epitaxial nitride and carbide heterojunctions are particularly challenging since very high temperatures are required during growth of these materials.

In general, the accuracy and reliability of these various band offset measurements can differ greatly, not just between different experimental techniques but even for measurements by the same method. Thus, techniques may have a precision of tens of meV, yet measurements of the same heterojunction by different researchers using the same technique can yield results differing by hundreds of meV. These discrepancies can be traced to variations in experimental procedure and sample interfaces as well as to uncertainties associated with the models underlying these experiments. Margaritondo and Perfetti estimate the typical accuracy of these measurements to be +0.1 eV [53]. Such precision is, in fact, not adequate to distinguish between theoretical models for specific interface structures.

20.4.3
Interface Dipoles

The data presented in the previous section permit us to assess how important the dipole contribution in Equation 20.17 is to heterojunction band offsets in general. This dipole term represents the difference between the electron affinity difference and the measured conduction-band offset for a given semiconductor couple.

20.4.3.1 Inorganic Semiconductors

For inorganic semiconductors, the data are summarized in Figure 20.28 (16, 88). The large deviations from the $\Delta E_C = \chi_A - \chi_B = \Delta\chi$ line demonstrates that there is no systematic trend. In general, therefore, (i) the "electron affinity rule" [38]

Table 20.3 Heterojunction valence and conduction-band discontinuities from photoemission (PH), capacitance–voltage (C–V), current–voltage (I–V), internal photoemission (IPS), photoluminescence spectroscopy (PL), or other techniques (OT).

Interface	ΔE_V	ΔE_C	$\Delta E_V/\Delta E_G$ (%)	Method	References	Comments
Si–Ge	0.28	0.16	–	PH, C–V	[53]	Average value: a-Si or c-Si on Ge(111), a-Ge or c-Ge on Si(111)
AlAs–Ge	0.86	0.67	–	PH	[53]	Average value: a-Ge or c-Ge on AlAs, AlAs on Ge(100)
AlAs–GaAs	0.34	0.51	60	PH, OT	[53]	Average value: GaAs on AlAs(100) and (110), AlAs on GaAs(100)
AlAs–ZnSe	0.15–1.15	0.30–0.61	0.38(0.77)	PH	[55]	ZnSe on AlAs(001) Se-rich to Zn-rich, respectively
$Al_xGa_{1-x}As$–GaAs	–	–	59	PH, C–V, I–V, IPS, OT	[53]	Average value: $x = 0.09$–0.7
$Al_xIn_{1-x}As$–GaAs	0.49	−0.52	–	PL	[56]	$x = 0.48$, staggered gaps
AlN–InN	1.52	4.0	0.38	PH	[57]	InN on AlN(0001)
AlN–InN	3.1	2.4	0.56	PH	[58]	InN on AlN(000-1)
AlSb–GaSb	0.4	0.5	–	PH	[59]	Superlattice
AlSb–GaSb	0.40	0.50	0.44	PH	[60]	GaSb on AlSb
AlSb–GaSb	0.45	0.45	0.5	PL	[61]	GaSb-AlSb multi-QW
AlSb–ZnTe	0.42	1.13	−0.27	PH	[61a]	ZnTe(100) on AlSb(100)
GaAs–Si	0.05	0.19	–	PH	[54]	a-Si on GaAs(110)
CdS–Ge	1.75	0.00	–	PH	[54]	a-Ge on cleaved CdS
CdS–InP	0.77	0.30	–	PH	[62]	c-CdS on InP(100)
CdS–Si	1.55	−0.24	–	PH	[54]	a-Si on cleaved CdS, staggered gaps

(continued overleaf)

Table 20.3 (continued)

Interface	ΔE_V	ΔE_C	$\Delta E_V/\Delta E_G$ (%)	Method	References	Comments
CdSe–Ge	1.30	−0.23	—	PH	[54]	a-Ge on cleaved CdSe, staggered gaps
CdSe–Si	1.20	−0.57	—	PH	[54]	a-Si on cleaved CdSe, staggered gaps
CdTe–GaAs	0.21	−0.13	—	PH	[63]	c-CdTe on cleaved GaAs(110)
CdTe–Ge	0.85	−0.08	—	PH	[54]	a-Ge on cleaved CdTe
CdTe–Si	0.75	−0.42	—	PH	[54]	a-Si on cleaved CdTe
CdTe–α-Sn	1.1	0.26	—	PH	[64]	α-Sn on CdTe(111)
CdTe–HgTe	0.13	1.16	—	PH, IR,OT	[53]	Average value
CuBr–GaAs	0.85	0.74	—	PH	[65]	—
CuBr–Ge	0.7	1.6	—	PH	[65]	—
CuGaSe$_2$–Ge	0.62	−0.33	—	PH	[66]	a-Ge on fractured CuGaSe$_2$, staggered gaps
CuInSe$_2$–CdS	0.79	0.31	0.71	PH	[67]	CdS on CuInSe$_2$(001) decapped
CuInSe$_2$–ZnSe	0.70	0.7	0.5	PH	[68]	ZnSe on CuInSe$_2$
CuInSe$_2$–Ge	0.48	−0.25	—	PH	[66]	a-Ge on fractured CuInSe$_2$, staggered gaps
CuInSe$_2$–Si	0.00	−0.21	—	PH	[66]	a-Si on fractured CuInSe$_2$
Cu$_2$Se–CdS	0.8	0.4	0.66	PH	[69]	RT PVD CdS on p-Cu$_{2-x}$Se(001) on GaAs(100)
CdTe–ZnSe	0.2	0.96	0.18	PH	[70]	ZnSe on CdTe(100)
CdTe–PbTe	0.135	1.145	0.18	PH	[71]	PbTe on CdTe(111)B
GaAs–CdTe	−0.10	0.25	−0.67	PH	[72]	Average CdTe(111) on GaAs(001)
GaAs–CdTe	0.08	−0.03	0.53	PH	[72]	Average CdTe(001) on GaAs(001)
GaAs–Ge	0.49	0.19	—	PH, I-V, OT	[53]	Average value: a-Ge and c-Ge on GaAs(110), c-Ge on GaAs(111)Ga, (100)Ga, (10)As, (111)As
GaAs–InAs	0.17	0.90	—	PH	[74]	GaAs on InAs(100)
GaAs–InN	0.94	1.66	−1.29	PH	[75]	InN on GaAs(111); Type II

GaAs–MgO	4.2	2.2	0.66	PH	[76]	MgO on GaAs(001)
GaAs–SrTiO$_3$	2.5	−0.6	1.32	PH	[77]	SrTiO$_3$ on GaAs(001)
GaAs–ZnO	2.39	−0.44	1.22	PH	[78]	ZnO on GaAs(111) on Al$_2$O$_3$; Type II
GaAs–ZnSe	0.52–1.01	0.75–0.26	0.41(0.80)	PH	[79]	ZnSe on GaAs(001) Se-rich to Zn-rich, respectively
GaN–AlN	0.5	2.3	0.18	PL	[80]	Fe impurity alignment
GaN–AlN	1.36	1.45	0.48	PH	[81]	AlN on GaN(0001) on SiC
GaN–InN	1.01	1.68	0.38	I-V,C-V, IPC	[82]	InN on GaN(0001)
GaN–InN	0.58	2.22	0.22	PH	[83]	InN on GaN(0001)
GaN–MgO	1.06	3.30	0.24	PH	[84]	MgO on GaN(0001)
GaN–ZnO	0.53	−0.53	0.53/∼0	PH, PL	[85]	RF sputter poly-ZnO on GaN(0001); Type II
GaN–ZnO	0.60	−0.60	0.60/∼0	PH, PL	[85]	RF sputter poly-ZnO on GaN(0001); Type II, higher O
GaP–Ge	0.80	0.77	–	PH	[54]	a-Ge on GaP(110)
GaP–Si	0.80	0.33	–	PH	[86]	a-Si or c-Si on GaP(110)
GaSb–AlSb	0.40	0.50	0.44	PH	[60]	AlSb on GaSb
GaSb–AlSb	0.39	0.51	0.43	PH	[61]	AlSb on GaSb(100)
GaSb–Ge	0.20	−0.20	–	PH	[54]	a-Ge on GaSb(110), staggered gaps
GaSb–Si	0.05	−0.49	–	PH	[54]	a-Si on GaSb(110), staggered gaps
GaSb–ZnTe	0.60	0.05	0.92	PH	[61]	ZnTe(100) on GaSb(100)
GaSe–Ge	0.83	0.55	–	PH	[87]	a-Ge on cleaved GaSe
GaSe–Si	0.74	0.20	–	PH	[87]	a-Si on cleaved GaSe
In$_y$Al$_{1−y}$As–In$_x$Ga$_{1−x}$As	0.22	0.47	0.32	PH	[88]	Average value; InGaAs on InAlAs(100); x=0.53, y=0.52

(continued overleaf)

Table 20.3 (continued)

Interface	ΔE_V	ΔE_C	$\Delta E_V/\Delta E_G$ (%)	Method	References	Comments
$In_xAl_{1-x}As-In_yGa_{1-y}As$	0.36	0.36	51	$C-V$, IPS, PL, $I-V$, OT	[53]	$x = 0.52, y = 0.53$
InAs–Ge	0.33	0.27	–	PH	[54]	a-Ge on InAs(110)
InAs–Si	0.15	0.01	–	PH	[54]	a-Si on InAs(110)
$In_xGa_{1-x}As-InP$	0.34	0.26	0.57	PH	[88]	Average value; InP on InGaAs(100); $x = 0.53$
InGaP–GaAs	0.24	0.21	0.53	$I-V$, IPS	[16]	Double HJ bipolar transistors
$In_xGa_{1-x}P-GaAs$	0.08	0.59	88	PH	[89]	$x = 0.48$
InN–MgO	1.59	5.54	0.22	PH	[90]	MgO(111) on InN on Al_2O_3
$InP-(In,Ga)_x(As,P)$	–	–	57	$C-V$, PL	[53]	Different quaternary compositions, lattice matched with InP, $E_G = 0.70-1.20$ eV
$InP-CaF_2$	6.2	4.56	0.60	PH	[91]	CaF_2 (001) on InP(001)
$InP-In_xGa_{1-x}As$	0.37	0.23	38	$C-V$	[50]	$x = 0.53$
InP–Ge	0.64	−0.04	–	PH	[54]	a-Ge on InP(110)
InP–Si	0.57	−0.41	–	PH	[54]	a-Si on InP(110)
$InP-SrF_2$	6.4	3.46	0.66	PH	[91]	SrF_2 (001) on InP(001)
InSb–Ge	0.0	−0.50	–	PH	[54]	a-Ge on InSb(110)
InSb–Si	0.0	−0.94	–	PH	[54]	a-Si on InSb(110)
MgO–ZnO	0.87	3.59	0.20	PH	[93]	ZnO on MgO(111)
PbTe–Ge	0.35	−0.42	–	PH	[94]	a-Ge on PbTe(110), staggered gaps
$Si(111)-CaF_2$	8.2	2.8	0.75	PH	[95]	Average CaF_2 on Si(111)
$Si-SrF_2$	8.2	1.8	0.82	PH	[95]	Average SrF_2 on Si(111)
$Si-SrTiO_3$	2.12	0.1	0.95	PH	[96]	$SrTiO_3$ on Si(001)

System					Comment
SiC–GaN	0.48	−0.11	C–V	[97]	GaN on SiC(0001) Type II
SiC–MgO	3.65	0.92	PH	[98]	MgO on 4H-SiC(100)8° → <11–20>
SiC–ZnO	1.61	−1.50	PH	[99]	ZnO on 4H-SiC(0001)8° → <11–20>; Type II
ZnO–CdS	1.2	−0.3	PH	[100]	Thermal CdS on poly-ZnO
ZnO–InN	0.82	1.85	PH	[101]	InN on ZnO(0001)
ZnS–Cu$_2$S	1.4	1.35	I–V	[102]	–
ZnSe–Ge	*1.40*	*0.51*	PH, I–V	[53]	Average value: a-Ge and c-Ge on cleaved ZnSe, c-ZnSe on GaAs(110)
ZnSe–Si	1.25	0.22	PH	[54]	a-Si on cleaved ZnSe
ZnSe–GaAs	*1.03*	*0.20*	PH	[53]	Average value: annealed c-ZnSe on Gas(110)
ZnSe–GaAs	1.25	0.05	PH	[103]	c-ZnSe on GaAs(100)
ZnSnP$_2$–GaAs	0.13	−0.03	OT	[104]	Staggered gaps
ZnTe–GaSb	0.34	1.25	PH	[105]	c-ZnTe on GaSb(110)
ZnTe–Ge	0.95	0.64	PH	[54]	a-Ge on cleaved ZnTe
ZnTe–Si	0.85	0.30	PH	[54]	a-Si on cleaved ZnTe

Figure 20.18 defines the sign convention. Average values are italicized. Updated from Margaritondo and Perfetti [53], with permission.

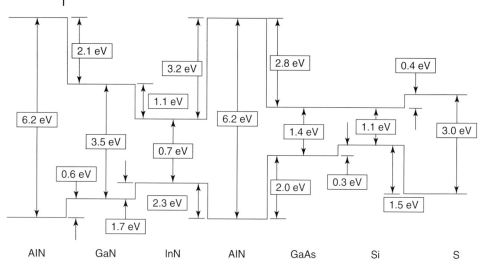

Figure 20.27 Approximate band lineups for WBG semiconductors with Si and GaAs based simply on electron affinity and bandgap differences. (Courtesy, J. Schetzina.)

fails as a predictive tool of heterojunction band offsets and (ii) the dipole term ($V_i + V_S^A - V_S^B$) plays a major role in determining these offsets.

20.4.3.2 Organic Semiconductors

Unlike inorganic semiconductors, the carrier concentrations in organic thin film semiconductors are typically quite low so that the surface space charge region can extend throughout the entire film. Hence, the organic–organic band diagrams typically include little or no band curvature. Figure 20.29 illustrates the various factors that can contribute to dipole formation for an adsorbed molecule on a surface [107]. Contributing factors include the following: (i) charge transfer between a nonpolar donor molecule and an organic acceptor or high work function metal, (ii) charge redistribution due to image charge in a (metal) film, (iii) a "pillow" effect that reduces wavefunction tailing out of the substrate, (iv) chemical bonding between the film and the molecule that rearranges the electron density as well as the near-surface film and the molecular structures themselves, (v) interface states that localize charge between the organic molecule and the film, and (vi) polar molecules or functional groups within a molecule that can produce dipole fields normal to the interface plane, depending on their orientation.

Charge transfer directly between the donor and the acceptor produces dipoles localized near the interface since conduction in the organic layers can be very low. The "pillow" effect represents a repulsion of the electron cloud and dipole extending out from the surface layer due to the presence of charge already present outside this surface. The net effect is a decrease in negative charge outside the surface, resulting in a positive dipole relative to that of the bare surface.

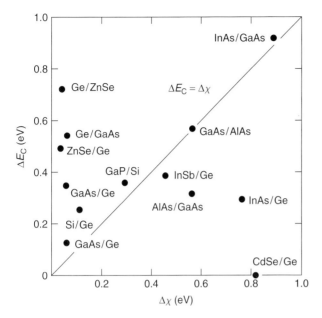

Figure 20.28 Test of the electron affinity rule. Each point signifies a heterojunction with the substrate first, the overlayer second. Deviations from the measured $\Delta E_C = \Delta \chi$ line indicate dipole contributions. (After Milnes and Feucht [106] as well as Bauer and Sang. [16])

Interface states can arise from one or more of the various factors described in Sections 4.3 and 4.4. For the low carrier concentrations usually present in films of organic molecules, image forces and wavefunction tailing are unlikely. Their presence at metal–organic interfaces gives rise to large dipole shifts in general.

Besides the mechanisms represented in Figure 20.29 for clean interfaces, exposure to air can have pronounced effects on interface dipoles. Besides introducing adventitious carbon, hydrogen, and oxygen as adsorbate layers, air exposure may oxidize organic surfaces and thus change the electronic structure near the surface significantly, for example, by introducing a new dielectric layer with properties unlike either organic. Such layers can easily produce vacuum level shifts.

Improved understanding of organic heterojunctions awaits more detailed measurements of band alignment and band bending at interfaces in well-characterized material systems.

20.4.4
Theories of Heterojunction Band Offsets

A variety of physical mechanisms provide the basis for theoretical models of heterojunction band offsets. These include: (i) tunneling approaches involving charge neutrality levels [108, 109], (ii) local bonding approaches based on tight-binding [110]

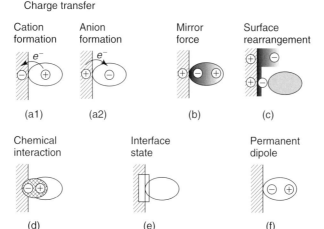

Figure 20.29 Mechanisms of dipole formation at organic heterojunctions. (a) Charge transport at the donor or the acceptor molecule substrate interfaces, (b) image force, (c) "pillow" effect, (d) chemical bonding, (e) interface states, and (f) oriented polar molecules or functional groups. (After Ishii et al. [107])

or pseudopotential [111] calculations of electrostatic potentials at the atomic-scale interface, (iii) an effective work function model based on new chemical phases and electron affinities at the atomic-scale interface, (iv) empirical deep-level schemes relating the band edges in the two semiconductors to transition-metal levels in each based on average dangling bond energies [112] or the vacuum level [113], and (v) various approaches based on dielectric electronegativity [114], superlattice band structure [115], and empirical values of the bulk optical dielectric and lattice constants [116]. This section deals with the most comprehensive of these approaches.

20.4.4.1 Charge Neutrality Levels

The charge neutrality levels developed for Schottky barrier formation can also be used to predict heterojunction band offsets. Here, the wavefunction tunneling across the interface defines equilibrium energies for each semiconductor that can be aligned. Figure 20.30a represents the complex band structure of a semiconductor in one dimension [117]. The left panel shows the usual free-electron-like energy bands $E(k)$ with the bandgap $E_C - E_V$ at wave vector $k = \pi/a$, while the right panel shows energy plotted versus $\text{Im}(k)$ for $\text{Re}(k) = \pi/a$. At an energy E_B, $\text{Im}(k)$ reaches a maximum, thereby minimizing wavefunction tunneling.

The crossover between conduction and valence band-derived states is defined as the *branch point* at which the Fermi level reaches equilibrium. Alignment of the branch points of the two semiconductors in a heterojunction therefore aligns their

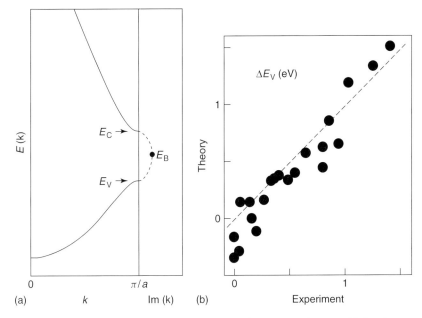

Figure 20.30 (a) Complex band structure in one dimension for a semiconductor with free-electron-like bands $E(k)$ (a) and energy versus Im(k) at Re(k) = π/a (b). The branch point E_B occurs at an energy where Im(k) reaches a maximum, that is, at the crossover between filled and empty gap states. (b) Comparison between experimental and theoretical values of ΔE_V derived from branch point alignment. (After Tersoff. [117])

band structures and yields values for ΔE_C and ΔE_V. Figure 20.30b shows the results of aligning the branch points of individual semiconductors. The agreement between theory and experiment is considerably improved over the electron affinity rule and, despite its relative simplicity, is in fact at least as predictive as any other approach. Nevertheless, deviations between theory and experiment are considerable, in many cases, several 100 meV. The charge neutrality approach takes into account only the bulk semiconductor band structure and ignores local bonding or other interactions at the interface.

20.4.4.2 Local Bond Approaches

Local bond approaches take interface-specific features into account explicitly. These include tight-binding pseudopotential calculations that determine the semiconductor band edges from hybridized atomic orbitals (HAOs). Thus, the valence-band energy maximum is given by

$$E_V = 1/2(\varepsilon_a + \varepsilon_c) + \left[1/4(\varepsilon_a - \varepsilon_c)^2 + (4E_{XX})^2\right]^{1/2} \tag{20.26}$$

where ε_a and ε_c are the anion and cation orbital energies, respectively, and

$$E_{XX} \approx -1.28 h/m^* d^2 \tag{20.27}$$

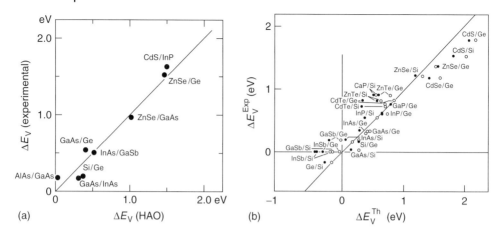

Figure 20.31 Experimental ΔE_V values versus ΔE_V derived from (a) the HAO method of Harrison (After Kroemer [118]) and (b) the HAO method corrected for lattice mismatch with an average lattice constant. (After Katnani. [119])

is an energy term that depends on the nearest anion–cation distance d and free electron mass m^*. This term incorporates the local potential differences at the interface. Figure 20.31a illustrates the comparison of experimental and theoretical ΔE_V values [118] based on the HAO method of Harrison [110]. The improved agreement shown versus that of Figure 20.30b stems from the incorporation of local potential differences at the atomic interface.

Figure 20.31b displays an expanded set of heterojunctions [119]. Here, the ΔE_V values have been corrected for lattice mismatch between semiconductors by the substitution of an average lattice constant for d in Equation 20.27 The improved agreement between experiment and theory for this expanded data set is evident. However, estimates of ΔE_V by this approach are only accurate to within 0.2 eV.

20.4.4.3 Empirical Deep-Level Schemes

Instead of the midgap neutrality levels discussed in Section 20.4.4.1, researchers have also considered the alignment of impurity levels as a method of determining band offsets. This approach is based on the nearly constant binding energies of deep-level transition-metal impurities calculated for different semiconductors. These energies are calculated with reference to either the vacuum level or the valence band. Figure 20.32 illustrates how the vacuum-related binding energies of the transition metals V, Cr, Mn, Fe, Co, and Ni are nearly the same, independent of the compound semiconductor [120]. Again, this reflects a bulk approach to band alignment that ignores interface interactions.

An analogous approach references these transition-metal energies to the valence bands of the semiconductors. Figure 20.33 illustrates the energy positions of the same transition metals relative to the valence bands of GaAs, InP, and GaP. This figure displays the band edges and thereby the band offsets of all three

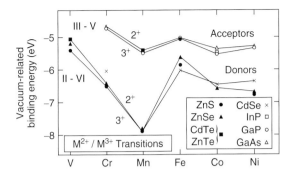

Figure 20.32 Universal binding energy curve for deep acceptors in III–V compounds and deep donors in II–IV compounds referenced to the vacuum level. (After M.J.Caldas et al. [120])

semiconductors relative to these deep-level valence band energies with an average error of ~100 meV or less.

More scatter is apparent for the energy positions of the same transition metals relative to the valence bands of the II–VI compounds displayed in Figure 20.34. The agreement is comparable to or less than that of other approaches. The utility of this approach is also limited by the availability of calculations for the semiconductor alloys that comprise lattice-matched heterojunctions.

20.4.5
Assessment of Theory Approaches

The results presented here indicate that several theory approaches show agreement with experimental measurements to within a precision of less than 0.2 eV. It is perhaps surprising that these models achieve comparable agreement despite being based on conceptually different physical phenomena. Nevertheless, an order of magnitude higher experimental and theoretical precision – approximately the thermal energy of ~25 meV at room temperature – is required in order to (i) distinguish between these approaches and (ii) to predict heterojunction band offsets within the tolerances necessary for electronic device design and fabrication.

20.4.6
Interface Contributions to Band Offsets

Besides the measurement of absolute band offset values, another strategy to identify the role of different physical mechanisms is the measurement of ΔE_V and ΔE_C as they vary with interface composition, structure, and growth conditions. This approach is possible with a combination of epitaxial growth and surface science measurements during the initial stages of heterojunction formation. Here, we present the findings of heterojunction experiments that vary (i) growth sequence, (ii)

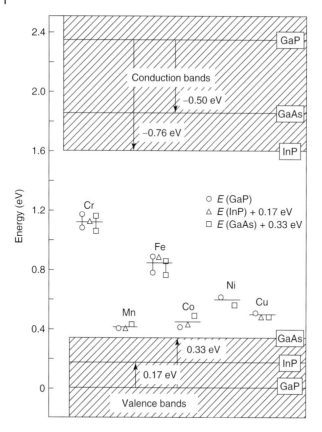

Figure 20.33 Universal binding energies referenced to the valence bands of III–V compound semiconductors. (After Langer and Heinrich. [112])

crystallographic orientation, (iii) surface reconstruction, (iv) interface composition, and (v) interfacial layers. This research reveals that the deviations from classical theory due to dipoles can be related to specific features of the microscopic interface. Furthermore, the control of these interface features at the atomic scale lends themselves to "band offset engineering" to improve practical electronic devices.

20.4.6.1 Growth Sequence

The dependence of band offset on growth sequence is an effective way to distinguish between physical models based on intrinsic properties of the semiconductors comprising the heterojunction, that is, *linear models*, versus extrinsic properties related to interface microscopic structure. The dependence of band offset on growth sequence can be expressed as the degree of commutativity and transitivity, that is, the linearity of electronic structure variations.

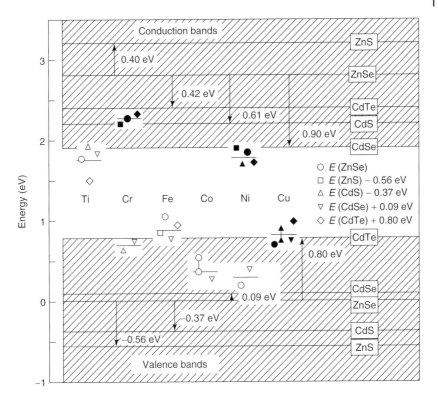

Figure 20.34 Universal binding energies referenced to the valence bands of II–VI compound semiconductors. (After Langer and Heinrich. [112])

For semiconductors A and B, linear models must satisfy the *commutativity* relation

$$\Delta E_V(A-B) + \Delta E_V(B-A) = 0 \tag{20.28}$$

for semiconductors A on B versus B on A as well as the *transitivity* relation

$$\Delta E_V(A-B) + \Delta E_V(B-C) + \Delta E_V(C-A) = 0 \tag{20.29}$$

Deviations from linearity correspond to interface dipole contributions.

The band offset measurements indicate that these linear relations are satisfied to varying degrees ranging from below experimental resolution to ~0.2 eV [51]. Isovalent semiconductors such as GaAs and AlAs appear to form heterojunctions that are linear to <0.1 eV. On the other hand, heterovalent semiconductors form junctions that are linear only to 0.2 eV or more. For example, ΔE_V(Ge/ZnSe) + ΔE_V(ZnSe/Ge) = 0.23 eV and ΔE_V(Ge/GaAs) + ΔE_V(GaAs/Ge) = 0.2 eV. An interface reaction is responsible for the nonlinear behavior of Ge with CuBr. Ge on CuBr(110) reacts strongly while CuBr on Ge exhibits epitaxial growth. The Ga/Ge(110) interface is also unstable with time and annealing, varying by 0.2–0.3 eV [121, 122].

Besides reactions, nonlinearities at heterovalent junctions may be due to atoms from different columns of the Periodic Table moving across the interface to minimize otherwise large charge buildup [30], a topic to be discussed further in Section 20.4.7.

20.4.6.2 Crystallographic Orientation

High-precision XPS measurements show that ΔE_V values vary systematically with reconstruction, that is, by bond termination and crystal face polarity. Nevertheless, these variations are small. For Ge on (111), (100), (110), and (111) oriented GaAs, Table 20.4 shows ΔE_V variations of only 0.17 eV. These measurements require considerable care since spectral linewidths were approximately 1.1–1.2 eV, yet it was possible to obtain measurement reproducibility of ±0.01 eV.

Table 20.4 shows that ΔE_V for Ge on GaAs varied according to the surface polarity [123]. Thus, ΔE_V increases steadily from the electropositive, Ga-terminated (111)Ga(2×2) surface to the electronegative As-terminated (111)As(1×1) face. This

Table 20.4 Crystallographic orientation dependence of straddling heterojunction valence-band offsets for Ge/GaAs interfaces.

Electropositive ↑	Substrate surface	Ge layer thickness (Å)	Γ(Ga3d) (eV)	Γ(Ga3d) (eV)	$(E_{Ga3d}^{GaAs} - E_{Ga3d}^{Ge})^b$	$\delta(\Delta E_v)_{AVE}$ (eV)
	(111)Ga (2×2)	13	1.17 ± 0.02	1.25 ± 0.01	−10.27	
						−0.085
		20	1.22 ± 0.02	1.26 ± 0.01	−10.31	
	(100)Ga c(8×2)	22	1.19 ± 0.02	1.25 ± 0.01	−10.22	−0.015
	(110)(1 × 1)	14	1.13 ± 0.01	1.29 ± 0.01	−10.20	0
		17	1.16 ± 0.01	1.27 ± 0.01	−10.21	
	(100)As	14	1.15 ± 0.02	1.25 ± 0.01	−10.17	+0.035
	(111)As (1 × 1)	13	1.21 ± 0.01	1.32 ± 0.01	−10.11	
						+0.10
Electronegative ↓		18	1.22 ± 0.01	1.28 ± 0.01	−10.10	

The Ge epitaxial layer thickness, Ge 3d and Ga 3d photoelectron linewidths, Ge 3d-Ga 3d core-level binding energy differences, the average variation in E_V relative to the (110) interface, and the ΔE_V value for eight different Ge-GaAs interfaces. After Grant et al. [51]

is consistent with the substrate's surface dipole contribution to the interface dipole term $(V_i + V_S^A - V_S^B)$ in Equation 20.17.

20.4.6.3 Surface Reconstruction: Band Bending versus Offsets

Different surface reconstructions on the same crystal orientation also produce different heterojunction band offsets. As illustrated in Figures 7.9 and 7.10, photoemission can monitor Fermi level movement due to band bending as a function of overlayer deposition. Figure 20.32a illustrates the Fermi level movement within the GaAs bandgap as a function of Ge overlayer coverage. The dashed line shows that E_F moves rapidly from a position near the GaAs conduction band to a position near midgap for Ge on cleaved GaAs(110). In contrast, E_F at the MBE-grown GaAs(100)–Ge interface exhibits more gradual movement to final positions that vary substantially with the surface reconstructions indicated [124]. For each (100) surface reconstruction, the band discontinuities obtained using both methods described in Figures 20.25 and 20.26 remain constant. Figure 20.35b shows the final position for the Fermi level position at the interface on a 10-Å scale.

As shown, the Fermi level position due to band bending varies by more than 0.3 eV, while the heterojunction band offsets do not change. This result is independent of the Ge overlayer's doping density. Ge overlayers doped with As during growth to vary their bulk Fermi level position again yield the same valence-band offset. These growth and in situ XPS studies demonstrate that the band offsets are not correlated with band bending within the semiconductors – an expected result since dipoles at the interface must be charge neutral, contributing no net charge

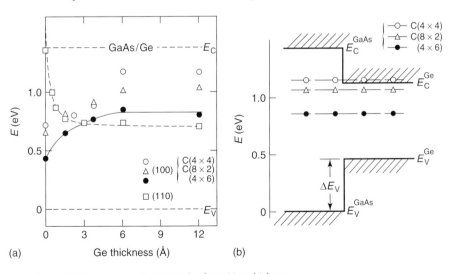

Figure 20.35 (a) $E_F - E_V$ versus Ge deposition thickness on MBE-grown GaAs(100) and melt-grown and cleaved GaAs(110) surfaces [124]. (b) Final E_F position in GaAs and Ge for each reconstruction. (After Chiaradia et al. [124])

that could induce band bending within the semiconductors' surface space charge regions.

20.4.6.4 Surface Reconstruction: Interface Bonding

Surface reconstruction and chemical composition can strongly affect whether interface states form as well as their energies and densities. A notable example is the lattice-matched ZnSe/GaAs heterojunction. C–V measurements show 2 orders-of-magnitude variation of the electrically active *interface state density* D_{it} as a function of the substrate GaAs(100) reconstruction measured by reflection high-energy electron diffraction (RHEED). Figure 20.36 shows the presence of two peaks in the density of midgap states that are the highest for $c(4\times4)$ and (2×4) reconstructions, which are As -rich. These state energies agree with those derived from Schottky barrier studies of melt-grown GaAs, to be discussed in Chapter 21.

In contrast, the Ga-rich (4×3) and annealed $c(4\times4)$ reconstructions are much lower [125]. These densities decrease in the order of decreasing As surface composition. This decrease in D_{it} is attributed to the different interface bonding: ZnAs-like bonding for As-rich reconstructions and GaSe-like bonding for Ga-rich conditions. The reduced interface state density results in (i) near-ideal C–V characteristics, (ii) E_F movement across the full GaAs bandgap, and (iii) E_F positions due to band bending that follow the doping dependence of E_F in ZnSe [126].

To conclude this section, interface preparation and chemical bonding can significantly alter heterojunction band offsets and band bending. Yet while localized states at these interfaces are sensitive to interface geometric and chemical structure,

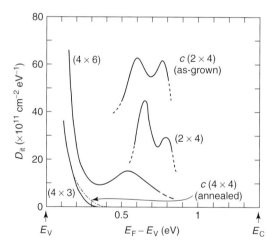

Figure 20.36 Density of interface states calculated from the capacitance-voltage characteristics for a variety of GaAs epilayer surface reconstructions prior to ZnSe epitaxial growth. Interface state densities decrease with decreasing As concentration, emphasizing the sensitivity of interface state densities to local atomic bonding [125].

the heterojunction band offsets do not vary accordingly. Rather the heterojunction band offsets can vary, but only with the introduction of interface dipoles. Furthermore, different surface orientations and chemical effects display larger effects than different surface reconstructions or growth sequences.

20.4.7
Theoretical Methods in Band Offset Engineering

Epitaxical growth of semiconductor heterojunctions provide theorists with models of well-defined atomic structures with which band offsets can be calculated. Nevertheless, such calculations are challenging because there is no absolute reference potential for the average lattice potential $V^{lattice}$(avg) due to the long-range Coulomb potential. Therefore, one can only define $V^{lattice}$(avg) to within a constant. Nevertheless, several methods have been developed that illuminate the variations in heterojunction band offsets with interface atomic structure and composition. Franciosi and Van de Walle have provided an authoritative review of these methods [127].

20.4.7.1 First-Principles Calculations
Self-consistent, first-principles calculations start with no experimental input. Density function theory establishes a ground state energy E from Schroedinger's equation for a single particle

$$(\hbar^2/2m)\nabla\psi^2 + V\psi = E\psi \tag{20.30}$$

where potential energy V consists of an electron–atom lattice potential in addition to an electron–electron exchange-correlation function of local charge density. The wavefunction is fixed by interface boundary conditions. The energy of the charge within the lattice of the semiconductor heterojunction is a function of atomic positions. Therefore, calculations of energy for different configurations of lattice atoms, including relaxations and reconstructions, can be useful as tools to identify stable structures with minimum energy.

20.4.7.2 Mathematical Approach
The mathematical approach to these calculations is a self-consistent solution of a variational problem. It involves the use of (i) periodic boundary conditions and (ii) average potentials in planes parallel to the interface. Condition (i) involves superlattice structures in which slabs of one semiconductor terminate on both sides with the other semiconductor. Such periodic boundary conditions are not only suitable for actual superlattices but also permit solutions for isolated interfaces as long as the two boundaries do not interact.

Condition (ii) involves an average potential

$$\overline{V}(z) = (1/S) \int V(x, y, z) \mathrm{d}x \mathrm{d}y \tag{20.31}$$

with area S for a unit cell in the x-y plane of the interface. Figure 20.37 illustrates the variation of this plane-averaged potential across a Si–Ge(001) interface consisting of four Si and four Ge layers. Potential maxima (minima) of the bulk lattice

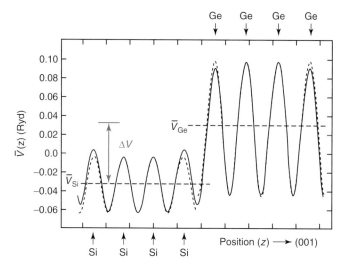

Figure 20.37 Average potential (in Rydbergs) as a function of position across four layers of Si and four layers of Ge. The band offset ΔV is the difference between $\overline{\overline{V}}_{Si}$ and $\overline{\overline{V}}_{Ge}$. (After Van de Walle and Martin. [128])

potential $V(z)$ shown as solid lines occur at (between) the atomic lattice positions. Periodic boundary conditions require that $\overline{V}(z)$ match at left and right boundaries. Near the boundaries, $\overline{V}(z)$ (dotted line) deviates only slightly from $V(z)$. Within one atomic layer, the bulk- and plane-averaged potential are approximately the same.

The average of planar-averaged potentials $\overline{\overline{V}}(z)$ across each semiconductor can be expressed as [128]

$$\overline{\overline{V}}(z) = (1/a)_{(z-a/2)} \int^{(z+a/2)} \overline{V}(z')dz' \qquad (20.32)$$

for length a of one period perpendicular to the interface. Figure 20.37 shows planar-averaged $\overline{\overline{V}}_{Si}$ and $\overline{\overline{V}}_{Ge}$ (dashed lines) across each layer as well as their separation ΔV, equivalent to the heterojunction band offset.

All-electron calculations that take into account both the core and the valence electrons can mimic the valence-band and core-level offsets measured by XPS as presented in Section 20.4.2.3. Pseudopotential calculations, which use averaged core potentials, do not provide valence-band offsets directly, but the averaged core and valence potentials provide average ΔV values.

ΔV can also be expressed in terms of the deviations between calculated electrons $\overline{\overline{n}}(z)$ per cell and the planar average electron density n_0 calculated as a function of position normal to the interface plane:

$$\Delta \overline{V} = \frac{e^2}{\varepsilon_0} \int z \left[\overline{\overline{n}}(z) - n_0\right] dz \qquad (20.33)$$

The term $\overline{\overline{n}}(z) - n_0$ becomes significant at the heterojunction, permitting direct calculation of $\Delta \overline{V}$ from the dipole moment set up by the unbalanced charges.

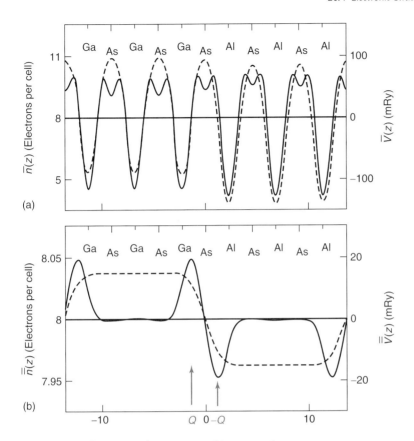

Figure 20.38 Planar (a) and macroscopic (b) average of the charge density per cell (solid line) and planar average electrostatic potential (dashed line) versus z normal to a GaAs–AlAs junction [129].

Figure 20.38a shows the electrons $\bar{n}(z)$ per cell (solid line) and the planar average electrostatic potential $V(z)$ (dashed line) versus position normal to a GaAs–AlAs heterojunction [129]. The fine structure shown for $\bar{n}(z)$ reflects the two-dimensional variations in atom density within single lattice periods. Within the GaAs lattice, the average $\bar{n}(z)$ and n_0 are equal, except at the GaAs side of the junction, where $\bar{n}(z)$ is higher than n_0 and at the AlAs side of the junction, where $\bar{n}(z)$ is lower than n_0. Figure 20.38b reflects this change in charge and potential at the interface in terms of a macroscopic average $\bar{\bar{n}}(z)$ per cell (solid line) and macroscopic planar average electrostatic potential $\bar{\bar{V}}(z)$ (dashed line).

Near the GaAs–AlAs heterojunction, the macroscopic planar average electron density $\underline{n}_0$ (dashed line) passes smoothly through the origin. However, $\simeq n(z) > \underline{n}_0$ near the interface on the GaAs side and $\simeq n(z) < \underline{n}_0$ on the AlAs side. The resultant charge densities Q and $-Q$ on opposite sides of the junction produce a dipole of potential ΔV defined by Equation 20.33.

20.4.7.3 Alternative Methods

Alternative approaches include (i) linear response theory, which model the difference between two semiconductors across their interfaces relative to an average, for example, GaAs and AlAs relative to $Al_{0.5}Ga_{0.5}As$ at a GaAs/AlAs interface; (ii) simplified Hamiltonians such as self-consistent tight-binding models. However, these cannot extract energies since parameters are fitted to bulk rather than interface properties; and (iii) linear models such as Anderson's electron affinity rule or charge neutrality levels, both of which take only the intrinsic properties of the semiconductor bulk into account.

20.4.8
Application to Heterovalent Interfaces

20.4.8.1 Polarity Dependence

The crystal orientation of the interface plane determines the local bonding and charge transfer between atoms at the heterojunction. In turn, this charge transfer determines whether an average electric field builds up away from the interface, rendering the junction unstable.

Figure 20.39a shows a nonpolar junction between Ge and GaAs with a (110) interface plane. All atoms are tetrahedrally coordinated. Double bonds in the figure represent two tetrahedral bonds separated by the usual 109° angle and projected

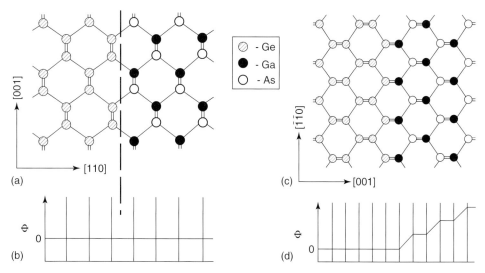

Figure 20.39 (a) Schematic lattice representation of a Ge–GaAs(110) heterojunction, viewed along the [110] direction with the [001] direction vertical. Dashed line signifies the interface plane. Every atomic plane parallel to the junction is neutral on average, corresponding to a nonpolar junction. (b) Potential versus lattice plane. (c) Ge–GaAs(001) heterojunction viewed along the [110] direction with the [110] direction vertical. The first atomic plane to the right of the junction consists of only negatively charged Ga atoms. (d) Potential versus lattice plane increasing from left to right due to nonzero electric field. (After Harrison et al. [30])

onto the plane of the figure. Note the equal numbers of Ge–Ga and Ge–As bonds. Inspection shows that each lattice plane is charge neutral. Thus, Figure 20.39b shows no potential change in planar-averaged $\overline{\overline{V}}$ from the Ge bulk into the GaAs bulk crystal.

Figure 20.39c shows a polar junction between Ge and GaAs with a (100) interface plane. The first atomic plane on the GaAs side of the junction consists entirely of Ga atoms, which are negatively charged. The potential averaged over planes parallel to the junction is obtained by integrating Poisson's equation from left to right. For this polar junction, Figure 20.39d shows that there is a nonzero average electric field in the GaAs due to charge accumulation at the junction and the dipoles between planes.

20.4.8.2 Interface Atomic Mixing

Atomic mixing can reduce or eliminate the net charge, electric field, and potential buildup at polar interfaces. Figure 20.40a shows a Ge–GaAs(001) heterojunction as shown in Figure 20.39c but with half of the Ga atoms replaced by Ge atoms in the first GaAs atomic layer. Figure 20.40b shows the change in potential in each GaAs layer that results. The average electric field is now 0. However, there is a potential change δ between the Ge and the GaAs average potentials. For lattice constant a, this dipole shift $\delta = e/(4a\varepsilon\varepsilon_0)$ is equal to the band offset. Assuming ε for GaAs, $2\delta = 0.73$ eV, while for ε of Ge, $2\delta = 0.50$ eV.

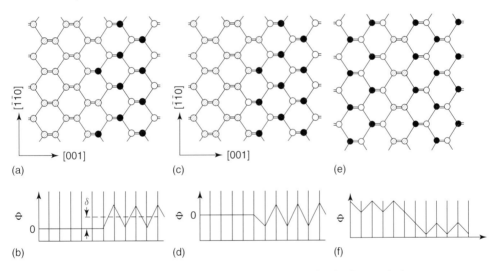

Figure 20.40 (a) The polar Ge–GaAs(100) heterointerface as in Figure 20.36c but with half the Ga atoms replaced by Ge in the first GaAs plane. (b) The resultant potential in GaAs and junction dipole shift in potential between GaAs and Ge. (c) The polar Ge–GaAs(100) junction with double-layer mixing – the first layer with one-fourth of As, the second three-fourth of Ga, with the remaining atoms Ge. (d) The resultant potential variation showing zero dipole shift. (e) A GaAs-2 layer Ge–GaAs(001) structure. (f) The potential change introduced by opposite charge exchange between Ge and GaAs on either side of the interlayer. (After Harrison [30])

Figure 20.40c illustrates the polar interface with double-layer mixing, here with the first layer containing 1 As atom for every three Ge atoms and the second layer with three Ga atoms to every one Ge atom. As Figure 20.40d shows, this geometry avoids any charge accumulation or dipole shift.

In general, charge buildup at polar interfaces can drive atomic mixing. Heterojunction sequences and orientations are desirable that avoid atomic mixing in order to obtain atomically abrupt interfaces.

20.4.8.3 Atomic Interlayers

Atomic interlayers can introduce band offsets between layers, even of the same semiconductor. Figure 20.40e shows a thin interlayer at a (100) polar interface. Here, a double layer of Ge between two GaAs lattices forms an abrupt, ideal interlayer. The potential variation in Figure 20.40f indicates that the charge exchange between these layers and their neighboring GaAs lattices introduces a large dipole shift. The sign of this dipole will depend on the GaAs surface termination prior to depositing the Ge layers. In this case, the Ge interlayer represents the introduction of a microscopic capacitor.

Figure 20.41a illustrates the planar-averaged potential for the GaAs-double-layer Ge–GaAs homojunction analogous to Figure 20.40e [130]. Extrapolating the dotted lines from the GaAs bulk lattice on either side of the interface, the dipole shift across the Ge layer is calculated to be 0.74 eV, a considerable shift that is nearly $1/2 E_G^{GaAs}$.

XPS measurements of valence-band offsets for the AlAs–Si–GaAs(100) and GaAs–Si–AlAs(100) heterojunctions ΔE_V appear in Figure 20.41b. They demonstrate that ΔE_V is tunable with the thickness of ordered Si interlayers. In both cases, the Si interlayer grows on an As-stabilized surface. ΔE_V is tunable continuously around the intrinsic value of 0.40 ± 0.07 eV (similar to Ge on GaAs, Figure 20.35b)

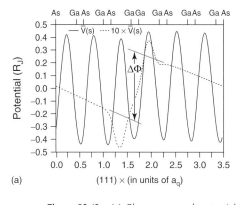
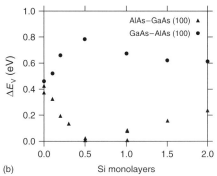

Figure 20.41 (a) Planar-averaged potential variation of a GaAs-2-layer Ge-GaAs homojunction along the (111) direction (solid line) and the unit cell-averaged potential showing the band offset (dotted line). (b) Valence-band offset for a GaAs-multilayer Si–GaAs heterojunction for AlAs–GaAs(100) (△) and GaAs–AlAs(100) (·) heterostructures as a function of ordered Si interlayer thickness. Band offset varies as a function of Si interlayer thickness. (After Sorba et al. [130])

from 0.02 to 0.78 eV. As with Ge interlayers, variable thicknesses of Si at the interface provide large and controllable changes in valence-band offsets.

Two additional points are worth noting. First, the sign of ΔE_V depends on the growth sequence. Second, the change in ΔE_V saturates at Si thicknesses of 0.5 monolayers. This is a prime example to illustrate how microscopic interlayers can produce macroscopic potential changes in semiconductor heterojunctions.

Elemental interlayers produce sizable changes in heterojunction band offsets in many other III–V and II–VI heterojunction, for example: (i) CdS heterojunctions including CdS–Ge with Al or Au [131], ZnSe(110)–Ge with Al or Au [130], GaP(110)–Si with Al or Au [130], GaAs(100)–Ge with Al [132], GaAs(100)–AlGaAs with donor or acceptor sheets [133], GaAs(100)–AlAs and GaAs(100) with Si [130], Ge(111)–Ge with Ga–As [134] and AlAs [135], and Si(111)–Si with Ga–P and Al–P [134].

The influence of growth parameters on band offsets leads to the following general conclusions: (i) heterovalent junctions exhibit wider band offset variations than isovalent junctions; (ii) nonpolar (110) heterovalent junctions exhibit ~0.2 eV variations; and (iii) polar (100) heterovalent junctions can vary by more than 0.5 eV. These features can be useful in obtaining desirable band offsets in specific heterojunctions.

20.4.9
Practical Band Offset Engineering

The design of a heterojunction for a particular application begins with low lattice mismatch and electronic properties of the specific constituent semiconductors. Band offset engineering involves control of the chemical bonding at the atomic interface as just described. This includes both the sequence of atom growth as well as their layer-by-layer composition. Furthermore, these semiconductors must

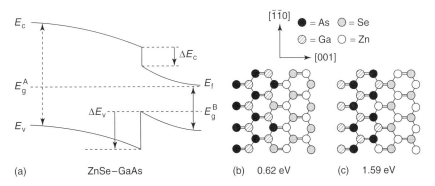

Figure 20.42 (a) p-ZnSe/n-GaAs heterojunction with band offsets illustrated. For the atoms designated in (b), ΔE_V varies in (c) from 0.62 eV at a Zn-rich interface to 1.59 eV for a Se-rich junction. (After Nicolini et al. [136])

remain thermodynamically stable at growth temperatures. Chemical instability leads not only to the formation of lattice defects in each material but also to cross-doping of atoms from one semiconductor inside the other.

The ZnSe/GaAs heterojunction illustrates these concepts of band offset engineering. For n-type GaAs with p-type ZnSe, both of which have zincblende lattice structure, the band diagram of the interface appears in Figure 20.42a [136]. As shown, $\Delta E_V > \Delta E_C$ for the lattice-matched heterojunction before changes in interface bonding and composition.

20.4.9.1 Spatially-Confined Nonstoichiometry

Calculations show that different near-interface atomic bonding can alter the ZnSe–GaAs junction dipole and band offsets by nearly a volt. Figure 20.42b,c indicates that ΔE_V can vary by nearly 1 eV depending on whether the interface bonding is Zn- or Se.

Figure 20.43a illustrates how these nonstoichiometric interfaces can be formed. By varying the beam pressure ratio (BPR) of Zn versus Se during ZnSe growth on As-stabilized GaAs by 2 orders of magnitude, Nicolini *et al.* could alter the XPS-measured ZnSe stoichiometry within the first 20–30 Å of the interface by over a factor of 3 [136].

Figure 20.43b shows that the effect of these BPR changes is to change the measured ΔE_V in the early growth stage ($\theta = 3$ Å) from 0.55 to 1.2 eV, certainly consistent with Figure 20.42b,c. The systematic increase in ΔE_V with R shows only a secondary dependence on the doping or surface reconstruction of the GaAs substrate.

Franciosi *et al.* showed that only the first 20 Å of ZnSe is needed in order to vary ΔE_V from 0.58 to 1.05 eV, emphasizing the importance of the spatially-confined stoichiometry on the macroscopic electronic transport.

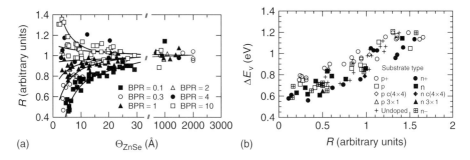

Figure 20.43 (a) Ratio R of Zn 3d versus Se 3d core-level intensities plotted as a function of ZnSe overlayer thickness θ on As-stabilized 2×4 reconstructed GaAs(100). (b) Corresponding valence-band offsets. (After Nicolini *et al.* [136])

20.4.9.2 Chemical Stability, Cross-Doping, and Interface States

The decrease in ΔE_V from ~0.9 to 0.55 eV permits more efficient hole transport from GaAs to ZnSe. However, the Se-rich stoichiometry required to achieve this effect degrades the chemical stability, increasing defects within the two semiconductors. This is illustrated by photoluminescence spectra of ZnSe/GaAs interfaces grown under stoichiometric and Se-rich conditions. The HeCd laser used here photoexcites the entire thickness of the ZnSe but rapidly attenuates upon passing into the narrower bandgap GaAs substrate. Figure 20.44a shows that a number of defect emissions below the 2.8 eV ZnSe bandgap are present with no annealing [137]. A new emission begins to appear at 1.9 eV for annealing temperatures of 400 °C. This emission is associated with Ga diffusion out of the GaAs into Zn vacancy sites within the ZnSe [138].

Figure 20.44b shows similar defect features to those in (a) except that the 1.9 eV peak now appears at temperatures as low as 300 °C. The emergence of this defect at significantly lower temperatures is consistent with the higher concentration of Zn vacancies expected for the Se-rich growth and the ability of diffusing Ga atoms to fill those vacancies. Hence, there is a trade-off between the decrease in ΔE_V with Se-rich growth versus the chemical stability of the heterojunction. Such electronic versus thermodynamic trade-offs must be considered in the design of heterojunctions.

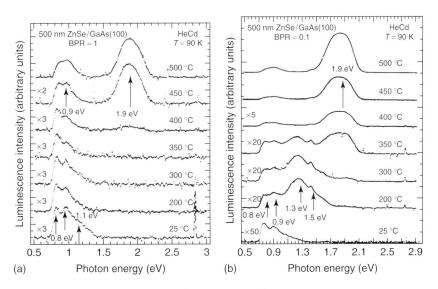

Figure 20.44 PL spectra versus annealing temperature for ZnSe–GaAs(001) heterostructures grown under (a) stoichiometric (BPR = 1) versus (b) Se-rich (BPR = 0.1) conditions. Annealing leads to a new 1.9–2.0 eV feature at temperatures above 300 °C. (After Raisanen et al. [137])

20.4.9.3 "Delta" Doping

Besides semiconductor or metal interlayers, dopant atoms provide another option for introducing an electric dipole at a microscopic heterojunction. Figure 20.45 illustrates this "delta doping" concept. Using MBE growth techniques, one can create near-monolayer sheets of ionized donor and acceptor charges within a few monolayers of and on opposite sides of an interface. This produces high electric field gradients that alter the band alignment.

Figure 20.45a shows a nested heterojunction with ΔE_C and ΔE_V indicated. The interface dipole shown in Figure 20.42b produces a voltage change $\Delta \Phi = \sigma d/\varepsilon$ between the two sides. The sign of this dipole results in downward band bending on the smaller gap side versus upward band bending on the higher gap side. If the distance d between the two sides of the dipole is only a few atomic layers, thermally activated electrons can tunnel through the triangular conduction band on the upward band-bending side. This effectively lowers ΔE_C by $e\Delta\Phi$. The overall effect of these changes tends to equalize ΔE_C and ΔE_V so that holes and electrons encounter nearly equal barriers to transport from the smaller to the larger heterojunction.

For AlGaAs p–i–n diodes, such dipoles can reduce ΔE_C by 0.1 eV. Reversal of the dipole sheets reverses the electric field so that ΔE_C increases and ΔE_V decreases. Asymmetric barriers are desirable for devices whose performance improves with single carrier injection, for example, gain in bipolar junction transistors.

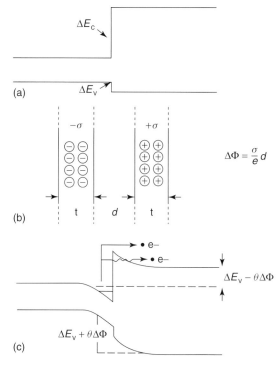

Figure 20.45 Type I heterojunction before (a) and after (c) introduction of a delta doped interface dipole (b) of magnitude $\Delta\Phi = \sigma d/\varepsilon$ with sheet charge $+\sigma$, thickness t, and separation d. The interface dipole effectively decreases ΔE_C and increases ΔE_V. (After Capasso. [133])

These heterojunction examples illustrate that (i) polar interfaces can give rise to large electric fields that render these structures unstable, (ii) atom exchange across the junction can minimize or eliminate these fields, and (iii) careful control of layer composition can provide a tool to "grow in" desired heterojunctions.

20.5 Summary

Semiconductor heterostructures are the basis of many current optoelectronic and microelectronic devices. It is possible to create lattice structures with a wide range of band structures and physical properties. The design of these properties depends on joining materials with similar lattice structures and controlled chemical bonding at their intimate junction.

Lattice-mismatched structures under strain give rise to dislocations and other morphological defects that degrade electronic properties. Lattice-strained heterostructures can be used to create carrier confinement channels with both high mobility and high carrier density. Likewise, strained films in superlattices enable the design of new band structures. The polarization fields induced by strain in these superlattices also have applications in nonlinear optics such as switches and logic elements. The bandgap in these structures can be engineered over wide ranges by controlling (i) layer composition, (ii) layer width, and (iii) strain.

Chemical reaction and diffusion must be minimized in order to maintain atomically abrupt interfaces. Semiconductor constituents that can act as dopants in their heterojunction counterparts are of particular concern.

Template structures between the semiconductors are useful as "bridges" between materials of significantly different lattice structures. They can also provide passivation, introduce new crystal orientations, improve morphology, and introduce electric fields and dipoles that can alter heterojunction band offsets.

Band offsets are fundamental to the transport and confinement of charge at heterojunctions. A variety of techniques are available to measure these band offsets, although each has limitations such that more than one technique is desirable to obtain unambiguous values. Experimental values offer precision limited to ≥ 0.1 eV, whereas 0.01 eV precision is needed to achieve predictive value.

The band offsets measured experimentally provide approximate support for several theoretical models despite their different physical bases. The failure of the electron affinity model underscores the importance of interface dipoles in the band structure alignment.

Many factors can contribute to these interface dipoles including (i) growth sequence, (ii) crystallographic orientation, (iii) surface reconstruction, (iv) interface composition, and (v) interfacial layers.

Band offset engineering makes use of these factors to obtain variations as large as 1 eV. The design of new heterojunctions may also involve trade-offs between electronic structure and themodynamical stability.

Problems

1. Using Vegard's law, calculate the critical thickness of $Al_{0.4}Ga_{0.6}N$ on GaN. Why do AlGaN/GaN heterojunctions for high electron mobility transistors typically have Al compositions in this range?
2. Calculate the critical thickness of GaN grown epitaxially on a ZnO wafer. (b) In terms of lattice matching, is ZnO better than SiC? (c) Besides lattice match, what other factors should be considered in using ZnO wafers as a large area template for GaN growth?
3. Sketch the band diagram for a heterojunction consisting of a p^+ semiconductor A with a bandgap $E_G = 2.7$ eV and an undoped n-type semiconductor B with bandgap $E_G = 1.4$ eV. Assume that $\Delta E_C = 2/3 \Delta E_G$, $\varepsilon_A = \varepsilon_B$ and, from continuity of the electric flux density at the interface, band bendings $V_{0A}/V_{0B} = N_B \varepsilon_B / N_A \varepsilon_A$ for doping densities N_A and N_B. Show ΔE_C, ΔE_V, the band lineups, the band bending, and relative depletion widths in each semiconductor.
4. Design a thin high mobility channel layer to be grown epitaxially on an InP substrate. What is the bandgap of this layer?
5. Design an epilayer structure grown epitaxially on a GaAs substrate that emits light at 5900 Å.
6. The space charges within the band-bending regions of a heterojunction are equal and opposite in sign. Assuming the depletion approximation for semiconductors A and B, show that the ratio of band bending V_{0A} in semiconductor A with doping N_A versus V_{0B} in semiconductor B with doping N_B is $V_{0A}/V_{0B} = N_B \varepsilon_B / N_A \varepsilon_A$. (Hint: use Equation 3.5)
7. Prove Equation 20.22 for a heterojunction comprised of semiconductor 1 with ε_1 and doping N_1 versus and semiconductor 2 with ε_2 and doping N_2. Assume the depletion approximation for the two semiconductors.
8. Calculate the required sheet doping to increase the valence-band offset by 0.1 eV across four monolayers at a GaAs–AlGaAs heterojunction. Assume $\varepsilon = \varepsilon(GaAs)$.
9. The gain in a p–n–p bipolar junction transistor (BJT) depends in part on the *emitter injection efficiency* γ, the ratio of majority versus majority plus minority carriers injected under electrical bias from an n-type emitter into a p-type base region. Minority carriers lower γ depending in part on the relative conduction versus valence-band offsets. (a) Give two examples of heterojunctions with offset differences that could be used to increase γ. (b) Calculate the ratio of thermal injection for electrons versus holes across one of these heterojunctions. (c) Draw a heterojunction band diagram to illustrate this concept.

References

1. Casey, H.C. Jr. and Panish, M.B. (1978) *Heterojunction Lasers, Part A: Fundamental Principles; Part B: Materials and Operating Characteristics*, Academic Press, New York.
2. Kuech, T.F., Wolford, D.J., Potoemski, R., Bradley, J.A., Kelleher, K.H., Yan, D., Farrell, J.P., Leser, P.M.S., and Pollak, F.H. (1987) *Appl. Phys. Lett.*, **51**, 505.
3. Cho, A.Y. (1985) in *Molecular Beam Epitaxy and Heterostructures*, NATA ASI Series, Vol. E87 (eds L.L.Chang and K. Ploog), Martinus Nijhoff, Dordrecht.
4. Mayer, J.W. and Lau, S.S. (1990) *Electronic Materials Science: for Integrated Circuits in Si and GaAs*, Macmillan Publishing Company, New York, p. 424.
5. Raisanen, A., Brillson, L.J., Goldman, R.S., Kavanagh, K.L., and Wieder, H.H. (1994) *J. Vac. Sci. Technol. A*, **12**, 1050.
6. Matthews, J.W. and Blakeslee, A.E. (1974) *J. Cryst. Growth*, **27**, 118.
7. Woodall, J.M., Kirchner, P.D., Rogers, D.L., Chisholm, M., and Rosenberg, J.J. (1988) *Mater. Res. Soc. Symp. Proc.*, **126**, 3.
8. Osbourn, G.C. (1986) *IEEE J. Quantum Electron.*, **QE-22**, 1677.
9. Osbourn, G.C., Gourley, P.L. II , Fritz, J., Biefeld, R.M., Dawson, L.R. and Zipperman, T.E. (1987) in *Semiconductors and Semimetals – Applications of Multiquantum Wells, Selective Dopings, and Superlattices*, Vol. 24 (ed. R.Dingle), Academic Press, New York.
10. Mailhiot, C. and Smith, D.L. (1986) *J. Vac. Sci. Technol.*, **4**, 997.
11. Patel, J.R., Golovchenko, J.A., Bean, J.C., and Morris, R.J. (1985) *Phys. Rev.*, **B 31**, 6884.
12. Durbin, S.M., Berman, L.E., Batterman, B.W., and Blakely, J.M. (1986) *Phys. Rev. Lett.*, **65**, 1227.
13. Dev, B.N., Materlik, G., Grey, F., Johnson, Rl.L., and Clausnitzer, M. (1986) *Phys. Rev. Lett.*, **57**, 3058.
14. Hoeven, J.J., Aarts, J., and Larsen, P.K. (1988) *J. Vac. Sci. Technol. A*, **7**, 5.
15. Chen, P., Bolmont, D., and Sébenne, C.A. (1984) *Thin Solid Films*, **111**, 367.
16. Bauer, R.S. and Sang, H.W. Jr. (1983) *Surf. Sci.*, **132**, 479.
17. Sarma, K., Dalby, R., Rose, K., Aina, O., Katz, W., and Lewis, N. (1984) *J. Appl. Phys.*, **56**, 2703.
18. Marschall, E.D., Lau, S.S., Palmstrøm, C.J., Sands, T., Schwartz, C.L., Schwartz, S.A., Harbison, J.P. and Florez, L.T. (1989) in *Materials Research Society Symposium Proceedings*, Vol. **148**, MRS, Pittsburgh, p. 163.
19. Chang, L.L. and Koma, A. (1976) *Appl. Phys. Lett.*, **29**, 138.
20. Laidig, W.D., Holonyak, N., Camras, M.D., Hess, K., Coleman, J.J., Dapkus, P.D., and Bardeen, J. (1981) *Appl. Phys. Lett.*, **38**, 776.
21. Li, D., Gonsalves, J.M., Otsuka, N., Qiu, J., Kobayashi, M., and Gunshor, R.L. (1990) *Appl. Phys. Lett.*, **57**, 449.
22. Qui, J., Qian, Q.-D., Gunshor, R.L., Kobayashi, M., Menke, D.R., Li, D., and Otsuka, N. (1990) *Appl. Phys. Lett.*, **56**, 1272.
23. Mackay, K.J., Allen, P.B., Herrenden-Harker, W., Williams, R.H., Williams, G.M., and Whitehouse, C.R. (1986) *Appl. Phys. Lett.*, **49**, 354.
24. Zahn, D.T.R., Williams, R.H., Golding, D.T., Dinan, J.H., Mackay, K.J., Geurts, J., and Richter, W. (1988) *Appl. Phys. Lett.*, **53**, 2409.
25. Woodall, J.M., Freeouf, J.L., Pettit, G.D., Jackson, T.N., and Kirchner, P. (1981) *J. Vac. Sci. Technol.*, **19**, 626.
26. Maissen, C., Mašek, J., Zogg, H., and Blunier, S. (1988) *Appl. Phys. Lett.*, **53**, 1608.
27. Cai, Z., Chen, Z., Goodrich, T.L., Harris, V.G. and Ziemer, K.S. (2007) *J. Cryst. Growth*, **307**, 321.
28. Tung, R.T. (1984) *Phys. Rev. Lett.*, **52**, 461.
29. Bringans, R.D. and Bachrach, R.Z. (1990) in *Synchrotron Radiation Research: Advances in Surface Science*, Chapter 12 (ed. R.Z. Bachrach), Plenum, New York.

30. Harrison, W.A., Kraut, E.A., Waldrop, J.R., and Grant, R.W. (1978) *Phys. Rev. B*, **18**, 4402.
31. Northrup, J.E. (1988) *Phys. Rev. B*, **27**, 9513.
32. Kroemer, H., Polasko, J., and Wright, S.J. (1980) *Appl. Phys. Lett.*, **36**, 763.
33. Petroff, P.M., Gossard, A.C., and Wiegmann, W. (1984) *Appl. Phys. Lett.*, **45**, 620.
34. LeGoues, F.K., Copel, M., and Tromp, R.M. (1989) *Phys. Rev. Lett.*, **63**, 1826.
35. Farrell, H.H., Tamargo, M.C., and de Miguel, J.L. (1991) *Appl. Phys. Lett.*, **58**, 355.
36. Streetman, B.G. and Banarjee, S. (2000) *Solid State Electronic Devices*, 5th edn, Prentice Hall, Upper Saddle River, p. 229.
37. Mailhiot, C. and Duke, C.B. (1986) *Phys. Rev. B*, **33**, 1118.
38. Anderson, R.L. (1962) *Solid-State Electron.*, **5**, 341.
39. Forrest, S.R. (1987) in *Heterojunction Band Discontinuities: Physics and Device Applications*, Chapter 8 (eds F.Capasso and G. Margaritondo), North-Holland, Amsterdam.
40. Lang, D.V. (1987) in *Heterojunction Band Discontinuities: Physics and Device Applications*, Chapter 9 (eds F.Capasso and G. Margaritondo), North-Holland, Amsterdam.
41. Hierro, A., Ringel, S.A., Hansen, M.A., Mishra, U., Denbaars, S., and Speck, J. (2000) *Appl. Phys. Lett.*, **76**, 3064.
42. Coluzza, C., Tuncel, E., Stachli, J.-L., Baudat, P.A., Margaritondo, G., McKinley, J.T., Ueda, A., Barnes, A.V., Albridge, R.G., Tolk, N.H., Martin, D., Morier-Genoud, F., Dupuy, C., Rudra, A., and Ilegems, M. (1992) *Phys. Rev. B*, **46**, 12834.
43. Dingle, R., Wiegmann, W., and Henry, C.H. (1974) *Phys. Rev. Lett.*, Vol. 33, 827.
44. Wolford, D.J. (1987) in *Heterojunction Band Discontinuities: Physics and Device Applications*, Chapter 6 (eds F. Capasso and G. Margaritondo), North-Holland, Amsterdam.
45. Wilson, B.A. (1988) *IEEE J. Quantum Electron.*, **QE-24**, 1763.
46. Menéndez, J. and Pinczuk, A. (1988) *IEEE J. Quantum Electron* **24**, 1698.
47. Albrektsen, O., Arent, D.J., Meier, H.P., and Salemink, H.W.M. (1990) *Appl. Phys. Lett.*, **57**, 31.
48. Abraham, D., Veider, L.A., Schönberger, Ch., Meier, H.P., Arent, D., and Alvarado, S.F. (1990) *Appl. Phys. Lett.*, **56**, 1564.
49. Smith, P.E., Goss, S.H., Gao, M., Hudait, M.K., Lin, Y., Ringel, S.A., and Brillson, L.J. (2005) *J. Vac. Sci. Technol. B*, **23**, 1832.
50. Forrest, S.R., Schmidt, P.-H., Wilson, R.B., and Kaplan, M.L. (1984) *Appl. Phys. Lett.*, **45**, 1199.
51. Grant, R.W., Kraut, E.A., Waldrop, J.R., and Kowalczyk, S.P. (1987) in *Heterojunction Band Discontinuities: Physics and Device Applications*, Chapter 4 (eds F. Capasso and G. Margaritondo), North-Holland, Amsterdam.
52. Kraut, E.A., Grant, R.W., Waldrop, J.R., and Kowalczyk, S.P. (1983) *Phys. Rev. B*, **28**, 1965.
53. Margaritondo, G. and Perfetti, P. (1987) in *Heterojunction Band Discontinuities: Physics and Device Applications*, Chapter 2 (eds F.Capasso and G. Margaritondo), North-Holland, Amsterdam.
54. Katnani, A.D. and Margaritondo, G. (1983) *Phys. Rev. B*, **28**, 1944.
55. Rubini, S., Milocco, E., Sorba, L., Pelucchi, E., Franciosi, A., Garulli, A., Parisini, A., Zhuang, Y., and Bauer, G. (2001) *Phys. Rev. B*, **63**, 155312.
56. Caine, E.J., Subbanna, S., Kroemer, H., Merz, J.L., and Cho, A.Y. (1984) *Appl. Phys. Lett.*, **45**, 1123.
57. King, P.D.C., Veal, T.D., Jefferson, P.H., McConville, C.F., Wang, T., Parbrook, P.J., Lu, H., and Schaff, W.J. (2007) *Appl. Phys. Lett.*, **90**, 132105.
58. Wu, C.L., Shen, C.-H., and Gwo, S. (2006) *Appl. Phys. Lett.*, **88**, 032105.
59. Tejedor, C., Calleja, J.M., Meseguer, F., Mendez, E.E., Chang, C.-A., and Esaki, L. (1985) in *Proceedings of the International Conference on the Physics of Semiconductors, San Francisco, CA* (eds J.A.Chadi and W.A. Harrison), Springer, Berlin, p. 559.

60. Gualtieri, G.J., Schwartz, G.P., Nuzzo, R.G., and Sunder, W.A. (1986) *Appl. Phys. Lett.*, **49**, 1037.
61. Menéndez, J., Pinczuk, A., Werder, D.J., Valladares, J.P., Chiu, T.H., and Tsang, W.T. (1987) *Solid State Commun.*, **61**, 1037. (a) Yu, E.T., Phillips, M.C., Chow, D.H., Collins, D.A., Wang, M.W., McCaldin, J.O. McGill, T.C. (1992) *Phys. Rev. B*, **46**, 13379.
62. Wilke, W.G., Seedorf, R., and Horn, K. (1990) *J. Cryst. Growth*, **101**, 620.
63. Yu, X., Raisanen, A., Haugstad, G., Ceccone, G., Troullier, N., and Franciosi, A. (1990) *Phys. Rev. B*, **42**, 1872.
64. Takatani, S. and Chung, Y.W. (1985) *Phys. Rev. B*, **31**, 2290.
65. Waldrop, J.R., Grant, R.W., Kowalczyk, S.P., and Kraut, E.A. (1985) *J. Vac. Sci. Technol. A*, **3**, 835.
66. Turowski, M., Margaritondo, G., Kelly, M.K., and Tomlinson, R.D. (1985) *Phys. Rev. B*, **31**, 1022.
67. Schulmeyer, T., Hunger, R., Klein, A., Jaegermann, W., and Niki, S. (2004) *Appl. Phys. Lett.*, **84**, 3067.
68. Nelson, A.J., Schwerdtfeger, C.R., Wei, S-H., Zunger, A., Rioux, D., Patel, R., and Höchst, H. (1993) *Appl. Phys. Lett.*, **62**, 2557.
69. Nelson, A.J. (1995) *J. Appl. Phys.*, **78**, 5701.
70. Nelson, A.J. (1995) *J. Appl. Phys.*, **78**, 2537.
71. Si, J., Jin, S., Zhang, H., Zhu, P., Qiu, D., and Wu, H. (2008) *Appl. Phys. Lett.*, **93**, 202101.
72. Bratina, G., Sorba, L., Antonini, A., Ceccone, G., Nicolini, R., Biasiol, G., Franciosi, A., Angelo, J.E., and Gerberich, W.W. (1993) *Phys. Rev. B*, **48**, 8899.
73. (a) Tersoff, J. (1984) *Phys. Rev. Lett.*, **52**, 465 ; (b) Tersoff, J. (1984) *Phys. Rev. B*, **30**, 4875.
74. Kowalczyk, S.P., Schaffer, W.J., Kraut, E.A., and Grant, R.W. (1982) *J. Vac. Sci. Technol.*, **20**, 705.
75. Zhang, R., Guo, Y., Song, H., Liu, Z., Yang, S., Wei, H., Zhu, Q., and Wang, Z. (2008) *Appl. Phys. Lett.*, **93**, 122111.
76. Lu, Y., Le Breton, J.C., Turban, P., Lépine, B., Schieffer, P., and Jézéquel, G. (2006) *Appl. Phys. Lett.*, **88**, 042108.
77. Liang, Y., Curless, J., and McCready, D. (2005) *Appl. Phys. Lett.*, **86**, 082905.
78. Zhang, P.F., Liu, X.L., Zhang, R.Q., Fan, H.B., Yang, A.L., Wi, H.Y., Jin, P., Yang, S.Y., Zhu, Q.S., and Wang, Z.G. (2008) *Appl. Phys. Lett.*, **92**, 012104.
79. Pellegrini, V., Börger, M., Lazzeri, M., Beltram, F., Paggel, J.J., Sorba, L., Rubini, S., Lazzarino, M., Franciosi, A., Bonard, J-M., and Ganiére, J-D. (1996) *Appl. Phys. Lett.*, **69**, 3233.
80. Baur, J. Maier, K. Kunzer, M. Kaufmann, U. Schneider, J. (1994) *Appl. Phys. Lett.*, **65**, 2211.
81. Waldrop, J.R. and Grant, R.W. (1996) *Appl. Phys. Lett.*, **68**, 2879
82. Wang, K., Lian, C., Su, N., Jena, D., and Timler, J. (2007) *Appl. Phys. Lett.*, **91**, 232117.
83. King, P.D.C., Veal, T.D., Kendrick, C.E., Bailey, L.R., Durbin, S.M., and McConville, C.F. (2008) *Phys. Rev. B*, **78**, 033308.
84. Chen, J.-J., Gila, B.P., Hlad, M., Gerger, A., Ren, F., Abernathy, C.R., and Pearton, S.J. (2006) *Appl. Phys. Lett.*, **88**, 104113.
85. Liu, H.F., Hu, G.X., Gong, H., Znag, K.Y., and Chua, S.J. (2008) *J. Vac. Sci. Technol. A*, **26**, 1462.
86. Perfetti, P., Patella, F., Sette, F., Quaresima, C., Capasso, C., Savoia, A., and Margaritondo, G. (1984) *Phys. Rev. B*, **230**, 4533.
87. Daniels, R.R., Margaritondo, G., Quaresima, C., Perfetti, P., and Levy, F.J. (1985) *J. Vac. Sci. Technol. A*, **3**, 479.
88. Waldrop, J.R., Kraut, E.A., Farley, C.W., and Grant, R.W. (1991) *J. Appl. Phys.* **69**, 372.
89. Hsieh, S., Hsieh, J., Patten, E.A., and Wolfe, C.M. (1984) *Appl. Phys. Lett.*, **45**, 1125. (a) Lee, T.W., Houston, P.A., Kumar, R., Yang, X.F., Hill, G., Hopkinson, M., and Claxton, P.S. (1992) *Appl. Phys. Lett.*, **60**, 474.
90. Zhang, P.F., Liu, X.L., Zhang, R.Q., Fan, H.B., Song, H.P., Wi, H.Y., Jiao, C.M. Yang, S.Y., Zhu, Q.S., and

Zhang, Z.G. (2008) *Appl. Phys. Lett.*, **92**, 042906.
91. Weiss, W. Wiemhöfer, H.D., and Göpel, W. (1992) *Phys. Rev. B*, **45**, 8478
92. Li, Y.F., Yao, B., Lu, Y.M., Li, B.H., Gai, Y.Q., Cong, C.X., Zhang, Z.Z. Zhao, D.X., Zhang, J.Y., Shen, D.Z., and Fan, X.W. (2008) *Appl. Phys. Lett.*, **92**, 192116.
93. Cerrina, F., Daniels, R.R., and Fano, V. (1983) *Appl. Phys. Lett.*, **43**, 182.
94. Cerrina, F., Daniels, R.R., Zhao, Te-Xiu, and Fano, V. (1983) *J. Vac. Sci. Technol. B*, **1**, 570.
95. Olmstead, M.A. and Bringans, R.D. (1990) *Phys. Rev.*, **41**, 8420.
96. Chambers, S.A., Liang, Y., Yu, Z., Droopad, R., Ramdani, J., and Eisenbeiser, Ki. (2000) *Appl. Phys. Lett.*, **77**, 1662.
97. Torvik, J.T. Leksono, M., Pankove, J.I., Zeghbroeck, B.V., Ng, H.M., and Moustakas, T.D. (1998) *Appl. Phys. Lett.*, **72**, 1371.
98. Zhang, B.L. Cai, F.F., Sun, G.S., Fan, H.B., Zhang, P.F., Wei, H.Y., Liu, X.L., Yang, S.Y., Zhu, Q.S., and Wang, Z.G. (2008) *Appl. Phys. Lett.*, **93**, 172110.
99. Fan, H.B., Sun, G.S., Yang, S.Y., Zhang, P.F., Zhang, R.Q., Wei, H.Y., Jiao, C.M., Liu, X.L., Chen, Y.H., Zhyu, Q.S., and Zhang, Z.G. (2008) *Appl. Phys.*, **92**, 192107.
100. Ruckh, M., Schmid, D., and Schock, H.W. (1994) *J. Appl. Phys.*, **76**, 5945.
101. Zhang, R. Zhang, P., Kang, T., Fan, H., Liu, X., Yang, S., Wei, H., Zhu, Q., and Wang, Z. (2007) *Appl. Phys. Lett.*, **91**, 162104.
102. Gorbik, P.P., Komashchenko, V.N., and Fedorus, G.A. (1980) *Sov. Phys. Semicond.*, **14**, 753.
103. Colbow, K.M., Gao, Y., Tiedje, T., Dahn, J.R., and Eberhardt, W. (1991) *J. Vac. Sci. Technol. A*, **9**, 2614.
104. Patten, E.A., Davis, G.D., Hsieh, S.J., and Wolfe, C.M. (1985) *IEEE Electron Device Lett.*, **EDL-6**, 60.
105. Yu, E.T., Phillips, M.C., McCaldin, J.O., and McGill, T.C. (1991) *J. Vac. Sci. Technol. B*, **9**, 2233.
106. Milnes, A.G. and Feucht, D.L. (1972) *Heterojunctions and Metal-Semiconductor Junctions*, Academic Press, New York.
107. Ishii, H., Sugiyama, K., Ito, E., and Seki, K. (1999) *Adv. Mat.*, **11**, 605.
108. Flores, F. and Tejedor, C. (1979) *J. Phys. C*, **12**, 731.
109. Tersoff, J. (1984) *Phys. Rev. Lett.*, **52**, 465; (b) Tersoff, J. (1984) *Phys. Rev. B*, **30**, 4874.
110. (a) Harrison, W.A. (1977) *Phys. Rev. Lett.*, **37**, 312; (b) Harrison, W.A. (1979) *ibid.*, **16**, 1492.
111. Frensley, W.R. and Kroemer, H. (1977) *Phys. Rev. B*, **16**, 2642.
112. Langer, J.M. and Heinrich, H. (1985) *Phys Rev. Lett.*, **55**, 1414.
113. Zunger, A. (1985) *Annu. Rev. Mater. Sci.*, **15**, 411.
114. Van Vechten, J.A. (1985) *J. Vac. Sci. Technol. B*, **3**, 1240.
115. Christensen, N.E., Gorczyca, I., Christensen, O.B., Schmid, U., and Cardona, M. (1990) *J. Cryst. Growth*, **101**, 318.
116. Jaros, M. (1988) *Phys. Rev. B*, **37**, 7112.
117. Tersoff, J. (1987) in *Heterojunction Band Discontinuities: Physics and Device Applications* (eds F.Capasso and G. Margaritondo), North-Holland, Amsterdam, pp. 1–57.
118. Kroemer, H. (1985) in *Molecular Beam Epitaxy and Heterostructures*, NATAO ASI Series E, Vol. 87 (eds L.L.Chang and K. Ploog), Martinus Nijhoff, Dordrecht, p. 331.
119. Katnani, A.D. (1987) in *Heterojunction Band Discontinuities: Physics and Device Applications*, Chapter 3 (eds F.Capasso and G. Margaritondo), North-Holland, Amsterdam.
120. Caldas, M.J., Fazzio, A., and Zunger, A. (1984) *Appl. Phys. Lett.*, **45**, 671.
121. Grant, R.W., Waldrop, J.R., Kowalczyk, S.P., and Kraut, E.A. (1985) *J. Vac. Sci. Technol. B*, **6**, 1295.
122. Grant, R.W., Waldrop, J.R., Kowalczyk, S.P., and Kraut, E.A. (1986) *Surf. Sci.*, **168**, 498.
123. Grant, R.W. and Harrison, W.A., (1988) *J. Vac. Sci. Technol. B*, **3**, 1295.

124. Chiaradia, P., Katnani, A.D., Sang, H.W. Jr., and Bauer, R.S. (1984) *Phys. Rev. Lett.*, **52**, 1246.
125. Qiu, J., Qian, Q.-D., Gunshor, R.L., Kobayashi, M., Menke, D.R., Li, D., and Otsuka, N. (1990) *Appl. Phys. Lett.*, **56**, 1272.
126. Olega, D.J. and Cammack, D (1990) *J. Cryst. Growth*, **101**, 546.
127. Franciosi, A. and Van de Walle, C.G. (1996) *Surf. Sci. Rep.*, **25**, 1–140.
128. Van de Walle, C.G. and Martin, R.M. (1986) *Phys. Rev.*, **34**, 5621.
129. Baroni, S., Resta, R., Baldereschi, B., and Peressi, M. (1989) *Spectroscopy of Semiconductor Microstructures*, Plenum, London, p. 251.
130. Sorba, L., Bratina, G., Ceccone, G., Antonini, A., Walker, J.F., Micovic, M., and Franciosi, A. (1991) *Phys. Rev. B*, **43**, 2450.
131. Niles, D.W., Tang, M., McKinley, J.T., Zanoni, R., and Margaritondo, G. (1988) *Phys. Rev. B*, **38**, 10949.
132. Katnani, A.D., Chiaradia, P., Cho, Y., Mahowald, P., Pianeta, P., and Bauer, R.S. (1985) *Phys. Rev.*, **32**, 4071.
133. Capasso, F., Cho, A.Y., Mohammed, K., and Foy, P.W. (1985) *Appl. Phys. Lett.*, **46**, 664.
134. McKinley, J.T., Hwu, Y., Koltenbah, B.E.C., and Margaritondo, G. Baroni, S. Resta, R. (1991) *J. Vac. Sci. Technol. A*, **9**, 917.
135. Marsi, M., La Rosa, S., Hwu, Y., Gozzo, F., Coluzza, C., Baldereschi, A., Margaritondo, G., McKinley, J.T., Baroni, S., and Resta, R. (1992) *J. Appl. Phys.*, **71**, 2048.
136. Nicolini, R., Vanzetti, L., Mula, G., Bratina, G., Sorba, L., Franciosi, A., Peressi, M., Baroni, S., Resta, R., Baldareschi, A., Angelo, J.E., and Gerberich, W.W. (1994) *Phys. Rev. Lett.*, **72**, 294.
137. Raisanen, A.D., Brillson, L.J., Vanzetti, L., Bonanni, A., and Franciosi, A. (1995) *Appl. Phys. Lett.*, **66**, 3301.
138. Gutowski, J., Presser, N., and Kudlek, G. (1990) *Phys. Status Solidi. A*, **120**, 11 and references therein.

Further Reading

Brillson, L.J. (1992) in *Basic Properties of Semiconductors*, Vol. 1, Chapter 7 (ed. P.T. Landsberg), North-Holland, New York.

Capasso, F. and Margaritondo, G. (eds) (1987) *Heterojunction Band Discontinuities: Physics and Device Applications*, North-Holland, Amsterdam.

Franciosi, A. and Van de Walle, C.G. (1996) Heterojunction band offset engineering. *Surf. Sci. Rep.*, **25**, 1–140.

21
Metals on Semiconductors

21.1
Overview

The most extensively studied electronic material interfaces are between metals and semiconductors. This reflects the importance that metal contacts to semiconductors have in electronics and the variety of phenomena that researchers have encountered in making such contacts. Unlike semiconductor–semiconductor heterojunctions that typically require epitaxical growth, metal contacts are usually not ordered and can be formed by a variety of methods. Thousands of scientific articles published over the last 70 years exemplify the continuing effort to understand and control these interfaces. In Chapters 3 and 4, we described the techniques used to measure the macroscopic electronic properties of metal–semiconductor interfaces and the various interface states that can alter their Schottky barriers. In this chapter, we examine these interface states and their effect on Schottky barriers in greater detail.

Researchers have used the ultrahigh vacuum (UHV) surface science techniques described in previous chapters to show the importance of atomic structure, chemical interactions, and native defects both at the intimate interface and extending into the semiconductor. We return to our model for the dipole contributions to Schottky barriers and the models developed to account for experimental results. We then present the experimental results available to evaluate and distinguish between these models. Surprisingly, several models can, to first order, account for selected Schottky barrier results despite quite different physical bases. We review the extrinsic phenomena that have an observable impact on Schottky barriers. This leads to a framework for describing Schottky barriers that requires both intrinsic and extrinsic factors. These factors contribute in varying degrees depending on (i) the intrinsic bulk metal and semiconductor properties, (ii) the crystal quality, (iii) chemical reaction and/or diffusion at or near the interface, and (iv) the conditions under which the interface is formed. Each of these factors can have a significant effect on charge transfer across the interface and the effective Schottky barrier height.

Surfaces and Interfaces of Electronic Materials. Leonard J. Brillson
Copyright © 2010 WILEY-VCH Verlag GmbH & Co. KGaA, Weinheim
ISBN: 978-3-527-40915-0

21.2
Metal–Semiconductor Interface Dipoles

As with the semiconductor–semiconductor junction described in Figure 20.19, the metal–semiconductor junction can be described in terms of multiple dipoles and internal potentials. Figure 21.1 illustrates the potential changes across the junction in terms of the metal internal potential S_M, semiconductor internal potential S_S, metal surface dipole V_M, semiconductor surface dipole V_S, and interface dipole V_i [1]. The surface dipole corresponds to the extension of electron wavefunction into vacuum as illustrated schematically in Figure 21.2. From Figure 21.1a, the metal work function Φ_M is

$$\Phi_M = V_M + S_M \tag{21.1}$$

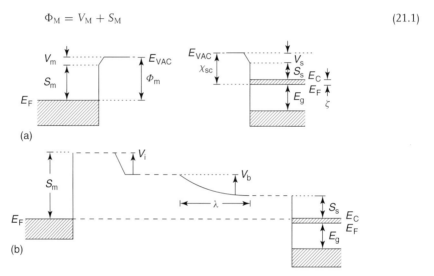

Figure 21.1 Potential distribution for (a) separated metal and semiconductor surfaces and (b) the metal–semiconductor interface. The observable work function Φ_M (electron affinity χ_{SC}) consists of a calculated internal potential S_M (S_S), plus a surface dipole V_M (V_S). The local interface dipole V_i plus the dipole V_b of the surface space charge region account for the difference in internal potentials when the metal and semiconductor are joined [1].

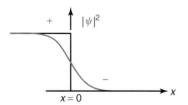

Figure 21.2 Wavefunction probability function (gray) extending out of the semiconductor or metal surface (black) ($x = 0$) into vacuum. Dipoles are required to confine electrons to the solid.

and the semiconductor electron affinity χ_S is

$$\chi_S = V_S + S_S + \varsigma \qquad (21.2)$$

where $\varsigma = E_F - E_C$ for filled states in the conduction band. In this picture, internal potentials are defined with respect to the highest occupied electron states.

Figure 21.1b illustrates the metal and semiconductor with their Fermi levels aligned. Now the surface dipoles V_S and V_M are replaced by an interface dipole V_i. As shown,

$$S_M = V_i + V_b + S_S \qquad (21.3)$$

where V_b is the semiconductor band bending and λ the width of the surface depletion region. V_b is the primary contribution to the Schottky barrier height Φ_{SB} since

$$\Phi_{SB} = V_b + (E_C - E_F) \qquad (21.4)$$

and $E_C - E_F$ is typically 0.1–0.2 eV or less. Solving for V_b and expressing the internal potentials in terms of measurable parameters

$$V_b = S_M - S_S - V_i \qquad (21.5)$$

$$= (\Phi_M - V_M) - (\chi_S - V_S - \varsigma) \qquad (21.6)$$

$$= (\Phi_M - \chi_S - \varsigma) + (V_S - V_M - V_i) \qquad (21.7)$$

As with semiconductor–semiconductor junctions, only the first term in Equation 21.7 is observable. The dipole term is inferred from the difference between $(\Phi_M - \chi_S - \varsigma)$ and V_b.

21.3 Interface States

21.3.1 Localized States

The localized states at interfaces can be categorized as follows: (i) intrinsic surface states whose properties depend only on the bulk semiconductor and metal, (ii) metal-induced gap states (MIGS) that derive from wavefunction tunneling of electrons from the metal into the semiconductor. These states are also intrinsic to the bulk properties of the junction constituents. (iii) Extrinsic states due to crystal contamination or other imperfections, and (iv) extrinsic states created by localized atomic bonding, interdiffusion, and chemical reaction. Intrinsic states of the clean semiconductor surface were discussed in Chapter 4 in light of indications that the extent of ionic bonding in the crystal lattice influences the extent of Fermi-level "pinning." Since these intrinsic surface states are outside the bandgap for most semiconductors *before* contact with metals and are replaced by other states *after* contact, they will not be discussed further.

21.3.2
Wavefunction Tailing (Metal-Induced Gap States)

Consider again the wavefunction of electronic states within the semiconductor bandgap. If present at the surface, its wavefunction must decay into both the semiconductor and the vacuum since neither has allowed states at these energies (see Figure 21.3a). Now consider the wavefunction tailing from a metal into a semiconductor. Wavefunctions of propagating states inside the metal can be represented by sinusoids as shown in Figure 21.3b. Inside the semiconductor, this wavefunction again decays within the forbidden gap. For simple metals on Si, such tails should extend ~10 Å into the semiconductor [2]. This produces a local maximum in electron density $|\psi|^2$ at the interface. Substitution of a featureless "jellium" of continuum states for the metal's electronic structure affords calculations that test the barrier sensitivity to different metal work functions for a wide range of semiconductors while neglecting any interface structural or chemical changes [3]. Such calculations display a weak dependence of Φ_{SB} on Φ_M for high dielectric constants of covalent semiconductors such as GaAs and Si and much stronger dependence for the more ionic semiconductors such as CdS, SiO$_2$, and ZnO [4, 5].

A related approach to such tunneling charge redistribution involves the existence of a charge neutrality level (CNL) in the semiconductor bandgap which minimizes the influence of the metal work function on the interface E_F position within

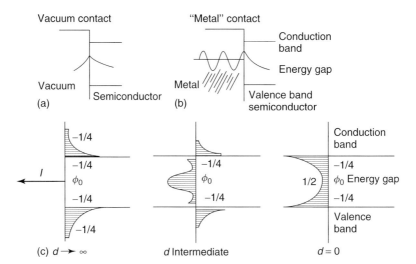

Figure 21.3 (a) Gap state wavefunction decay into vacuum and semiconductor. (b) Propagating metal wavefunction decay into semiconductor at gap state energies. (c) Metal-induced gap states (MIGS). The exponentially decaying behavior leads to the continuous density of interface states shown for a one-dimensional model of a covalent semiconductor–metal interface, where Φ_0 defines the *charge neutrality energy* of the semiconductor and d is the metal–semiconductor separation [6].

the semiconductor bandgap [6, 7]. As shown in Figure 21.3c, a new density of states due to the metal wavefunction is compensated for by a decrease in the semiconductor valence- and conduction-band density of states. Charge neutrality defines the energy below which the interface and valence-band densities of states compensate each other locally. This compensation results from a sum rule stating that the integrated density of states at any site, both filled and empty, is a constant [8]. Addition of the metal to the semiconductor surface changes the potential or boundary conditions, introducing new gap states which alter the occupancy of states in the local conduction and valence bands according to their conduction- or valence-band character [7]. The effective midgap energy between these states then determines an E_F position and local charge redistribution required to maintain charge neutrality across the interface. Figure 21.3c illustrates the initial conduction- and valence-band densities without a metal, the formation of discrete gap states at intermediate distances, and the formation of a continuum of states at the intimate contact. Here, the transfer of charge to or from the metal sets up a local dipole that tends to restore equilibrium of the metal E_F and CNL and to render Φ_{SB} insensitive to Φ_M.

Figure 21.4 represents the influence of the CNL on Fermi-level stabilization. In Figure 21.4, the CNL is defined such that [6]

$$E_F - CNL = S(-CNL + \Phi_M) \tag{21.8}$$

where CNL is the charge neutrality level and S is a dimensional parameter that can be related to semiconductor dielectric properties.

One can see the effect of parameter S on the Schottky barrier calculation by considering its limiting values in Equation 21.9 If the induced dipole $D = 0$, so that the interface is Schottky-like, then $S = 1$ and $E_F = \Phi_M$. On the other hand, if $S = 0$, then $E_F = CNL$ and the Fermi level is pinned at the CNL. Figure 21.4 also shows

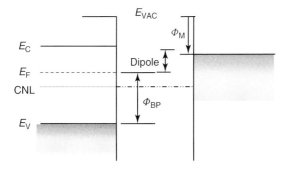

Figure 21.4 The CNL represents the cross over energy between the weighted average of conduction and valence bands. Charge tunneling into the semiconductor forms a dipole that accounts for the difference between metal work function and Fermi level at the semiconductor–metal interface.

that $(\Phi_M - CNL) = D + (E_F - CNL) = D + S(\Phi_M - CNL)$, so that [6, 9]

$$D = (1 - S)(\Phi_M - CNL) \tag{21.9}$$

For a given distribution of gap states, S can be expressed as

$$S = 1/\left[1 + \left(4\pi e^2 D(E_S)\delta/A\right)\right] \tag{21.10}$$

where $D(E_S)$ = density of states at energy E_S, δ is the distance into the semiconductor of the states, and A is the area. Example 21.1 provides an illustrative calculation of these parameters.

Example 21.1. Given a work function $\Phi_M(Au) = 5.1\,eV$, $\Phi_M(Cu) = 4.65\,eV$, $\Phi_{BP}(Au) = 0.34\,eV$, and $\Phi_{BP}(Cu) = 0.46\,eV$, calculate (a) S and (b) CNL for Si.
ANSWER: (a)

(a) $S = \Delta\Phi_{BP}/\Delta\Phi_M = (0.46 - 0.34)/(5.1 - 4.65) = 0.27$

(b) $E_F - CNL = (\Phi_M - CNL) + (E_F - \Phi_M) = S(\Phi_M - CNL)$

so that $(\Phi_M - CNL) = (\Phi_M - E_F)/(1 - S)$
For Au, $\Phi_{BP}(Au) = 0.34$ and $E_V = \chi + E_G = 4.05\,eV + 1.11\,eV = 5.16\,eV$,

so that $E_F = E_V - \Phi_{BP} = 5.16 - 0.34 = 4.82\,eV$

and $(\Phi_M - CNL) = (5.1 - (E_V - 0.34))/(1 - 0.27)$
$= (5.1 - (5.16 - 0.34))/0.73 = 0.28/0.73 = 0.38\,eV$

CNL $= \Phi_M - 0.38 = 5.1 - 0.38 = 4.72\,eV$, which is $E_V - CNL = 5.16 - 4.72 = 0.44\,eV$ above the valence band. It is worth noting that the CNL position is quite sensitive to the parameter S, which depends on a linear slope fit to barrier height data.

For $S \sim 0.1 - 0.3$ and $E_F - CNL \sim 0.1\,eV$, the interface dipole D can be on the order of 1 eV. Calculations of metal wavefunction tunneling yield large local dipole charge densities of $\sim 10^{14} - 10^{15}\,cm^{-2}\,eV^{-1}$. Calculation of the complex band structure branch points, that is, where the gap states cross over from valence- to conduction-band character, appear for many common semiconductors in Table 21.1.

Table 21.1 Charge neutrality levels (CNL in eV) relative to the valence band E_V from [7] and references therein.

Semiconductor	CNL	Semiconductor	CNL
Si	0.36	AlSb	0.45
Ge	0.18	GaAs	0.07
AlP	1.27	InSb	0.01
GaP	0.81	ZnSe	1.70
InP	0.76	MnTe	1.6
AlAs	1.05	ZnTe	0.84
GaAs	0.50	CdTe	0.85
InAs	0.50	HgTe	0.34

21.3.3
Charge Transfer, Electronegativity, and Defects

The magnitude of charge transfer and the resultant dipole depend sensitively on the theoretical boundary conditions assumed for charge penetration. Local atomic bonding leads to variations in CNL that depend on the metal–anion versus metal–cation bonding [10]. Experimentally, E_F "pinning" energies exhibit a monotonic dependence on the electronegativity of adsorbates for both Si and GaAs. The Miedema electronegativity X used here derives from a semiempirical fit with thermodynamic solubility data [11] and is related to the Pauling [12] electronegativity [13]. For metals on Si, the monotonic dependence shown in Figure 21.5a extends over half the Si bandgap and includes both metals that have reacted with Si to form silicides as well as nonreacting metals [14].

For metals on GaAs, the solid line shown in Figure 21.5b signifies the Φ_{SB} variation due to tunneling charge transfer as a function of $X_{ad} - X_{GaAs}$ and no defects [15]. Here, the CNL for $X_{ad} - X_{GaAs} = 0$ corresponds to a barrier height of ~ 0.92 eV. The dashed line indicates the E_F position with the largest density of defects compatible with the available data at $E_C - 0.65$ eV. The deviations from the solid line are large, indicating that the E_F stabilization energies depend on the relative density of extrinsic versus intrinsic gap states. As shown later, these extrinsic states can be related to both the semiconductor bulk as well as the metal–semiconductor interface.

21.3.4
Additional Intrinsic Pinning Mechanisms

Besides intrinsic surface states and metal-induced gap states, other proposed pinning mechanisms include (i) dangling bond hybrid orbitals that shift electrostatically to keep E_F pinned at a constant energy in the bandgap [17], (ii) a "negative-U" model of negative electron correlation between the interface electrons due to atomic disorder and electron localization [3], a narrowing of the bandgap due to reduction of exchange-correlation contributions to the bandgap [18], and combinations of metal-induced gap states and defects. However, all these intrinsic mechanisms require abrupt interfaces with no additional chemical phases. In fact, only a few such interfaces are known (see Chapter 18), and all exhibit a dependence on interface atomic structure.

21.3.5
Extrinsic States

In contrast to intrinsic mechanisms, there are a wide range of extrinsic effects that can impact charge transfer at metal–semiconductor interfaces. These can be grouped according to their surface or bulk origin.

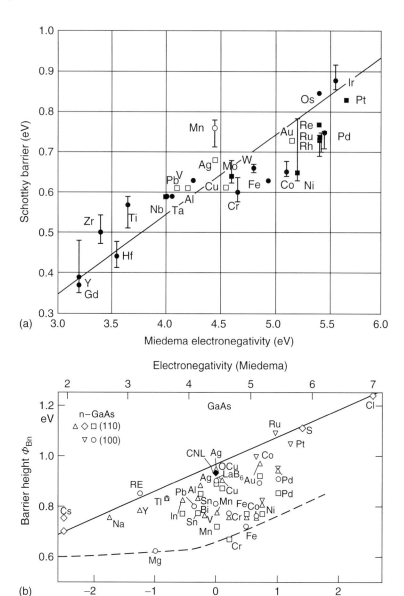

Figure 21.5 (a) Φ_{SB} versus Miedema electronegativity for transition-metal silicides and nonreacting metals on n-type Si. Data points correspond to silicides (■ and ●) and nonreacting metals (□) [14]. (b) Φ_{SB} versus Miedema electronegativity X difference between adatoms and GaAs substrate with $X_{GaAs} = 4.45$ for metal–GaAs contacts and for adsorbates on GaAs(110) cleaved surfaces. The deviations in Φ_{SB} from the solid line indicate high densities of interface defects [13, 15].

21.3.5.1 Surface Imperfections and Contaminants

Lattice imperfections and contaminants on semiconductor surfaces can introduce localized states that trap charge. They can be categorized as (i) chemisorbed species, (ii) chemical contamination, (iii) structural imperfections, and (iv) point defects. Some of the earliest Schottky barrier studies showed that surface chemisorbed species produce new electronic states in the semiconductor bandgap [19, 20]. More recent soft X-Ray Photoemission Spectroscopy (SXPS) studies provide an elegant example relating adsorbate energies to their ionization potentials. Deposition of adsorbates in submonolayer amounts on clean, cleaved GaAs(110) surfaces at low temperature, where metallic clustering and/or new chemical bonding is retarded, exhibit "overshoot" movements of E_F due to charge exchange involving adsorbate donor or acceptor states [21]. Figure 21.6 shows the E_F dependence on metal coverage and its relation to the proposed adsorbate level located an energy E_d above E_V [16]. For example, the midgap donor level in p-type GaAs shown in Figure 21.6a introduces additional positive charge at the surface, increasing the downward band bending and raising E_F closer to E_D. The "overshoot" represents the dominant effect of the adsorbate at coverages below which metallization or reaction move E_F to its final position. Figure 21.6b illustrates a linear dependence of E_d obtained from the "overshoot" energy on the adsorbate's first ionization potential for an array of adsorbate-induced donors (o) and acceptors (•) on GaAs(110). The two-stage E_F movement reflects the submonolayer dependence on the electron affinity (EA) of the isolated atoms but the ultimate dominance of a second mechanism at multilayer coverages.

Chemical contamination at the monolayer level can itself affect E_F stabilization strongly. Clean surfaces with and without exposure to the atmosphere exhibit qualitatively different E_F movements and band bending with subsequent metallization. For example, air exposure prior to Au deposition on InGaAs alters both

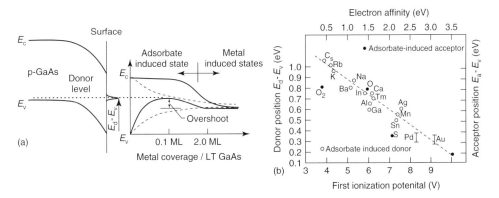

Figure 21.6 (a) SXPS-measured E_F movement versus metal coverage for adsorbates on GaAs(110) at room temperature (dashed line) and low temperature (solid line). (b) Adsorbate-induced donor (acceptor) level position obtained from the initial E_F movements on low-temperature p-type (n-type) GaAs as a function of the adatom's first ionization energy E_d (electron affinity) [16].

the premetallized band bending and the final Schottky barrier height [22], with air-exposed values in agreement with Φ_{SB} for diodes prepared under low vacuum conditions [23]. Air exposure also degrades the Φ_{SB} stability of metal/GaAs(110) diodes with annealing [24].

Structural imperfections such as lattice step edges due to crystal cleavage or misorientation as well as near-surface point defects can introduce new states into the semiconductor bandgap that can "pin" the Fermi level. Thus, cleavage steps can introduce high ($> 10^{14}$ cm^{-2}) densities of states at poorly cleaved surfaces [26]. Indeed such steps accounted for early reports of intrinsic surface states on UHV-cleaved surfaces. Atoms at step edges are also more electrically and chemically active than terrace atoms due to their lower bond coordination. Thus, intentionally stepped, vicinal surfaces of GaAs(100) exhibit a quantitative correlation between the density of electrically active interface states deep in the semiconductor bandgap and the density of exposed, chemically active sites [27].

Point defects are also electrically active and include interstitials and vacancies introduced by high-energy (MeV) electrons and ions as well as during sample cleaning, for example, by ion bombardment or by reactive ion etching.

21.3.5.2 Bulk States

Imperfections within the semiconductor bulk can also affect metal–semiconductor interface properties. These include (i) impurities, (ii) native point defects, and (iii) extended defects.

Impurities Impurities within the semiconductor lattice can introduce deep levels that alter the surface space charge region [28]. Such impurities may be introduced during the crystal growth process, for example, C in GaAs grown by molecular organic chemical vapor deposition (MOCVD). Impurities can also diffuse into semiconductors during the processes of metallization or thermal annealing, for example Au diffusion into GaAs during contact formation. Impurities and their complexes with native defects in III–V and II–VI compound semiconductors can introduce multiple deep levels with energies throughout the bandgap. Noble and near-noble metals such as Au, Ag, Cu, and In can diffuse rapidly into semiconductors at relatively low temperatures since they do not react strongly with the lattice atoms [29]. Impurity diffusion is higher in more ionic semiconductors, which have larger lattice constants and higher concentrations of vacancies.

Similarly, nonstoichiometry introduces electrically active sites. Such nonstoichiometry can be intentional in order to control morphological defects. A notable example is the excess As used in the melt growth of GaAs. Figure 21.7 shows that the dislocation density in GaAs grown by a Horizontal Bridgman technique decreases by orders of magnitude to reach a minimum at an optimum melt stoichiometry that is slightly As-rich [25]. This effect is associated with the charge state of native defects and their tendency to coalesce into dislocations. The excess As in such melt-grown bulk GaAs also introduces native point defects such as As$_{Ga}$ antisites that have concentrations orders of magnitude higher than GaAs grown by vapor-phase, molecular-beam, or liquid-phase epitaxy [30].

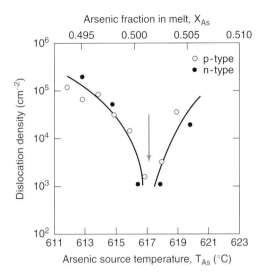

Figure 21.7 Dislocation density versus As source temperature and melt stoichiometry for lightly doped n-type and p-type GaAs [25].

Native Defects Native defects within the semiconductor bulk lattice produce many different trap levels in semiconductors with densities that vary by orders of magnitude depending on the growth method. Such variations are readily observed by luminescence or trap spectroscopies in crystals of the same semiconductor grown by different techniques or sources and even in different crystals grown by the same source. Furthermore, such imperfections may segregate to the near-surface region where their density can increase by orders of magnitude. Thermal processing will enhance such defect or impurity redistribution.

Extended Defects Extended surface or bulk imperfections such as misfit dislocations, antiphase domain boundaries, and combinations of native defects produced during growth or processing can be electrically active. These defects can produce electrostatic band bending both in as well as normal to the interface plane. For example, misfit dislocations at the junction between InGaAs and GaAs exhibit depletion regions extending radially outward, consistent with line charge densities. Figure 21.8 illustrates the three-dimensional nature of this band bending with the InGaAs/GaAs interface in the x–z plane and the dislocation lines parallel to the z-axis [31]. A plan view (along the y axis) of a similar InGaAs/GaAs interface appears in Figure 20.6. As the misfit density increases with increasing In concentration, the dislocation density increases so that the carrier depletion regions begin to overlap. The result is a "pinch-off" of undepleted regions and a change in the average barrier height across the interface plane.

Interfacial stress produced by plastic deformation, chemical reaction, or even metallization can also introduce electrically active sites that are extended in nature.

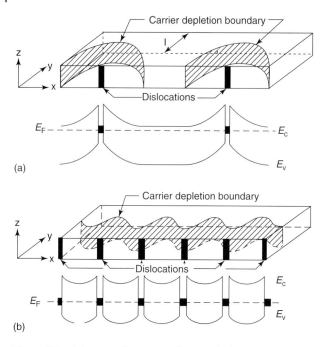

Figure 21.8 Schematic illustration of carrier depletion region surrounding line charges within dislocations at an InGaAs/GaAs interface [31].

These extrinsic bulk and surface features emphasize the importance of crystal quality and surface preparation, respectively, in Schottky barrier formation. Deviations from ideal-crystal properties on an atomic, nanometer, or macroscopic scale contribute electrically active sites that can strongly affect charge transfer and band bending at metal–semiconductor interfaces.

21.3.6
Interface-Specific Extrinsic States

There is now strong evidence that extrinsic states produced by chemical interaction at the metal–semiconductor interface can dominate Schottky barrier formation. These interface-specific states include localized states due to metal–semiconductor bonding and interdiffusion, defect formation, and new phase formation. Each of these interactions is sensitive to the kinetics and thermodynamics of the chemical interactions at the intimate metal–semiconductor junction. These interface-specific features are common to all semiconductors.

21.3.6.1 Interface Reaction and Diffusion
The role of metal–semiconductor bonding and interdiffusion is supported by a diverse array of phenomenological correlations. For Si, there exists a linear

correlation between barrier heights of transition-metal silicide/Si junctions and the heats of formation, that is, enthalpy changes, of the transition-metal silicide. This can be interpreted as the strength of chemical bond charge transfer and bond hybridization between metal and Si atoms [33] (see Figure 4.14). Barrier heights for transition metals on Si also show a dependence on the eutectic temperature of the transition-metal silicide/Si system, as shown in Figure 4.15 [34]. This correlation reflects a measured thermodynamic parameter, that is, the minimum reaction temperature, and thereby the enthalpy and the bond strength of the interfacial layer. Again, macroscopic barrier heights appear related directly to properties of an extrinsic layer formed between metal and semiconductor.

Compound semiconductors also exhibit a strong correlation between Φ_{SB} and interface thermodynamics. Figure 4.17 displayed a qualitative transition in Φ_{SB} for metals on compound semiconductors according to an interface heat of reaction ΔH_R defined by Equation 4.9 [35]. This macroscopic barrier height measured by internal photoemission exhibits a transition from low to high Φ_{SB} that occurs at a critical value ΔH_R^C observed by low energy electron loss spectroscopy as described in Section 10.4. The same chemical transition occurs, regardless of semiconductor ionicity. This systematic behavior suggests that chemical stability factors into the Φ_{SB} dependence. Indeed, Figure 4.16 shows that the index of interface behavior defined by Equation 4.1 exhibits the same behavior with semiconductor heat of formation as with ionicity. Thus, semiconductors with the lowest chemical stability and thus the strongest tendency to form interfacial layers exhibit the weakest Schottky barrier dependence on metal work function or ionicity.

This variation of metal–semiconductor interfaces with stability and reactivity manifests itself in atomic-scale diffusion experiments. Figure 8.7 showed the ratio of outdiffused In and P from InP through thin metal overlayers [36]. First of all, this plot demonstrates that In and P outdiffuse into metal overlayers. Such decomposition and outdiffusion is a general phenomenon at metal–semiconductor interfaces.

Secondly, more anions than cations diffuse through unreactive metals such as Au, Ag, Cu, and Pd. Conversely, outdiffusion is cation-rich for reactive metals such as Al, Ni, and Ti. Similar results are obtained for the reactive metals Cr and Mn [37]. This behavior reflects the "chemical trapping" of anions by the reactive metals and is characteristic of III–V compound semiconductors in general.

Finally, the stoichiometry of outdiffusion correlates with the magnitude of Φ_{SB}, with unreactive metals producing high barriers and reactive metals creating low barriers. This result links the outdiffusion of atoms from the interface region with the Fermi-level stabilization energy. Since atomic vacancies in the semiconductor lattice are typically electrically active, this outdiffusion phenomenon shows that atomic movements across the metal–semiconductor interface create electrically active sites that stabilize E_F.

21.3.6.2 Atomic Structural and Geometric Effects

Even subtle local bonding effects can impact Schottky barrier formation. As illustrated in Chapter 4, the detailed atomic positions of metal atoms at the semiconductor interface can strongly influence the density of states in the bandgap region. Thus, Figure 18.4 illustrated two epitaxial variants of the $NiSi_2/Si$ interface rotated by $180°$ about the surface normal [32]. Despite this relatively small difference in bonding, Figure 21.9 shows I–V measurements that differ by over 2 orders of magnitude in forward current, corresponding to $\Phi_{SB} = 0.64$ versus $0.78\,eV$. Such differences could be complicated by differences in structural perfection rather than epitaxy. However, the intrinsic nature of these $NiSi_2/Si$ differences has been confirmed experimentally and justified on the basis of tunneling charge transfer [38, 39]. Similar structure versus Φ_{SB} differences are observable for $Pb/Si(111)$ epitaxial interfaces [40]. For both these Ni– and Pb–Si epitaxial interfaces, the abrupt metal–semiconductor interfaces are free of extended chemical reactions, interdiffusion, and lattice disorder, and their significant differences in Φ_{SB} emphasize how even single monolayer bonding can change macroscopic electronic properties.

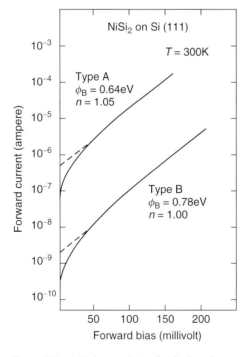

Figure 21.9 I–V characteristics for diodes where rectifying contacts are type-A and type-B $NiSi_2$ on $Si(111)$. Both exhibit ideality factors near unity and significant Φ_{SB} differences [32].

21.3.6.3 Chemisorption-Induced Effects

Adsorbate chemisorption can create electrically active defects at semiconductor surfaces. In this process, the energy released by chemisorption breaks surface bonds of the semiconductor, forming defects with energy levels in the bandgap that can "pin" the Fermi level [41, 43–47]. In addition to the modifications in surface bonding induced by the condensation process, the atoms can form clusters that release sufficient energy to overcome the activation barriers for breaking bonds within the semiconductor surface [48]. The introduction of inert intermediate layers provides a test for this bond breaking picture since condensation then occurs at the Xe rather than the semiconductor surface. Indeed, Xe layers between metal condensates and the GaAs(110) surface result in different band bending versus metal condensation directly on GaAs [49].

The work of Spicer and coworkers in the 1980s showed that a variety of adsorbates on cleaved GaAs(110) produced Fermi-level pinning at energies that were independent of the adsorbate. Using photoemission spectroscopy (PES) techniques (described in Section 7.8.4), they monitored band bending as a function of adsorbate coverage to determine the final E_F position in the bandgap. The pinning positions for n-type (o) and p-type (Δ) GaAs appear in Figure 21.10a for the elements indicated. For the room temperature 1.42 eV GaAs bandgap, these various adsorbates indicate two pinning energies at 0.65 and 0.5 eV above the n- and p-type valence band maximum (VBM), respectively [41].

This commonality of E_F pinning position for different adsorbates as diverse as O and various metals led to a *"unified defect model"* based on chemisorption-induced creation of the same defect, regardless of adsorbate. Chemisorption on the cleaved GaAs surface induced a rapid E_F movement to a final position within a few angstroms of adsorbate coverage, suggesting a local interaction confined to the outer semiconductor. Subsequent photo-spin resonance measurements showed a nearly identical pair of energy levels at 0.52 and 0.75 eV above E_{VBM} at 8 K (E_G^{GaAs} = 1.52 eV) that were associated with bulk As_{Ga} antisite defects, termed *EL2* [42, 51] (see Figure 21.10b). This close correspondence between native defects common to bulk GaAs grown by Czochralski or Horizontal Bridgman techniques showed that the E_F pinning induced by chemisorbed species on the GaAs(110) surfaces could be directly associated with native point defects rather than any interface-specific bonding.

Theoretical studies of deep-level defects in compound semiconductors have centered on vacancies [52–54] and antisites [55, 56] and their trends with particular metal or semiconductor compositions. While early Schottky barrier models incorporated surface anion and cation vacancies [41, 45], later calculations showed that antisite defects have lower activation energies than simple vacancy or interstitial defects in III–V compounds [57, 58]. In general, a variety of defect or defect plus tunneling models are capable of accounting for the narrow Φ_{SB} range widely reported for cleaved GaAs.

The correspondence between n- and p-type E_F stabilization energies shown in Figure 21.10a and EL2 energy levels in Figure 21.10b requires As_{Ga} concentrations significantly higher than Ga_{As}, as expected for crystals grown with an excess of As.

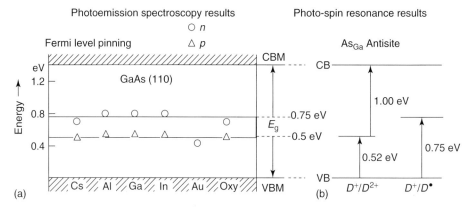

Figure 21.10 (a) Photoemission spectroscopy measurements of E_F pinning levels at the cleaved GaAs (110) surface for various UHV-deposited elements [41]. (b) Photoinduced electron spin resonance measurements of GaAs and their relation to E_F pinning [42].

For high enough defect concentrations, E_F movement away from the n-type GaAs conduction band with increasing chemisorption would then stabilize at the higher defect level in Figure 21.10b, while E_F movement away from the p-type GaAs valence band would stabilize at the lower energy. Significantly, the melt-grown cleavage material widely used by many researchers in the 1970s and 1980s to measure GaAs E_F positions was grown on the As-rich end of the GaAs bulk phase diagram as shown in Figure 21.7. Thus, high densities of antisite defects resident in the bulk provide a plausible explanation for the midgap E_F stabilization observed at cleaved, melt-grown GaAs surfaces. It also highlights the role that crystal quality can play in Schottky barrier formation.

21.3.6.4 Interface Chemical Phases

New chemical phases at metal–semiconductor interfaces represent another extrinsic factor in Schottky barrier formation. As described in Section 18.2, chemical interactions can lead to interfacial phases with work functions and dielectric properties that differ substantially from those of either metal or semiconductor. For metals on Si, a correlation between Φ_{SB} and an alloy work function based on a common metal–Si_4 stoichiometry suggests the presence and dielectric effect of such an interphase [59].

For metals on III–V compound semiconductors, there is considerable evidence for excess anions segregated to the interface [36, 60, 61]. Such anion layers would then establish an "effective work function" such that various metals would appear to have the same local work function with band bending determined by charge transfer between the semiconductor and the interfacial phase rather than the metal itself [50, 62]. Schottky barriers reported for Au on many semiconductors supports this model. Figure 21.11 shows E_F positions on an absolute energy scale with respect to the valence and conduction bands of different III–V and pseudobinary alloy

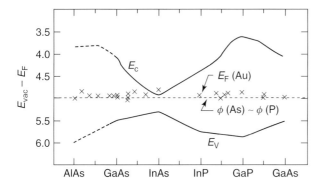

Figure 21.11 E_F positions for Au contacts to various III–V compounds and alloys plotted on an absolute energy scale. E_F "pinning" energies lie at the anion work function energy, suggesting that an anion-rich interphase dominates the interface charge transfer [50].

compound semiconductors. The nearly constant interface Fermi-level position for relatively unreactive Au contacts to these arsenides and phosphides all lie close to the dashed line $E_{VAC} - E_F$, corresponding to As and P work functions of 4.8–5.1 eV. The interface states associated with As-rich phases may also be considered chemisorption states at submonolayer coverages that evolve into metallic edges of As clusters, precipitates or macroscopic phases. The small variations in Φ_{SB} shown in Figure 21.11 can be interpreted in terms of changing interface anion composition. An advantage of such interfacial models is that the new phases can in principle be verified directly by microscopic observation.

Lattice disorder at the intimate metal–semiconductor junction may also be considered to be a new interfacial phase. For example, fluctuations in bond angle, bond length, as well as the presence of stress and irregularity boundaries can produce a distribution of bonding and antibonding states with energies around a characteristic hybrid orbital energy [64, 65]. This disorder introduces dangling bonds with a characteristic continuum of energies in the bandgap and a minimum density near an empirical hybrid energy. Such a *disorder-induced gap state* (DIGS) model can account for insulator–GaAs and metal–GaAs "pinning."

21.3.6.5 Organic Semiconductor–Metal Dipoles

Interfaces are a critical factor in charge injection into and transport through organic semiconductors. Many of the mechanisms that affect the metal–organic semiconductor interface are similar to their inorganic counterparts [66]. As with inorganic semiconductors, there is a breakdown of the vacuum-level alignment rule due to interface dipoles. Deviations from the Schottky–Mott limit can be as large as 1 eV for many metal–organic junctions. Charge injection and transport are more complex, the latter involving thermally activated intermolecular hopping or band movement, depending on the organic's crystallinity. Furthermore, space-charge-limited, trap-limited, or injection-limited models are typically used to describe movement across the interfaces involving these low dielectric constant, nearly charge-depleted materials. Finally, these transport and injection processes can not always be separated experimentally.

The morphology and chemistry of metal–organic interfaces can have a strong impact on their electrical properties. Because organic materials are soft, the processes used to form their interfaces can greatly affect the junction morphology, leading to chemical reaction or diffusion, depending on the sequence of deposition. Figure 21.12 illustrates this difference for the interface between Mg and tris(8-hydroxyquinoline)aluminum (Alq_3). This interface is particularly significant for organic light emitting diodes (OLEDs) since Alq_3 is used in electron transport and green light emission, while an Mg : Ag alloy is a good electron injector for Alq_3 [63].

For Alq_3 on Mg, exponential attenuation of the Mg with coverage indicates an abrupt interface, whereas the slower attenuation of the Alq_3 C 1s peak by Mg signifies a more diffuse interface with Mg penetrating the organic matrix. Notwithstanding this difference, both interfaces exhibit similar photoemission features, chemical shifts, states in the Alq_3 gap, and E_F relative to the *highest occupied molecular orbital* (HOMO). I–V measurements indicate identical injection for the two junctions. This result is consistent with a chemically induced density of gap states that pins E_F at the same energy, regardless of deposition sequence. Conversely, any broad distribution of interface states due to chemical reaction and diffusion does not play a major role in charge injection across these interfaces.

A counter example is the $F_{16}CuPc$ (Pc, pthalocyanine)/Au system in which the deposited molecules do not react with the metal and do not induce gap states in the organic material. On the other hand, Au deposition on $F_{16}CuPc$ leads to doping or a reaction with the organic and E_F that is 0.5–0.6 eV lower relative to the organic HOMO [67].

Even without strong chemical interactions, interface dipoles can form due to the change in electron wavefunction tunneling into a dielectric medium rather than vacuum, for example, the "pillow" effect described in Section 20.4.3.2.

Electrical doping has proven to be an effective method to achieve efficient charge injection across metal–organic interfaces. Extensive studies of both organic and inorganic dopants have already led to significant improvements in OLED technology [68, 69]. Figure 21.13 illustrates the *lowest unoccupied molecular orbital* (LUMO) and HOMO for ZnPc, tetrafluoro-tetracyanoquinodimethane (F_4-TCNQ), and N,N'-diphenyl-N,N'-bis(1-naphthyl)-1,1'-biphenyl-4,4'-diamine (α-NPD) with their vacuum levels aligned. Valence-band photoemission combined with inverse PES provides HOMOs and LUMOs, respectively, for the same organic surface in UHV.

To dope the host molecule p-type, an electron must transfer to the dopant. This transfer requires that the EA of the dopant and the ionization energy (IE) of the host molecule by close in energy. As shown, electron transfer occurs from either ZnPc or α-NPD HOMO to the highly electronegative F_4-TCNQ LUMO, resulting in hole doping of the host. Figure 21.14 illustrates the effect of this doped film.

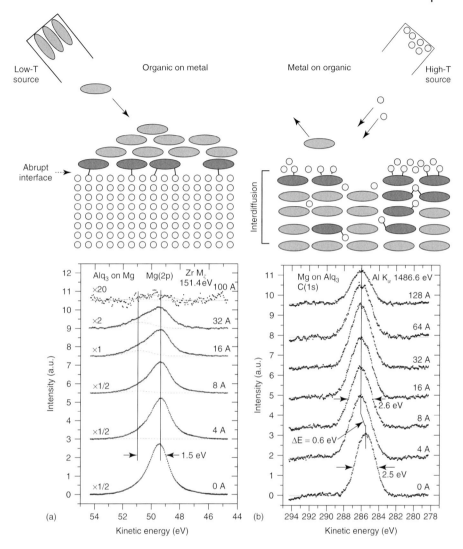

Figure 21.12 (a) Low-temperature deposition of Alq$_3$ on Mg (top) and Mg 1s core level versus Alq$_3$ coverage. Note the reacted Mg component (dashed) localized at the buried interface. (b) High-temperature deposition of Mg on Alq$_3$ (top) and C 1s core level versus Mg coverage [63].

Au deposited on the undoped film produces an interface with $E_F - E_V$ constant (flat bands) extending away from the interface, an interface dipole of 0.86 eV, and a Φ_{SB} of 1.24 eV. With a 0.5% F$_4$-TCNQ doping, the bands bend down with a depletion width of 20–40 Å, the interface dipole is 0.68 eV, and Φ_{SB} decreases to 1.10 eV.

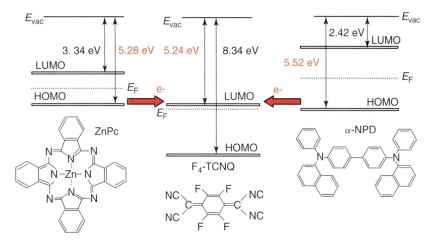

Figure 21.13 Ionization energy (IE) and electron affinity (EA) of ZnPc, F_4-TCNQ, and α-NPD with their molecular structures. The energy-level alignment between F_4-TCNQ LUMO and both ZnPc and α-NPD HOMOs enables efficient electron transfer as indicated [66].

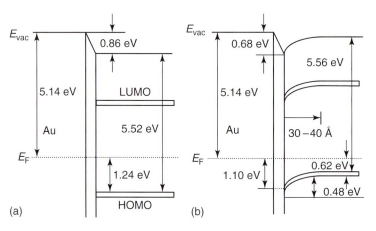

Figure 21.14 Schematic energy band diagram of the molecular levels near the interface between Au and (a) undoped α-NPD and (b) α-NPD: 0.5% F_4-TCNQ. Along with p-type band bending, doping reduces the electron affinity and the hole barrier height [66].

Even though the interface chemical interaction constrains the position of E_F in the gap, tunneling through the now p-type bent band region enables tunneling and an increase of hole injection by nearly 4 orders of magnitude. Thus, interface dipoles have a dominant effect on charge injection at metal–organic molecule interfaces that is difficult to control with the intrinsic materials alone.

On the other hand, molecular doping can be effective in overcoming these limitations. This continues to motivate the development of efficient and stable n- and p-type molecular dopants that can be inserted into molecular organic host matrices.

21.4
Self-Consistent Electrostatic Calculations

The preceding sections described numerous models that can account for the Fermi-level stabilization in a narrow range of energies. However, *independent of a particular physical mechanism*, one can gauge the effect of localized charge states on semiconductor band bending. As described in Chapter 4, the presence of acceptors and donors at the interface creates dipoles with potential V_i that alter the overall band bending V_b according to Equation 21.7. The charge states of these electrically active sites will depend upon the E_F position relative to the energy levels of these donors and acceptors in the bandgap. In turn, the total interface charge depends upon which donor and/or acceptor sites of this type are present, their densities, and their relative positions in the bandgap. The relationship between V_b and metal work function Φ_M will exhibit a characteristic dependence or "fingerprint" of the detailed energetics and position of these localized states. Figure 21.15 illustrates the V_b versus Φ_M dependence for various densities of different charge-center densities for both negative and positive U ordering of three-charge centers (i.e., positive, neutral, negative) near both n- and p-type semiconductor surfaces [70]. Depending on the exchange–correlation interaction potential U for such centers, acceptor energies lie above or below the donor energies. Acceptor states lie above donor states in energy for positive U and vice versa for negative U. In the calculation shown, charge centers are situated 10 Å below the semiconductor surface. Charge-center levels are at energies $E_D = E_G/3$ and $E_A = 2E_G/3$ above the valence band for $U > 0$ versus energies $E_A = E_G/3$ and $E_D = 2E_G/3$ above the valence band for $U < 0$. Figures 21.15a,b show that E_F versus Φ_M deviates from Schottky-like behavior for outside the energy range between donor and acceptor states. Figures 21.15c,d are characteristic of the E_F "pinning" behavior, where V_b is constant over a wide range of Φ_M. Many of the physical models of Schottky barrier formation already described are based on this case.

Here, the neutral state is unstable with respect to the charged donor and acceptor states, and electrons are exchanged in pairs between D^+ and A^- at E_F positions between the two. For near-surface densities of negative-U centers on the order of 10^{14} cm^{-2}, charge exchange between the metal and semiconductor moves almost entirely into these centers so that almost no change in V_b takes place. For example, Figure 21.15c shows that charge-center densities of 10^{13} cm^{-2} or less permit a V_b variation of >1 V for $4.2 < \Phi_M < 5.4$ eV, whereas V_b is nearly constant for densities of 10^{14} cm^{-2} or more. The qualitatively different functional dependences in Figure 21.15 provide an easy way to recognize the various charge-center combinations. In general, this self-consistent approach can

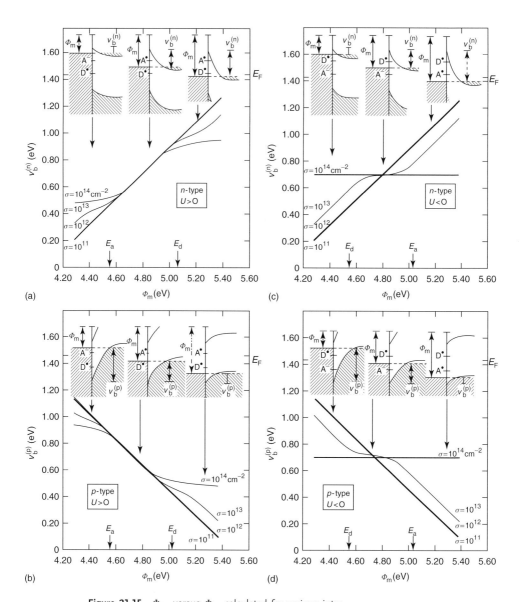

Figure 21.15 Φ_{SB} versus Φ_M calculated for various interface charge-center densities in the case of a 3-charge-state center with either positive or negative U for both n- and p-type GaAs. n-type, positive U (a); p-type, positive U (b); n-type, negative U (c); p-type, negative U (d). (After Duke and Mailhiot [70]).

gauge the effect of different combinations of charge centers on the variation of V_b with Φ_M, independent of the specific physical mechanism that produces the localized electronic state.

An example can illustrate the self-consistent method of calculating the energy at which the Fermi level stabilizes.

Example 21.2. Consider an abrupt metal–semiconductor junction with metal work function $\Phi_M = 5.4$ eV, semiconductor bulk Fermi-level position $E_C - E_F = 0.1$ eV, electron affinity $\chi = 4.2$ eV, and relative dielectric constant $\varepsilon = 10$. A donor-level density $\sigma = 10^{13} \text{cm}^{-2}$ resides 0.9 eV below the conduction-band edge. Assume that the Fermi level at the interface is $E_C - E_F = \Phi_M - \chi = 5.4 - 4.2 = 1.2$ eV. The density of positively charged donors is given by $\sigma^+ = \sigma_0/[1 + \exp[(E_F - E_d)/k_B T]] = 10^{13}/[1 + \exp[(0.9 - 1.1)/0.0259 \text{ eV}]] \sim 10^{13} \text{cm}^{-2}$, that is, the donors are fully charged at room temperature. Assuming the positively charged acceptor and its image charge on the metal side are 1.0 nm apart, calculate the potential drop across the interface due to interface dipole and the corresponding interface Fermi level.

$$q\Delta V = q^2 \frac{\sigma_a}{\varepsilon} d = \frac{1.6 \times 10^{-19} \times 10^{13}}{10 \times 8.85 \times 10^{-14}} \times 1 \times 10^{-7} = 0.18 \text{eV}$$

The interface Fermi level is then raised by 0.18 eV toward E_C.

$E_C - E_F = 1.2 - 0.18 = 1.02$ eV and $E_F - E_d = 1.02 - 0.90 = 0.12$ eV $\gg k_B T$, so that the donors remain nearly fully charged. Thus, the Fermi level stabilizes in the vicinity of the defect level.

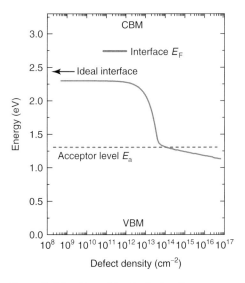

Figure 21.16 E_F position in the band gap versus acceptor defect density. E_F is "pinned" near E_a for defect densities above 10^{12} cm^{-2}.

Figure 21.17 The Schottky Barrier proponents at a scientific meeting held at Stanford University in the early 1980s. (Courtesy R. Ludeke, with permission).

For work functions $\Phi_M < \chi + 0.9$, the donors are not charged, the dipole contribution is zero, and the Schottky barrier behaves classically. For $\Phi_M - \chi \sim E_C - E_d$, one must take the occupancy of the donor level into account. For Φ_M even higher than 5.4 eV, the dipole contribution ΔV remains fixed and $E_C - E_F = \Phi_M - \chi + \Delta V$ increases proportional to Φ_M but offset by ΔV. See, for example, Figure 21.15c. Similarly, increasing defect density increases the dipole contribution. Figure 21.16 illustrates the Fermi-level dependence on defect density for an acceptor level located 2.0 eV below E_C for $\Phi_M = 5.1$ eV, $E_C - E_F = 0.15$ eV in the bulk, and $\varepsilon = 8.2$. See Problem 8.

In general, the Fermi-level position, defect charge occupancy, and dipole contribution must be established self-consistently, for example, by computer iteration.

21.5
Fermi-Level Pinning Models

The diverse array of physical mechanisms described above lead to intrinsic and extrinsic interface states, each of which can account for the narrow range of energies reported for strongly "pinned" metal–semiconductor junctions. While none of these various approaches identified a unique mechanism responsible for the deviations from Schottky–Mott behavior, they stimulated considerable controversy in the scientific literature and at scientific meetings. Figure 21.17 presents a memento from one of the many surface and interface conference proceedings at which these models were discussed. As sketched by Dr Rudy Ludeke during one such meeting, this cartoon attempts to convey some of the animated discussion characteristic of Schottky barrier sessions as each proponent of a Schottky barrier model attempted to "shoot down" their competitors' models. While the various competitors shall remain unnamed, the reader may be able to identify the characters from their models.[1]

While a number of physical models can account for the narrow range of energies observed in strongly "pinned" junctions, there is not yet a physical framework for *predicting* Schottky barrier height with any precision. To distinguish various models, we need a more refined description of the localized interface states and how they form. Several experimental advances now make it possible to characterize interface state energies, relative densities, charge character, and spatial distribution on a nanometer or even an atomic scale. Higher quality crystal growth enables researchers to avoid complications due to bulk defects that affect the Fermi-level movement. Furthermore, studies of wide bandgap semiconductors such as ZnO, GaN, and SiC make it possible to vary Schottky barriers over much wider energy ranges, thereby permitting easier distinction among physical models. These and other developments may soon help researchers reach a predictive framework of Schottky barrier formation.

21.6
Experimental Schottky Barriers

Schottky barrier measurements accumulated over the past 60 years provide researchers and technologists with a starting point for selecting particular combinations of metals and semiconductors. This section attempts to summarize these results concisely.

---

1) Hints: Jack Dow (defect theory) played football for Notre Dame, this author had a mustache, and observations of interface arsenic argued for the effective work function model, underscored by Jerry Woodall's comment that "If it looks like a duck and quacks like a duck, then maybe it is a duck"!

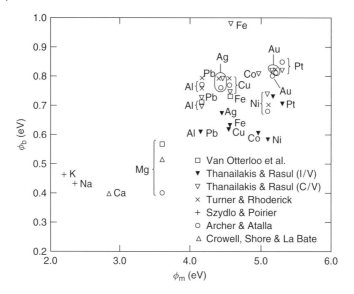

Figure 21.18 *I–V* and *C–V* measurements of Φ_{SB} for elemental metals on cleaved Si(111) surfaces [71].

21.6.1
Metals on Si and Ge

There have been considerable efforts to study metal–Si Schottky barriers since Si is the engine of the microelectronics industry.

21.6.1.1 Clean Surfaces

Schottky barriers of elemental metals to clean Si surfaces obtained by cleavage under near-UHV conditions are represented in Figure 21.18 [71] according to the references indicated [73–80]. Work functions were measured in UHV by Thanailakis and Rasul [73] and are nearly the same as reported by Michaelson [81].

Table 21.2 summarizes the barrier height measurements of elemental metals on clean Si by *I–V*, *C–V*, and internal photoemission spectroscopy (IPS). There is good agreement between *I–V* and IPS values, with the exception of Al, for which oxidation may account for some differences. There is good agreement among *C–V* values, but they are consistently higher than *I–V* values for the same diodes. Higher *C–V* values are typical for many semiconductors due to inhomogeneities within individual diodes. This is because smaller Φ_B patches within a diode lower the effective barrier measured by transport measurements. See Section 3.4. In general, the results of Table 21.2 and Figure 21.18 show that large metal work functions produce high barriers while low work function metals lead to lower barriers. However, the variation in Φ_{SB} values is only ~0.2 eV in Table 21.2 and ~0.4 eV in Figure 21.18 for Φ_M values extending over ~1 eV. This range

Table 21.2 Comparison of Φ_{SB} values (in eV) for elemental metals on clean, cleaved Si as measured by I–V, C–V, and internal photoemission spectroscopy.

Metal	Metal work function	I–V [73]	IPS [73]	C–V [73]	C–V [75]	C–V [77]	Other C–V	Other I–V	Other IPS
Pt	5.30	0.71	0.71	0.82	–	–	–	–	–
Au	5.10	0.73	0.73	0.82	0.82	0.81	0.82 [79]	0.68 [80]	
Ni	5.10	0.59	0.59	0.74	0.70	0.68	–	–	–
Co	4.97	0.61	0.61	0.81	–	–	–	–	–
Cu	4.55	0.62	0.62	0.75	0.79	0.77	–	–	–
Fe	4.58	0.63	0.63	0.98	–	–	0.71–0.76 [74]	–	0.7 [74]
Ag	4.41	0.68	0.68	0.79	0.79	0.76	–	0.71–0.83 [80]	0.70–0.81 [80]
Pb	4.25	0.61	0.61	0.72	0.79	–	–	–	–
Al	4.17	0.61	0.61	0.70	0.76	0.77	0.72 [74]	–	0.70 [74]
Mg	–	–	–	–	–	–	0.55–0.58	–	0.56

Adapted from [71] with permission.

is similar to that shown in Figure 3.3, which includes chemically prepared Si surfaces.

Rather more pertinent to microelectronics, Figure 4.15 summarizes the leading metal–silicide Schottky barriers to Si. Since the silicide reaction establishes a new interface away from any initial surface contamination, these Φ_{SB} values represent clean metallic interfaces. Again, the 0.45 eV range of Φ_{SB} values is comparable to that for UHV-cleaved surfaces.

21.6.1.2 Etched and Oxidized Surfaces

Elemental metals on etched and oxidized Si surfaces produce Φ_{SB} values that follow trends similar to those shown in Figure 21.18 [75]. Φ_{SB} values for specific metals are within 0.1 eV of their cleaved Si counterparts. As in Figure 21.18, there is a trend of increasing Φ_{SB} with increasing work function but with considerable scatter around this trend line. For both cleaved and etched Si, Φ_{SB} varies significantly even for a particular metal. For example, in Figure 21.18, barriers for Al range over nearly 0.2 eV, from 0.6 to 0.8 eV, while Al barriers on etched Si range from 0.5 to 0.72 eV. Similar variations are apparent for Cu on cleaved and etched Si. These wide variations in barrier height for the same metal on Si indicate the strong effect of different surface cleaning treatments prior to metal deposition and oxidation during deposition.

Furthermore, these contacts all exhibit *aging* effects, that is, changes in barrier height with annealing or with exposure to various gas ambients. Such aging effects can be due to (i) interface chemical interdiffusion or reaction, (ii) diffusion of O or

Table 21.3 Φ_{SB} for n- and p-type Si.

Barrier (eV)	Ag	Al	Au	Cu	Ni	Pb
Φ_{SB}^n	0.56	0.50	0.81	0.69	0.67	0.41
Φ_{SB}^p	0.54	0.58	0.34	0.46	0.51	0.55
$\Phi_{SB}^n - \Phi_{SB}^p$	1.10	1.08	1.15	1.15	1.18	0.96

Adapted from [83].

H through the metal film or into their periphery, or (iii) chemical reaction involving adventitious O, OH, or C on the Si prior to metallization. These chemically altered junctions can well be inhomogeneous and exhibit barriers that depend sensitively on the details of the treatment used to prepare the junctions. For more discussion of these variations in elemental metal–Si Schottky barriers, see Rhoderick and Williams [71].

21.6.1.3 N versus P-type Barriers

Metal–semiconductor contacts that follow the Schottky–Mott model expressed by Equation 3.2 have barriers that are simply $\Phi_{SB}{}^n = \Phi_M - \chi_{SC}$ for n-type and $\Phi_{SB}{}^p = \Phi_M - \chi_{SC} - E_G$ for p-type. The sum of these barriers should then just equal the bandgap, that is, $\Phi_{SB}{}^n - \Phi_{SB}{}^p = E_G$ (taking the difference in sign into account. In other words, the Fermi level reaches the same position in the bandgap whether the semiconductor is n-type or p-type.

Table 21.3 shows the results of such measurements for six elemental metals on n- and p-type Si [83]. In all but one case, the sum of these barriers approximately equals the 1.11-eV bandgap of Si at room temperature.

One also expects this result for junctions whose Fermi level is "pinned" at one energy within the bandgap. This holds true for both a MIGS model as well as a defect model. A counter example is found for the GaAs used in Figure 21.10, where two "pinning" levels are evident. Here, the reader should appreciate the distinction between barrier height as opposed to band bending. For the "unpinned" Schottky barriers, the band bending qV_B equals the barrier height minus the Fermi-level separation from the band edge in the bulk, for example,

$$\Phi_{SB}{}^n = qV_B^n + E_C^{bulk} - E_F \tag{3.3}$$

for n-type semiconductors and

$$\Phi_{SB}{}^p = qV_B^p - \left(E_V^{bulk} - E_F\right) \tag{21.11}$$

so that

$$\Phi_{SB}^n - \Phi_{SB}{}^p = qV_B^n - qV_B^p + \left(E_C^{bulk} - E_F\right) - \left(E_V^{bulk} - E_F\right) \tag{21.12}$$

again taking into account the opposite signs of n- and p-type band bending. For heavily doped semiconductors, the bulk terms are small so that the band bendings approximate the barrier heights. For "pinned" Fermi levels, this is not the case. In this case, an interface dipole forms that further separates band bending from barrier height. Hence,

$$\Phi_{SB} = \phi_M - \chi_{SC} - \Delta\chi \tag{21.13}$$

for n-type semiconductors. The difference between qV_B and Φ_{SB} after taking bulk doping into account thus permits one to differentiate between "pinned" and "unpinned" behavior.

In contrast to Si, there have been very few studies of Schottky barrier formation for Ge. This lack of interest stems from the fact that Ge does not form a robust oxide suitable for metal-oxide–Ge microelectronics. Barriers for metals deposited on vacuum-cleaved, n-type Ge are limited to Au (0.45 eV) and Al (0.48 eV) [84]. However, this situation is now changing. There is now renewed interest in Ge because of its much higher mobilities compared to Si for high-speed microelectronics and the advent of new high-K dielectrics that can replace the Ge oxide.

21.6.1.4 Si, Ge Summary

The data presented in this section illustrates that Schottky barriers to Si follow a general trend with work function but with considerable scatter from the expected Schottky–Mott relationship. Furthermore, results for the same metal on Si vary considerably, reflecting differences in surface and interface preparation. The deviations from ideal behavior illustrated in Figure 21.18 can be understood by one or more of the various effects already described in Section 21.3.

It is worth noting that the behavior of these elemental metals on Si is in fact a moot point for technologists, who routinely pattern contacts using reactive metals that form silicides. Such contacts can circumvent surface contamination or lattice damage – difficult-to-control extrinsic features that are evident from surface science experiments and which can alter barrier heights. Furthermore, reactive metal contacts promote good adhesion and control over the physical dimensions of the interface region.

21.6.2
Metals on III–V Compound Semiconductors

Contacts between metals and III–V compound semiconductors are of great importance for microelectronics and optoelectronics technology because of their applications in high-speed switching, amplification, light emitting diodes, lasers, and solar cells – to name but a few. The ability to control Schottky barrier heights and to produce highly ohmic contacts has become increasingly important as electronics moves to ever higher speed, power, and energy efficiency. III–V compound semiconductors have distinct advantages over Si in terms of their direct bandgaps, high mobilities, and the ability to create materials with unique band structures on a quantum scale. At the same time, their interfaces have enabled the discovery of many fundamental phenomena and insights into the nature of solids, recognized by many Nobel Prizes. Nevertheless, these semiconductors present serious challenges because Fermi-level "pinning" at their interfaces can be quite pronounced. Thus, both science and technology have motivated the extensive study of metal/III–V compound interfaces for the past 40 years.

21.6.2.1 GaAs(110) Pinned Schottky Barriers

Among III–V compounds, GaAs has been the semiconductor most thoroughly studied, in large part due to its many high-speed device applications. These studies have involved many of the techniques described in previous chapters. Most macroscopic electrical measurements of contacts prepared in air have yielded only the narrow range of Schottky barriers characteristic of "pinning." Here, we present Schottky barrier results obtained using surface science techniques under more controlled surface conditions.

Clean Surfaces The GaAs(110) surface obtained by bulk crystal cleavage in UHV provides a clean surface that is free of surface states in the bandgap. This result is evident from SXPS measurements of the Fermi-level movement as a function of metal coverage. Figure 21.19 shows that the initial E_F position before metal deposition is near the conduction-band minimum E_{CBM} for n-type and near the valence band maximum E_{VBM} for p-type GaAs [72].

These starting points indicate that any localized states (i) intrinsic to the surface, (ii) related to surface imperfections, or (iii) due to contamination must have densities below the $\sim 10^{12}$ cm^{-2} densities that affect E_F positions in, for example, Figure 21.15. These surface densities translate to approximately one part per thousand surface atoms – below the detection limit of most surface science techniques.

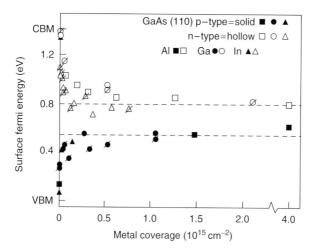

Figure 21.19 SXPS measurements of E_F versus metal coverage on UHV-cleaved melt-grown n-type (hollow points) and p-type (solid points) GaAs(110). At submonolayer coverages, E_F moves to a narrow midgap range of energies with different n- and p-type values [72].

21.6 Experimental Schottky Barriers

Table 21.4 Φ_{SB} for metals on clean n-GaAs(110) and (100) surfaces.

	(110)				(100)			
	Newman et al. [86]		Mead and Spitzer [84]		Waldrop [87]		Other	
Metal	I–V	C–V	C–V	IPS	I–V	C–V	I–V	C–V
Cu	0.87	0.94–1.08	0.83–0.90	0.82	0.96	0.96	0.85	0.85 [88]
Pd	0.85	0.88	–	–	0.91	0.93	–	–
Ag	0.85–0.90	0.95–0.99	0.90–0.95	0.88	0.90	0.89	0.82 [89]	1.03 [89]
–	–	–	–	–	–	–	0.90 [88]	0.88 [88]
Au	0.92	0.99–1.05	0.93–0.98	0.90	0.89	0.87	0.88 [89]	0.98 [89]
–	–	–	–	–	–	–	0.94 [88]	0.95 [88]
Al	0.80–0.85	0.84–0.93	0.78–0.92	0.80	0.85	0.84	0.85 [90]	0.87 [90]
–	–	–	–	–	–	–	0.78 [91]	0.78 [91]
Ti	–	–	–	–	0.83	0.83	–	–
Mn	0.72	0.75	–	–	0.81	0.89	–	–
Pb	–	–	–	–	0.80	0.91	–	–
Bi	–	–	–	–	0.77	0.79	–	–
Ni	0.77	0.82	–	–	0.77	0.91	–	–
Cr	0.67	0.72	–	–	0.77	0.81	–	–
Fe	–	–	–	–	0.72	0.75	–	–
Mg	–	–	–	–	0.62	0.66	–	–
Pt	–	–	0.90–0.98	0.86	–	–	–	–
Be	–	–	0.82	0.81	–	–	–	–
Sn	0.77	0.82	0.68–0.74	–	–	–	0.79 [88]	0.80 [88]
Ba	–	–	0.94	–	–	–	–	1.05 [90]
In	–	–	–	–	–	–	0.76 [88]	0.77 [88]
Co	–	–	–	–	0.76	0.86	–	–
Sb	–	–	–	–	–	–	0.73 [91]	0.74 [91]

Adapted from [71].

With the deposition of metal atoms in submonolayer amounts, E_F for both n- and p-type move rapidly to final "pinning" positions in the GaAs bandgap. Figure 21.19, therefore, suggests a Fermi-level "pinning" mechanism that involves only the outermost semiconductor layer.

Pinned Transport Measurements Transport measurements for metals on similar GaAs(110) surfaces in UHV are consistent with the submonolayer coverage results. Table 21.4 tabulates results from several research groups for diodes on both bulk-grown GaAs(110) and molecular beam epitaxy (MBE-)grown GaAs(100) [86–91]. The MBE results are for surfaces chemically etched and heat treated in UHV.

These transport measurements show only a weak dependence of Φ_{SB} on metal work function with barriers than change by only ~0.2–0.3 eV, varying in the range of $\sim 0.75 \leq \Phi_{SB} \leq 1.05$ eV. As before, C–V barriers exceed I–V barriers by $\sim$50–100 meV, while C–V and IPS measurements are in closer agreement. In general, Φ_{SB} measurements of the same metal on both n- and p-type GaAs yield $\Phi_{SB}^n - \Phi_{SB}^p = E_G$. While Table 21.4 suggests E_F "pinning" in a range of 0.2–0.3 eV, these results should be viewed with caution since the (110) results can be influenced by the high densities of As antisites in the bulk, while the (100) results are from surfaces with an oxide desorbed thermally in vacuum.

Oxidized Surfaces There have been many measurements of Schottky barriers for air-cleaved or chemically etched GaAs. These produce a wider range of barriers than those of UHV-cleaved (100) GaAs. These Φ_{SB} extend between $\sim$0.65 and 1 eV, similar to the oxidized and oxide-desorbed (100) results in Table 21.4. This larger range could be due to the influence of an intermediate oxide. The effect of such oxides will depend on the reactivity of the particular metal deposited over it, such that metal oxides can form for elements such as Mg, Al, Ti, and Cr while little or no interaction may take place for such metals as Au.

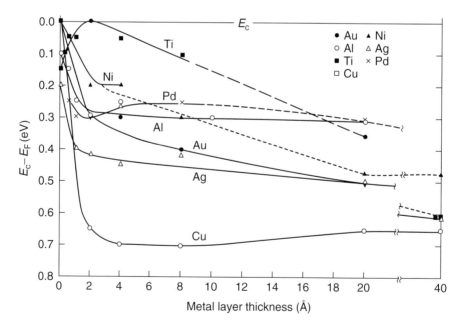

Figure 21.20 SXPS measurements of E_F versus metal coverage on UHV-cleaved n-type InP(110). E_F moves to a 0.5 eV range of energies with over 20 Å thicknesses of Au, Al, Ti, Cu, Ni, Ag, and Pd [82].

21.6.3
InP(110) Unpinned Schottky Barriers

21.6.3.1 Clean Surfaces

The semiconductor InP offers a contrast to the results for cleaved (110) or thermally desorbed (100) GaAs surfaces. While transport studies suggested a narrow Φ_{SB} range, other results have revealed Fermi-level behavior much less characteristic of E_F "pinning." Figure 21.20 shows the Fermi-level movement of UHV-cleaved n-type InP as a function of deposited thickness for a wide variety of metals [82]. Unlike Figure 21.19, E_F stabilizes at a range of energies extending nearly 0.5 eV. Thus, $E_C - E_F$ for Ni is 0.2 eV versus 0.65 eV for Cu. Furthermore, E_F movement occurs much more slowly, in some cases requiring over 20 Å to reach its final equilibrium value. Finally, the E_F movement of each metal can vary significantly. For Ni and Al, E_F reaches its final value within a few angstroms, while Au and Ag require considerably thicker coverages.

Figure 21.20 indicates that the initial chemisorption of metals on clean InP surfaces does not create sufficient states in the bandgap to "pin" the Fermi level and that metal-specific interactions determine the E_F movements to their final values. Similar metal-specific E_F behavior and wide E_F ranges are observed for both n- and p-type InP(110) as well as with Xe interlayers [98].

21.6.3.2 Macroscopic Measurements

$I-V$ and $C-V$ measurements of Schottky barriers for clean, UHV-cleaved InP exhibit a similar range. Table 21.5 again shows larger Φ_{SB} values of $C-V$ versus $I-V$ as well as considerable scatter between $I-V$ measurements, indicating the effect of interface-specific chemical differences described earlier [99–106]. The

Table 21.5 Macroscopic Schottky barrier heights for metals on clean, cleaved n-InP.

Metal	$I-V$ [99]	$I-V$ [100–103]	$C-V$ [100–103]	$I-V$
Ag	0.54	0.42	0.47	0.515 [104]
Cr	0.45	–	–	–
Cu	0.42	–	0.49	–
Au	0.42	0.43	0.50	0.43 [104]
Pd	0.41	0.41	–	–
Mn	0.35	0.35	–	–
Sn	0.35	0.35	–	Ohmic-0.4 [105]
Al	0.325	Ohmic	–	Ohmic [104], 0.22 [106]
Ni	0.32	Ohmic	–	–
Ga		0.6	–	–
In	–	–	–	0.35 [104]

Adapted from [71].

overall Φ_{SB} range extends from 0.6 to 0.22 eV or less, reflecting the difficulty of I–V measurements for barriers of only a few hundred meV. The Φ_{SB} reported by Williams et al. exhibit the transition between reactive and unreactive metals shown in the Figure 8.7 inset.

Metals on UHV-cleaved p-type InP satisfy the $\Phi_{SB}{}^n - \Phi_{SB}{}^p = E_G$ complementarity, analogous to metals on GaAs [107]. As with GaAs, contacts to air-exposed or intentionally oxidized InP exhibit considerable scatter between laboratories [71], again reflecting differences in surface chemistry before and during metal deposition.

21.6.4
GaN Schottky Barriers

The wide bandgap semiconductor GaN and its alloys are of high interest for lasers, light emitting diodes, and transistors. Schottky barrier measurements are almost entirely for diodes on polar surfaces since almost all GaN is grown in epitaxial thin film formed on these surfaces. Table 21.6 lists barrier heights for elemental metals on n- and p-type GaN surfaces prepared by a number of different process treatments. Barriers for different metals range from 0.58 to 1.65 eV over a work function range of 1.4 eV. However, there is considerable scatter for the same metal on GaN prepared by different methods and measured by different techniques [108–121].

In aggregate, the barrier heights increase with increasing work function, spanning a range of ∼1.4 eV for a 4-eV range of metal work function. However, this work function dependence is not strong, likely due to surface damage by the various surface treatments used to prepare diodes. The effect of these chemical treatments is evident from the large scatter in barrier heights for the same metal, for example, 0.75 eV for Pt or 0.4 eV for Au. In contrast, the 1.56 eV barrier measured for the UHV-cleaved nonpolar ($\bar{1}2\bar{1}0$) surface is close to that expected from $\Phi_{Au} - \chi_{GaN} = 5.1 - 3.5 = 1.6$ eV. Nevertheless, maximum GaN barrier heights appear limited to 1.65 eV for Pt, the metal with the largest work function, notwithstanding the $\Phi_{Pt} - \chi_{GaN} = 5.65 - 3.5 = 2.15$ eV barrier expected.

GaN Schottky barriers manifest a number of other interface-specific features as follows: (i) Φ_{SB} increases of ∼0.18 eV for Au/n- and p-type GaN diodes are observed with cryogenic metal deposition, indicating that different metal deposition conditions can alter barriers [122]; (ii) Chemical treatments that minimize or prevent reformation of native oxides can reduce AuPt/n-GaN barriers by 0.1–0.2 eV [123]; (iii) Decreases in depletion widths can occur due to unintentional near-surface donors, for example, point defects near dislocations, that reduce GaN Φ_{SB}'s substantially [124]; (iv) GaN Schottky barriers exhibit an orientation dependence. Pd on GaN(0001) barriers are 0.24–0.3 eV higher than on ($11\bar{2}0$) faces, attributed to the absence of polarization-induced surface charges on nonpolar surfaces [125]. The orientation dependence of Pt/GaN diodes in Table 21.6 is similar in magnitude.

Table 21.6 Schottky barrier heights for elemental metals on GaN.

Metal	$q\Phi_{SB}^n$ (eV)	$q\Phi_{SB}^p$ (eV)	Surface treatment	Measurement technique	References
Pt	1.65	0.92	ISA	XPS, UPS	[108]
Pt	1.1 (0001)	–	CCA	I–V, C–V	[109]
Pt	0.9 (000$\bar{1}$)	–	CCA	I–V, C–V	[109]
Pt	1.6	1.4	CCA	SXPS	[110]
Pt	1.2	–	NVC	XPS, UPS, I–V, C–V	[111]
Pt	1.57/1.43	–	CCA/EDEP	C–V/I–V	[112]
Ni	1.4	1.9	CCA	SXPS	[110]
Ni	0.9/0.8	–	CCA	C–V/I–V	[113]
Ni	–	2.4	CCA	I–V	[114]
Ni	1.11/1.15	–	CCA	C–V/I–V	[115]
Ni	0.85/0.79	–	CCA/AG	I–V	[116]
Pd	1.9	1.5	CCA	SXPS	[110]
Pd	0.94/0.91	–	CCA	C–V/I–V	[117]
Au	1.32	1.21	ISA	XPS, UPS	[108]
Au	0.9	1.9	CCA	SXPS	[110]
Au	0.9	–	NVC	XPS, UPS, I–V, C–V	[111]
Au	1.03/1.03	–	CCA	C–V/I–V	[115]
Au	1.25	–	CCA	XPS	[118]
Au	1.56 ($\bar{1}2\bar{1}0$)	–	UHV cleave	SET	[121]
Re	1.06	–	CCA	C–V/I–V	[119]
Fe	0.85	–	CCA	I–V	[120]
Cr	0.58/0.53	–	CCA	C–V/I–V	[115]
Ti	1.40	1.83	ISA	XPS, UPS	[108]
Ti	0.6	2.3	CCA	SXPS	[110]
Al	0.9	0.92	ISA	XPS, UPS	[108]
Al	0.8	2.5	CCA	SXPS	[110]
Ag	0.82	–		I–V	[120]
Ag	0.6	–	NVC	XPS, UPS, I–V, C–V	[111]
Ag	0.95 (01$\bar{1}$0)	–	UHV cleave	SET	[121]
Pb	0.73	–	GFA	I–V	[120]
Mg	0.77	1.92	ISA	XPS, UPS	[108]
Cs	0.14	–	GFA	I–V	[120]

Surface treatments: As-grown (AG), UHV N$^+$ ion sputter and anneal (ISA), Ga flux and anneal (GFA), chemically cleaned and air-exposed (CCA), NH$_3$-based vapor clean (NVC), electrochemical deposition (EDEP), ultrahigh vacuum cleavage (UHV cleave).

Measurement Techniques: X-ray photoemission spectroscopy (XPS), UV photoemission spectroscopy (UPS), capacitance–voltage (C–V), current–voltage (I–V), internal photoemission spectroscopy (IPS), secondary electron threshold (SET).

21.6.4.1 Other Binary III–V Semiconductors

There are far fewer Schottky barrier measurements for other binary III–V compound semiconductors. For GaP ($E_G = 2.26$ eV), Φ_{SB}^n ranges from 1.0 to 1.4 eV [126]. For GaSb ($E_G = 0.7$ eV), Φ_{SB}^n exhibits 0.55 eV barriers, independent of metal [46]. For InAs ($E_G = 0.36$ eV), E_F for metals on UHV-prepared surfaces can range

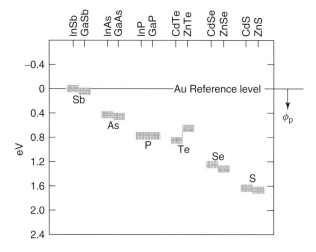

Figure 21.21 Φ_{SB} values for Au on common III–V and II–VI compound semiconductors [85].

over 0.6 eV and depend sensitively on air exposure [127]. For InSb ($E_G = 0.18$ eV), Au and Ag produce $\Phi_{SB}{}^n$ are 0.17 eV and 0.18 eV, respectively [84]. For these narrow-gap semiconductors, it is difficult to distinguish between interface state models since the range of E_F movement is very limited.

McCaldin, McGill, and Mead [85] noted that Au barrier heights measured by I–V, C–V, and IPS methods appeared to follow a *"common anion rule"* such that $\Phi_{SB}{}^p = E_G - \Phi_{SB}{}^n$ is constant for a given anion. Thus, in Figure 21.21, $\Phi_{SB}{}^p$ appears constant between InSb and GaSb, InAs and GaAs, plus InP and GaP. This relation is relevant only for Au since Φ_{SB} can vary widely for different metals on a given semiconductor. Nevertheless, it is consistent with the interfacial phase models of the interface such as the effective work function model.

21.6.5
Ternary III–V Semiconductors

As epitaxial growth has extended to more semiconductors, the "common anion rule" has interested researchers concerned with contacts to ternary alloy semiconductors. Chief among such alloys is the $Ga_xAl_{1-x}As$ system, used widely in laser diodes and other optoelectronic devices. Figure 21.22 illustrates $\Phi_{SB}{}^n$ values for $Ga_xAl_{1-x}As(100)$ predicted theoretically and measured experimentally. While early measurements for Au on chemically etched surfaces indicated agreement with the common anion rule for $0 \leq x \leq 0.4$ [92], later measurements showed a monotonic increase across the full alloy series [93]. $\Phi_{SB}{}^n$ values for Al deposited directly on MBE-grown $Ga_xAl_{1-x}As$ show a similar increase starting at $x \approx 0.4$, where the alloy changes from a direct to an indirect gap semiconductor [94].

The experimental results show a clear disagreement with the common anion rule [97]. Theory based on anion vacancies [95] versus antisites [96] shows the best

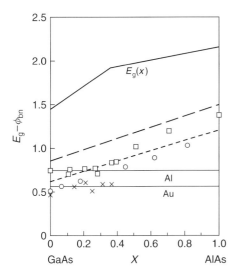

Figure 21.22 Φ_{SB}^n results for Al and Au on $Ga_xAl_{1-x}As(100)$ across the alloy series: Au (x) [92] and (o) [93] and Al (□) [94]. (- - -) lines signify theory based on anion vacancies [95], solid lines based on common anion rule [96], (— — —) based on antisite defects. (After Robinson [97]).

agreement with experiment. The disagreement between Au Φ_{SB} results may be due to differences in interface preparation, particularly since the Al constituent reacts strongly with any oxygen present during growth.

SXPS measurements of $\Phi_{SB}{}^n$ for metals deposited on clean $Ga_xIn_{1-x}As(100)$ in UHV also show strong disagreement with the common anion rule as well as with results for air-exposed surfaces. Figure 21.23a shows the E_F stabilization energies for metals and Ge on $Ga_xIn_{1-x}As$ for $0 \leq x \leq 1$ [128]. The corresponding $\Phi_{SB}{}^n = E_C - E_F$ scale in proportion to the metal work functions, with Au the highest and Al the lowest. The results also disagree with results on the same alloy series which support the common anion rule [130], but for which the $Ga_xIn_{1-x}As$ surfaces were air-exposed before metallization. Figure 21.23b emphasizes the impact of air exposure on these results. The E_F movement versus Au coverage for clean versus air-exposed InAs(100) reaches a high barrier for the clean surface, whereas E_F at the air-exposed surface stabilizes in the conduction band. Such results emphasize the impact of different surface preparation.

Schottky barriers on P- or Sb-based alloys are limited. For $Au/n-Al_xGa_{1-x}Sb$ junctions, $E_G - \Phi_{SB}{}^n$ increases with increasing x [131]. For $Au/GaAs_xP_{1-x}$, $E_G - \Phi_{SB}{}^n = 0.55$ eV so that $\Phi_{SB}{}^p$ increases with P content, in contradiction to the common anion rule [132]. Similarly, $\Phi_{SB}{}^p$ values for Au on p-type $Ga_yIn_{1-y}As_{1-x}P_x$ for $0 \leq x \leq 1$ increase with x such that $\Phi_{SB}{}^p$ is relatively constant.

III–V ternary alloys incorporating N, particularly $Al_xGa_{1-x}N$, are of high technological interest because of their high thermal stability and high mobilities in two-dimensional electron gas (2DEG) device structures. Schottky barrier studies of

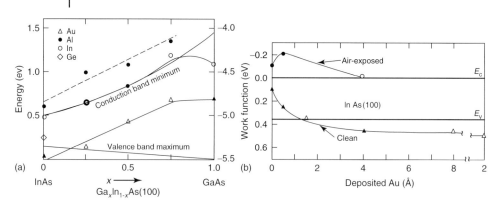

Figure 21.23 (a) E_F stabilization energies relative to the $Ga_xIn_{1-x}As$ conduction and valence-band edges across the alloy series for Au, Al, In and Ge. (b) The different behavior of Au on air-exposed versus clean $Ga_xIn_{1-x}As$ [128].

GaN and $Al_xGa_{1-x}N$ clean surfaces are challenging because bulk crystals are not generally available and, unless metallization takes place in the growth chamber, it is difficult to prepare Schottky barriers on epitaxial surfaces without air exposure. This difficulty is compounded by the high reactivity of Al at the $Al_xGa_{1-x}N$ surface with oxygen. Nevertheless, metal/$Al_xGa_{1-x}N$ barriers are observed to increase nonlinearly with increasing Al content. For example, Φ_{SB}^n (Au) increases from ~1 to ~2 eV for $x = 0$ to 0.4, respectively [133], while Φ_{SB}^n (Ni) increases from ~0.8 to 1.15 eV for $x = 0$–0.25, respectively. These values can increase by enhancing effective barrier heights using the piezoelectric effect, for example, Ni/$Al_{0.25}Ga_{0.75}N$ barriers as high as 1.89 eV [134]. Chemical treatments can also alter the semiconductor Al mole fraction near the interface and the resultant barrier heights [135]. In addition, metals to strained $Al_xGa_{1-x}N$/GaN 2DEG layers can act in opposition to the expected work function dependence for bulk $Al_xGa_{1-x}N$. With decreasing work functions of 5.26, 5.15, and 4.96 eV, Ir, Ni, and Re Schottky contacts to strained $Al_{0.25}Ga_{0.75}N$ are 1.12, 1.27, and 1.68 eV, respectively. This is attributed to a decrease of 2DEG sheet carrier concentration as Φ_M decreases [136]. In general, both interface chemistry and strain can affect III–V ternary alloy Schottky barriers strongly.

21.6.6
Metals on II–VI Compound Semiconductors

There have been relatively few Schottky barrier studies on II–VI compound semiconductors until recently. Early work focused on device applications such as photodetectors, phosphors for television screens, and gas sensors. They involved vacuum-cleaved [137] or chemically etched surfaces [138] of bulk single crystals which exhibited a wide range of barrier heights for different metals and for the same metal on surfaces prepared by different techniques. As with Group

IV and III–V compound semiconductors, $\Phi_{SB}{}^n$ increases with Φ_M and there is considerable scatter between measurements of nominally identical systems due to surface preparation or other factors. However, no simple dependence of $\Phi_{SB}{}^n$ on Φ_M or χ_{SC} is evident.

21.6.6.1 Sulfides, Selenides, and Tellurides

Studies of different metals on CdS, CdSe, ZnS, ZnSe, and ZnTe revealed that these interfaces varied are not abrupt and that their barriers correlated thermodynamically [139]. See Section 4.4.3. UHV-cleaved CdTe Schottky barriers also exhibited large barrier height variations [140].

21.6.6.2 ZnO: Dependence on Metals

More recently, attention has turned to the ZnO interface because of its promise as a high efficiency light emitter. As with other wide bandgap semiconductors, ZnO provides a test bed for different Schottky barrier models since E_F can move over a larger range. Table 21.7 lists $\Phi_{SB}{}^n$ for various metals on ZnO surfaces prepared by different methods [4, 129, 141–165].

Table 21.7 shows that $\Phi_{SB}{}^n$ ranges from 1.2 eV down to ohmic, depending on the metal and on surface treatment. Reliably p-type ZnO is still under development. A wide $\Phi_{SB}{}^n$ energy range is observed even for the same metal, for example, Pt, Au, and Ta. This strong dependence on surface preparation indicates that extrinsic factors such as crystal quality and surface treatment have a large effect on ZnO barrier heights. In general, $\Phi_{SB}(C-V) \geq \Phi_{SB}(I-V)$, with the exception of Ag oxide. Similarly, with the exception of Ag, low work function metals such as In, Al, and Ti yield low Φ_{SB}'s. (Ag oxidizes easily, producing high barrier heights that depend on the degree of oxidation.) These results indicate that tunneling lowers $\Phi_{SB}(I-V)$ except where interfacial oxide layers form.

On the other hand, high work function metals exhibit lower than expected $\Phi_{SB}{}^n$. Since $\chi_{ZnO} = \sim 4.2$ eV, $\Phi_{SB}{}^n$ should be $\Phi_M - \chi_{ZnO} = 5.65 - 4.2 = 1.45$ eV for Pt, whereas measured Φ_{SB} are limited to 0.96 eV for air-exposed surfaces and 0.75 eV for high vacuum-cleaved surfaces. Wide barrier variations are evident for other high work function metals such as Pd and Au with various air-exposed treatments. Highest Au $\Phi_{SB}{}^n$ were measured for remote oxygen plasma (ROP)-cleaned surfaces of low defect (LD) density ZnO. Highest Pd Φ_{SB} were measured for hydrogen peroxide-oxidized surfaces. The increased barrier with oxidized interfacial layers indicates a dipole contribution to Φ_{SB}. Similarly, the blocking contact formed at the Ta–ZnO interface indicates the formation of a Ta oxide layer [129, 144]. Indeed, cathodoluminescence (CL) spectra of the Ta–ZnO interface reveal the presence of a Ta_2O_3-like layer. Conversely, H diffusion can form a dipole at the Pd–ZnO interface that lowers Φ_{SB} [143].

21.6.6.3 ZnO: Dependence on Native Point Defects

While many barrier variations can be ascribed to interface preparation, other extrinsic factors appear to play a role. Only recently have factors such as crystalline quality, impurity content, and native point defect concentration become serious

Table 21.7 $\Phi_{SB}n$ measurements by various techniques versus orientation and crystal quality for metals on ZnO surfaces prepared by different cleaning methods.

Metal	$q\Phi_{SB}n(eV)$	Ideality factor	Surface treatment	Measurement technique	References
Pt (5.65)	0.75 nonpolar	–	Vac-cleave	IPS	[4]
Pt	0.42 (000$\bar{1}$)	3.45	HD, ROP1A	I–V	[145]
Pt	Ohmic (000$\bar{1}$)	NA	LD, ATMA	I–V	[145]
Pt	0.39 (000$\bar{1}$)	1.00	LD, ROP1A	I–V	[145]
Pt	0.61 (0001)	1.70	PLD, AG	I–V	[150]
Pt	0.85, 0.73 (0001)	1.77	ATMA, LA	C–V, I–V	[151]
Pt	0.96 (0001)	1.1	OCA	I–V	[152]
Pt	0.6 (0001)	3.1	OCA	I–V	[152]
Pt	0.70 (0001)	1.5	UVOA	I–V	[157]
Pt	0.93, 0.89 (0001)	1.15	HPA	C–V, I–V	[161]
Pt	0.55 (0001)	2.0	CCA	I–V	[162]
Pt	0.72, 0.68 (000$\bar{1}$)	1.2	CCA	C–V, I–V	[162]
Ir (5.27)	0.65 (000$\bar{1}$)	2.62	HD, ROP1A	I–V	[145]
Ir	0.69 (000$\bar{1}$)	1.58	HD, ROP1A	I–V	[129]
Ir	0.54 (000$\bar{1}$)	1.66	LD, ATMA	I–V	[139]
Ir	0.64 (000$\bar{1}$)	1.36	LD, ROP1A	I–V	[129]
Pd (5.12)	0.68 nonpolar	–	Vac-cleave	IPS	[4]
Pd	0.59, 0.61, 0.60	–	Acid etch	IPS, C–V, I–V	[141]
Pd	0.73, 0.53 (0001)	1.3	ROP2	C–V, I–V	[143]
Pd	0.68, 0.61 (000$\bar{1}$)	1.2	ROP2	C–V, I–V	[143]
Pd	1.14, 0.81 (0001) PLD	1.49	AG	C–V, I–V	[153]
Pd	0.74 (000±1)	2.0	Acetone, Acid	I–V	[154]
	0.60 (000 ± 1)	1.4	etch	I–V	[154]
Pd	0.83 (000$\bar{1}$)	1.03	HPA	I–V	[155]
Pd	1–1.2 (000$\bar{1}$)	1.8	HPA	I–V	[156]
Pd	0.55 (0001)	2.0	CCA	C–V, I–V	[162]
Pd	0.59, 0.59 (000$\bar{1}$)	1.2	CCA	I–V	[162]
Au (5.1)	0.65 nonpolar	–	Vac-cleave	I–V, C–V	[4]
Au	0.645, 0.67, 0.66 -	–	CCA	IPS	[141]
Au	0.67, 0.60 (000$\bar{1}$)	1.86, 1.03	ROPHT, ROPRT	IPS, C–V, I–V	[142]
Au	0.71 (0001)	1.17	ROPHT, ROPRT	I–V	[160]
	0.60 (000$\bar{1}$)	1.03	–	I–V	–
Au	ohmic (0001)	NA	LD, ATMA		[4]
Au	1.2, 0.81 (0001)	1.2	LD, ROP2	I–V	[143]
Au	1.07, 0.77 (000$\bar{1}$)	1.3	LD, ROP2	C–V, I–V	[143]
Au	0.48 (000$\bar{1}$)	1.30	LD, ROP1A	C–V, I–V	[129]
Au	ohmic (000$\bar{1}$)	NA	LD, ROP1, 650°C	I–V	[144]
Au	0.43 (000$\bar{1}$)	3.57	HD, ROP1A	I–V	[129]
Au	0.46 (000$\bar{1}$)	1.56	LD, ATMA	I–V	[129]
Au	0.48 (000$\bar{1}$)	1.30	LD, ROP1A	I–V	[145]
Au	Ohmic (0001)	NA	ATMA	I–V	[146]
Au	0.63 (0001)	1.15	HPA	I–V	[146]

(*continued overleaf*)

Table 21.7 (Continued)

Metal	qΦ_{SB}n(eV)	Ideality factor	Surface treatment	Measurement technique	References
Au	0.65 (0001)	–	ATMA	I–V	[147]
Au	0.66 (0001) ZnO : N	1.8	OCA	C–V	[149]
Au	0.37 (000$\bar{1}$) ZnO : N	3.5	OCA	I–V	[149]
Au	0.71, 0.70 (0001)	1.4	CCA	I–V	[162]
Au	0.70, 0.69 (000$\bar{1}$)	1.1	CCA	I–V, C–V	[162]
PEDOT : PSS (5.0)	0.9, 0.7 (0001)	1.2	HT, CCA	C–V, I–V	[165]
Ti (4.33)	<0.3 nonpolar	–	Vac-cleave	I–V, C–V	[4]
Ti	Ohmic (0001)	NA	OCA		[158]
Cu (4.65)	0.45 nonpolar	–	Vac-cleave	I–V	[4]
Al (4.28)	0.0 nonpolar	–	Vac-cleave	I–V	[4]
Ag (4.26)	0.68 nonpolar	–	Vac-cleave	I–V	[4]
Ag	0.92, 0.89(11$\bar{2}$0)	1.33	AG	I–V	[148]
Ag	0.69 (0001)	–	ATMA	IPS	[147]
Ag	0.84 (11$\bar{2}$0)	1.5	O_2 plasma	C–V, I–V	[159]
Ag	0.80, 0.78 (0001)	1.2	CCA	C–V	[162]
Ag	1.11, 1.08 (0001)	1.08	CCA highest	I–V	[163]
Ag	0.80, 0.77 (000$\bar{1}$)	1.1	CCA	C–V, I–V	[162]
Ag	0.99, 0.97 (000$\bar{1}$)	1.06	CCA highest	I–V, C–V	[163]
Ag oxide (~5.0)	1.20, 0.93 (0001)	1.03	HT, OCA, OPS	C–V, I–V	[164]
Ag oxide	0.99, 0.79 (000$\bar{1}$)	1.04	HT, OCA, OPS	I–V, C–V	[164]
Ag oxide	1.03 (0001)	1.14	MG, OCA, OPS	I–V, C–V	[164]
Ag oxide	0.98, 0.89 (000$\bar{1}$)	1.10	MG, OCA, OPS	I–V, C–V	[164]
Ag oxide	1.02 M-plane	1.10	MG, OCA, OPS	I–V	[164]
Ta (4.25)	ohmic (000$\bar{1}$)	NA	ROP1A	I–V, C–V	[144]
Ta	blocking (000$\bar{1}$)	NA	LD, ROP1A, 350°C	I–V	[144]
Ta	blocking (000$\bar{1}$)	NA	LD, ROP1A, 550°C	I–V	[144]
Ta	ohmic (000$\bar{1}$)	NA	HD, ROP1A	I–V	[144]
	Leaky (000$\bar{1}$)	NA	HD, ROP1A, 550°C	I–V	[144]
In (4.12)	<0.3 nonpolar	–	Vac-cleave	I–V	[4]

Crystal quality: Low defect (LD); high defect (HD), Pulsed laser deposited (PLD), hydrothermal (HT), melt-grown (MG).

Surface treatment: Vacuum cleaved in stream of evaporating metal (Vac-cleave); H_3PO_4, HCl, deionized water (Acid etch); chemically cleaned and air-exposed (CCA); laser annealed (LA); remote oxygen plasma cleaned at high (ROPHT) temperature, plus room temperature reexposure (ROPRT); room temperature remote oxygen plasma for one hour and air exposure (ROP1A) or for two hours without breaking vacuum (ROP2); hydrogen peroxide and air-exposed (HPA); acetone, trichloroethylene, methanol and air-exposed (ATMA); organic clean and air-exposed (OCA); UV ozone and air-exposed (UVOA), oxygen-plasma sputtered (OPS), as-grown (AG).

Measurement technique: Internal photoemission spectroscopy (IPS); current-voltage (I–V), capacitance-voltage (C–V).

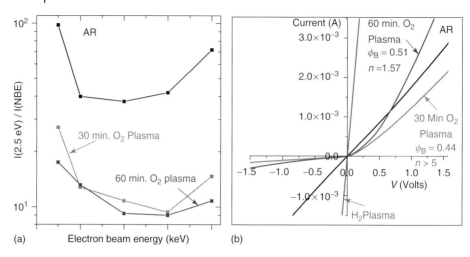

Figure 21.24 (a) DRCLS ratio $I(2.5\,\text{eV})/I(3.36\,\text{eV})$ versus depth of peak excitation. Defect densities increase toward the free surface. (b) I–V characteristics of Au/ZnO(000$\bar{1}$) diodes on these surfaces change from (i) as received and chemically etched (AR), (ii) with 30 minute ROP cleaning, (iii) 60 minute ROP cleaning, and (iv) subsequent H plasma exposure [129].

considerations. Several studies indicate that metals on Zn-polar surfaces yield higher Φ_{SB} than on O-polar surfaces. This can be attributed to higher subsurface defect concentrations for O-polar crystals [143]. Furthermore, there is a qualitative difference between reactive versus nonreactive metal–ZnO interfaces. Metals such as Al, Ta, Ti, and Ir that react with ZnO to form oxides produce defects associated with oxygen vacancies, whereas metals such as Au, Pd, and Pt that form eutectics with Zn produce defects associated with zinc vacancies. Depending on the density of defects and the extent of reaction in the subsurface region, reactive metals can form either ohmic or blocking contacts [129, 144]. A new strategy to avoid interface chemical interaction is to use high work function and chemically inert conducting polymers to form high barriers [165].

To illustrate the effect of native point defects on I–V features, Figure 21.24 shows how contacts to ROP-cleaned ZnO(000$\bar{1}$s) surfaces can change from ohmic to Schottky-like as surface impurities and subsurface native point defects are removed. The contributions of surface contamination, subsurface H, and native point defects can be separated at different stages of this conversion process. Depth-resolved cathodoluminescence spectroscopy (DRCLS) into a ZnO surface shows that deep-level "green" defect emission at ∼2.5 eV increases proportionally as excitation occurs closer to the surface. Figure 21.24a illustrates this increase in the ratio of this defect emission intensity $I(2.5\,\text{eV})$ versus near-band-edge (NBE) emission intensity $I(3.36\,\text{eV})$ versus depth of peak excitation rate. It also shows that remote plasma treatment decreases this ratio strongly within the outermost few tens of nanometers [144].

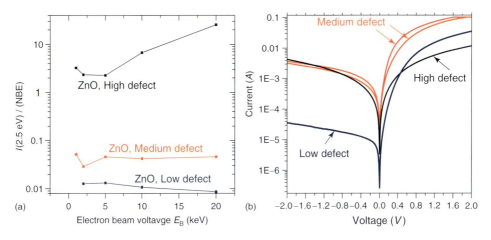

Figure 21.25 (a) Defect-to-NBE intensity ratio versus depth for "high," "medium," and "low" defect density ZnO. (b) Corresponding I–V characteristics [129].

Furthermore, this defect ratio continues to decrease even after 30-minute ROP processing, which X-ray photoemission spectroscopy (XPS) and low-energy electron diffraction (LEED) measurements show cleans the surface completely. Figure 21.24b shows the effect of these treatments. Au diodes on the as-received surface are ohmic. With 30-minute ROP surface cleaning, the I–V characteristic becomes asymmetric with $\Phi_{SB}{}^n = 0.44$ eV and ideality factor $n > 5$. Additional ROP processing further increases rectification: $\Phi_{SB}{}^n$ increases to 0.51 eV while n decreases to 1.51. Such studies show that subsurface defects rather than surface contaminants determine the Schottky barrier changes.

Hydrogen within ZnO can act as donors, as predicted theoretically [168] and observed experimentally [169, 170], which in turn can lead to heavy n-type doping that narrows the surface space charge region and promotes tunneling. This is evident from the ohmic contact (straight line) formed on ZnO exposed to a remote H plasma in Figure 21.24b.

Figure 21.25 further illustrates how the density of native point defects within the ZnO crystal before metallization strongly affects Schottky barriers. Figure 21.25a shows $I(2.5\,eV)/I(3.36\,eV)$ DRCLS intensity ratios for ROP-cleaned ZnO crystals with "high," "medium," and "low" defect concentrations. This panel shows that the defect ratios can vary by over 2 orders of magnitude for different ZnO crystals [129].

Figure 21.25b shows that the reverse currents vary by 2 orders of magnitude for Au contacts to these surfaces. "High" defect ZnO shows nearly ohmic behavior, "medium" defect ZnO shows only weak rectification, while the strongest rectification appears for the "low" defect ZnO. In general, $\Phi_{SB}{}^n$ are higher for "low" versus "high" defect ZnO crystals with correspondingly lower ideality factors [129]. The direct correlation between reverse current leakage and defect density underscores the importance of using low defect crystals in Schottky barrier research.

21.6.6.4 ZnO: Dependence on Polarity

Schottky barriers to clean ZnO surfaces also exhibit a pronounced dependence on the Zn- versus O-polar surfaces. DRCLS, $C^{-2}-V$, and deep-level transient spectroscopy (DLTS) studies show that Zn-face diodes have higher barrier heights, lower subsurface deep-level defects, and lower subsurface free carrier concentrations than O-face diodes of the same ZnO crystal. This polarity dependence correlates to defect emission, traps, and interface chemistry [143].

Finally, native point defects created at the metal–ZnO interface produce large changes in barrier height. Figure 16.15 illustrated the creation of new native point defects at the Au and Al–ZnO interfaces [129]. Along with the increase in defect emission measured with DRCLS at the intimate metal/ZnO contact, the corresponding $I-V$ characteristics show increased leakage or current blockage, depending on whether or not the metal forms an interfacial oxide [144]. These ZnO results demonstrate that native point defects resident in the bulk crystal as well as similar defects produced by chemical interaction at the ZnO interface have a strong effect on Schottky barrier formation.

21.6.7
Metals on IV–IV, IV–VI, and III–VI Compound Semiconductors

Schottky barriers for the IV–IV compound semiconductor SiC can vary considerably with not only surface preparation but also crystal polytype and surface polarity. Itoh and Matsunami have tabulated extensive results for 3C-, 4H-, and 6H-SiC polytypes with n- or p-type doping, and for Si and C surface polarity [172]. Among these polytypes, 4H–SiC is the most commonly used, with applications in high-power, high-temperature electronics. Table 21.8 summarizes results for n- and p-type 4H–SiC with both Si and C surface polarity. The SiC surfaces were

Table 21.8 Schottky barrier heights of metal/4H–SiC interfaces.

Metal	Surface polarity	$q\Phi_{SB}I - V$(eV)	$q\Phi_{SB}C - V$(eV)	$q\Phi_{SB}$IPS(eV)	Ideality factor
Au	C	1.80	2.10	2.07	1.20
Au	Si	1.73	1.85	1.81	1.08
Ni	C	1.60	1.90	1.87	1.08
Ni	Si	1.62	1.75	1.69	1.02
Ti	C	1.16	1.30	1.25	1.04
Ti	Si	0.95	1.17	1.09	1.03

Adapted from [172].

chemically cleaned and hydrogen-terminated prior to metallization. In general, 4H– and 6H–SiC exhibit 0.2–0.3 eV higher barriers on C- versus Si-polar surfaces. Regardless of surface polarity, barrier heights lie within ∼0.8 eV ranges compared with Φ_M variations of ∼1 eV. Porter and Davis have tabulated Schottky barriers for 3C and 6H polytypes [173]. Both 6H–SiC tabulations show higher barriers for metals on C- versus O-polar surfaces as well.

There are only a few Φ_{SB} results for the IV–IV semiconductor diamond due in part to the difficulty of doping this wide gap semiconductor to obtain conductive material. Early work indicated Φ_{SB}^p values of either 1.35 or 1.73 eV for Au [84, 174]. Al, Ni, Au, and Pt diodes on (001) diamond annealed in vacuum at temperatures up to 1200 °C yielded Φ_{SB}^p = 0.95, 0.8, 0.57, and 0.2 eV, respectively. The corresponding $S \sim 0.6$ indicates a substantial reduction in surface state density [175]. Diamond (111) Φ_{SB}^p values exhibit a weaker ($S \sim 0.3$) dependence on Φ_M for surfaces prepared by a wet-chemical oxidation followed by annealing in Ar or vacuum, suggesting a dependence on crystal orientation [176]. Recent SXPS band-bending measurements of clean (001) diamond prepared by UHV annealing yield Φ_{SB}^p = 1.05 eV for Al with an ideality factor of 1.4. *In situ* annealing at temperatures above 755 K induces Al–C bonding that reduces Φ_{SB}^p steadily until bulk carbide formation above 1020 K drives the contact ohmic [177]. See [177(a)] for a review of these diamond contacts. Thus, surface chemical termination, orientation, and chemical bonding can all influence diamond Schottky barriers.

Limited Schottky barrier studies of IV–VI compounds include PbTe, useful for near-IR photodetectors. Low work function Pb and Sn contacts produce Ohmic contacts to n-PbTe and rectifying barriers to p-PbTe [178], consistent with a Schottky–Mott model. Φ_{SB} studies of III–VI compound GaSe display similar characteristics to those described by an interface heat of reaction in Figure 4.17, namely, unreactive metals that produce sizable variations in barrier height versus reactive metals for which Φ_{SB} showed little variation [179].

21.6.8
Compound Semiconductor Summary

The experimental results described in this section demonstrate several important conclusions:

1) First and foremost, localized states clearly affect Schottky barriers for most if not all semiconductors.
2) Specific surfaces, surface treatments, reactions, and annealing all affect the Schottky barrier for a particular metal–semiconductor combination.
3) As a result, there is at present no simple theoretical model that predicts Schottky barriers *a priori*.

21.7
Interface Passivation and Control

Schottky barriers can be controlled on both a macroscopic and an atomic scale. Even without a predictive Schottky barrier model, technologists have developed mesoscopic and macroscopic techniques to control contact properties in real devices. With the increasing use of epitaxy to fabricate microelectronic and optoelectronic devices, researchers have also identified interface passivation and control methods on an atomic scale.

21.7.1
Macroscopic Methods of Contact Formation

Figure 21.26 summarizes several macroscopic methods of contact formation. The top panel shows the typical 0.8 eV thermionic emission barrier of metals to

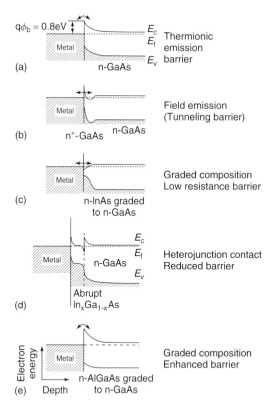

Figure 21.26 Schematic band diagrams illustrating a metal–GaAs junction before (top panel) and after the use of macroscopic methods to decrease (middle three panels) or increase (bottom panel) the effective barrier height [166, 167].

melt-grown GaAs. With increased doping, the depletion region for this barrier decreases in width so that field emission increases. This contact then becomes a tunneling barrier [166, 167].

Graded compositions can also produce low-resistance barriers. Contacts to air-exposed InAs are Ohmic since, as Figure 21.23b shows, E_F stabilizes in the conduction band. Table 20.2 shows that ΔE_C between GaAs and InAs is only 0.17 eV so that a graded InGaAs layer between GaAs and a metal effectively eliminates the n-type barrier [181]. A uniform and abrupt $In_xGa_{1-x}As$ layer also reduces the effective barrier since the $\exp[-qV_{SB}/k_B T]$ transport increases for two small barriers versus a single large one.

Graded composition can also increase barrier heights. The bottom panel illustrates a metal–GaAs contact with a graded $Al_xGa_{1-x}As$ surface layer that increases the bandgap. Since $\Delta E_C = 0.34$ eV at the AlAs–GaAs heterojunction, Φ_{SB}^n can increase by several tenths of electron volts depending on the Al alloy composition.

21.7.2
Processing Contacts

Localized states themselves provide another method to reduce barrier heights. Thus, lattice disorder produced mechanically, thermally, chemically or with particle bombardment can lead to high densities of localized states distributed in energy across the bandgap through which transport can occur. Figure 21.27 illustrates the introduction of recombination centers that effectively reduces the Schottky barrier by localized state conduction through rather than over the barrier [171]. The defects created at metal–ZnO interfaces, for example, Figure 16.15 are examples of lowered barriers due to localized states induced chemically. Mechanical scraping while soldering metal to semiconductors is a coarse method sometimes used to form low-resistance contacts, not only to create lattice disorder but also to punch through any native insulating oxide layer.

21.7.2.1 Elemental Metals on GaAs

There has been extensive research on the metallurgy of Schottky and Ohmic contacts between elemental metals and GaAs. Table 21.9 illustrates the complexity of GaAs interface composition, depending on the thermodynamics of the specific metal [184].

This table illustrates several general features: (i) new phases form with increasing temperatures and annealing times; (ii) the corresponding barrier heights can change, depending on the phases that form; (iii) the new phases can assume different compositions and morphologies; and (iv) particular morphologies can produce ohmic or rectifying contacts as well as macroscopic decomposition of the metallization. The Mo–GaAs contact illustrates the sequence of phases that can form at elevated temperature. The Au–GaAs interface shows large Φ_{SB}^n decreases with diffusion, defect formation, and changes in doping. Annealing the Ta–GaAs contact produces a number of different mesoscopic morphologies.

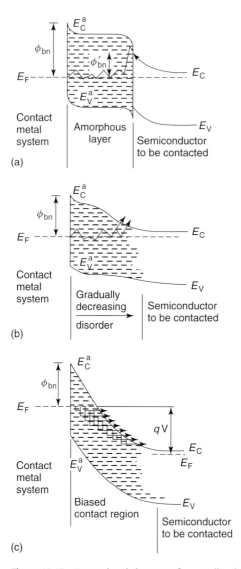

Figure 21.27 Energy band diagrams for metallized (a) abrupt, (b) graded, and (c) biased amorphous/crystalline junction. Schematic dashes and arrows indicate localized states in the mobility gap and the electron paths between them, respectively [171].

Table 21.9 Barrier heights and chemical interactions for elemental metal contacts to GaAs.[a]

Metallization	Temperature (°C)	Φ_{SBn} (eV)	Analysis technique	Comments on contact degradation
Al	RT	~0.7	µAES	–
	200	~0.8	SIMS, XRD	Slight Ga and As outdiffusion
	250	~0.9	SEM	Ga outdiffusion, Al indiffusion
	350	~0.87	TEM	More Ga outdiffusion, slight As outdiffusion
	450–500	~0.8	RBS	More Ga and As outdiffusion, Al indiffusion 20–40 nm. GaAs + Al precipitates near GaAs surface, Al–As crystallites at interface.
	550	–	–	More Ga and As outdiffusion
Ag	RT	–	SEM	–
	300	Poor ohmic for p$^+$	–	–
	400–750	–	–	Faceted holes below contact in GaAs
Au	RT	~0.9	RBS channeling	–
	200–300	0.7	AES, EPM	Ga outdiffuses to surface to form Ga_2O_3; low concentration of As in Au film, Au diffuses slightly into GaAs (40–150 nm)
	350–450	0.5	XRD	More Ga outdiffusion; As outdiffusion and evaporation. Au–Ga (21 at% Ga) phase with Au segregation. Nonuniform Au diffusion into GaAs; Au precipitates extend 300 nm into GaAs producing high dislocation density
	>450	–	SIMS, TEM	

(continued overleaf)

Table 21.9 (Continued)

Metallization	Temperature (°C)	$\Phi_{SB}n$(eV)	Analysis technique	Comments on contact degradation
Cr	RT-300	–	RBS, SEM	No reaction
	425	–	SIMS	2–3 at%Ga outdiffusion into Cr
	550	–	–	Intermetallic formation Cr_2X (X = Ga and/or As), Cr indiffusion
Mo	RT-450	–	RBS, SIMS,	Stable – no reaction
	–	–	SEM, AES, XRD	Bubbles on film surface due to thermal expansion
	600/30 min	Rectifying	AES, SIMS	No interdiffusion
	700/30 min	Ohmic	SEM, XRD	Mo_5As_4 forms
	800/30 min	Ohmic	–	Mo_5As_4 major phase; Mo_2As_3, Mo_3Ga and Mo_3GaAs_2 also detected
	1000/30 min	Rectifying	–	Mo_3GaAs_2 major phase
Ni	250	–	RBS	Formation of Ni_2GaAs
	400	–	SEM, XRD, AES	Ni_2GaAs decomposes to form NiAs and β'-NiGa
Pd	RT	–	–	–
	250	–	UPS, XPS, RBS	Pd_2GaAs formation
	>350	–	XRD, AES, TEM	Pd_2GaAs decomposes, PdGa forms
Pt	RT	~0.9	RBS, AES SIMS, XRD, TEM	–
	300–350	0.86	–	$PtAs_2$ and PtGa form. Layered structure Pt/Pt_3Ga/Pt–$PtGa_2$/$PtAs_2$/GaAs also reported
	500	1.0	–	PtGa–$PtGa_2$/$PtAs_2$/GaAs forms
	600	–	–	Brittle alloy forms

Metal	Temperature (°C)	Φ_{SBn}	Techniques	Comments
Ta	RT-300	–	RBS, SIMS	Stable – no reaction
	425–450	–	μAES, EPM	Ta indiffusion ~200–300 nm, Ga outdiffusion
	500	–	–	More Ga and As outdiffusion; Ta indffusion and possible TaX formation at interface (X–Ga and/or As)
Ti	RT	~0.82	RBS, SIMS	TiAs forms
	450	~0.84	XRD, EPM	–
	500–550	0.80	AES	TiAs, Ti$_5$As$_3$, Ti$_2$Ga$_3$, Ti$_5$Ga$_4$ intermetallics reported
W	RT-500	~0.65 to ~0.67	RBS, SIMS, SEM, EDS	Stable – no reaction; slight Φ_{SBn} increase with temperature, layers tend to peel off (T > 400 °C) due to thermal expansion mismatch
	800/10 min	0.73	XRD, RBS	W–GaAs interaction occurs
	950/10 min	0.74	–	Lateral diffusion of implanted dopants in self-aligned MESFET structures. Less spread if gates parallel to [01$\bar{1}$] vs. [0$\bar{1}\bar{1}$].
	900/3–30 min		–	

[a] Adapted from [184].

Table 21.10 Multilayer metal contacts to GaAs.

Metallization	Temperature (°C)	Φ_{SBn}(eV)	Analysis technique	Comments on contact degradation
Au/Pt	350	–	AES	Ga, As outdiffuse; Ga edge leads As edge; Au diffuses through Pt
	500	–	–	More Ga than Pt in Au; Au/Pt–Ga/Pt–Ga–As/GaAs
Au/W	RT-500	–	XRD	Stable – no reaction
Ni/Au	RT-350	–	RBS, EPM	No Ni accumulation at Au/GaAs interface
Pt/Ni	RT-200	0.88	XRD	No reaction
	380–480	~0.95	–	GaPt, PtAs$_2$, Ga$_2$Pt formed
	>500	–	–	Ga$_2$Pt, PtAs$_2$, GaPt Φ_{SB} degraded
Pt/Ti	RT-350	~0.81	RBS	No reaction
	500	~0.86	XRD	Pt$_3$Ga/TiAs/PtAs$_2$/GaAs formed
Pt/W	RT-500	~0.7	XRD	No reaction
Au/Pt/Ti	RT 250	–	–	Ga outdiffusion into Ti, slight As outdiffusion, Pt diffusion through Ti and into GaAs
Ti/W/Si	As deposited	0.75	RBS, SIMS	Composition close to Ti–W silicide
	850/1 h	0.85	–	No interaction observed
	800/10 min	–	–	Used for activating implants for self-aligned MESFETs

Material	Condition		Technique	Observation
$TaSi_2$	As deposited	–	XRD, AES, TEM, RBS	Amorphous with $R_s \sim 32\Omega\ cm^{-2}$; high leakage current due to damage induced by sputter deposition
	450/2 min	0.79	–	Amorphous; withstands alloy cycle for ohmic contacts in MESFETs
	500/30 min	–	–	Crystallization starts and R_s decreases
	800/20 min	0.78	–	Crystallization $R_s \sim 8\Omega\ cm^{-2}$; no observable interdiffusion
$Au/\alpha Ni - Nb$	As deposited	0.51	AES, XRD	Amorphous $Ni_{0.45}Nb_{0.55}$ alloy film
	400/2 h	–	–	Au diffusion; Ni–As reactions at Ni–Nb/GaAs interface; soft reverse I–V
$Au/\alpha Ta - Ir$	As deposited	0.79	–	Amorphous $Ta_{0.56}Ir_{0.44}$ alloy
	500/24 h	–	–	No observable interdiffusion
	600/24 h	–	–	Some Ga outdiffusion
	700/24 h	–	–	Complete chemical intermixing
$Al/Ti/GaAs$	400	0.75	XRD, AES	Al_3Ti, GaAlAs formation at interface
$Au/Ti-W$	As deposited	–	SEM-EDS	Leaky due to damage induced by sputter deposition
	700/10 min	0.8	–	No degradation
	800/10 min	0.9	–	Silver color appears; poor breakdown characteristics
	860/10 min	–	–	Surface morphology unchanged
	As deposited	0.65	AES	Leaky due to damage induced by sputter deposition
	500/24 h	0.7	–	–
	600/15 h	0.75	–	Diode becomes leaky; Au detected throughout Ti–W

(continued overleaf)

Table 21.10 (Continued)

Metallization	Temperature (°C)	Φ_{SBn}(eV)	Analysis technique	Comments on contact degradation
Au/αW – Si	400/16 h	–	RED, AES, SEM, Electron channeling	Amorphous W–Si layer; no interdiffusion; $D < 3 \times 10 - 18 \text{ cm}^2 \text{ s}^{-1}$ for Au, Ga, and As
Au/Mo/Ti	As deposited	0.8	SEM-EDS	–
	500/10 min	0.8	–	TiAs forms
	550/10 min	0.9	–	Ga detected at surface; silver color
Ti/W	As deposited	0.6	RBS, SIMS	–
	750/15 min	0.68	–	Ti–W–GaAs interacts starts
	850/15 min	0.5	–	Ga, As outdiffusion; Ti indiffusion
	As deposited	0.8	SEM-EDS	–
	860/10 min	0.9	–	–
Ti/W/Ti	As deposited	0.8	SEM-EDS	High leakage current due to damage induced by sputter deposition, lower leakage currents for evaporated metallization
	500/10 min	0.74	–	–
	680/10 min	0.8	–	–
W/Si/W	As-deposited	–	–	Weak channeling pattern; polycrystalline W near surface; amorphous region forms at W/Si interface
	500/4 h	–	–	Metallization becomes amorphous
Ti–W/Si/Ti–W	500/4h	–	RED, AES, SEM, Electron channeling	Metallization becomes amorphous

Adapted from [184].

W–GaAs junctions exemplify how thermal lattice mismatch leads to layer peeling that destroys the electrical contact.

21.7.2.2 Metal Multilayers on GaAs

Multiple metal layers can serve to control diffusion and alloy formation at semiconductor contacts. A summary of useful metal multilayers appears in Table 21.10. For the metallizations shown, the sequence of deposition is from right to left. Multiple metal layers are commonly used to achieve not only the desired ohmic or rectifying electrical contact but also to obtain (i) good metal–semiconductor adhesion, (ii) chemical stability at the elevated temperatures encountered during device operation, and (iii) minimum interdiffusion of constituents that would otherwise introduce defects and doping changes. Layers in direct contact with the semiconductor are commonly reactive, for example, Ti, for good adhesion. Outer layers are often relatively unreactive metals, for example, Au, to promote easy wire bonding without intervening oxides. As with elemental metal layers, these metallurgical interactions extend more than one to two monolayers since thinner layers are difficult for Rutherford backscattering spectroscopy (RBS) and X-ray diffraction (XRD) to detect.

Table 21.10 shows that metal multilayers can in general provide greater chemical stability at higher temperatures than elemental metal contacts. Several of these contacts, for example, Pt/W and Pt/Ti, are stable to several hundred degrees celsius and even remain rectifying at higher temperatures when reactions become evident.

Metal alloys have also proven useful in forming rectifying contacts to GaAs. Table 21.11 lists a number of alloys that maintain high Schottky barriers even with rapid thermal annealing at relatively high temperatures [185, 186].

Refractory metal contacts such as ZrN retain high barrier heights and ideality factors near unity at temperatures as high as 850 °C. Barrier height increases at elevated temperatures are attributed to N incorporation into GaAs near the metal/GaAs interface. Similarly, the Al–Ga exchange reaction between Al and GaAs produces an $Al_xGa_{1-x}As$ interfacial layer that increases the effective barrier. Also, such interface reactions produce diffusion barriers that serve to inhibit further interdiffusion. As already discussed in Sections 8.1 and 18.2, surface science techniques enable detection of such interfacial reactions at the monolayer

Table 21.11 Φ_{SB} values for alloy metals at high temperatures.

Metallization	Temperature (°C)	$\Phi_{SB}n$	Analysis technique	References
ZrN	900, 10 s	0.92	I–V, C–V	[185]
TiN	800, 10 s	0.80	I–V, C–V	[185]
NbN	800, 10 s	0.73	I–V, C–V	[185]
NiAl	650, 20 s	0.99	I–V	[186]
WN	800, 10 s	0.95	I–V	[186]

scale even at room temperature. Thus, interface chemistry provides a useful tool to achieve barrier enhancement as pictured in Figure 21.26.

21.7.2.3 Useful Metallizations for III–V Compound Ohmic Contacts

A number of useful metallizations are available to form ohmic contacts to III–V compound semiconductors. These are listed in Table 21.12 [187].

Elemental contact materials typically have low or high work functions, depending on the semiconductor's n- or p-type nature, respectively. To achieve melting with the lowest temperature annealing, alloy compositions are matched to the lowest melting point eutectic of their binary phase diagram. Interdiffusion at these junctions can also produce high concentrations of dopant atoms or native point defects that promote tunneling and hopping transport through any barriers otherwise formed.

As circuit features shrink and device speeds increase, ohmic contacts require lower contact resistance. In general, an RC circuit with equivalent resistance R and capacitance C has a time constant related to its cutoff frequency f_C as $\tau = RC = 1/2\pi f_C$. The resistance of circuit elements increases in inverse proportion to contact area, which shrinks quadratically as scaling decreases. Thus, for submicron high electron mobility transistors (HEMTs), contact resistivities of $10^{-6} \Omega$ cm^2 or less are needed. Table 21.13 presents representative metallizations developed to achieve resistances in this range for GaAs [188–193]. Here again, interdiffusion is used to promote high doping densities within the semiconductor's near-surface region. In addition, the short anneal times are intended to minimize the depth of such diffusion since the semiconductor-active regions themselves have thicknesses on a submicron scale.

21.7.3
Atomic-Scale Control

Just as atomic interlayers can modify heterojunction band offsets (see Section 20.4.9), new interface chemical layers that are only monolayers thick can also control Schottky barriers. Methods to produce these chemically distinct interphases include (i) reactive metals, (ii) less-reactive buffer layers, (iii) wet-chemical cleaning or bonding, and (iv) gas–surface interactions.

21.7.3.1 Reactive Metal Interlayers

We have already seen how monolayers of reactive metals can alter the outdiffusion of semiconductor constituents. Recall, for example, Figure 8.7 and the effect of different metals on GaAs or InP outdiffusion. The native point defects generated by anions or cations leaving the semiconductor create electrically active sites with high-enough densities to alter Φ_{SB}. For Au on cleaved, bulk GaAs(110) surfaces, single monolayers of Al decrease Φ_{SB} measured electrically by 0.1–0.2 eV [180]. The high defect densities present in the melt-grown GaAs bulk limit these barrier decreases. Yb, Sm, and Sm–Al interlayers on MBE-grown GaAs(100) produce nearly 0.2 eV I–V and C–V changes as well [194].

Table 21.12 Ohmic contact technology for III–V and mixed III–V compound semiconductors.

III–V	E_G (eV)	Type	Contact material	Technique	Alloy temperature (°C)
AlN	5.9	Semi-i	Si	Preform	
		Semi-i	Al, Al–In	Preform	1500–1800
		Semi-i	Mo, W	Sputter	1000
AlP	2.45	N	Ga–Ag	Preform	500–1000
AlAs	2.16	n, p	In–Te	Preform	150
		n, p	Au	Preform	160
		n, p	Au–Ge	Preform	700
		N	Au–Sn	Preform	–
GaN	3.36	Semi-i	Al–In	Preform	–
GaP	2.26	P	Au–Zn(99 : 1)	Preform, evaporation	700
		P	Au–Ge	Preform	–
		N	Au–Sn(62 : 38)	Preform	360
		N	Au–Si(98 : 2)	Evaporation	700
GaAs	1.42	P	Au–Zn(99 : 1)	Electroless, evaporation	600
		P	In–Au(80 : 20)	Preform	–
		N	Au–Ge(88 : 12)	Evaporation	–
		N	In–Au(90 : 10)	Evaporation	350–450
		N	Au–Si(94 : 6)	Evaporation	550
		N	Au–Sn(90 : 10)	Evaporation	300
		N	Au–Te(98 : 2)	Evaporation	300
GaSb	0.72	P	In	Preform	500
		n	In	Preform	–
InP	1.35	p	In	Preform	–
		n	In, In–Te	Preform	350–600
		N	Ag–Sn	Preform, evaporation	350–600
InAs	0.36	N	In	Preform	–
			Sn–Te(99 : 1)	Preform	–
InSb	0.17	N	In	Preform	–
		N	Sn–Te(99 : 1)	Preform	–
$GaAs_{1-x}P_x$	1.42–2.16	P	Au–In	Electroplate	400–450
		p	Au–Zn	Evaporation	–
		P	Al	Evaporation	500
		N	Au–Ge–Ni	Evaporation	500
		N	Au–Sn	Evaporation, electroless	450–485
		N	Au–Si	Evaporation	–
$Ga_{1-x}In_xSb$	0.7–0.17	N	Sn–Te	Evaporation	–
$Al_xGa_{1-x}P$	2.31–2.45	N	Sn	Preform	–
$Ga_{1-x}In_xAs$	1.47–0.35	N	Sn	Preform	–
$InAs_xSb_{1-x}$	0.17–0.35	N	In–Te	Preform	–

After Sze [187].

Table 21.13 Very low resistance ohmic contacts to GaAs.

Metallization	Annealing temperature (°C)	Annealing time (s)	Annealing ambient	ρ_C (cm^2)	N_D (cm^{-3})	Analysis	Comments
Ge/P/W/Au	800	5	N_2	4×10^{-6}	1×10^{17}	AES, STEM, EDS	Can withstand rapid thermal annealing schedules used for ion implant activation [188]
Ge/Mo/W/Au	850	10	N_2	5×10^{-6}	1×10^{17}	AES, STEM, EDS	Can withstand rapid thermal annealing schedules used for ion implant activation [188]
Ge/Pd	275–450	30	H_2–N_2	2×10^{-6} – 8×10^{-6}	8×10^{17} – 10×10^{17}	RBS	Solid-phase epitaxy to form metal/Ge (n + epi) GaAs (n$_i$[100]) [189] heterostructure
In,Zn,Ti/Pt/Au	400, 440	5, 5	H_2–N_2	1.7×10^{-7}	P	AES, SIMS	In, Zn dopant diffusion prior to refractory metallization [190]
In/Pt	300–500	—	N_2	2×10^{-6}	1.5×10^{17}	SIMS, SEM, RBS	InAs/GaAs heterojunction forms [191]
Si delta doping	—	—	—	6×10^{-6}	10^{18}	SEM	Delta-doped barrier [192]
Au/Ge	ArF, KrF laser	—	Ar	6.2×10^{-7}	0.45×10^{18} – 1.8×10^{18}	AES	Excimer laser annealing interdiffused Ge, Ga [193]

Interlayers have a larger effect at metal-II–VI compound semiconductor interfaces. Figure 21.28 displays I–V characteristics for 150-Å thick Au layers on UHV-cleaved CdS($10\bar{1}0$) surfaces. Without any interlayer, the Au–CdS($10\bar{1}0$) junction is strongly rectifying with a barrier height measured by I–V and C–V of 0.8 eV. Submonolayer deposits of Al on the CdS surface prior to Au deposition reduce the barrier height incrementally, as indicated by the labeled curves, such that the contact becomes ohmic with a 2 Å thick Al interlayer. Analogous changes are also observed for Au/Al/CdTe [195] and Al/Yb/CdTe [196] interfaces.

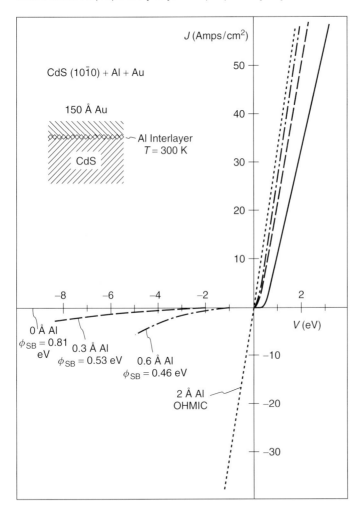

Figure 21.28 Atomic-scale control of macroscopic device characteristics. Room temperature I–V characteristics of 150 Å thick Au diodes on UHV-cleaved CdS($10\bar{1}0$). With increasing submonolayer Al interlayer thickness, Φ_{SB} decreases steadily from 0.8 eV to Ohmic. The inset shows a cross-sectional schematic diagram of the interlayer structure [180].

Photoemission measurements provide a microscopic explanation for these changes in terms of chemical interdiffusion [198]. As with III–V compounds, reactive metals such as Al alter the stoichiometry of II–VI anion versus cation outdiffusion. For CdS, Al retards Cd outdiffusion, resulting in excess cation concentrations below the CdS interface. This cation–anion imbalance results in a highly n-type CdS near-surface layer. With high carrier concentration, the width of the surface space charge region decreases and tunneling through the barrier dominates carrier transport. Thus, only one to two atomic layers at the intimate metal–semiconductor interface are sufficient to produce large barrier height changes at the macroscopic level.

21.7.3.2 Less-Reactive Buffer Layers

Less-reactive adsorbate layers can affect macroscopic barriers either by (i) introducing dopants or (ii) preventing chemical changes. An example of (i) is H_2S on UHV-cleaved InP(110), which reduces the Au/InP Φ_{SB} from 0.5 to <0.3 eV [199]. This is attributed to S indiffusion and surface doping that produce shallow donor levels and increased tunneling through the barrier.

Examples of (ii) are Xe buffer layers that reduce interface disruption by dissipating the energy associated with metal condensation and clustering on semiconductor surfaces. With subsequent thermal desorption of the Xe, the metal comes in contact with the semiconductor, producing junctions with significantly different barrier heights. For different metals on UHV-cleaved p-GaAs with Xe interlayers, E_F ranges from 0.37 to 0.62 eV above E_V, whereas E_F is "pinned" at $E_V + 0.5$ eV without a Xe interlayer [200]. Similarly, E_F stabilization is centered at $E_C - 0.32 + 0.1$ eV for metals on UHV-cleaved n-GaAs with Xe interlayers versus the $E_C - 0.7$ eV position typically reported. Xe interlayers produce even larger effects for metals on InP [201] and ZnSe [202].

21.7.3.3 Semiconductor Interlayers

Elements that either form semiconductor interlayers or alter the near-surface doping are very effective methods to control Schottky barrier heights. Thus, nanometer thicknesses of the chalcogens S, Se, and Te permit MBE-grown GaAs(100) barrier changes of >0.6 eV for different metals [182, 183]. Metals that react strongly with these chalcogens form the smallest n-type Φ_{SB}, whereas metals that do not react form the highest barriers – in agreement with Figure 4.17.

Combinations of Au, Ge, and Ni layers on liquid encapsulated Czochralski (LEC) GaAs(100) permit E_F variations than range from 0.25 to 0.9 eV. In this case, different metallurgical phases form in contact with the GaAs that either changes the substrate doping due to Ge indiffusion or that produces a multiinterface metal/Ge/GaAs contact as in the heterojunction contact pictured schematically in Figure 21.26.

Intentionally doped semiconductor interlayers also produce large Φ_{SB} changes. Figure 21.29 illustrates the E_F stabilization energies for (i) Ge doped with As or P and (ii) Si doped with As or Ga deposited *in situ* on GaAs(100) with particular growth treatments [182, 183]. Ge deposition alone produces E_F stabilization in the

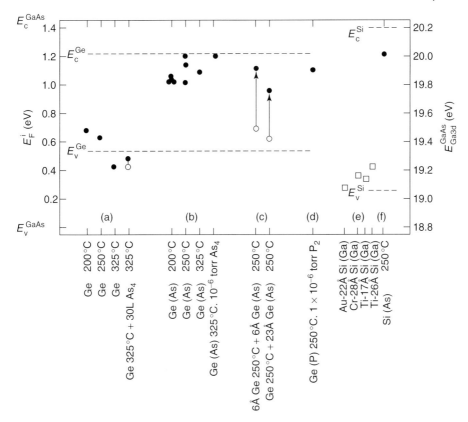

Figure 21.29 Wide energy range of XPS-measured E_F positions within the GaAs(100) bandgap for a variety of doped Ge and Si interlayers and metal overlayers. Unless otherwise noted, overlayer thicknesses are ~10 Å and the depositions intended to incorporate As or P in the overlayer occurred in a 10^{-7} torr background pressure of As_4 or P_2: (a) Ge overlayers deposited in vacuum; (b) Ge overlayers deposited in As_4; (c) Ge overlayers initially deposited in vacuum but completed with deposition in As_4; (d) Ge overlayer deposited in P_2; (e) Si overlayer deposited with a Ga monolayer in ~2 × 10^{-6} torr H_2 background pressure; and (f) Si overlayer deposited in As_4. Dashed lines signify the Ge and Si band edges derived from measured band offsets. E_F positions range across energies corresponding to the Ge and Si bandgap energies [182, 183].

lower part of the GaAs bandgap, whereas E_F for Ge doped n-type with As stabilizes at energies up to 0.7 eV higher. Similarly, Si doped p-type with Ga stabilizes near the GaAs valence band, while Si doped n-type with As stabilizes 1.1 eV higher. The dashed lines serve as guides to the eye and correspond to the band-edge separations of Ge (E_G = 0.7 eV) and of Si (E_G = 1.11 eV), whose positions were determined from XPS band offset measurements [203].

Figure 21.29 indicates that the Ge or Si interface layers move E_F according to their effective work functions, which decrease or increase according to whether these interlayers are doped n-type or p-type, respectively, that is, E_F is near the interlayer E_C or E_V. Si interlayers at Al/GaAs interfaces grown under Ga- or As-rich background conditions also produce Φ_{SB} values that extend from 0.3 to 1.04 eV [205].

In general, these results show that (i) work functions of the deposited layers dominate the energy of Fermi-level stabilization and (ii) E_F is not "pinned" at these interlayer–GaAs interfaces.

21.7.4
Wet-Chemical Treatments

21.7.4.1 Photochemical Washing

Wet-chemical treatments can also passivate semiconductor surfaces to reduce surface and interface state densities. An appealingly simple method to reduce states in the GaAs bandgap is photochemical washing of MBE-grown GaAs(100) wafers with deionized water coupled with above-bandgap illumination [197]. This treatment decreases the band bending and recombination velocity of the air-exposed GaAs by dissolving and removing surface As and As oxides. Here, the surface recombination velocity (SRV) is gauged by the intensity of band-edge photoluminescence since (i) surface states in the bandgap promote subgap recombination that competes with bandedge emission and (ii) band bending separates photogenerated electrons and holes, further reducing band-to-band recombination. Likewise, the decrease in surface states permits greater Fermi-level movement within the gap, as measured by $C–V$ techniques.

Figure 21.30 illustrates the experimental setup for the photochemical washing procedure. Band bending and surface recombination are monitored continuously

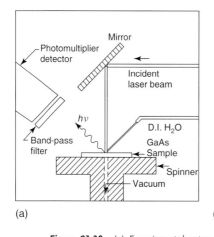
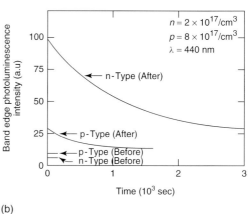

Figure 21.30 (a) Experimental setup for the unpinning procedure and the photoluminescence measurements. The n- or p-type GaAs spins on a vacuum chuck spinner under Ar^+ ion laser illumination while bathed in 18 MΩ deionized H_2O. A photomultiplier and 1.4 eV band-pass filter monitors the GaAs band-edge luminescence intensity. (b) Band-edge photoluminescence intensity excited by a 440 nm laser versus time after photochemical washing [197].

while the surface is washed. Reexposure to air causes the band-edge photoluminescence to decrease over time to levels at or above the original values before washing. Capacitors formed by immediately spinning polymethylmethacrylate (PMMA) on to the washed surface before the photoluminescence degraded displayed the characteristic shape of a metal–insulator–semiconductor (MIS) capacitor at high frequency with a clear transition from accumulation to depletion. This C–V behavior indicated "unpinned" E_F movement and surface state densities reduced by orders of magnitude to $\sim 10^{11}$ cm^{-2}. Short circuit photocurrent measurements with metal contacts formed on photo-washed surfaces exhibited reduced series resistance and band bending [206].

The photowashing experiment demonstrates that the Fermi level can be "unpinned" at GaAs surfaces, that the GaAs surface slowly forms new gap states spontaneously with air exposure, and that these new states are associated with water-soluble As and As oxides. Besides its fundamental scientific significance, the procedure has value for GaAs technology since it demonstrates the feasibility of improved GaAs metal–insulator field effect transistors (MISFETs).

21.7.4.2 Inorganic Sulfides

Inorganic sulfides can provide large reductions in SRV while passivating surfaces to prevent reoxidation. Skromme *et al.* treated GaAs grown by organometallic vapor-phase epitaxy with Na$_2$S to increase band-edge photoluminescence by nearly 3 orders of magnitude, comparable to values achieved by passivating GaAs with an AlGaAs cap [208]. Yablonovitch *et al.* used radiofrequency (RF) induction to probe transient surface conductivity and extract SRV values [209]. As Figure 21.31 shows, SRV for unpassivated Si, GaAs, and InGaAs surfaces are $\sim 10^6$–10^7 cm s^{-1} [204].

Along with photowashing, sulfides such as Na$_2$S, NaOH, and semiconductor epilayers reduce SRV by over 6 orders of magnitude. The reduction in interface state density for these sulfides is attributed to the formation of Ga chalcogenides which rehybridize dangling bond states at the surface, moving their energy levels out of the bandgap [212].

21.7.4.3 Thermal Oxides and Hydrogen

Thermal oxides and hydrogen treatment have similar effects on SRV for Si surfaces. Indeed, the reduction of interface state density between Si and SiO$_2$ using gas and temperature treatments was a prime goal of the Si microelectronics industry for decades in order to improve speed and reduce power. Aqueous etching using HF solutions with pH adjusted to eliminate surface roughness produce microscopically smooth surfaces with SRV values as low as <1 cm s^{-1} [213]. Here, H terminates the Si dangling bonds of a (111) surface with a homogeneous monohydride passivation layer. The termination persists long enough for surfaces to remain passivated during transfer from solution to vacuum or other nonoxidizing environment.

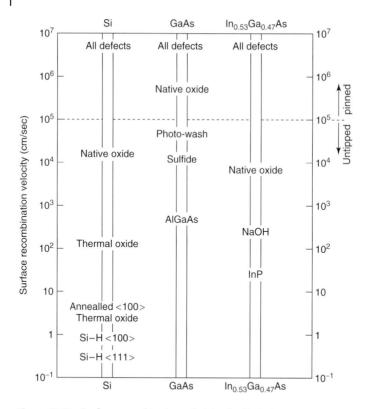

Figure 21.31 Surface recombination velocities for Si, GaAs, and $In_{0.53}Ga_{0.47}As$ epitaxial layers on GaAs(100) for various surface treatments. Low SRV values correspond to reduced interface state densities and unpinned surface Fermi levels. (After Yablonovitch and Gmitter [204]).

In general, removing electrically active species, forming stable overlayers that hybridize surface atomic bonds, and preventing reoxidation are effective procedures to avoid E_F "pinning" and high surface recombination.

21.7.5
Semiconductor Crystal Growth

Semiconductor crystal growth enables many atomic-scale approaches to the control of Schottky barriers. Researchers have used stoichiometry, surface misorientation, and the epitaxical growth of binary alloys to demonstrate the primary effect of extrinsic structural properties on the electronic interface.

21.7.5.1 Variations in Stoichiometry

Layer-by-atomic layer crystal growth permits the control of stoichiometry, purity, point defect and dislocation density, precipitates, and their associated physical properties. Stoichiometric semiconductor crystals with low native defect and dislocation densities are now widely available with epitaxial growth methods. The importance of these high-quality semiconductor crystals for controlling Schottky barriers is evident from a comparison of Figure 21.19 with Figure 21.32 [207]. There is a qualitative difference in the metal-induced E_F movements of melt- versus MBE-grown GaAs. Figure 21.19 showed that metals on UHV-cleaved metal-grown GaAs(110) produced E_F "pinning" within a narrow energy range with only submonolayer coverages. Furthermore, the same metal on n- versus p-type GaAs stabilized at different energies, indicative of "pinning" by two energy levels. In contrast, Figure 21.32 shows that metals on As-decapped, MBE-grown GaAs(100) move E_F over several monolayers to a wide range of energies in the bandgap with matching n- and p-type values. IPS confirms these energies. The much wider range indicates much lower interface states in MBE GaAs. Figure 21.19 is consistent with the orders of magnitude larger native point defect densities in melt-grown GaAs because of its As-rich stoichiometry.

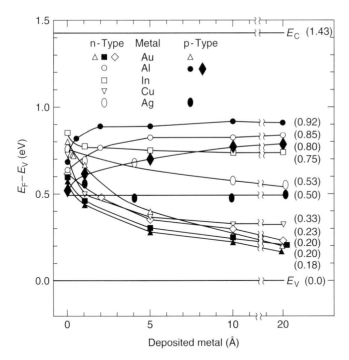

Figure 21.32 Qualitative difference in metal-induced E_F movements within the GaAs bandgap as measured by SXPS. Unlike metals on UHV-cleaved metal-grown GaAs(110), metals on As-decapped, MBE-grown GaAs(100) move E_F over several monolayers to a wide range of energies with matching n- and p-type values [207].

21.7.5.2 Misorientation/Vicinal Surfaces

Because E_F can move over a wide energy range, MBE-grown GaAs interfaces with metals are a model system to test the effect of different interface states. By simply changing the crystal growth orientation, one can introduce controlled densities of interface states. As Figure 17.3 shows, off-axis growth results in step edges with well-defined densities of dangling bonds. The self-consistent electrostatic model pictured in Figure 21.15 then provides a tool to extract energies, densities, and types of gap states that affect the relationship between semiconductor barrier height and metal work function.

Figure 21.33a shows the application of this model to different metals on MBE-grown GaAs. Here, SXPS measurements of rigid core level shifts on As-decapped vicinal GaAs provide the Φ_{SB} values [210, 211]. The straight line indicates the Φ_{SB} versus Φ_M dependence without interface states. Interface state density depends on orientation and increases with vicinal angle. Figure 21.33a shows little change for Au barriers with density changes, whereas Al barriers increase monotonically with the density of dangling bond sites. The family of curves corresponds to different interface state densities of a level at $E_V + 0.6$ eV and a (constant) background density of states at $E_V + 0.2$ eV. The Figure 21.33a

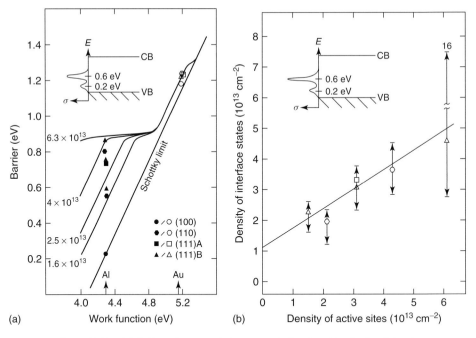

Figure 21.33 (a) Self-consistent analysis of Φ_{SB} versus Φ_M and (b) interface state densities versus active structural sites for Al (closed symbols) and Au (open symbols) on vicinal GaAs(100) surfaces. Φ_{SB} increases monotonically with misorientation angle and direction. The family of density curves for the 0.6 eV and (constant density) 0.2 eV states pictured in the insets yield the interface state density values shown in (b) [210, 211].

inset represents states at these energies schematically. States at both energies are required to account for the changes in measured barriers with interface state density. Indeed, CL spectra reveal emissions involving states at these energies.

Figure 21.33b compares the density of interface states obtained from Figure 21.33a with the density of chemically active edge sites calculated from the Figure 17.3 picture of vicinal steps. The linear dependence and slope indicates ~2/3 electron per active structural site and an extrapolated density of 1×10^{13} cm^{-2} for the [100]-oriented surface. scanning tunneling microscopy (STM) measurements of kink sites at vicinal GaAs(100) surfaces have densities sufficiently high to account for the observed electrical effects.

Hence, metals on clean vicinal GaAs(100) surfaces illustrate how discrete interface states calculated self-consistently and observed spectroscopically correlate with densities of structural imperfections introduced and controlled on an atomic scale.

21.7.5.3 Epitaxical Growth of Binary Alloys on Compound Semiconductors

Semiconductor crystal growth also provides a method to distinguish between the effects of strain and misorientation. Epitaxical growth of binary alloys on compound semiconductors permits one to adjust the interface strain from compressive to tensile by varying the alloy composition and thereby the lattice mismatch. Thus, in Figure 21.34, strain at the GaAs(100) interface with semimetallic alloy Sc$_x$Er$_{1-x}$As ($0 \leq x \leq 1$) ranges from compressive to tensile as ScAs composition increases

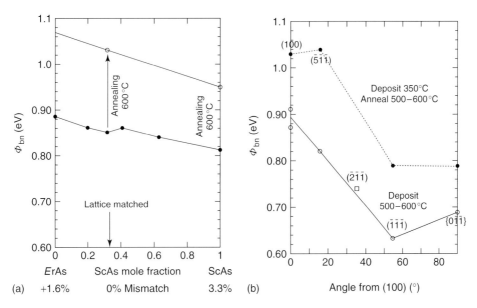

Figure 21.34 Φ_{SB} of Sc$_x$Er$_{1-x}$As metal alloys on GaAs(100) versus (a) lattice mismatch and annealing and (b) interface misorientation and annealing. The weak dependence on lattice mismatch contrasts with the strong dependence on misorientation and annealing. (After Palmstrøm et al. [214]).

[214]. Diodes formed with different composition and lattice mismatch show only a weak (<0.1 eV) Φ_{SB} dependence, whereas annealing and misorientation angle with the same lattice-matched composition produce Φ_{SB} changes greater than 0.4 eV. For the 90° range of orientations shown, interface structures include different amounts of {100} and {111} ledges and steps for {h11}-type, $h > 1$, surfaces and different stacking sequences between {100} and {111} $Sc_xEr_{1-x}As$ ($0 \leq x \leq 1$). Indeed, barriers of such interfaces may be macroscopic averages, and even larger Φ_{SB} variations may exist within microscopic domains.

As with vicinal surfaces, it is possible to vary Φ_{SB} over large energy range for the same metal–semiconductor contact by controlling interface structure at the atomic scale [215].

21.8
Summary

In this chapter, we have examined the intrinsic and extrinsic mechanisms that can produce localized electronic states and the interface dipoles that impact Schottky barriers at metal–semiconductor interfaces. While different intrinsic models can account for several general trends at ideal interfaces, a wide variety of extrinsic physical mechanisms exist that have major effects on Schottky barrier formation.

Schottky barriers measured experimentally extend over wide energy ranges for UHV-clean semiconductors with low-bulk native defect densities, while Φ_{SB} ranges are much more limited for semiconductors of poorer surface cleanliness and crystal quality.

Schottky barriers can be controlled by both macroscopic and microscopic techniques. A variety of passivation methods are now available to reduce extrinsic effects. UHV and epitaxial growth methods have advanced our ability to obtain metal–semiconductor interfaces with a high level of control and reproducibility.

In turn, these methods have enabled large barrier height variations by changing interface structure at the atomic level. The success of these techniques is evidence that localized bonding and composition has a primary impact on interface electronic structure.

Problems

1. Calculate S and CNL for n-type InP given $\Phi_{SB}^n(Au) = 0.6$ eV, and $\Phi_{SB}^N(Al) = 0.3$ eV.
2. According to the S and CNL in Problem 1, what would be the n-type barrier height for Cu?
3. For a Si(111) surface and a density of interface states equal to one electron per atom distributed across the bandgap, calculate S. Assume depth $\delta = 1.5$ Å.

4. For Au and In on melt-grown GaAs with barriers $\Phi^n_{SB}(\text{Au}) = 0.9$ eV and $\Phi^n_{SB}(\text{In}) = 0.65$ eV, calculate their respective dipoles with the CNL given in Table 21.1 and $S = 0.1$.

5. Figure 21.7 illustrates the carrier depletion region surrounding line charges within misfit dislocations at a semiconductor heterojunction. See also, for example, Figure 20.7. These depletion regions may limit transport through a metal contact to this heterojunction, depending on their spacing. Poisson's equation in cylindrical coordinates yields (Ref. 21.31) $V_D = q\Phi(r_0) = qNr_s^2/2\varepsilon[\ln(r_s/r_0) - 0.5]$, where V_D is the potential at the dislocation core, N is the semiconductor carrier concentration, r_s is the radius of the depletion boundary around the dislocation, and r_0 is the effective radius of the dislocation. (a) Calculate r_s for GaAs with $N = 2 \times 10^{16}$ cm^{-3} and $V_D = 0.8$ eV. (b) Calculate r_0 for this potential and doping. (c) How important is this effect for a dislocation density of 6×10^{10} cm^{-2}? For 10^8 cm^{-2}? (d) Is it possible to increase the bulk carrier density to obtain barriers equivalent to reducing dislocation density of 6×10^{10} to 10^8 cm^{-2}?

6. Defects at metal–semiconductor interfaces can introduce excess free charge that can change the effective barrier height by tunneling through a narrowed depletion region. (a) For a Au–InP diode with $\Phi_{SB} = 0.5$ eV, calculate the probability for electrons at E_{FM}, the Fermi level of the metal, to tunnel under reverse bias $V_R = 1$ V. (b) How much does this tunneling probability increase due to a change in n-type carrier concentration from $N_D = 10^{15}$ cm^{-3} to $N_D = 5 \times 10^{17}$ cm^{-3}? The InP electron effective mass $m^*/m_0 = 0.077$. Hint: Approximate the reverse-biased depletion region by a triangular barrier, for which the transmission probability is given by $T = \exp(4E_B^{3/2}\sqrt{2\,m^*}/3qh\mathcal{E})$, where $E_B = E_C - E_{FM}$ at the interface and the base width is $\sim \frac{1}{3}$ W, the depletion width.

7. Calculate the dipole contribution to the Schottky barrier height of GaAs for a filled surface acceptor density $\sigma_a^- = 10^{13}$cm^{-2} located 10 Å below the surface.

8. At abrupt Au/n-ZnO ($N_D = 10^{16}$ cm^{-3}) interfaces, there are Zn vacancies, which act as compensating acceptors. The conduction-band density of states $N_C = 4 \times 10^{18}$ cm^{-3} and $\varepsilon_{\text{average}}/\varepsilon_0 = 8.2$. The density of charged acceptors is

$$\sigma^-(E_F) = \frac{\sigma_a}{1 + 2\exp[(E_F - E_a)/k_B T]}$$

where E_a is the energy level of the acceptor (2.0 eV below CBM), E_F is the Fermi level, k_B is Boltzmann's constant, and T is the absolute temperature. Assuming the negatively charged acceptor and its image charge on the metal side are 1.0 nm apart, calculate the potential drop across the interface due to interface dipole and the corresponding interface Fermi level with (a) $\sigma_a = 10^{12}$cm^{-2} and (b) $\sigma_a = 10^{13}$cm^{-2} respectively. Will the Fermi level be pinned around $\sigma_a = 10^{14}$cm^{-2}? Why?

9. Calculate and plot barrier heights versus metal work function from $\Phi_M = 3 - 6$ eV for ZnO with $E_C - E_F = 0.15$ eV in the bulk and a donor level located 0.9 eV below the conduction-band edge with densities 10^{12}, 10^{13}, and 10^{14} cm^{-2}.
10. Calculate and plot the interface Fermi-level position in the ZnO bandgap as a function of acceptor density for the Au–ZnO interface with $E_a = 2.0$ eV as shown in Figure 21.16.
11. Consider an intrinsic n-type ZnO crystal with defect levels due to oxygen vacancy donors at $E_V + 2.5$ eV and zinc vacancy acceptors at $E_C - 2.1$ eV. (a) Calculate the classical Schottky barrier height for Pt, Ir, and Ta assuming no interface states. (b) Calculate the corresponding Schottky barriers assuming interface chemical reactions can create additional native point defects and/or interfacial layers. Use $\chi(\text{ZnO}) = 4.2$ eV.
12. Sketch a band diagram to illustrate an Ohmic contact between a metal and n-type GaAs using InAs and a graded InGaAs intermediate layer.
13. For an Ohmic contact to an AlGaN high electron mobility transistor, calculate the maximum contact resistivity to permit a cutoff frequency of 50 GHz. Assume an AlGaN thickness of 23 nm and a 30% Al composition. Neglect channel resistance.

References

1. Duke, C.B. and Mailhiot, C. (1985) *J. Vac. Sci. Technol. B*, **3**, 1170.
2. Heine, V. (1965) *Phys. Rev.*, **138**, A1689.
3. Schlüter, M. (1982) *Thin Solid Films*, **93**, 3 and references therein.
4. Mead, C.A. (1966) *Solid State Electron.*, **9**, 1023.
5. Kurtin, S., McGill, T.C., and Mead, C.A. (1969) *Phys. Rev. Lett.*, **22**, 1433.
6. (a) Tejedor, C., Flores, F., and Louie, E. (1977) *J. Phys. C*, **10**, 2163; (b) Flores, F. and Tejedor, C. (1987) *J. Phys. C: Solid State Phys.*, **20**, 145.
7. (a) Tersoff, J. (1984) *Phys. Rev. Lett.*, **52**, 465; (b) Tersoff, J. (1984) *Phys. Rev. B* **30**, 4875; (c) Capasso, F. and Margaritondo, G. (eds) (1987) *Heterojunction Band Discontinuities: Physics and Device Applications*, North-Holland, Amsterdam, p. 44.
8. Applebaum, J.A. and Hamann, D.R. (1974) *Phys. Rev. B*, **10**, 4973.
9. Mönch, W. (1986) in *Festkörperprobleme – Advances in Solid State Physics*, Vol. 26 (ed. P. Grosse), Vieweg, Braunschweig, p. 67.
10. Flores, F. and Tejedor, C. (1979) *J. Phys. C*, **12**, 731.
11. Miedema, A.R., de Chątel, P.F., and de Boer, F.R. (1980) *Phys. B*, **100**, 1.
12. Pauling, L.N. (1960) *The Nature of the Chemical Bond*, Cornell University, Ithaca.
13. Mönch, W. (1989) *Appl. Surf. Sci.*, **41/42**, 128.
14. Schmid, P.E. (1985) *Helv. Phys. Acta*, **58**, 371.
15. Mönch, W. (1988) *Phys. Rev. B*, **37**, 7129.
16. Kahn, A., Stiles, K., Mao, D., Horng, S.F., Young, K., McKinley, J., Kilday, D.G., and Margaritondo, G. (1989) in *Metallization and Metal-Semiconductor Interfaces*, NATO ASI Series B, Vol. 195 (ed. I.P. Batra), Plenum, New York, p. 163.

17. Harrison, W.A. (1976) *Phys. Rev. Lett.*, **37**, 312.
18. Inkson, J.C. (1974) *J. Vac. Sci. Technol.*, **11**, 943.
19. Brattain, W.H. and Bardeen, J. (1953) *Bell Syst. Tech. J.*, **82**, 1.
20. Goodwin, T.A. and Mark, P. (1972) *Progress in Surface Science*, Vol. 1, Pergamon, New York, p. 1.
21. Stiles, K., Kahn, A., Kilday, D.G., and Margaritondo, G. (1987) *J. Vac. Sci. Technol. B*, **5**, 987.
22. (a) Brillson, L.J., Slade, M.L., Viturro, R.E., Kelly, M.J., Tache, N., Margaritondo, G., Woodall, J.M., Kirchner, P.D., and Wright, S.L. (1986) *Appl. Phys. Lett.*, **48**, 1458; (b) Brillson, L.J. Slade, M.L., Viturro, R.E., Kelly, M.J., Tache, N., Margaritondo, G., Woodall, J.M., Kirchner, P.D., and Wright, S.L. (1986) *J. Vac. Sci. Technol. B*, **4**, 919.
23. Kajiyama, K., Mizushima, Y., and Sakata, S. (1973) *Appl. Phys. Lett.*, **23**, 458.
24. Newman, N., Lilienthal-Weber, Z., Weber, E.R., Washburn, J., and Spicer, W.E. (1988) *Appl. Phys. Lett.*, **53**, 145.
25. Lagowski, J., Gatos, H.C., Aoyama, T., and Lin, D.G. (1984) *Appl. Phys. Lett.*, **45**, 680.
26. Huijser, A. and van Laar, J. (1975) *Surf. Sci.*, **52**, 202.
27. Chang, S., Brillson, L.J., Kime, Y.J., Rioux, D.S., Pettit, G.D., and Woodall, J.M. (1990) *Phys. Rev. Lett.*, **64**, 2551.
28. Milnes, A.G. (1973) *Deep Impurities in Semiconductors*, Wiley-Interscience, New York.
29. Shaw, D. (1973) *Atomic Diffusion in Semiconductors*, Plenum, New York.
30. Mirceau, A. and Bois, D. (1979) *Inst. Phys. Conf. Ser.*, **46**, 82.
31. Woodall, J.M., Pettit, G.D., Jackson, T.N., Lanza, C., Kavanagh, K.L., and Mayer, J.W. (1983) *Phys. Rev. Lett.*, **51**, 1783.
32. Tung, R. (1984) *Phys. Rev. Lett.*, **52**, 461.
33. Andrews, J.M. and Phillips, J.C. (1975) *Phys. Rev. Lett.*, **35**, 56.
34. Ottaviani, G., Tu, K.N., and Mayer, J.W. (1980) *Phys. Rev. Lett.*, **44**, 284.
35. Brillson, L.J. (1978) *Phys. Rev. Lett.*, **40**, 260.
36. Brillson, L.J., Brucker, C.F., Katnani, A.D., Stoffel, N.G., and Margaritondo, G. (1981) *Appl. Phys. Lett.*, **38**, 784.
37. Kendelewicz, T., Newman, N., List, R.S., Lindau, I., and Spicer, W.E. (1985) *J. Vac. Sci. Technol. B*, **3**, 1206.
38. Fujitani, H. and Sasano, S. (1990) *Phys. Rev. B*, **42**, 696.
39. Das, G.P., Blöchl, P., Andersen, O.K., Christensen, N.E., and Gunnarson, O. (1989) *Phys. Rev. Lett.*, **63**, 1168.
40. Heslinga, D.R., Weitering, H.H., van der Werf, D.P., Klapwijk, T.M., and Hibma, T. (1990) *Phys. Rev. Lett.*, **64**, 1589.
41. Spicer, W.E., Lindau, I., Skeath, P., and Su, C.Y. (1980) *Phys. Rev. Lett.*, **44**, 420.
42. Weber, E.R., Ennen, H., Kaufmann, V., Windscheif, J., Schneider, J., and Wasinki, T. (1982) *J. Appl. Phys.*, **53**, 6140.
43. Mark, P., Chang, S.C., Creighton, W.F., and Lee, B.W. (1975) *CRC Crit. Rev. Solid State Sci.*, **5**, 189.
44. Mark, P., So, E., and Bonn, M. (1977) *J. Vac. Sci. Technol.*, **14**, 865.
45. Wieder, H.H. (1978) *J. Vac. Sci. Technol.*, **15**, 148.
46. Lindau, I., Chye, P.W., Garner, C.M., Pianetta, P., Su, C.Y., and Spicer, W.E. (1978) *J. Vac. Sci. Technol.*, **15**, 1337.
47. Williams, R.H., Varma, R.R., and Montgomery, V. (1979) *J. Vac. Sci. Technol.*, **16**, 1418.
48. Zunger, A. (1981) *Phys. Rev. B*, **24**, 4372.
49. Weaver, J.H. and Waddill, G.D. (1981) *Nature*, **251**, 1444 and references therein.
50. Freeouf, J.L. and Woodall, J.M. (1981) *Appl. Phys. Let.*, **39**, 727.
51. Spicer, W.E., Lilienthal-Weber, Z., Weber, E.R., Newman, N., Kendelewicz, T., Cao, R.K., McCants, C., Mahowald, P., Miyano, K., and Lindau, I. (1988) *J. Vac. Sci. Technol. B*, **6**, 1245.
52. Daw, M.S. and Smith, D.L. (1979) *Phys. Rev. B*, **20**, 5150.
53. Daw, M.S. and Smith, D.L. (1980) *Appl. Phys. Lett.*, **8**, 690.

54. Srivistava, G.P. (1979) *Phys. Status Solidi B*, **93**, 761.
55. Allen, R.E. and Dow, J.D. (1981) *J. Vac. Sci. Technol.*, **19**, 383.
56. Allen, R.E. and Dow, J.D. (1982) *Phys. Rev. B*, **25**, 1423.
57. Van Vechten, J.A. (1975) *J. Electrochem. Soc.*, **122**, 423.
58. Van Vechten, J.A. (1980) Materials, properties and preparation, in *Handbook on Semiconductors*, Vol. 3 (ed. S.P. Keller), North-Holland, Amsterdam, p. 1.
59. Freeouf, J.L. (1980) *Solid State Commun.*, **33**, 1059.
60. Brillson, L.J. (1982) *Surf. Sci. Rep.*, **2**, 123 and references therein.
61. Lilienthal-Weber, Z., Gronsky, R., Washburn, T., Newman, N., Spicer, W.E., and Weber, E.R. (1986) *J. Vac. Sci. Technol. B*, **4**, 912.
62. Woodall, J.M. and Freeouf, J.L. (1981) *J. Vac. Sci. Technol.*, **19**, 794.
63. Shen, C., Kahn, A., and Schwartz, J. (2001) *J. Appl. Phys.*, **89**, 449.
64. Hasegawa, H. and Ohno, H. (1986) *J. Vac. Sci. Technol. B*, **5**, 1130.
65. Hasegawa, H., He, L., Ohno, H., Sawada, T., Haga, T., Abe, Y., and Takahashi, H. (1987) *J. Vac. Sci. Technol. B*, **5**, 1097.
66. Kahn, A., Koch, N., and Gao, W. (2003) *J. Polymer Sci.*, **41**, 2529.
67. Shen, C. and Kahn, A. (2001) *J. App. Phys.*, **90**, 4549.
68. Greczynski, G., Fahlman, M., Salaneck, W.R., Johansson, N., do Santos, D.A., and Bredas, J.L. (2000) *Thin Solid Films*, **363**, 322.
69. Koch, N., Rajagopal, A., Zojer, E., Ghijsen, J., Cripsin, X., Pourtois, G., Bredas, J.L., Johnson, R.L., Pireaux, J.J., and Leising, G. (2000) *Surf. Sci.*, **454**, 1000 and references therein.
70. Duke, C.B. and Mailhiot, C. (1985) *J. Vac. Sci. Technol. B*, **3**, 1170.
71. Rhoderick, E.H. and Williams, R.H. (1988) *Metal-Semiconductor Contacts*, Oxford Science, Oxford, pp. 49–84.
72. Spicer, W.E., Lindau, I., Skeath, P., and Su, C.Y. (1980) *J. Vac. Sci. Technol.*, **17**, 1019.
73. Thanalaikas, A. and Rasul, A. (1976) *J. Phys. C: Solid-State Phys.*, **9**, 337.
74. Van Otterloo, J.D. and Gerritsen, L.J. (1978) *J. Appl. Phys.*, **49**, 723.
75. Turner, M.J. and Rhoderick, E.H. (1968) *Solid-State Electron.*, **11**, 291.
76. Szydlo, N. and Poirier, R. (1973) *J. Appl. Phys.*, **44**, 1386.
77. Archer, R.J. and Atalla, M.M. (1963) *Ann. N.Y. Acad. Sci.*, **101**, 697.
78. Crowell, C.R., Shore, H.B., and La Bate, E.E. (1965) *J. Appl. Phys.*, **36**, 3843.
79. Varma, R.R., McKinley, A., Williams, R.H., and Higginbotham, I.G. (1977) *J. Phys. D: Appl. Phys.*, **10**, L171.
80. Van Otterloo, J.D. and de Groot, J.G. (1976) *Surf. Sci.*, **57**, 93.
81. Michaelson, H.B. (1977) *J. Appl. Phys.*, **48**, 4729.
82. Brillson, L.J., Brucker, C.F., Katnani, A.D., Stoffel, N.G., Daniels, R., and Margaritondo, G. (1982) *J. Vac. Sci. Technol.*, **21**, 564.
83. Smith, B.L. and Rhoderick, E.H. (1971) *Solid-State Electron.*, **14**, 71.
84. Mead, C.A. and Spitzer, W.G. (1964) *Phys. Rev.*, **134**, A713.
85. McCaldin, J.M., McGill, T.C., and Mead, C.A. (1976) *J. Vac. Sci. Technol.*, **13**, 802.
86. Newman, N., van Schilfgaarde, M., Kendelewicz, T., Williams, M.D., and Spicer, W.E. (1985) *Phys. Rev. B*, **33**, 1146.
87. Waldrop, J.R. (1984) *J. Vac. Sci. Technol. B*, **2**, 445.
88. Smith, B.L. (1969) Schottky barriers on p-type silicon, Ph.D. Thesis, Manchester University.
89. Ismail, A., Palau, J.M., and Lassabatere, L. (1984) *Rev. Phys. Appl.*, **19**, 205.
90. Svensson, S.P. and Andersson, T.G. (1985) *J. Vac. Sci. Technol. B*, **3**, 760.
91. Missous, M., Rhoderick, E.H., and Singer, K.E. (1986) *Electron. Lett.*, **22**, 241.
92. Gol'dberg, Y.A., Rafiev, T.Y., Tsarenkov, B.V., and Yakoulev, Y.P. (1972) *Sov. Phys. Semicond.*, **6**, 398.
93. Best, J.S. (1972) *Appl. Phys. Lett.*, **34**, 522.
94. Okamoto, K., Wood, C.E., and Eastman, L.F. (1981) *Appl. Phys. Lett.*, **38**, 636.

95. Daw, M.S. and Smith, D.L. (1981) *Solid-State Commun.*, **37**, 205.
96. Allen, R.E. and Dow, J.D. (1981) *J. Vac. Sci. Technol.*, **19**, 383.
97. Robinson, G.Y. (1985) in *Physics and Chemistry of III-V Compound Semiconductor Interfaces*, Chapter 2 (ed. C. Wilmsen), Plenum, New York and references therein.
98. Vitomirov, I.M., Aldao, C.M., Waddill, G.D., Capasso, C., and Weaver, J.H. (1990) *Phys. Rev. B*, **41**, 8465.
99. Newman, N., van Schilfgaarde, M., Kendelewicz, T., Williams, M.D., and Spicer, W.E. (1985) *Phys. Rev. B*, **33**, 1146.
100. Williams, R.H., Varma, R.R., and McKinley, A. (1977) *J. Phys. C: Solid-State Phys.*, **10**, 4545.
101. Williams, R.H., Montgomery, V., and Varma, R.R. (1978) *J. Phys. C: Solid-State Phys.*, **11**, L735.
102. Williams, R.H., Dharmadasa, I.M., Patterson, M.H., Maani, C., and Forsythe, N.M. (1986) *Surf. Sci.*, **168**, 323.
103. Williams, R.H., McLean, A.B., Evans, D.A., and Herrenden-Harker, W.G. (1986) *J. Vac. Sci. Technol. B*, **4**, 966.
104. Ismail, A., Palau, J.M., and Lassabatere, L. (1984) *Rev. Phys. Appl.*, **19**, 205.
105. Humphreys, T.P., Hughes, G.I., McKinley, A., Cunningham, E.C., and Williams, R.H. (1985) *Surf. Sci.*, **152**, 1222.
106. Slowik, J.H., Brillson, L.J., and Richter, H.W. (1986) *J. Vac. Sci. Technol. B*, **4**, 974.
107. Newman, N., van Schilfgaarde, M., and Spicer, W.E. (1987) *Phys. Rev. B*, **35**, 6298.
108. Wu, C.I. and Kahn, A. (1998) *J. Vac. Sci. Technol. B*, **16**, 2218.
109. Karrer, U., Ambacher, O., and Stutzmann, M. (2000) *Appl. Phys. Lett.*, **77**, 2012.
110. Rickert, K.A., Ellis, A.B., Kim, J.K., Lee, J.-L., Himpsel, F.J., Dwikusuma, F., and Kuech, T.F. (2002) *J. Appl. Phys.*, **92**, 6671.
111. Tracy, K.M., Hartlieb, P.J., Einfeldt, S., Davis, R.F., Hurt, E.H., and Nemanich, R.J. (2003) *J. Appl. Phys.*, **94**, 3939.
112. DeLucca, J.M., Mohney, S.E., Auret, F.D., and Goodman, S.A. (2000) *J. Appl. Phys.*, **88**, 2593.
113. Liu, Q.Z., Yu, L.S., Deng, F., Lau, S.S., and Redwing, J.M. (1998) *J. Appl. Phys.*, **84**, 881.
114. Shiojima, K., Sugahara, T., and Sakai, S. (1999) *Appl. Phys. Lett.*, **74**, 1936.
115. Kalinina, E.V., Kuznetsov, N.I., Dmitriev, V.A., Irvine, K.G., and Carter, C.H. Jr. (1996) *J. Electron. Mater.*, **25**, 831.
116. Lewis, L., Corbett, B., Mahony, D.O., and Maaskant, P.P. (2007) *Appl. Phys. Lett.*, **91**, 162103.
117. Guo, J.D., Feng, M.S., Guo, R.J., Pan, F.M., and Chang, C.Y. (1995) *Appl. Phys. Lett.*, **67**, 2657.
118. Maffeis, T.G.G., Simmonds, M.C., Clark, S.A., Peiro, F., Haines, P., and Parbrook, P.J. (2002) *J. Appl. Phys.*, **92**, 3179.
119. Venugopalan, H.S. and Mohney, S.E. (1998) *Appl. Phys. Lett.*, **73**, 1242.
120. (a) Kampen, T.U. and Mönch, W. (1997) *Appl. Surf. Sci.*, **117/118**, 388; (b) Rizzi, A. and Lüth, H. (2002) *Appl. Phys. Lett.*, **80**, 530.
121. Walker, D.E. Jr., Gao, M., Chen, X., Schaff, W.J., and Brillson, L.J. (2006) *J. Electron. Mater.*, **35**, 581.
122. Wang, H.-T., Jang, S., Anderson, T., Chen, J.J., Kang, B.S., Ren, F., Voss, L.F., Stafford, L., Khanna, R., Gila, B.P., Pearton, S.J., Shen, H., LaRoche, J.R., and Smith, K.V. (2006) *Appl. Phys. Lett.*, **89**, 122106.
123. Cao, X.A., Pearton, S.J., Dang, G., Zhang, A.P., Ren, F., and Van Hove, J.M. (1999) *Appl. Phys. Lett.*, **75**, 4130.
124. Hasegawa, H. and Oyama, S. (2002) *J. Vac. Sci. Technol. B*, **20**, 1647.
125. Kim, H., Lee, S.-N., Park, Y., Kwak, J.S., and Seong, T.-Y. (2008) *Appl. Phys. Lett.*, **93**, 132105.
126. Cowley, A.M. (1966) *J. Appl. Phys.*, **37**, 3024.
127. Brillson, L.J., Slade, M.L., Vitturo, R.E., Kelly, M.K., Tache, N., Margaritondo, G., Woodall, J.M., Kirchner, P.D., Pettit, G.D., and Wright, S.L. (1986) *J. Vac. Sci. Technol. B*, **4**, 919.
128. Brillson, L.J., Slade, M.L., Vitturo, R.E., Kelly, M.K., Tache, L., Margaritondo,

G., Woodall, J.M., Kirchner, P.D., Pettit, G.D., and Wright, S.L. (1986) *J. Vac. Sci. Technol. B*, **4**, 919.

129. Brillson, L.J., Mosbacker, H.L., Hetzer, M.J., Strzhemechny, Y., Jessen, G.H., Look, D.C., Cantwell, G., Zhang, J., and Song, J.J. (2007) *Appl. Phys. Lett.*, **90**, 102116.

130. Kajiyama, K., Mizushima, Y., and Sakata, S. (1973) *Appl. Phys. Lett.*, **23**, 458.

131. Chin, R., Milano, R.A., and Law, H.D. (1980) *Electron. Lett.*, **16**, 626.

132. Rideout, V.L. (1974) *Solid-State Electron.*, **17**, 1107.

133. Dumont, J., Monroy, E., Muñoz, E., Caudano, R., and Sporken, R. (2001) *J. Cryst. Growth*, **230**, 558.

134. Yu, E.T., Dang, X.Z., Yu, L.S., Qiao, D., Asbeck, P.M., Lau, S.S., Sullivan, G.J., Boutros, K.S., and Redwing, J.M. (1998) *Appl. Phys. Lett.*, **73**, 1880.

135. (a) Bradley, S.T., Goss, S.H., Hwang, J., Schaff, W.J., and Brillson, L.J. (2004) *Appl. Phys. Lett.*, **85**, 1368; (b) Bradley, S.T., Goss, S.H., Hwang, J., Schaff, W.J., and Brillson, L.J. (2005) *J. Appl. Phys.*, **97**, 084502.

136. Lin, Z., Lu, W., Lee, J., Liu, D., Flynn, J.S., and Brandes, G.R. (2003) *Appl. Phys. Lett.*, **82**, 4364.

137. Spitzer, W.G. and Mead, C.A. (1963) *J. Appl. Phys.*, **34**, 3061.

138. Goodman, A.M. (1963) *J. Appl. Phys.*, **24**, 329.

139. Brillson, L.J. (1981) *Thin Solid Films*, **89**, 461.

140. Dharmadasa, I.M., Herrenden-Harker, W.G., and Williams, R.H. (1986) *Appl. Phys. Lett.*, **48**, 1082.

141. Neville, R.C. and Mead, C.A. (1970) *J. Appl. Phys.*, **41**, 3795.

142. Coppa, B.J., David, R.F., and Nemanich, R.J. (2003) *Appl. Phys. Lett.*, **82**, 400.

143. Dong, Y., Fang, Z.-Q., Look, D.C., Cantwell, G., Zhang, J., Song, J.J., and Brillson, L.J. (2008) *Appl. Phys. Lett.*, **93**, 072111.

144. Mosbacker, H.L., Zgrabik, C., Hetzer, M.J., Swain, A., Look, D.C., Cantwell, G., Zhang, J., Song, J.J., and Brillson, L.J. (2007) *Appl. Phys. Lett.*, **91**, 072102.

145. Mosbacker, H.L., El Hage, S., Gonzalez, M., Ringel, S.A., Hetzer, M., Look, D.C., Cantwell, G., Zhang, J., Song, J.J., and Brillson, L.J. (2007) *J. Vac. Sci. Technol. B*, **25**, 1405.

146. Gu, Q.L., Cheung, C.K., Ling, C.C., Ng, A.M.C., Djurišić, A.B., Lu, L.W., Chen, X.D., Fung, S., Beling, C.D., and Ong, H.C. (2008) *J. Appl. Phys.*, **103**, 093706.

147. Polyakov, A.R., Smirnov, N.B., Kozhukhova, E.A., Vdovin, V.I., Ip, K., Norton, D.P., and Pearton, S.J. (2003) *J. Vac. Sci. Technol. A*, **21**, 1603.

148. Sheng, H., Muthukumar, S., Emanetoglu, N.W., and Lu, Y. (2002) *Appl. Phys. Lett.*, **80**, 2132.

149. Oh, D.C., Kim, J.J., Makino, H., Handa, T., Cho, M.W., Yao, T., and Ko, H.J. (2005) *Appl. Phys. Lett.*, **86**, 042110.

150. Ip, K., Heo, Y.W., Baik, K.H., Norton, D.P., Pearton, S.J., Kim, S., LaRoche, J.R., and Ren, F. (2004) *Appl. Phys. Lett.*, **84**, 2835.

151. Oh, M.-S., Hwang, D.-K., Lim, J.-H., Choi, Y.-S., and Park, S.-J. (2007) *Appl. Phys. Lett.*, **91**, 042109.

152. Endo, H., Sugibuchi, M., Takahashi, K., Goto, S., Sugimura, S., Hane, K., and Kashiwaba, Y. (2007) *Appl. Phys. Lett.*, **90**, 121906.

153. von Wenckstern, H., Biehne, G., Rahman, R.A., Hochmuth, H., Lorenz, M., and Grundmann, M. (2006) *Appl. Phys. Lett.*, **88**, 092102.

154. von Wenckstern, H., Kaidashev, E.M., Lorenz, M., Hochmuth, H., Biehne, G., Lenzner, J., Gottschalch, V., Pickenhain, R., and Grundmann, M. (2004) *Appl. Phys. Lett.*, **84**, 79.

155. Grossner, U., Gabrielsen, S., Børseth, T.M., Grillenberger, J., Kuznetsov, A.Y., and Svensson, B.G. (2004) *Appl. Phys. Lett.*, **85**, 2259.

156. Schifano, R., Monakhov, E.V., Grossner, U., and Svensson, B.G. (2007) *Appl. Phys. Lett.*, **91**, 193507.

157. Ip, K., Gila, B.P., Onstine, A.H., Lambers, E.S., Heo, Y.W., Baik, K.H., Norton, D.P., Pearton, S.J., Kim, S., LaRoche, J.R., and Ren, F. (2004) *Appl. Phys. Lett.*, **84**, 5133.

158. Yang, H.S., Norton, D.P., Pearton, S.J., and Ren, F. (2005) *Appl. Phys. Lett.*, **87**, 212106.
159. Liang, S., Sheng, H., Liu, Y., Huo, Z., Lu, Y., and Shen, H. (2001) *J. Cryst. Growth*, **225**, 110.
160. Coppa, B., Fulton, C.C., Kiesel, S.M., Davis, R.F., Pandarinath, C., Burnette, J.E., Nemanich, R.J., and Smith, D.J. (2005) *J. Appl. Phys.*, **97**, 103517.
161. Kim, S.-H., Kim, H.-K., and Seong, T.-Y. (2005) *Appl. Phys. Lett.*, **86**, 112101.
162. Allen, M.W., Alkaisi, M.M., and Durbin, S.M. (2006) *Appl. Phys. Lett.*, **89**, 103520.
163. Allen, M.W., Miller, P., Reeves, R.J., and Durbin, S.M. (2007) *Appl. Phys. Lett.*, **90**, 062104.
164. Allen, M.W., Durbin, S.M., and Metson, J.B. (2007) *Appl. Phys. Lett.*, **91**, 053512.
165. Nakano, M., Tsukazaki, A., Gunji, R.Y., Ueno, K., Ohtomo, A., Fukumura, T., and Kawasaki, M. (2007) *Appl. Phys. Lett.*, **91**, 142113.
166. Murakami, M., Kim, H.J., Shih, Y.-C., Price, W.H., and Parks, C.C. (1989) *Appl. Surf. Sci.*, **41/42**, 195.
167. Sands, T., Palmstrøm, C.J., Harbison, J.P., Keramidas, V.G., Tabatabaie, N., Cheeks, T.L., Ramesh, R., and Silberberg, Y. (1990) *Mat. Sci. Rep.*, **5**, 99.
168. Van de Walle, C.G. (2000) *Phys. Rev. Lett.*, **85**, 1012.
169. Strzhemechny, Y.M., Mosbacker, H.L., Look, D.C., Reynolds, D.C., Litton, C.W., Garces, N.Y., Giles, N.C., Halliburton, L.E., Niki, S., and Brillson, L.J. (2004) *Appl. Phys. Lett.*, **84**, 2545.
170. Hoffman, D.M., Hofstaetter, A., Leiter, F., Zhou, H., Henecker, F., and Meyer, B.K. (2002) *Phys. Rev. Lett.*, **88**, 045505.
171. Sebestyen, A. (1982) *Solid State Electron.*, **25**, 543.
172. Itoh, A. and Matsunami, H. (1997) *Phys. Status Solidi A*, **162**, 389.
173. Porter, L.M. and Davis, R.F. (1995) *Mater. Sci. Eng. B*, **34**, 83.
174. Glover, G.H. (1973) *Solid-State Electron.*, **16**, 973.
175. Chen, Y.G., Ogura, M., and Okushi, H. (2004) *J. Vac. Sci. Technol. B*, **22**, 2084.
176. Ri, S.-G., Takeuchi, D., Tokuda, N., Okushi, H., and Yamasaki, S. (2008) *Appl. Phys. Lett.*, **92**, 112112.
177. (a) Evans, D.A., Roberts, O.R., Vearey-Roberts, A.R., Langstaff, D.P., Twitchen, D.J., and Schwitters, M. (2007) *Appl. Phys. Lett.*, **91**, 132114, (b)Evans, D.A., Roberts, O.R., Williams, G.T., Vearey-Roberts, A.R., Bain, F., Evans, S., Langstaff, D.P., and Twitchen, D.J., (2009) *J. Phys.: Condens. Matter* **21**, 364223.
178. Nill, K.W., Walpole, J.N., Calawa, A.R., and Harman, T.C. (1970) *Proceedings of the Conference on the Physics of Semimetals and Narrow-Gap Semiconductors*, Pergamon, Oxford, p. 383.
179. Hughes, G.J., McKinley, A., and Williams, R.H. (1983) *J. Phys. C: Solid-State Phys.*, **16**, 2391.
180. Brucker, C.F. and Brillson, L.J. (1981) *Appl. Phys. Lett.*, **39**, 67.
181. Woodall, J.M., Freeouf, J.L., Pettit, G.D., Jackson, T., and Kirchner, P. (1981) *J. Vac. Sci. Technol.*, **19**, 626.
182. Waldrop, J.R. and Grant, R.W. (1987) *Appl. Phys. Lett.*, **50**, 250.
183. Waldrop, J.R. and Grant, R.W. (1988) *Appl. Phys. Lett.*, **52**, 1794.
184. Palmstrøm, C.J. and Morgan, D.V. (1985) in *Gallium Arsenide* (eds M.J. Howes and D.V. Morgan), John Wiley & Sons, Inc., New York.
185. Zhang, L.C., Cheung, S.K., Liang, C.L., and Cheung, N.W. (1987) *Appl. Phys. Lett.*, **50**, 447.
186. Sands, T., Chan, W.K., Chang, C.C., Chase, W.W., and Keramidas, V.G. (1988) *Appl. Phys. Lett.*, **52**, 1338.
187. Sze, S.M. (1985) *Semiconductor Devices*, 2nd edn, John Wiley & Sons, Inc., New York.
188. Chen, C.L., Mahoney, L.J., Woodhouse, J.D., Finn, M.C., and Nitishin, P.M. (1987) *Appl. Phys. Lett.*, **50**, 1179.
189. Marshall, E.D., Chen, W.X., Wu, C.S., Lau, S.S., and Kuech, T.F. (1985) *Appl. Phys. Lett.*, **47**, 298.
190. Shealy, J.R. and Chinn, S.R. (1985) *Appl. Phys. Lett.*, **47**, 410.
191. Marvin, D.C., Ives, N.A., and Leung, M.S. (1985) *J. Appl. Phys.*, **58**, 2659.

192. Shubert, E.F., Cunningham, J.E., Tsang, W.T., and Chiu, T.H. (1986) *Appl. Phys. Lett.*, **49**, 292.
193. Imanaga, S., Kawai, H., Kajiwara, K., Kaneko, K., and Watanabe, N. (1987) *J. Appl. Phys.*, **62**, 2381.
194. Hirose, K., Tsuda, T., and Mizutani, T. (1989) *Appl. Surf. Sci.*, **41/42**, 174.
195. Williams, R.H. and Patterson, M.H. (1982) *Appl. Phys. Lett.*, **40**, 484.
196. Shaw, J.L., Viturro, R.E., Brillson, L.J., and LaGraffe, D. (1988) *Appl. Phys. Lett.*, **53**, 1723.
197. Offsey, S.D., Woodall, J.M., Warren, A.C., Kirchner, P.D., Chappell, T.I., and Pettit, G.D. (1986) *Appl. Phys. Lett.*, **48**, 475.
198. Brucker, C.F. and Brillson, L.J. (1981) *J. Vac. Sci. Technol.*, **18**, 787.
199. Montgomery, V., Williams, R.H., and Srivastava, G.P. (1981) *J. Phys. C*, **14**, L191.
200. Waddill, G.D., Vitomirov, I.M., Aldao, C.M., and Weaver, J.H. (1989) *Phys. Rev. Lett.*, **62**, 1568.
201. Vitomirov, I.M., Aldao, C.M., Waddill, G.D., Capasso, C., and Weaver, J.H. (1990) *Phys. Rev. B*, **41**, 8465.
202. Vos, M., Aldao, C.M., Aaustuen, D.J.W., and Weaver, J.H. (1989) *Phys. Rev. Lett.*, **62**, 1568.
203. Waldrop, J.R., Kraut, E.A., Kowalczyk, S.P., and Grant, R.W. (1983) *Surf. Sci.*, **132**, 513.
204. Yablonovitch, E. and Gmitter, T.J. (1988) *Proc. Electrochem. Soc.*, **88-20**, 207.
205. Costa, J.C., Williamson, F., Miller, T.J., Beyzavi, K., Nathan, M.I., Mui, D.S.L., Strite, S., and Morkoc, H. (1991) *Appl. Phys. Lett.*, **58**, 382.
206. Woodall, J.M., Kirchner, P.D., Freeouf, J.L., McInturff, D.T., Melloch, M.R., and Pollak, F.H. (1993) *Proc. R. Soc., Philos. Trans. R. Soc. Lond. A*, **344**, 521.
207. Brillson, L.J., Viturro, R.E., Shaw, J.L., Mailhiot, C., Tache, N., McKinley, J.T., Margaritondo, G., Woodall, J.M., Kirchner, P.D., Pettit, G.D., and Wright, S.L. (1988) *J. Vac. Sci. Technol. B*, **6**, 1263.
208. Skromme, B.J., Sandroff, C.J., Yablonovitch, E., and Gmitter, T. (1987) *Appl. Phys. Lett.*, **51**, 2022.
209. Yablonovitch, E., Sandroff, C.J., Bhat, R., and Gmitter, T. (1987) *Appl. Phys.*, **51**, 439.
210. Chang, S., Brillson, L.J., Kime, Y.J., Rioux, D.S., Pettit, G.D., and Woodall, J.M. (1990) *Phys. Rev. Lett.*, **64**, 2551.
211. Chang, S., Vitomirov, I.M., Brillson, L.J., Rioux, D.S., Kirchner, P.D., Pettit, G.D., and Woodall, J.M. (1991) *Phys. Rev. B*, **44**, 1391.
212. Nelson, R.J., Williams, J.S., Leamy, H.J., Miller, B.J., Casey, H.C. Jr., Parkinson, B.A., and Heller, A. (1980) *Appl. Phys. Lett.*, **38**, 76.
213. Higashi, G.S., Chabal, Y.J., Trucks, G.W., and Raghavachari, K. (1990) *Appl. Phys. Lett.*, **56**, 656.
214. Palmstrøm, C.J., Cheeks, T.L., Gilchrist, H.L., Zhu, J.G., Carter, C.B., and Nahory, R.E. (1990) in *Electronic, Optical and Device Properties of Layered Structures* (eds J.R. Hayes, M.S. Hybertson, and E.R. Weber), Materials Research Society, Pittsburgh, p. 63.
215. Tung, R.T. (1991) *Appl. Phys. Lett.*, **58**, 2821.

Further Reading

Brillson, L.J. (1982) *Surf. Sci. Rep.*, **2**, 123.
Brillson, L.J. (1992) Surfaces and interfaces: atomic-scale structure, band bending, and band offsets, *Handbook on Semiconductors, Basic Properties of Semiconductors*, Vol. 1, Chapter 7, Elsevier, New York.
Brillson, L.J. (ed.) (1993) *Contacts to Semiconductors*, Noyes Publications, Park Ridge.
Kahn, A., Koch, N., and Gao, W. (2003) *J. Polym. Sci. B: Polym. Phys.*, **41**, 2529–2548.
Mönch, W. (2001) *Semiconductor Surfaces and Interfaces*, Springer-Verlag, Berlin.
Rhoderick, E.H. and Williams, R.H. (1988) *Metal-semiconductor Contacts*, Oxford Science, Oxford.
Tung, R.T. (1992) *Phys. Rev.*, **45** (23), 13509–13523.

22
The Future of Interfaces

22.1
Current Status

Our understanding of electronic surfaces and interfaces has evolved from the model of an atomically abrupt boundary represented in most solid-state textbooks to a more complex picture that depends on quantum mechanics, chemistry, and materials science. For the metal–semiconductor contact, Figure 22.1 illustrates this evolution from abrupt junction to atomically thin interface dipoles to extended interfaces.

The abrupt metal–semiconductor interface pictured in Figure 22.1a represents the Schottky barrier Φ_{SB} as simply the difference between metal work function Φ_M and semiconductor electron affinity χ_{SC}. In turn, the semiconductor band bending qV_B depends only on Φ_{SB} and the semiconductor bulk $E_C - E_F$ position.

Because most semiconductors fail to exhibit this classical behavior, it was necessary to introduce an interface dipole pictured in Figure 22.1b that accounted for part or all of the potential difference between metal and semiconductor but which was atomically thin in order to render this layer transparent to tunneling. This picture has been the basis for many physical models of Schottky barrier formation and the motivation for thousands of research studies.

As surface science techniques have refined our understanding of semiconductor interface structure, new phenomena have emerged to indicate the importance of interface atomic bonding on a monolayer scale as well as chemical reaction and diffusion on a nanometer scale. This now leads to the more general picture of the semiconductor interface shown in Figure 21.1c. Here the interface is no longer abrupt. The junction between metal and semiconductor includes not only a dipole layer but also a possible reacted phase with unique dielectric properties as well as an altered band-bending region within the semiconductor. This altered semiconductor region represents possible metal indiffusion and doping or semiconductor atom outdiffusion and defect formation.

Any new dielectric phases change the potential difference at the semiconductor boundary, altering the charge transfer. Interdiffusion would change the free carrier concentration within the surface space charge region, altering the band profile and balance between thermionic and tunneling charge transport. Likewise, the formation of localized states extending into the surface space charge region

Surfaces and Interfaces of Electronic Materials. Leonard J. Brillson
Copyright © 2010 WILEY-VCH Verlag GmbH & Co. KGaA, Weinheim
ISBN: 978-3-527-40915-0

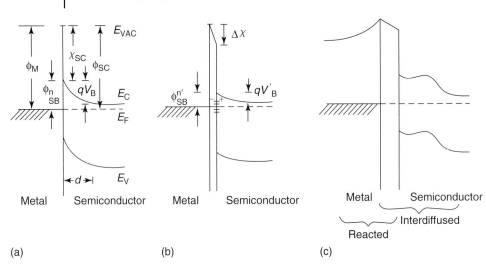

Figure 22.1 Evolution of a metal–semiconductor interface from (a) abrupt junction to (b) atomically thin interface dipoles plus band bending to (c) extended interfaces.

can promote trap-assisted hopping or tunneling transport, both of which reduce the effective barrier height. Figure 22.2 illustrates this competition of several interface charge transport mechanisms. Here thermionic emission of carriers over the Schottky barrier is augmented by tunneling through the barrier and hopping transport through states within the bandgap. Impurities and/or defects that increase carrier density inside the surface space charge region will decrease the barrier width, thereby increasing the contribution of tunneling to the overall current. In a similar manner, increasing defect concentrations near the semiconductor surface enable charge movement through the semiconductor depletion region, especially at defect densities at which their gap state wave functions begin to overlap. If enough defects are introduced, for example, by near-surface segregation, metal–semiconductor reaction, or surface mechanical damage, the otherwise Schottky barrier pictured in Figure 22.2 becomes ohmic. Indeed, Chapter 21 provided examples of chemically induced defect states and their corresponding barrier-lowering effects. Conversely, interface reactions can introduce dielectric layers that, merely at a few atomic-layer thicknesses, are sufficient to block transport in either direction. The chemical interactions evident in Chapters 4, 18 and 21 demonstrate that such interactions are common for metal–semiconductor interfaces. The contributions of these extended interface features to the electronic properties of a specific junction will depend sensitively on the thermodynamics and kinetics associated with the materials and the processes used to join them.

The evolution from abrupt to extended interface pictured in Figure 22.1 encompasses semiconductor heterojunctions as well since the interface region can have not only atomically thin dipoles but also extended interface chemical phases.

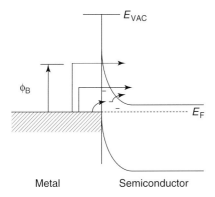

Figure 22.2 Competing charge transport mechanisms: thermionic emission, tunneling, and hopping transport through defect levels in the bandgap.

While the extended interface represents greater complexity, the physical mechanisms that determine its structure and properties provide a platform on which to develop new electronics technology.

This chapter illustrates several developing areas of solid-state electronics that incorporate surfaces and interfaces. Many new materials have emerged that present exciting opportunities for future devices. As with the generations of devices that have come before, the control of surfaces and interfaces are integral to their development.

22.2
Current Device Applications and Challenges

Up to this point, we have viewed the control of electronic surfaces and interfaces in terms of their Schottky barrier heights, ohmic contacts, and heterojunction band offsets. The following are among their many uses that are sought:

- high barrier heights for (i) solar cells to increase their open circuit voltage, (ii) diode switches to improve rectification, (iii) transistor gate electrodes to suppress current leakage, (iv) field effect transistors to improve detection sensitivity;
- low barrier heights for ohmic contacts to all electronic devices to minimize parasitic voltage, heat generation, and power loss;
- adjustable heterojunction band offsets to (i) improve carrier confinement and radiative recombination efficiency at quantum well laser, (ii) increase emitter efficiency in bipolar junction transistors, and (iii) tune confinement energies within quantum-confined devices such as tunnel diodes and two-dimensional electron gas transistors.

An example of tuned confinement energies is the resonant tunnel diode (RTD). Figure 22.3a schematically illustrates the conduction band profile of an RTD and

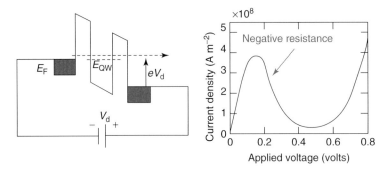

Figure 22.3 (a) Schematic illustration of a resonant tunnel diode. (b) Current injection increases then decreases with increasing applied voltage V_d as E_F^1 passes through alignment with E_{QW}.

its connection to a biasing circuit. Current increases when applied voltage V_d aligns the quantum-confined energy level E_{QW} with E_F^1 of the injecting electrode. As V_d increases further, the levels no longer align, the transition probability for tunneling decreases, and the current drops. Because tunneling times are so short, one can achieve subpicosecond response times and terahertz resonant frequencies for applied voltages in this negative resistance region. The importance of the interfaces in such RTDs is twofold: (i) the current injection efficiency depends on carriers tunneling through the confinement barrier region without scattering or recombining at interface gap states and (ii) the energy alignment of E_F at a given voltage with the quantum-confined level depends in part on the conduction band offset. Control of both interface perfection and band offset require careful materials design and growth.

Another device application that incorporates multiple interfaces is an avalanche detector. Figure 22.4a illustrates a graded heterostructure with large ΔE_C and small ΔE_V at a series of interfaces [1]. The composition of each layer in this device is graded to produce a low bandgap (E_{g1}) at one surface and a high bandgap (E_{g2}) at the other over a thickness l Photons $h\nu$ incident on the biased staircase photomultiplier generate electron–hole pairs that move in opposite directions under bias. As shown in Figure 22.4b, electrons experience a kinetic energy increase ΔE_C upon crossing each heterojunction, producing secondary electrons by impact ionization that produces an electron avalanche. No impact ionization occurs for holes due to the much lower ΔE_V, resulting in lower noise than in conventional avalanche detectors.

Doping and step grading with multiple interfaces prove useful in achieving low resistance across heterojunctions with large band offsets. Figure 22.5 shows how these techniques work [2]. n-Type doping of both sides of the single heterojunction in Figure 22.5a lowers E_C of the high bandgap semiconductor and produces a thin barrier region through which electrons can tunnel. Figure 22.4b shows the effect of combining such doping with graded layers. Now the tunnel barriers are

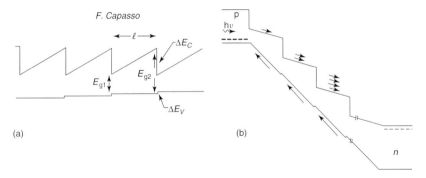

Figure 22.4 Staircase solid-state photomultiplier consisting of graded heterostructures with (a) no applied bias and (b) under forward bias. Arrows indicate carrier multiplication due to impact ionization of electrons but not holes [1].

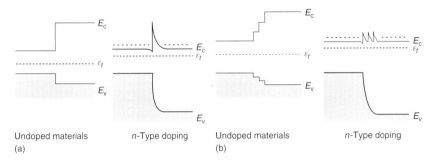

Figure 22.5 Effect of (a) n-type doping and (b) graded layers with n-type doping to reduce the resistance of a high ΔE_C heterojunction [2].

reduced in height and width, resulting in lower resistance. Here again, control of heterojunction band offsets is central to this method.

Low-resistance contacts are more difficult to achieve as bandgap energies increase. For example, integrated heterostructures such as p-on-n $Al_{0.5}Ga_{0.5}N$ diodes grown on SiC (for improved lattice matching) have a high barrier to electron flow, $\Delta E_C = 1.4\,eV$, as pictured in Figure 22.6a for $E_G(Al_{0.5}Ga_{0.5}N) = 4.9\,eV$. n-Type doping of the AlGaN produces the conduction band spike that retards electron flow.

On the other hand, p-type SiC forms a good ohmic contact to p-type $Al_{0.5}Ga_{0.5}N$ (Figure 22.6b) since ΔE_V is low and ΔE_C is no longer a factor. SiC is used for the p-type contact since metals with sufficiently high work functions for such wide bandgap semiconductors are not available. Thus the growth sequence of p-SiC/p-$Al_{0.5}Ga_{0.5}N$/n-$Al_{0.5}Ga_{0.5}N$ in Figure 22.6b permits low-resistance contacts while the reverse p–n junction sequence does not.

Low-resistance n-type contacts becomes challenging for semiconductors such as AlN with extremely high (6.2 eV) bandgaps. One such device application is

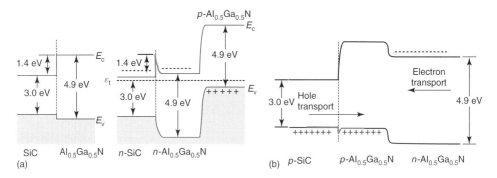

Figure 22.6 Contacts to an $Al_{0.5}Ga_{0.5}N$ p–n diode with (a) high electron barrier at an n-SiC/n-$Al_{0.5}Ga_{0.5}N$ versus a (b) p-SiC/p-$Al_{0.5}Ga_{0.5}N$ junction. (After Schetzina [2]).

the negative electron affinity (NEA) emitter already described in Section 18.3.3.4. AlN is a true NEA semiconductor with conduction band minimum above the vacuum level at its surface. To emit electrons, the AlN must be n-type and its SiC substrate must be ohmic. This presents the same high barrier to electrons shown in Figure 22.6a. Here again, the method of graded layers with n-type doping pictured in Figure 22.6b, in this case graded $Al_xGa_{1-x}N$, reduces the AlN barrier to provide the backside ohmic contact.

22.3
New Directions

There are many new directions for surface and interface science and technology. Chief among fundamental issues is the *prediction* of Schottky barrier heights rather than the rationalization of measured barriers after the fact. Intrinsic wave function tunneling models are successful in accounting for Schottky barriers involving non-reactive metals to within 0.1–0.2 eV. On the other hand, the wide range of chemical reactions, interdiffusion, and defect formation known for metal–semiconductor interfaces in general requires a more comprehensive framework.

Interface studies of wide bandgap semiconductors such as ZnO, GaN, and SiC can yield new insights into this problem since the large energy range of E_F movement observed under different experimental conditions permits researchers to resolve differences between physical mechanisms. At present, results suggest that more than one mechanism plays a role at metal–semiconductor interfaces in general.

Another fundamental issue is the prediction and control of heterojunction band offsets. Here again, the band alignment depends on multiple physical mechanisms – the bulk band structures of the constituents, the detailed interface atomic bonding, and any chemical reaction or interdiffusion. Such interdiffusion is now apparent at monolayer thicknesses even for otherwise abrupt heterojunctions.

The altered interface stoichiometry, interface dipole, and band offsets depend sensitively on the sequence and duration of deposition steps used in the transition between growing one semiconductor and the next. This sensitivity is at once a problem but also an opportunity to control band offsets. To develop the techniques required to control band offsets for specific heterojunctions, researchers must combine epitaxial growth with electronic, chemical, and structural characterization – all with monolayer precision.

22.3.1
High-K Dielectrics

An important and challenging topic of interface research involves high permittivity (high-K) dielectrics. With the ever-shrinking dimensions of complementary metal-oxide-semiconductor (CMOS) transistors, Moore's Law (doubling transistor chip density every 18 months) requires gate insulators that for SiO_2 are too thin to prevent current leakage via tunneling. To avoid this tunneling, technologists have turned to higher K insulators such as HfO_2, ZrO_2, and their alloys, which increase K from 3.9 for SiO_2 to, for example, 30 for HfO_2, increasing transistor capacitance proportionally for the same thickness. The challenge of this transition away from SiO_2 is that alternative dielectrics introduce more bulk and interface traps. The microelectronics industry required over 50 years to reach the current state of the art with interface state densities below 10^{10} cm^{-2}.

Furthermore, these traps depend sensitively on the process conditions used to fabricate the metal–oxide–semiconductor film structure. An example of this sensitivity appears in Figure 22.7. The capacitor structure consists of a 10-nm Mo film deposited by plasma vapor deposition (PVD) on 4-nm HfO_2 deposited by atomic-layer deposition on clean Si wafers [3].

The unannealed metal–oxide–Si structure displays pronounced defect emissions within the oxide structure at 3.35, 3.53, and 3.9–4.3 eV that can be associated with HfO_2 oxygen vacancies in different charge states [4, 5]. With increasing excitation depth, defect emissions appear at 2–2.5 eV that resemble SiO_2-related nonbonding oxygen hole centers (NBOHCs) and positively charged O vacancies (termed *E′ centers*). These SiO_2-related features suggest Hf silicate formation at the HfO_2/Si interface.

A forming gas anneal (FGA) increases all these defect emissions, particularly the SiO_2-related emissions, consistent with increased HfO_2 crystallinity and silicate formation at the HfO_2/Si interface. Rapid thermal annealing (RTA) (1000 °C, 5 seconds N_2) decreases these defect emissions dramatically, particularly the silicate features both before and after FGAs. Indeed, such high-temperature RTAs improve Si/SiO_2 [6] and Si/HfO_2 [7] interface abruptness, reduce dangling bonds and interface state densities, and thereby reduce interface-localized cathodoluminescence (CL) emissions [8].

The identity of defects within high-K dielectrics (and electronic materials in general) can be probed using electron spin resonance (ESR). For HfO_2, ESR reveals $E′$

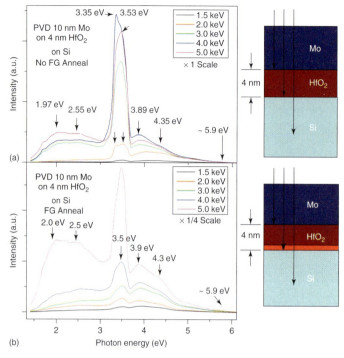

Figure 22.7 (a) DRCL spectra of PVD-deposited Mo on HfO$_2$/Si without annealing with multiple deep levels below 5.5 eV bandgap. The 2–2.5 eV features increase relative to the dominant 3.35/3.53 eV doublet with increasing excitation depth. (b) PVD-deposited Mo on HfO$_2$/Si with an FGA shows increased 2 and 2.5 eV intensities due to silicate formation. (After Walsh et al. [3]).

centers present in the interfacial layer of hafnium-oxide-based metal–oxide–silicon structures [9]. Detection of such states within individual gate structures is challenging due to small number of spins involved. However, new techniques that make use of transistor gating enable ESR to characterize such defects in fully processing metal gate HfO$_2$ metal-oxide semiconductor field effect transistors (MOSFETs) [10].

The ability to probe defects in ultrathin films sandwiched between other materials combined with theoretical analysis of their origin holds promise for rapid refinement of new materials, film structures, and thermal processing to achieve low interface state densities.

22.3.2
Complex Oxides

Complex oxides have emerged as an exciting new area of condensed matter research because of their intrinsic functional properties of ferroelectricity, ferromagnetism, superconductivity, spintronics, and multiferroic behavior. Furthermore, heterostructures of these materials can exhibit unique properties that depend critically on their interface composition, bonding, and lattice structure. A leading example is the change in physical properties with lattice strain induced by lattice mismatch. Thus, it is possible to achieve room temperature ferroelectricity in strained $SrTiO_3$, a material that is not normally ferroelectric at any temperature [11]. To obtain this effect, Schlom et al. developed a new substrate material, $DyScO_3$ for the growth of $SrTiO_3$ under uniform biaxial tensile strain. The high dielectric constant of nearly 7000 at high frequency (10 GHz) and its strong dependence on electric field are useful for microwave and other dielectric applications.

Researchers are developing new composite materials that couple ferroelectric and ferromagnetic behavior, adding a new degree of freedom in the design of electromagnetic devices [12]. For example, heterostructures consisting of nanopillars of a ferromagnetic phase embedded in a ferroelectric matrix employ three-dimensional epitaxy to achieve mechanical coupling between the two phases. Such architectures can enable devices that convert electric fields into magnetic fields and vice versa. Here, interface chemical structure plays a role in translating strain between phases effectively.

A striking electronic example is the quasi-two-dimensional electron gas reported at the epitaxial heterostructure interface between insulating oxides. Figure 22.8 illustrates how one can use epitaxy to control the termination layer and thereby the polarity discontinuity between two insulating perovskite oxides, $LaAlO_3$ and $SrTiO_3$ [13]. The reflection high-energy electron diffraction (RHEED) oscillations demonstrate the introduction of an SrO layer on the otherwise TiO_2-terminated $SrTiO_3$ surface to change the interface termination of $LaAlO_3$ from $(LaO)^+$ to $(AlO2)^-$. The resulting extra half hole or electron, respectively, per two-dimensional unit cell leads to an insulating or a highly conducting (mobility $<10^4 \, cm^2 V^{-1} s^{-1}$) interface sheet. This effect is analogous to modulation doping in conventional semiconductors and opens the way to create low-dimensional charge states using oxide heteroepitaxy controlled at the atomic level.

Extending this concept, Thiel et al. tuned the conducting quasi-two-dimensional electron gas formed at the insulating dielectric interface by altering the thickness of the $LaAlO_3$ sheets on a unit cell level [14]. Native point defects such as oxygen vacancies may play an important role at such interfaces [15], requiring atomic-scale methods such as transmission electron microscopy and electron loss spectroscopy [16] along with nanoscale CL spectroscopy to examine their contributions.

The temperature and electric field dependence of the complex oxides presents another dimension to Schottky barrier research. Band bending at metal–semiconductor interfaces depends on the semiconductor permittivity, normally assumed to be constant. This is no longer the case for complex oxides since the permittivity of the dielectric can change by orders of magnitude with electric field. Since electric field changes with position inside a surface space charge region, a description of the transport properties requires a self-consistent analysis of the barrier profile [17].

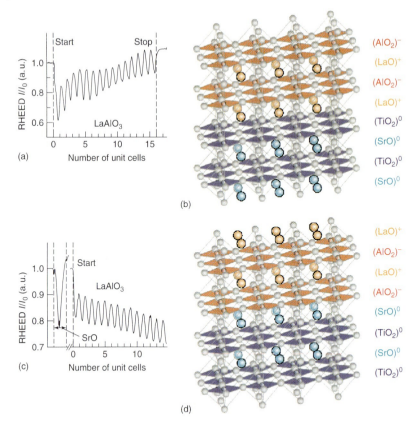

Figure 22.8 Growth and schematic models of the two possible interfaces between LaAlO$_3$ and SrTiO$_3$ in the (001) orientation. (a) RHEED intensity oscillations of the specular reflected beam for the growth of LaAlO$_3$ directly on the TiO$_2$-terminated SrTiO$_3$ (001) surface. (b) Schematic of the resulting (LaO)$^+$/(TiO$_2$)^0 interface, showing the composition of each layer and the ionic charge state of each layer. (c) RHEED oscillations for the growth of LaAlO$_3$, after a monolayer of SrO was deposited on the TiO$_2$-terminated SrTiO$_3$ surface. (d) Schematic of the resulting (AlO$_2$)$^-$/(SrO)^0 interface. (After Ohtomo and Hwang [13]).

Thus far, magnetic properties have not been discussed. In general, the complex transition-metal oxides form interfaces whose physical properties depend on the interplay among charge, spin, and orbital degrees of freedom [18]. Understanding and controlling these interactions can lead to advances in high-temperature superconductivity and colossal magnetoresistance.

22.3.3
Spintronics

The use of electron spin to carry information has given rise to a new technology for spin transport electronics termed *spintronics*. Spin adds a new degree of freedom

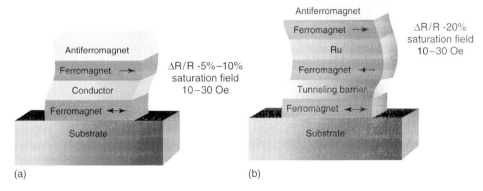

Figure 22.9 Spin-dependent transport architectures. (a) Spin valve. (b) Magnetic tunnel junction. (After Wolf et al. [19]).

to conventional charge-based electronic devices as well as increased speed, energy consumption, and device integration density [19]. One of the major challenges for this technology is the transport of spin-polarized carriers across heterointerfaces.

Spin-polarized injection from ferromagnetic layers into semiconductors commonly employs Schottky tunnel barriers [20–22]. Interfacial layers with strong magnetic anisotropy enable spin injection with preferential orientation [23], increased injected spin polarization [24], and spin injection into Si [25].

At present, spin-based electronic devices include spin valves and magnetic tunnel junctions. See, for example, Figure 22.9. Spin valves involve two ferromagnetic layers sandwiching a thin nonmagnetic metal such as copper with one the two magnetic layers "pinned" (with an adjoining antiferromagnetic layer) such that its magnetization direction remains in a fixed orientation for moderate applied magnetic fields. The other magnetic layer is "free" to change magnetization with the application of a small magnetic field. When the ferromagnetic directions are antiparallel, the resistance between them increases. When they are parallel, the resistance of the spin valve decreases.

A magnetic tunnel junction involves a "pinned" and "free" layer separated by an insulating layer. In both devices, the layer separating the two magnetic layers is only a few nanometers thick so that even monolayer perturbations of the otherwise abrupt interfaces between each layer could easily dominate the charge transport and spin. Among factors that could degrade spin injection are (i) charged lattice sites due to native point or morphological defects (e.g., dislocations), (ii) interface roughness due to steps or interfacial phase formation, both of which change the local potential, (iii) interface compounds that alter the ferromagnetic metal, tunnel barrier, or semiconductor. Such extrinsic effects can increase charge scattering, reducing spin polarization transferred across the interface.

Spintronics research has led to rapid improvements in the polarization, efficiency, orientation, and coherence of spin injection that are already impacting the magnetic storage industry. Electronic interfaces are integral to these advances.

22.3.4
Nanoscale Circuits

The drive toward electronics miniaturization has led to research on nanoscale circuits that incorporate novel materials and structures. Chief among these have been nanowires, nanotubes, molecules, and monolayer sheets such as graphene. Ohmic contacts to such materials are desirable, yet current–voltage characteristics are, in general, nonlinear. Furthermore, conventional Schottky models are not always adequate to describe this behavior but instead require additional size-dependent considerations such as low intrinsic doping, charge traps, or depletion widths at the contacts that are larger than the length scale of the entire structure. For example, charge traps on the surface of a nanowire can deplete the wire of charge such that injected carriers are not effectively screened [26]. This can lead to *space charge-limited current* (*SCLC*) in which injected carriers set up fields that govern the injection of additional charge. The resultant $I-V$ characteristics are nonlinear, nonexponential, and obey a $I \propto V^2$ relationship.

Thus SCLC accounts for the symmetric, nonlinear electrical transport in GaN nanowires shown in Figure 22.10 [27]. With initial contact, the $I-V$ characteristics are rectifying and exponential under forward bias, typical of an injection-limited contact. With additional pressing, the $I-V$ curves become characteristic of SCLC transport. Indicative of its general nature, lithographic contacts to multiple nanowires appear to follow the same functional relationship ascribed to SCLC as individual tip contacts [27]. Similar behavior is expected for many other nanostructures since (i) electrostatic screening is poor in high aspect ratio structures, (ii) dopant incorporation is difficult to achieve, and (iii) surface states and charge traps can further deplete the nanostructures of carriers. One such example is an ultrathin membrane. When semiconductor crystals are made into membranes as thin as a few nanometers, they can behave like traditional soft matter while maintaining their single-crystal structure without defects - even after substantial deformation. In this regime, the entire structure is a surface or interface [28]. These size-dependent effects pose a challenge to nanoscale circuit development that surface and interface research is well suited to address.

Theoretical studies are now beginning to reveal the unique aspects of contacts to graphene, a zero-gap material with very high electron mobility, quantization of its conductivity, and a zero-energy anomaly in the quantum Hall effect. Metal–graphene interactions can be shown to describe the doping of graphene phenomenologically [29]. Covalent bonding of graphenelike layers to a SiC substrate induces a finite gap and doping [30]. Barrier lowering with oxidation of graphene-oxide flakes suggests that this material's bandgap can be tuned by changing the degree of oxidation [31]. Again, the unique features of graphene and related materials open a new area of interface science and engineering.

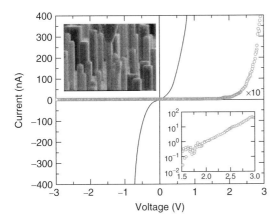

Figure 22.10 Upper left inset shows the image of tip-contacting nanowire. *I–V* characteristics before (symbols) and after (line) pressing the tip into the nanowire. Bottom right inset shows *I–V* before pressing on a log scale [27].

22.3.5
Quantum-Scale Interfaces

Quantum-scale structures have increased in importance with their use in optoelectronics, solar energy conversion, and sensing. The optical properties of quantum devices depend on their size, shape, composition, and strain. Engineering device structures must also take into account the interface between the quantum structures and their surrounding media.

Scanning tunneling microscopy (STM) and scanning tunneling spectroscopy (STS) are providing new insights into these interfaces. Thus the electronic structure of an InAs quantum dot (QD) embedded in a GaAs matrix exhibits features characteristic of quantum-confined conduction and valence band states shown in Figure 22.11a [32]. Conductance spectra of an individual dot display multiple peaks corresponding to quantized conduction band states of the QD. Figure 22.11b shows that these spectra can change with position within this single dot (left), corresponding to the bright areas within the dot shown at right. The small bright areas for hole states H_1, H_2, and H_3 show much less wave function extension that electron states E_1, E_2, and E_3. Furthermore, these hole states are not centered within the dot but rather localized at its upper interface. This difference between electron and hole states can give rise to local dipoles across the dot. It is attributed to the lower confining potential for holes and its variations with local chemical composition, charged defects, and strain. Such studies underscore the importance of quantum-scale interfaces in developing uniform, well-controlled optoelectronic properties.

Theoretical methods are needed to refine energy calculations of defects and impurities in semiconductors. These methods are particularly important for extrinsic states near interfaces, where they can affect electronic barriers. Experimental

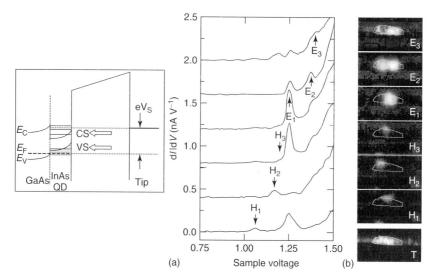

Figure 22.11 (a) Band diagram showing conduction (CS) and valence (VS) QD states that contribute to tunneling current with positive bias relative to E_F. (b) Spatially resolved tunneling spectra at left measured in different parts of an individual QD (outlined in white) at 5 K that were bright at the I–V energies indicated by arrows [31].

studies of impurity atoms at well-defined distances from surfaces or within well-defined quantum-scale structures can help distinguish between current local state models.

Deep-level techniques such as deep-level transient spectroscopy (DLTS) and deep-level optical spectroscopy (DLOS) that provide quantitative information on trap densities and cross sections can be correlated with near-surface probes such as depth-resolved cathodoluminescence spectroscopy (DRCLS) to obtain calibrated measurements of defect or impurity densities versus depth on a nanometer scale. Such calibrated spectra can then be extended to interface length scales too short for conventional capacitance techniques.

22.4
Synopsis

This book has attempted to convey the richness and usefulness of electronic surface and interface science. Conclusions drawn for each class of surface or interface are summarized here:

- Semiconductor surfaces and interfaces
  – Surface geometric structure differs from the bulk lattice in complex ways.

- New growth and characterization techniques demonstrate the importance of thermodynamics and kinetics.
- Surface electronic structure depends sensitively on specifics of reconstruction; localized states in the bandgap relate to atomic bonding.
- Overlayer–semiconductor interfaces
  - Adsorption can induce surface geometric changes.
  - Chemical reaction, diffusion, new phase formation can occur, depending on local atomic bonding and processing.
  - Metal–semiconductor localized states and band bending are influenced by extrinsic phenomena.
  - Schottky barriers are controllable by atomic-scale techniques.
- Semiconductor–semiconductor interfaces
  - Geometric structure is defined by crystal lattice parameters and imperfections generated.
  - Diffusion and reaction can occur, depending on surface bonding and stoichiometry.
  - Atomic-layer template structures can control nature of epitaxy and growth.
  - Techniques are available for nanoscale state-of-the-art analysis.

Electronic surfaces and interfaces provide fertile ground for future research and development. On a fundamental level, research can lead to new understanding of the electronic material structure and charge transfer on an atomic scale. On an applied level, such work holds the promise of electronic device structures on the smallest scale possible.

References

1. Capasso, F. (1987) in *Heterojunction Band Discontinuities, Physics and Device Application* (eds F. Capasso and G. Margaritondo), North-Holland, Amsterdam, pp. 399–450.
2. (a) Schetzina, J.F. (1996) *Mater. Res. Soc. Symp. Proc.*, **395**, 903; (b) Schetzina, J.F. (1997) Integrated heterostructures of group III-V Nitride semiconductor materials including epitaxial ohmic contact, non-nitride buffer layer and methods of fabricating same, U.S. Patent number 5,679963, Oct. 21, 1997, (U.S. Patent Office, Washington, DC).
3. Walsh, S., Fang, L., Schaeffer, J.K., Weisbrod, E., and Brillson, L.J. (2007) *Appl. Phys. Lett.*, **90**, 052901.
4. Xiong, K., Robertson, J., Gibson, M.C., and Clark, S.J. (2005) *Appl. Phys. Lett.*, **87**, 183505.
5. Strzhemechny, Y.M., Bataiev, M., Tumakha, S.P., Goss, S.H., Hinkle, C.L., Fulton, C.C., Lucovsky, G., and Brillson, L.J. (2008) *J. Vac. Sci. Technol. B*, **26**, 232.
6. Lee, D.R., Parker, C., Hauser, J.R., and Lucovsky, G. (1995) *J. Vac. Sci. Technol. B*, **13**, 1788.
7. Stemmer, S., Li, Y., Foran, B., Lysaght, P.S., Streiffer, S.K., Fuoss, P., and Seifert, S. (2003) *Appl. Phys. Lett.*, **83**, 3141.
8. Schäfer, J., Young, A.P., Brillson, L.J., Niimi, H., and Lucovsky, G. (1998) *J. Vac. Sci. Technol. A*, **73**, 791.
9. Ryan, J.T., Lenahan, P.M., Bersuker, G., and Lysaght, P. (2007) *Appl. Phys. Lett.*, **90**, 173513.
10. Campbell, J.P., Lenahan, P.M., Cochrane, C.J., Krishnan, A.T., and

Krishnan, S. (2007) *IEEE Trans. Device Mater. Rel.*, **7**, 540.

11. Haeni, J.H., Irvin, P., Chang, W., Uecker, R., Reiche, P., Li, Y.L., Choudhury, S., Tian, W., Hawley, M.E., Craigo, B., Tagantsev, A.K., Pan, X.Q., Streiffer, S.K., Chen, L.Q., Kirchoefer, S.W., Levy, J., and Schlom, D.G. (2004) *Nature*, **430**, 758.

12. Zheng, H., Wang, J., Lofland, S.E., Ma, Z., Mohaddes-Ardabili, L., Zhao, T., Salamanca-Riba, L., Shinde, S.R., Ogale, S.B., Bai, F., Viehland, D., Jia, Y., Schlom, D.G., Wuttig, M., Roytburd, A., and Ramesh, R. (2004) *Science*, **303**, 661.

13. Ohtomo, A. and Hwang, H.Y. (2004) *Nature*, **427**, 423.

14. Thiel, S., Hammerl, G., Schmehl, A., Schneider, C.W., and Mannhart, J. (2006) *Science*, **313**, 1942.

15. Siemons, W., Koster, G., Yamamoto, H., Harrison, W.A., Lucovsky, G., Geballe, T.H., Blank, D.H.A., and Beasley, M.R. (2007) *Phys. Rev. Lett.*, **98**, 196802.

16. Muller, D.A., Nakagawa, N., Ohtomo, A., Grazul, J.L., and Hwang, H.Y. (2004) *Nature*, **430**, 657.

17. Susaki, T., Kozuka, Y., Tateyama, Y., and Hwang, H.Y. (2007) *Phys. Rev. B*, **76**, 155110.

18. Tokura, Y. and Nagaosa, N. (2000) *Science*, **288**, 462.

19. Wolf, S.A., Awschalom, Dd., Buhrman, R.A., Daughton, J.M., von Molnár, S., Roukes, M.L., Chtchelkanova, A.U., and Treger, D.M. (2001) *Science*, **294**, 1488.

20. Zhu, H.J., Ramsteiner, M., Kotial, H., Wassermeier, M., Schönherr, H.-P., and Ploog, K.H. (2001) *Phys. Rev. Lett.*, **87**, 016601.

21. Hanbicki, A.T., Jonker, B.T., Iskos, G., Kioseoglou, G., and Petrou, A. (2002) *Appl. Phys. Lett.*, **80**, 1240.

22. Adelmann, C., Lou, X., Strand, J., Palmstrøm, C.J., and Crowell, P.A. (2005) *Phys. Rev. B*, **71**, 121301.

23. Adelmann, C., Hilton, J.L., Schultz, B.D., McKernan, S., Palmstrøm, C.J., Lou, X., Chiang, H.S., and Crowell, P.A. (2006) *Appl. Phys. Lett.*, **89**, 1125.

24. van't Erve, O.M.J., Kioseoglou, G., Hanbicki, A.T., Li, C.H., Jonker, B.T., Mallory, R., Yasar, M., and Petrou, A. (2004) *Appl. Phys. Lett.*, **84**, 4334.

25. Jonker, B.T., Kioseoglou, G., Hanbicki, A.T., Li, C.H., and Thompson, P.E. (2007) *Nat. Phys.*, **3**, 542.

26. Léonard, F. and Talin, A.A. (2006) *Phys. Rev. Lett.*, **97**, 026804.

27. Talin, A.A., Léonard, F., Swartzentruber, B.S., Wang, X., and Hersee, S.D. (2008) *Phys. Rev. Lett.*, **101**, 076802.

28. Cavallo, F., and Lagally, M.G. (2010) *Soft Matter*, DOI: 10: 101039/b916582g.

29. Giovannetti, G., Khomakov, P.A., Brocks, G., Karpan, V.M., Van den Brink, J., and Kelly, P.J. (2008) *Phys. Rev. Lett.*, **101**, 026803.

30. Yang, H., Baffou, G., Mayne, A.J., Comtet, G., Dujardin, G., and Kuk, Y. (2008) *Phys. Rev. B*, **78**, 041408.

31. Wu, X., Sprinkle, M., Li, X., Ming, F., Berger, C., and de Heer, W.A. (2008) *Phys. Rev. Lett.*, **101**, 026801.

32. Urbieta, A., Grandidier, B., Nys, J.P., Deresmes, D., Stiévenard, D., Lemaî, A., Patriarche, G., and Niquet, Y.M. (2008) *Phys. Rev. B*, **77**, 155313.

Appendices

Appendix 1: Glossary of Commonly Used Symbols

A, **B**	Primitive vectors
a_0	Lattice constant (Å)
b	Impact parameter (Å, cm)
C	Capacitance (F), concentration
cpd	Contact potential difference (V)
d	Depletion width, crystal thickness (Å, μm, cm)
D	Grid spacing (Å, μm), induced dipole, overlayer metal thickness (Å, μm)
d_{opt}	Optical density (dB)
D_q	Diffusion coefficient
$d\Omega$	Differential solid angle (steradian)
E	Energy (J, eV)
E_B	Core level binding energy (eV)
E_F	Fermi level (eV)
E_G	Bandgap energy level (eV)
E_K	Kinetic energy (J)
f	Vibration frequency (Hz), lattice mismatch
F	Flux of photons (s^{-1} cm^{-2}), diffusion flux (mol s^{-1}cm^{-2}), free energy (J), unit area (cm^2)
$f(E)$	Fermi function
f_C	Cut-off frequency (Hz)
f_{depol}	Fluctuating dipole field
f_P	Probability of electron emission
f_q	Ratio of total J with versus without tunneling
g	Vector of the Bravais lattice
G	Gibbs free energy (J), shear modulus
$g_\square$	Sheet conductance (Ω^{-1})
g_i	Statistical weight
H_F	Heat of formation (J mol^{-1})
H_R	Heat of reaction (J mol^{-1})
h	Film thickness (Å, μm)
h_c	Critical thickness at a heterojunction (Å, μm)
I	Ionization energy, ion intensity, mean excitation energy loss (eV)

Surfaces and Interfaces of Electronic Materials. Leonard J. Brillson
Copyright © 2010 WILEY-VCH Verlag GmbH & Co. KGaA, Weinheim
ISBN: 978-3-527-40915-0

Appendix 1: Glossary of Commonly Used Symbols

I_D Source–drain current (A)
I_T Tunneling current (A)
J Current density (A cm^{-2})
J_R Photocurrent per absorbed photon per unit area (A cm^{-2})
k Wave vector (rad μm^{-1})
K Kinematic factor
$\mathbf{k}$ Vector momentum (g m s^{-1})
L Scattering length (Å, μm), channel length (Å, μm)
l Layer thickness (Å, μm)
m Molecular mass (g)
M Tunneling matrix element
m^* Carrier effective mass (g)
M_I Incident particle mass (g)
M_{if} Matrix element between initial and final states
n Ideality factor
n   Number of molecules, refractive index of a solid, bulk charge density (C cm^{-2})
N Atomic density, semiconductor carrier concentration (cm^{-3})
n_0 Equilibrium carrier concentration (cm^{-3})
N_C Conduction band density of states (cm^{-3})
p Momentum (g m s^{-1})
P Dipole moment (V cm)
P Pressure (Pa)
P_0 Incident power (W)
R Rate of photoexcitation, transition probability per unit time (s^{-1})
r Rate of charge flow
R_B Bohr–Bethe range (nm, μm)
r_{min} Closest approach of particle–solid scattering (Å)
S Index of interface behavior, enthalpy (J), internal potential (V)
S_{eff} Effective pumping speed (l s^{-1})
S_j Scattering efficiency
S_P Pumping speed (l s^{-1})
$[S]$ Backscattering energy loss factor
T Temperature (K), energy transfer (J)
U_k Bloch function
V Applied voltage (V), volume (m^3), potential (V), Coulomb repulsion energy (J)
υ Photon frequency (Hz)
v Velocity (cm s^{-1})
$<v>$ Average thermal velocity (cm s^{-1})
$V(z)$ Lattice potential (eV)
V_B Semiconductor band bending (V)
v_D Effective diffusion velocity (m s^{-1})
v_R Recombination velocity (m s^{-1})
v_{TR} Transverse acoustic mode velocity (m s^{-1})
W Depletion layer width (Å, μm)
W_A Transition probability
X Electronegativity

Y	Young's modulus
Z	Atomic number, channel width (Å, μm)
α	Optical absorption coefficient
Δr_c	Coherence length (Å, μm)
ΔV	Band offset (eV)
$\varepsilon, \varepsilon_0, \varepsilon_S$	Permittivity, free space permittivity, static permittivity (F cm^{-1})
ε_i	Effective dielectric constant ($\varepsilon = \varepsilon_i \varepsilon_0$)
κ	Imaginary part of complex refractive index
η	Real part of complex refractive index
λ	Wavelength (μm)
μ	Electron mobility (cm^2 V^{-1} s^{-1}), Poisson ratio
$\mathcal{E}$	Applied electric field (V m^{-1})
ρ	Density of states (cm^{-3}), density (kg m^{-3})
$\rho(x)$	Charge density of a region (C cm^{-2})
σ	Conductivity (S m^{-1}), photoelectron cross section (cm^{-2}), strain/dislocation energy (J mol^{-1})
σ_e	Ionization cross section, Auger electron yield
τ	Lifetime (s), reacted layer thickness (Å, μm)
Φ, Φ_M, Φ_{SB}	Potential, metal work function, Schottky barrier (eV)
χ	Electron affinity (J mol^{-1}), dielectric susceptibility
Ψ	Wave function
ω	Frequency (rad s^{-1})
ω_X	Fluorescence yield
Γ_e	Energy of a grid of edge dislocations (J, eV)

Appendix 2: Table of Acronyms

2DEG	Two-dimensional electron gas
α-NPD	N,N′-diphenyl-N,N′-bis(1-naphthyl)-1,1′-biphenyl-4,4′-diamine
AES	Auger electron spectroscopy
AFM	Atomic force microscopy
AG	As-grown
amu	Atomic mass unit
AP	Antiphase
AREDC	Angle-resolved energy distribution curve
ARPES	Angle-resolved photoelectron spectroscopy
ATMA	Acetone, trichloroethylene, methanol, and air-exposed
BEEM	Ballistic electron energy microscopy
BEP	Beam equivalent pressure
BJT	Bipolar junction transistor
BPR	Beam pressure ratio
BZ	Brillouin zone
CNL	Charge neutrality level
CBE	Chemical beam epitaxy
CCA	Chemically cleaned air exposed
CFS	Constant final state spectroscopy
CIS	Constant initial state spectroscopy
CL	Cathodoluminescence
CLS	Cathodoluminescence spectroscopy
CMA	Cylindrical mirror analyzer
CRS	Confocal Raman Spectroscopy
$C-V$	Capacitance–voltage
CW	Continuous wave
CZ	Czochralski
DAP	Donor–acceptor pair
DAS	Dimer adatoms and stacking fault
DIGS	Disorder-induced gap state
DLOS	Deep-level optical spectroscopy
DLTS	Deep-level transient spectroscopy
DOS	Density of states

Surfaces and Interfaces of Electronic Materials. Leonard J. Brillson
Copyright © 2010 WILEY-VCH Verlag GmbH & Co. KGaA, Weinheim
ISBN: 978-3-527-40915-0

DRCLS	Depth-resolved cathodoluminescence spectroscopy
DSIMS	Dynamic secondary ion mass spectrometry
EBIC	Electron-beam-induced current
EDEP	Electrochemical deposition
EDC	Energy distribution curve
EDXS	Energy dispersive X-ray spectroscopy
EELS	Electron energy loss spectroscopy
ELS	Electron loss spectroscopy
ESCA	Electron spectroscopy for chemical analysis
ESR	Electron spin resonance
ESRF	European Synchrotron Radiation Facility
F_4-TCNQ	Tetrafluoro-tetracyanoquinodimethane
FCC	Face-centered cubic
FEL	Free electron laser
FGA	Forming gas anneal
FIM	Field ion microscopy
FWHM	Full width at half maximum
FZ	Float zone
GFA	Ga flux and anneal
HAO	Hybridized atomic orbital
HB	Horizontal Bridgman
HD	High defect
HEMT	High electron mobility transistor
HOMO	Highest occupied molecular orbit
HPA	Hydrogen peroxide and air exposed
HREELS	High-resolution electron energy loss spectroscopy
HRTEM	High-resolution transmission electron microscopy
HSA	Hemispherical analyzer
HT	Hydrothermal
HV	High vacuum
IC	Inner cylinder
ID	Insertion device
IE	Ionization energy
IP	In-phase
IPES	Inverse photoemission spectroscopy
IPS	Internal photoemission spectroscopy
IR	Infrared
IS	Injection source
ISA	Ion sputter and anneal
ISS	Ion beam scattering spectrometry
$I\text{–}V$	Current–voltage
KPFM	Kelvin force probe microscopy
LA	Laser annealed
LAPS	Laser-excited photoemission spectroscopy
LEC	Liquid-encapsulated Czochralski
LD	Low defect
LDOS	Local density of states
LEED	Low-energy electron diffraction
LEEM	Low-energy electron microscopy

LEELS	Low-energy electron loss spectroscopy
LEPD	Low-energy positron diffraction
LFM	Lateral force microscopy
LPE	Liquid-phase epitaxy
LO	Longitudinal optical
LUMO	Lowest unoccupied molecular orbit
MBE	Molecular beam epitaxy
MFM	Magnetic force microscopy
MG	Melt grown
MIGS	Metal-induced gap states
MIS	Metal–insulator–semiconductor
MISFET	Metal–insulator field effect transistor
ML	Monolayer
MOCVD	Molecular organic chemical vapor deposition
MOSFET	Metal-oxide semiconductor field effect transistor
NBE	Near band edge
NBOHC	Nonbonding oxygen hole centers
NEA	Negative electron affinity
NTP	nonthermal plasma
NVC	NH_3-based vapor clean
OC	Outer cylinder
OCA	Organic clean and air exposed
OLED	Organic light-emitting diode
OMCVD	Organometallic chemical vapor deposition
OPS	Oxygen–plasma sputtered
Pc	Pthalocyanine
PES	Photoemission spectroscopy
PL	Photoluminescence
PLD	Pulsed laser deposited
PLS	Photoluminescence spectroscopy
PMMA	Polymethylmethacrylate
PVD	Plasma vapor deposition
QCM	Quartz crystal monitors
QD	Quantum dot
RBS	Rutherford backscattering spectroscopy
RE	Rare earth
RF	Radio frequency
RGA	Residual gas analyzer
RHEED	Reflection high-energy electron diffraction
rms	Root-mean-square
ROP	Remote oxygen plasma
ROP1A	Room temperature remote oxygen plasma for 1 h and air exposure
ROP2	Room temperature remote oxygen plasma for 2 h
ROPHT	Remote oxygen plasma cleaned at high temperature
ROPRT	Remote oxygen plasma cleaned plus room temperature re-exposure
RPAN	Remote plasma-assisted nitride
RPAO	Remote plasma-assisted oxide
RS	Raman scattering

RSS	Raman scattering spectroscopy
RT	Room temperature
RTA	Rapid thermal annealing
RTD	Resonant tunnel diode
SCLC	Space charge-limited current
SE	Spectroscopic ellipsometry
SEM	Scanning electron microscope
SET	Secondary electron threshold
SEXAFS	Surface extended X-ray absorption fine spectroscopy
SIMS	Secondary ion mass spectroscopy
SLS	strained layer superlattice
SOI	Silicon-on-insulator
SPC	Surface photoconductivity
SPS	Surface photovoltage spectroscopy
SPV	Surface photovoltage
SRS	Surface reflectance spectroscopy
SRV	Surface recombination velocity
SS	Surface states
SSIMS	Static secondary ion mass spectrometry
STM	Scanning tunneling microscopy
STS	Scanning tunneling spectroscopy
SXAS	Soft X-ray absorption spectroscopy
SXPS	Soft X-ray photoemission spectroscopy
TED	Transmission electron diffraction
TEM	Transmission electron microscopy
TEXRD	Total external X-ray diffraction
TLM	Transmission line measurement
TM	Transition metal
TRCI	Time-resolved charge injection
TOF-SIMS	Time-of-flight secondary ion mass spectrometry
TSCAP	Thermally stimulated capacitance
TSP	Titanium sublimation pump
UHV	Ultrahigh vacuum
UPS	Ultraviolet photoemission spectroscopy
UV	Ultraviolet
UVOA	UV ozone and air-exposed
VBM	Valence band maximum
VCSEL	Vertical cavity surface emitting laser
VPE	Vapor phase epitaxy
VS	Valence states
WBG	Wide bandgap
XCLS	Cross-sectional cathodoluminescence spectroscopy
XKPFM	Cross-sectional Kelvin probe force microscopy
XPS	X-ray photoemission spectroscopy
XRD	X-ray diffraction
YL	Yellow emission

Appendix 3: Table of Physical Constants and Conversion Factors

Avogadro's number, N_A	6.023×10^{23} atoms/mol[1]
Bohr radius, a_0	5.29×10^{-9} cm
Boltzmann constant, k_B	1.38×10^{-23} J/molecule K[1]
	$8.62 \times 10^{-5\,\text{eV}}$ J/molecule K[1]
$k_B T$ at room temperature	0.0259 eV
Electron charge, e	1.602×10^{-19} C
Fine structure constant, α	$\sim 1/137$
Planck's constant, h	6.626×10^{-34} J s
	$4.13566733(10) \times 10^{-15}$
Reduced Planck's constant, $\hbar$	1.054×10^{-34} J s
Rest electron mass, m_e	$9.10938215 \times 10^{-31}$ kg
Speed of light	2.998×10^{10} cm/s[1]
Vacuum permittivity, ε_0	8.854×10^{-14} F/cm[1]
	8.854×10^{-12} F/cm[1]
1 Å (Angstrom)	10^{-8} cm
1 nm (nanometer)	10^{-7} cm
1 µm (micron)	10^{-4} cm
1 eV	1.602×10^{-19} J
e^2	14.4 eV Å
$e^2/2a_0$	13.58 eV

Surfaces and Interfaces of Electronic Materials. Leonard J. Brillson
Copyright © 2010 WILEY-VCH Verlag GmbH & Co. KGaA, Weinheim
ISBN: 978-3-527-40915-0

Appendix 4: Semiconductor Properties

	EG (eV)	χ (eV)[a]	Structure	A_0 (Å)	ε_s	Density (g cm^{-3})	Melting point (°C)	Enthalpy (kJ mol^{-1})[b]
Si	1.11	i 4.05	D	5.43	11.8	2.33	1415	0.0
Ge	0.67	i 4.0	D	5.65	16	5.32	936	0.0
C	5.5	i 0.3	D	3.5597	5.70	3.515	3850	0.0
SiC (3C)	2.36	i 4.12	ZB	4.3596	9.72	3.21	~3100 @ 35 atm	66.9
SiC (6H)	3.0	i 3.7	W	a = 3.0806 c=15.11733	9.66 10.03	3.21	~3100 @ 35 atm	66.9
SiC (4H)	3.2	i 3.2	W	a = 3.0730 c = 10.053	9.66 10.03	3.21	~3100 @ 35 atm	66.9
BN	4.5–5.5	i 4.5	W	a =2.55 c = 4.17	6.8 5.06	3.48	1400	252.3
BN	6.1–6.4	i 4.5	ZB	3.615	7.1	3.45	1400@4 GPa	252.3
BN	4.0–5.8	d 4.5	H	a = 2.5–2.9 c = 6.66	6.85 5.06	2.0–2.28	1400	252.3
AlN	6.2	d 0.6	W	a = 3.112 c = 4.982	8.5	3.23	2750@100 atm N_2	318.4
AlP	2.45	i 3.98	ZB	5.46	9.8	2.4	2000	164.4
AlAs	2.16	i 3.5	ZB	5.66	10.9	3.60	1740	116.3
AlSb	1.6	i 3.6	ZB	6.14	11	4.26	1080	50.2
GaN	3.39	d 3.5	W	a = 3.189 c = 5.186	8.9	6.15	2530	109.6
GaN	3.2	d 4.1	ZB	4.52	9.7	6.15	2530	109.6
GaP	2.26	i 3.65	ZB	5.45	11.1	4.13	1467	102.5
GaAs	1.43	d 4.07	ZB	5.65	13.2	5.31	1238	74.1
GaSb	0.7	d 4.06	ZB	6.09	15.7	5.61	712	43.9
InN	0.7[c]	d	ZB	a = 3.533 c = 5.693	15.3	6.81	~750@ 10^3 atm N_2	

Surfaces and Interfaces of Electronic Materials. Leonard J. Brillson
Copyright © 2010 WILEY-VCH Verlag GmbH & Co. KGaA, Weinheim
ISBN: 978-3-527-40915-0

	EG (eV)	χ (eV)[a]		Structure	A₀ (Å)	ε_s	Density (g cm⁻³)	Melting point (°C)	Enthalpy (kJ mol⁻¹)
InP	1.35	d	4.35	ZB	5.87	12.4	4.79	1070	75.3
InAs	0.36	d	4.90	ZB	6.06	14.6	5.67	943	57.7
InSb	0.18	d	4.72	ZB	6.48	17.7	5.78	525	30.5
CdS	2.42	d	4.79	W	4.137	8.9	4.82	1475	149.4
CdSe	1.73	d		W	4.30	10.2	5.81	1258	144.8
CdTe	1.58	d	4.28	ZB	6.482	10.2	6.20	1098	100.8
ZnS	3.6	d	3.82	W	5.409	8.9	4.09	1650	205.2
ZnSe	2.7	d	4.09	ZB	5.671	9.2	5.65	1100	163.2
ZnTe	2.25	d	3.53	ZB	6.101	10.4	5.51	1238	119.2
ZnO	3.39	d	4.2–4.57	W	$a = 3.253$ $c = 5.213$	7.8 8.75	5.67	~2000	350.5
PbS	0.37	i	3.3	R	5.936	17.0	7.6	1119	98.3
PbSe	0.27	i	–	R	6.147	23.6	8.73	1081	100.0
PbTe	0.29	i	–	R	6.452	30	8.16	925	68.6
MgO	7.8	d	0.85	R	4.2112	9.83	3.58	2800	601.6

All values are at 300 K.

Structure: D, diamond; ZB, zinc blende (T_d^2 – F43m); W, wurtzite (C_{6v}^4 P6₃mc); R, rocksalt (Oh₅-Fm3m); H, hexagonal (D₆ᵥ P6₃mmc).

ε_r, relative static dielectric constant; $\varepsilon = \varepsilon_r \varepsilon_0$; for W or H, top value = ⊥c axis, bottom value = ∥ c axis.

[a] Preferred values of electron affinity from among literature values.

[b] Enthalpy values $-\Delta H_{298}$ from Kubachewski, O., Alcock, C.B., and Spencer, P.J. (1993) *Materials Thermochemistry*, 6th edn, Pergamon Press, New York.

[c] Wu, J., Walukiewicz, W., Yu, K.M., Arger, J.W. III, Haller, E.E., LU, H., Schaff, W.J., Saito, Y., and Nanishi, Y. (2002) *Applied Physics Letters*, **80**, 3967.

SiC, BN, GaN, and InN values from Levinshtein, M., Rumyantsev, S., and Shur, M. (eds) (2001) *Properties of Advanced Semiconductor Materials*, John Wiley & Sons, Inc., New York.

BN gap parameters from: http://www.ioffe.ru/SVA/NSM/Semicond/BN/bandstr.html.

GaSb χ from: http://www.ioffe.ru/SVA/NSM/Semicond/GaSb/basic.html.

SiC ZB and 6H χ from Internet.

AlP χ from Internet: http://www.rpi.edu/~Schubert/Educational-resources/Materials-Semiconductors-III-V-phosphides.pdf.

AlAs χ from Internet: http://www.worldscibooks.com/phy_etextbook/2046/2046_chap1_1.pdf.

AlSb χ from Internet: http://adsabs.harvard.edu/abs/1965PhRv..139.1228F.

PbS χ from Internet: http://arxiv.org/ftp/cond-mat/papers/0412/0412307.pdf.

InAs χ: Takahashi, T., Kawamukai, T., Ono, S., Noda, T., and Sakaki, H. (2000) *Japanese Journal of Applied Physics* **39**, 3721.

InSb χ: Adachi, S. (2005) *Properties of Group-Iv, III–V and II–VI Semiconductors*, John Wiley & Sons, p. 196.

MgO values from: http://www.oxmat.co.uk/Crysdata/mgo.htm, also de Boer, P.K., and de Groot, R.A. (1998) *Journal of Physics: Condensed Matter*, **10**, 10241.

MgO χ: Electron affinities of the alkaline earth chalcogenides, *Journal of Applied Physics*, **45**, 47 (1979).

Diamond χ: http://pubs.acs.org/doi/full/10.1021/cm801752j?cookieSet=1; Qi D., Gao X., Wang L., Chen S., Loh K. P. and Wee A. T. S. (2008) *Chemistry of Materials*, **20** (21), 6871–6879.

Appendix 5: Table of Preferred Work Functions

Element	Work function	Element	Work function	Element	Work function
Ag	4.26	Hf	3.9	Rh	4.98
Al	4.28	Hg	4.49	Ru	4.71
As	3.75	In	4.12	Sb	4.55
Au	5.1	Ir	5.27	Sc	3.5
B	4.45	K	2.30	Se	5.9
Ba	2.7	La	3.5	Si	4.85
Be	4.98	Li	2.9	Sm	2.7
Bi	4.22	Lu	3.3	Sn	4.42
C	5.0	Mg	3.66	Sr	2.59
Ca	2.87	Mn	4.1	Ta	4.25
Cd	4.22	Mo	4.6	Te	4.95
Ce	2.9	Na	2.75	Th	3.4
Co	5.0	Nb	4.3	Ti	4.33
Cr	4.5	Nd	3.2	Tl	3.84
Cs	2.14	Ni	5.15	U	3.63
Cu	4.65	Os	4.83	V	4.3
Eu	2.5	Pb	4.25	W	4.55
Fe	4.5	Pd	5.12	Y	3.1
Ga	4.2	Pt	5.65	Zn	4.33
Gd	3.1	Rb	4.16	Zr	4.05
Ge	5.0	Re	4.96		

Adapted from H.B. Michaelson, "The work function of the elements and its periodicity," J. Appl. Phys. 48, 4729 (1977) with permission.

Surfaces and Interfaces of Electronic Materials. Leonard J. Brillson
Copyright © 2010 WILEY-VCH Verlag GmbH & Co. KGaA, Weinheim
ISBN: 978-3-527-40915-0

Appendix 6: Derivation of Fermi's Golden Rule

Fermi's Golden Rule provides the rate at which atomic or electronic transitions take place between two states. It applies to a wide range of optical and electronic processes for which the initial and final states can be described by wave functions. The various surface and interface analysis techniques may involve one or more atomic transitions. The rates are calculated from probabilities determined by transition matrix elements in quantum mechanical, first-order perturbation theory. This appendix follows the derivation presented in L.C. Feldman and J.C. Mayer, *Fundamentals of Surface and Thin Film Analysis* (Prentice Hall, NJ, 1986), pp. 189–207.

For a Hamiltonian,

$$H = H_0 + H' \qquad (A6.1)$$

where H_0 is a *Hamiltonian* for which the Schrödinger equation can be solved and H' represents an additional potential due to, for example, an applied electric field. Then, H_0 satisfies the time-dependent Schrödinger equation:

$$i\hbar \frac{\delta \psi_0}{\delta t} = H_0 \psi_0 \qquad (A6.2)$$

for a wavefunction

$$\psi_0 = u(x, y, z) \, e^{-iE_0 t/\hbar} \qquad (A6.3)$$

or, more generally,

$$\psi_0 = \sum_n a_n^0 u_n^0 e^{-iE_n^0 t/\hbar} \qquad (A6.4)$$

where the unperturbed state described by H_0 has a set of energy eigenvalues such that

$$H_0 u_n^0 = E_n^0 u_n \qquad (A6.5)$$

For an orthogonal set of eigenvectors u_n and prefactors a_n^0 are independent of time. Then,

$$H\psi = i\hbar \frac{\delta \psi}{\delta t} = (H_0 + H') \psi \qquad (A6.6)$$

Surfaces and Interfaces of Electronic Materials. Leonard J. Brillson
Copyright © 2010 WILEY-VCH Verlag GmbH & Co. KGaA, Weinheim
ISBN: 978-3-527-40915-0

and

$$\psi = \sum_n a_n(t) u_n^0 e^{-iE_n^0 t/\hbar} \qquad (A6.7)$$

Prefactors $a_n(t)$ now depend on time since they describe the perturbed system.

Substituting Equation A6.7 into Equation A6.6, multiplying by the complex conjugate u_s^{0*} and using the orthornormality of the wave functions,

$$\frac{da_s}{dt} = \frac{-i}{\hbar} \sum_n a_n(t) H'_{sn} e^{i(E_s^0 - E_n^0)t/\hbar} \qquad (A6.8)$$

where

$$H'_{sn} = \int u_s^{0*} H' u_n^0 d\tau \qquad (A6.9)$$

integrated over all space.

For a small perturbation, the time variation of $a_n(t)$ is slow; then, $a_n(t) \cong a_n(0)$ and

$$a_s(t) - a_s(0) \cong \frac{-i}{\hbar} \sum_n a_n(0) \int_0^t H_{sn}'(t) e^{i\omega_{sn}t} dt \qquad (A6.10)$$

where

$$\hbar \omega_{sn} = E_s^0 - E_n^0 \qquad (A6.11)$$

For simplicity, consider the special case for which the system is in state n at $t = 0$. Then $a_n(0) = 1$ and all other $a_n(t) = 0$. Then

$$a_s(t) = i/\hbar \int_0^t H'_{sn}(t) e^{i\omega_{sn}t} dt \quad \text{for } s \neq n \qquad (A6.12)$$

Equation A6.12 shows that $H'(t)$ can induce transitions from state n to any other state s, and the probability of finding the system in state s is $|a_s(t)|^2$.

For H' independent of time,

$$a_s(t) = -H'_{sn} \frac{(e^{i\omega_{sn}t} - 1)}{\hbar \omega_{sn}} \qquad (A6.13)$$

and

$$|a_s(t)|^2 = 4 \left(H'_{sn}\right)^2 \sin^2\left(\frac{\omega_{sn}t}{2}\right) / (\hbar^2 \omega_{sn}^2) \qquad (A6.14)$$

For many techniques, the result of perturbation is a final state particle in the continuum, that is, a free particle. In this case, an explicit final state can be replaced by a density of final states. Such a density is defined as the number of energy levels per unit energy interval dE. The *transition probability* $P(t)$ for discrete states is given as

$$P(t) = \sum_s |a_s(t)^2| = 4 |H'_{sn}|^2 \sum \sin^2\left(\frac{\omega_{sn}t}{2}\right) (\hbar^2 \omega_{sn}^2) \qquad (A6.15)$$

Appendix 6: Derivation of Fermi's Golden Rule

For a continuum, the summation is replaced by an integral so that

$$P(t) = 4\left|H'_{sn}\right|^2 \int_{-\infty}^{\infty} \rho(E_s) \sin^2 \frac{(\omega_{sn}t/2)}{(\hbar^2\omega_{sn}^2)} d\hbar\omega_{sn} \qquad (A6.16)$$

The major contribution to this integral occurs at $\omega_{sn} = 0$, similar to a delta function. Hence, $\rho(E_s) \sim \rho(E_n)$. Furthermore, the integral in Equation A7.16 is of the form $\int \sin^2(\alpha x)/x^2 = \pi\alpha$, where $\alpha = t/2\,h$.

Therefore, the integral reduces to

$$P(t) = 4\left|H'_{sn}\right|^2 \rho(E_n)(\pi t/2\hbar)$$

and the rate of transition $R = dP(t)/dt$ in units of s^{-1} is equal to

$$R = (2\pi/\hbar)\,\rho(E_n)\left|H'_{sn}\right|^2 \qquad (A6.17)$$

This is Fermi's Golden Rule as presented as Equation 7.9. It incorporates the density of states and the probability of transitions between initial and final states.

Appendix 7: Derivation of Photoemission Cross Section for a Square Well

Fermi's Golden Rule provides the basis for calculating the energy-dependent cross section in photoemission shown in Equation A6.17. Such calculations require the wave function of the electron in an unperturbed potential plus a perturbation potential. As a simple example to illustrate this dependence, Fermi's Golden Rule is used to calculate the photoemission cross section for electrons ejected from a bound state of energy E_B in square well into vacuum with kinetic energy $\hbar\omega - E_B$. This derivation follows that presented in L.C. Feldman and J.C. Mayer, *Fundamentals of Surface and Thin Film Analysis* (Prentice Hall, NJ, 1986), pp. 190–193.

Figure A 7.1 illustrates such a square well with a wave function corresponding to its lowest allowed energy state. The perturbation $H'(x, t) = H'(x)e^{i\omega t} = eEx\, e^{i\omega t}$ represents an electric field that is uniform in space with a wavelength $\lambda > a_0$ but harmonic in time, for example, a photon's electric field of frequency ω. According to Equation A6.17,

$$R = |<\psi_f|H'|\psi_i>|^2\, 2\pi\rho(E)/\hbar = |H_{fi}|^2\, \frac{2\pi\rho(E)}{\hbar} \quad (A7.1)$$

where ψ_i is the bound state in the potential well and ψ_f is a one-dimensional plane wave $= e^{ikx}$. This assumes a system of dimension L with periodic boundary conditions such that $\psi(x_0) = \psi(x_0 + L)$ and the normalized $\psi_f = e^{ikx}/\sqrt{(2L)}$. Here, $E = \hbar^2 k^2/2m$ for a free particle so that the periodic boundary conditions are as follows:

$$KL = \left(\frac{2mE}{\hbar^2}\right)^{1/2} L = 2\pi N \text{ for } -\infty < x < \infty \quad (A7.2)$$

The density of final states $\rho(E)$ is the number of states with energy between E and $E + \Delta E$ so that, from Equation A7.2, $\rho(E)\Delta E = \Delta N = \Delta kL/2\pi$ and

$$\rho(E) = \left(\frac{\Delta k}{\Delta E}\right)\left(\frac{L}{2\pi}\right) \quad (A7.3)$$

$$= 1/2 \left(\frac{2m}{\hbar^2}\right)^{1/2} E^{-1/2} \left(\frac{L}{2\pi}\right) \quad (A7.4)$$

$$= \left(\frac{2m}{E}\right)^{1/2} \frac{L}{2\pi\hbar} \quad (A7.5)$$

multiplying Equation A7.4 by a factor of 2 to account for negative and positive N.

According to Schrödinger's equation, the spatial component $u(x)$ of the wave function ψ_i must satisfy

Surfaces and Interfaces of Electronic Materials. Leonard J. Brillson
Copyright © 2010 WILEY-VCH Verlag GmbH & Co. KGaA, Weinheim
ISBN: 978-3-527-40915-0

Appendix 7: Derivation of Photoemission Cross Section for a Square Well

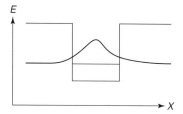

Figure A7.1 Square well potential and lowest energy wave function.

$$\left(\frac{-\hbar^2}{2m}\right)\frac{d^2 u(x)}{du^2} = -E_B u(x)$$

leading to an exponential form of ψ_i, which must decay exponentially. To simplify ψ_i and the calculation of $<\psi_f|H'|\psi_i>$ for a finite depth well, one can extend ψ_i to $x = 0$ in order to take $\psi = e^{-ik|x|}$ over all x, not just outside the well. Inside the well, this exponential form also approximates the harmonic form of the true wave function, except at the well center. Normalized, $\psi_i \cong \sqrt{k_i} e^{-k_i|x|}$ so that

$$|H_{fi}| = |<\psi_f|H'(x,t)|\psi_i>| = |<\psi_f|e\mathcal{E}x|\psi_i>| \tag{A7.6}$$

$$= \frac{e\mathcal{E}}{\sqrt{(2L)}} \int_{-\infty}^{\infty} e^{ikx} x \psi_i(x) dx \tag{A7.7}$$

$$= \frac{e\mathcal{E}k_i^{1/2}}{\sqrt{(2L)}} \int_{-\infty}^{\infty} e^{(ik-k_i)x} x\, dx \tag{A7.8}$$

$$= \frac{e\mathcal{E}\sqrt{(k_i/2L)}\,(4i\,k\,k_i)}{\left(k^2 + k_i^2\right)^2} \tag{A7.9}$$

Therefore, the transition rate for $\hbar^2 k_i^2/2m = E_B$ from Equation A7.1

$$R = |H_{fi}|^2 2\pi \rho(E)/\hbar$$

$$= \left(\frac{2\pi}{\hbar}\right)\left(\frac{L}{2\pi}\right)\left(\frac{2m}{E}\right)^{1/2}\left(\frac{e^2\mathcal{E}^2 k_i}{(k^2+k_i^2)^4\, 2L}\right)(16k^2 k_i^2)^4 \tag{A7.10}$$

$$= \left(\frac{16e^2\mathcal{E}^2}{2\hbar^2}\right)\left(\frac{2m}{E}\right)^{1/2}\left(\frac{2mE_B}{\hbar^2}\right)^{1/2}$$

$$\times \left(\frac{2m}{\hbar^2}\right)^2 \frac{EE_B}{\left[\left(\frac{2m}{\hbar^2}\right)^4 (E+E_B)^4\right]} \tag{A7.11}$$

$$= \left(\frac{8e^2\mathcal{E}^2}{\hbar^3}\right)\left(\frac{E_B}{E_B}\right)^{1/2}\left[\frac{(EE_B)}{(E+E_B)^4}\right]\left(\frac{2m}{\hbar^2}\right)\left(\frac{\hbar^2}{2m}\right)^2 \tag{A7.12}$$

$$= \left(\frac{4e^2\mathcal{E}^2\hbar}{m}\right)\frac{E_B^{3/2} E^{1/2}}{(E+E_B)^4} \tag{A7.13}$$

For $\hbar\omega \gg E_B$ such that $E + E_B \sim E = \hbar\omega$,

$$R = \left(4e^2\mathcal{E}^2\hbar/m\right) E_B^{3/2}/E^{7/2} \tag{A7.14}$$

which yields Equation 7.18.

Index

a

absorption 141, 260, 276
absorption coefficient 258, 259
absorption edge 259
absorption effects 298
absorption threshold spectroscopies 137
accelerator 187
acceptors 45
activation barrier E_{act} 329, 350
activation energy 350
adatom 328, 329
– metal 329
admittance spectroscopy 29
adsorbate 181, 221, 275, 310, 327, 329, 352, 358
– geometric structure 327
adsorbate chemisorption 461
adsorbate layers 506
adsorbate ordering 358
adsorbate-induced donor (acceptor) 455
adsorbate–semiconductor sensors 365
adsorbate–surface charge transfer 365
adsorbed trimers 227
adsorption 314, 329, 330, 378, 380
– metal 329
adsorption sensors 380
advacancies 310
AES depth profiles 166
AES depth profiling 164
Airy functions 266
all-electron calculations 430
alloy composition 384, 385
alloy spikes 343
amorphous 344
amu 210
analyzers, electron energy 122
– cylindrical mirror analyzer (CMA) 123

– hemispherical analyzer (HSA) 123
– retarding grid 122
angular resolved photoemission spectroscopy 42, 137, 138
anion vacancies 482, 483
antibonding 348, 349
antisite defects 461, 483
antisites 482
arsenic clusters 345, 463
atomic bonding 436
atomic force microscopy (AFM) 13, 246
atomic interlayers 434
atomic mass unit 200
atomic morphology 237
atomic number 87, 100, 105
atomic positioning 252
atomic resolution 239
atomic vacancies 459
atomic weight a 175, 289
atomic number Z 289
atomic-scale control 502
attempt frequency 367, 368
Auger depth profiling 163
Auger electron 86, 280
Auger electron energies 158
Auger electron spectroscopy (AES) 83, 151, 153, 215
Auger electron yield σ_e 156
Auger excitation process 156, 157
Auger transition energy 160
Auger transition probability 153
Auger transition rates 154
Auger versus X-ray Yields 154
automotive sensors 372
avalanche detector 526
average 431
average dangling bond energies 420
average electrostatic potential 431

average ionization energy 288
average lattice potential 429
average potential 429, 430, 433
Avogadro's number 175
azimuthal angle ϕ_P 138

b

back-bond states 44
backscattered electrons 280
backscattering energy loss factor 188
backscattering geometry 186
bakeout 69
ballistic electron emission microscopy (BEEM) 249, 250
ballistic electron energy microscopy (BEEM) 13
band bending 11, 19, 24, 38, 115, 118, 266, 268, 270, 346, 350, 351, 365–367, 369, 372, 375, 379, 380, 402, 403, 438
band discontinuities 403
– lattice-mismatched 403
band lineups 418
band offsets 423, 434, 439, 525
– interface contributions 423
"band offset engineering" 424, 429, 435
– spatially-confined 436
band offset results 412
band-bending 9, 20
band-bending effects 365
band-bending region 523
bandgap 5, 267, 270, 384, 385, 388
bandgap energy 259
bands 44
– model, tight-binding 44
Barns 105
barrier height 24, 56, 59, 272
barrier height measurements 472
beam decomposition 294
beam equivalent pressure (BEP) 223
beam lines 129, 130
BEEM 251, 408
– dislocations 251
– heterojunction band offsets 251
– quantum wells 251
BEEM spectrum 251
binary alloys 513
binary phase diagrams 56
binding energy 94, 101, 347, 350, 425
BioFET 375
biological sensors 380
biosensors 376, 377
bipolar junction transistors 525
black body radiation 93
Bloch function 39, 40

Bloch wave 41, 42
Bohr radius 95, 105
Bohr–Bethe range R_B 289, 290, 304
Boltzmann constant 67
Boltzmann distribution 309
bond density 319
bond hybridization 459
bond rotation model 233, 242
bond termination 323, 426
bond-rotated termination 218
bonding 11, 348, 349
– ionic and covalent 12
– ionic versus covalent 11
Bragg reflection 335
branch points 420
Bravais lattice 216
"bridge" layers 399
brightness 122, 131, 132, 138
brilliance 122
Brillouin zone 177
buckling 49
buckyballs 210
buffer layers 399, 506
built-in potential 403, 405
built-in voltage V_{bi} 406
bulk conductivity 367
bulk diffusion 341
bulk-loss function 171
Burgers vector 389, 391
buried interfaces 279, 293

c

cantilever 247
capacitance 248, 271, 377, 406
capacitance-sensitive force microscopy 246
capacitance–voltage 28, 413
capacitor 269
capture cross section 273
carrier concentration 379
carrier confinement 525
carrier mobility 24
cascade 280, 291
cathodoluminescence (CL) 86, 280, 529
cathodoluminescence spectroscopy (CLS) 83, 152, 279
Cerenkov radiation 129
change transfer 351
channeling and blocking 190
channeltron 200
– quadrupole 200
charge density 22, 174
charge exchange 119
charge neutrality energy 450

charge neutrality level   419, 420, 432, 450–452
charge transfer   20, 134, 329, 345, 350, 365, 367–371, 376, 377, 380, 418, 459, 463
– per molecule   377
– rate   380
charge transfer, electronegativity   453
charge transport   525
charge tunneling   451
charging   115, 117
ChemFET   375
chemical   377, 380
chemical and biosensors   375
chemical beam epitaxy (CBE)   314
chemical bonding   112, 132, 143, 370, 420
chemical diffusion   208
chemical etch treatments   321
chemical etching   324
chemical identification   99
chemical interaction   192, 300
chemical interdiffusion   334, 473, 506
chemical maps   207
chemical potential   341
chemical properties   336
chemical reaction   60, 439, 449
chemical reactions   143, 330, 399
chemical reactivity   137, 315, 319
chemical shift   112, 114, 115
chemical stability   437
chemical structure   311
"chemical trapping"   459
chemisorption   140, 329, 337, 347, 349, 350, 367, 461
chemisorption energy levels   349
chemisorption sites   328
cleavage   46, 462
cleaving   89
CLS equipment   295
clustering   330
coefficient of interface behavior   60
coherence length   231, 306
coincidence superlattice   220
collection angle   139
colossal magnetoresistance   532
Common-Anion Rule   61, 482
commutativity   425
compensating voltage   268
complementary metal-oxide-semiconductor (CMOS)   529
complex dielectric constant   258
complex dielectric permittivity   171
complex oxides   530, 531
complex refractive index   257
complex wave vector   257
compression   394
compression ratio   71
condensation   336, 337, 461
condensation energy   330
conducting polymers   379
conduction band   5
conduction band spike   527
conduction-band density of states (DOS)   406
conductivity   25, 242, 267, 369, 374, 377
conservation of energy, conservation of momentum   185
constant final state (CIS) photoemission   137
constant final state spectroscopy (CFS)   42, 140
constant initial state spectroscopy (CIS)   42, 141
constant-current STM images   242
contact   492
contact electrification   9
– triboelectricity   9
contact potential   268, 271
contact potential difference   23
contact potential difference Δcpd   23, 271
contact resistance ρ_C 27, 301
core level   114
core level binding energy E_B 87
core level shifts   118
core-level transitions   170
Coster–Kronig transitions   155, 157
Cottrell cloud   322
Coulomb field   183
Coulomb force   147
counterions   376
coupled modes   181
crack   315
cracking   73
– baffle   73
critical photon energy   131
critical thickness   266, 343, 387, 389, 390, 395
critical thickness h_c 392, 394
cross section   96, 187
cross-doping   437
cross-sectional STM   244
crystal face polarity   426
crystal growth   311, 314, 510
crystal orientation   399, 400, 426, 432
crystal oscillators   371
– magnetic field   371
crystal symmetry   265
crystallographic orientation   426, 439
Cs ion sputtering   203
current density   25
current density J   267
current–voltage   25, 413

cylindrical mirror analyzer (CMA) 156
Czochralski 10, 312, 461
Czochralski CZ growth 311

d
2DEG layers 375
dangling bond 50, 307, 308, 319, 512
dangling bond hybrid orbitals 453
dangling bond states 44
de Broglie relation 183, 214
deep-level optical spectroscopy (DLOS) 536
deep-level transient spectroscopy (DLTS) 29, 406, 490, 536
defect densities 514
defect model 474
defect states 300, 524
defects 13, 310, 453
"Delta" doping 438
density of states (DOS) 96, 97, 241, 246
depletion approximation 440
depletion layer width 22
– bulk charge density 22
depletion region 28
depletion width 277
depolarization 353
depopulation 276
deposit thickness monitor 378, 379
deposition monitors 79
– acoustic mode velocity 79
– crystal frequency constant 79
– flux density J 81
– quartz crystal monitor 80
– quartz crystal monitors (QCMs) 79
– vibration frequency f 79
deposition sources 77
– crystal growth 78
– Knudsen cell 78
– metallization 77
– plasma generator 78
depth of maximum energy loss rate U_0 290
depth profile 210
depth profiling 167
depth sensitivity 132
depth-resolved cathodoluminescence spectroscopy (DRCLS) 152, 293, 536
desorption 314, 356
desorption energy 350
desorption induced electron transition 86
detection limit 197, 198, 203, 270
– dynamic 198
diamond 491
dielectric 258
dielectric constant 263, 284, 396, 469, 524, 531

dielectric electronegativity 420
dielectric function 276
dielectric permittivity 171, 175, 272
dielectric response theory 171
dielectric susceptibility χ 264
differential reflectance spectroscopy 260
diffraction pattern 213, 215
diffusion 12, 143, 315, 317, 324, 338, 372, 380, 439, 458
diffusion barriers 343, 345, 501
diffusion coefficient 341
diffusion couple 341
diffusion flux F 342
diffusion potential 28
diffusion rate 342
diffusion velocity 26, 342
diffusion-controlled growth 342
dimer 49, 219, 228, 309
dimer-adatom-stacking fault 222
dimerization 219
diode switches 525
dipole 23, 24, 116, 172, 173, 274, 353, 355, 418, 419, 427, 438, 451, 453
dipole control structures 401
dipole effect 347
dipole fields 171
dipole formation 353, 420
dipole moment 347, 353
dipole shift 434
dislocation 266, 323, 392, 393, 456–458, 533
dislocation energy σ_D 390
disorder-induced gap state (DIGS) 463
dispersion relation 177
displacement current 269
displacement field vector 171
displacement vector $\mathbf{D}$ 172
dissociation 314
dissociation energy 350
domain 233, 306, 328
domain boundaries 227
donor-acceptor-pair 301
donors 45
doping profiles 406
double layer 20
DRCLS 488
dynamic secondary ion mass spectrometry (DSIMS) 199
dynamic SIMS 207
dynamical LEED 214
dynamical models 218

e
edge dislocations 319, 321, 323, 387, 389, 391
effective barrier height 492, 524

effective dielectric constant   179
effective mass   26
effective work function model   420
effective work functions   462, 507
Einstein's relation for   341
electric dipole   134
electric field vector   171
electric polarization   371
electrochemical potential   351
electrochemical reactivity   373
electromagnetic radiation   257
electron accumulation   366
electron affinity   98, 203, 319, 346, 351, 357, 404, 408, 420, 449, 469
electron affinity (EA)   466
electron affinity rule   405, 412, 419, 421, 432
electron beam spectroscopies   151
electron beam-induced current (EBIC)   293
electron charge   24
electron density   175, 285
electron energy analyzer   125
electron energy loss spectroscopy (EELS)   42, 85, 169, 180
electron escape depth   89
electron impact cross section   150
electron loss spectroscopy (ELS)   152
electron momentum   215
electron rest mass   95
electron scattering length   88, 89, 285
electron storage rings   129
electron thermal velocity v_T 273
electron-electron collisions   149, 291
electron-electron interactions   52
electron–electron impact cross section   165
electron–hole pair   280, 288
electronegativity χ 38, 61
electronegativity difference ΔX 38, 39
electronic "nose"   374
electronic conductors   379
electronic configuration   99
electronic devices   4, 301
– gas sensor   5
– solar cell   4, 5
electronic materials' surfaces   305
electronics technology   2
– drain   2
– source   2
– annealing   3
– boule   3
– diffusion   2, 3
– dopant   2
– implantation   3
– integrated   4

– interconnects   4
– localized charge   3
– localized states   3
– metal-on-insulator   3
– outdiffusion   3
– reactions   3
– silicide   2
– source   3
– source–gate–drain   2
– transistor   2
electroreflectance   86
electroreflectance spectroscopy   261
electrostatic imaging   247
elemental detection and analysis   166
elemental identification   160
elemental metal   334
elemental metals on GaAs   493
ellipsometric   315
ellipsometry Raman scattering   86
emitter efficiency   525
emitter injection efficiency   440
empirical deep-level schemes   420, 422
energy dispersive X-ray spectroscopy (EDXS)   86, 152
energy distribution curve (EDC)   96, 97, 139, 353
energy distribution curve $N(E)$ 411
energy gaps   387
energy loss   283, 291
energy transfer   150
energy transfer rate   171
enthalpy   57, 459
entropy   338
epitaxial   314
epitaxial growth   327, 337, 383
epitaxical interfaces   334
epitaxical metal–semiconductor applications   335
epitaxical overlayers   331
– elemental metal   331
epitaxical silicides   331
epitaxy   314
equilibrium carrier concentration   28
equilibrium phase diagram   341, 345
ESR   530
etch pit   321
etch rate   319, 322
etching   318
– electrolytic   318
– etching   318
– ion bombardment   318
– melt etching   318
– plasma   318
– reactive ion   318

etching (contd.)
- sputter   318
- thermal   318
- wet chemical   318
European Synchrotron Radiation Facility (ESRF)   122
eutectic   57
eutectic temperature   343, 459
evaporation   317
Everhart-Hoff   289
Ewald sphere   216, 231
exchange reaction   134, 330
exchange-correlation   52, 453
excitation depths   280
excitation energy   298
excitation sources   119
- near-UV   119
- UV discharge   119
excited electron spectroscopies   85
exciton   259
extended defects   456, 457
extrinsic states   449, 453, 458
extrinsic surface states   37

f

facets   227–230, 310, 323
Fermi level   4, 24
Fermi level movements   132, 143
Fermi level stabilization   137, 451, 459
Fermi's Golden Rule   96, 100, 141, 154
Fermi-level pinning   55, 449, 476, 477
- GaAs oxidized surfaces   478
- GaAs pinned transport measurements   477
Fermi-level pinning models   471
ferroelectric matrix   531
ferroelectricity, ferromagnetism   530
ferromagnetic   533
field effect   24
field effect principle   371, 375, 380
field effect transistor   377
field effect transistors   525
field ion microscopy   86
fine structure constant   95
first-principles calculations   429
float-zone (FZ) method   311, 312, 313
fluorescence   141
fluorescence yield   155
fluorescent X-rays   152
flux   100
forming gas anneal (FGA)   529
Fowler theory   30
Frank Van der Merwe growth   337
Franz–Keldysh effect   397

free carrier lifetime τ 292
free energy F 351
free space permittivity   396
frequency   93
Fuchs–Kliewer phonons   172

g

GaAs   16, 495, 498, 501
- barrier heights   495
- chemical interactions   495
- multilayer metal contacts   498
GaAs reconstructions   223
GaAs(110)   476
- clean surfaces   476
GaAs, InP   16
GaN   481
- Schottky barrier heights   481
GaN Schottky Barriers   480
GaP   16
gas adsorption   369
gas sensor   365, 373, 375, 380
gauges   76
- "Nude" ion gauges   77
- diaphragm gauges   76
- ionization gauges   76
- pirani gauges   76
Gauss' equation   353
Ge   11, 475
- Schottky barrier formation   475
generation rate G   292
geometric structure   305, 306
Gibbs free energy ΔG 57, 341
glancing incidence electron gun   294
graded compositions   493
graded layer   394, 528
grain boundaries   344, 374
granular gas sensors   374, 380
graphene   534
growth   317
growth sequence   439
growth template   394

h

4H–SiC   490
- Schottky barrier heights   490
Hamiltonian   349
Hartree potential   53
heat of formation   58
heat of reaction ΔH_R 59, 61, 62, 137, 341
heats of formation   56, 62, 135, 330
heats of fusion   337
helium backscattering   86
HEMT   301
heterojunction   13, 403

– float-zone   13
– horizontal Bridgman   13, 14
– molecular beam epitaxy   14
– molecular beam epitaxy (MBE)   13
– offsets   13
– pulling   13
heterojunction band diagrams   403
heterojunction band offsets   402, 418, 419, 423, 528
– theories   419
heterovalent interfaces   432
HfO_2 529, 530
HfO_2, ZrO_2 529
high electron mobility transistor (HEMT)   375, 502
highest occupied molecular orbital (HOMO)   464
high vacuum   68
high-frequency cutoff   302
high-K dielectrics   529
high-purity   10
– crystal-growth   10
high-resolution electron energy loss spectroscopy (HREELS)   170
high-resolution transmission electron microscopy (HRTEM) electron ELS (EELS)   170
Hooke's law   247
hopping transport   502
horizontal Bridgman   312, 456, 461
horizontal Bridgman growth   311
hybridization   218
hybridized atomic orbitals (HAOs)   421
hydrogen   489, 509
hydrogen adsorption   368
hydrogen catalysis   350
hydrogen diffusion   379, 485
hydrogenic atom   105
– wavefunction   105
hydrogenic binding energy   105

i

$I-V$, $C-V$ 473
ideal gas law   67
ideal Schottky barriers   20
ideality factor   26
II–VI compound semiconductor   46, 61
II–VI compound semiconductors Schottky barriers   484
III–V compound semiconductor   46, 61
III–V compound semiconductor reconstructions   223
image force correction   28
image forces   350

image-force dielectric constant   31
image-force lowering   31, 33
imaging SIMS   206
impact ionization   283
impact parameter   147, 281, 282
impact scattering   169
impurities   376, 456
impurity diffusion   456
impurity-induced disorder   398
in-phase (IP)   227
incident beam energy E_B 288
incommensurate   220
incubation   341
index of interface behavior S   38
inelastic collision cross section   286
inelastic scattering   264, 286
injection source   129
inorganic sulfides   509
InP(110) unpinned Schottky barriers   479
– $C-V$ 479
– clean surfaces   479
– InP $I-V$ 479
– measurements   479
insertion device   130
insulating   117
integrated circuits   11
interband transitions   170
interconnects   343
interdiffusion   132, 136, 143, 167, 193, 344, 345, 397, 528
interdiffusion, defect formation   458
interface atomic mixing   433
interface behavior   58
interface bonding   333
interface broadening   209
interface charge   406
interface charge-center   468
interface chemical interaction   466
interface chemical phases   462, 524
interface composition   439
interface defects   294
interface dipole   400, 402, 404, 405, 412, 427, 448
interface electronic transitions   177
interface heat of reaction   60, 491
interface layer   179, 181
interface misorientation   513
interface reaction   181, 458
interface state density D_{it} 428, 510
interface state transitions   170
interface states   37, 300, 420, 428, 437, 449, 513
– density   428

interface states (*contd.*)
– interface state models   37
– surface reconstructions   428
interface thermodynamics   459
interfaces   1, 293, 523
interfacial dielectric constants   181
interfacial layers   439, 533
interfacial reactions   180
interference   142
interlayer   136, 434, 505
intermetallic phases   334
intermixing   398
internal photoemission spectroscopy (IPS)   29, 30, 406, 459, 473
internal potential   404
– metal S_M 448
– semiconductor S_S 448
internal reflection spectroscopy   262
intrinsic surface states   37, 48, 52, 449
intrinsic surface-state models   44
inverse photoemission spectroscopy (IPES)   47, 49, 86, 141, 408
ion milling   89
ion neutralization   86
ion pump   72
ion scattering   86
ion yield   203, 204
ion-induced luminescence   86
ionic   38
ionic conductors   379
ionicity   459
ionization cross section   150
ionization efficiency   203
ionization energy (IE)   95, 286, 287, 288, 464, 466
ionization potential   98, 203, 319, 455
IR vibrational mode spectroscopy   263
IV–IV, IV–VI, and III–VI compound Schottky barriers   490

j
jellium   53, 450

k
Kelvin force probe microscopy (KPFM)   247
Kelvin probe   269–271, 355
kinematic factor   185, 188, 190
kinetic   310, 339, 341
kinetic energy E_K 87
kink   227, 307, 310, 323
knife edges   69
Knudsen   314
Kramers–Kronig   257

l
Langmuir   68
Langmuir–Blodgett films   379
laser   292
lateral force microscopy (LFM)   248
lattice constant   335, 384, 385, 387, 394, 422
lattice match   384, 385
lattice matching   334, 384, 386, 527
lattice mismatch   386, 389, 390, 392, 399, 439, 513, 531
lattice parameters   334
lattice relaxation energy   140
lattice spacing   215
lattice spacing of   214
lattice strain   384
LEED equipment   215
LEED kinematics   216
LEED pattern   219, 221
LEED reconstruction   227
Lever Rule   346
LFM   249
light scattering   407
linear models   424
linear response theory   432
– macroscopic   431
– planar   431
liquid encapsulated Czochralski (LEC) GaAs(100)   506
liquid phase   311
liquid phase epitaxy (LPE)   13, 313, 314
lithography   141
local bonding   419, 421
local density of states (LDOS)   242
localized atomic bonding, interdiffusion   449
localized states   12, 23, 24, 402, 449, 476, 491, 523
lowest unoccupied molecular orbital (LUMO)   464
low energy electron microscopy (LEEM)   306
low-energy electron diffraction (LEED)   43, 213, 295, 306, 354, 489
low-energy electron loss   86
low-energy electron loss spectroscopy (LEELS)   87
low-resistance contacts   527
luminescence   407, 508

m
macroscopic   492
magnetic anisotropy   533
magnetic domains   249
magnetic force microscopy (MFM)   248
magnetic sector   199, 202, 203

Index | 565

magnetic tunnel junction   533
mass balance equation   314
mass flow controllers   315
mass spectrometer   197, 199, 200
maximum energy loss per unit depth   289
MBE chamber   315, 316
mean excitation energy loss   289
mean free path   286
measurements   25
mechanical vibration   237
metal alloys   501
metal epitaxy   333
metal silicides   331
metal-induced gap state   55, 449, 450
metal-induced interface states   275
metal-induced states   37
metal-induced surface states   274
metal-oxide semiconductor field effect
    transistors (MOSFETs)   377, 530
metal–insulator field effect transistors
    (MISFETs)   509
metallic alloys   334
metallurgical reactions   192
metastability   346
microelectronics   192, 529
microspectroscopy   141
Miedema electronegativity X 453, 454
MIGS   55
MIGS model   474
minibands   395
miscibility gaps   385
misfit dislocations   389–391, 398
misfit dislocations, antiphase domain
    boundaries   457
misorientation   307, 510
misorientation angle   512
misorientation/vicinal surfaces   512
missing dimer   226
missing dimer model   226
mixed conductors   379
mixing   434
mobility   341
MOCVD   315
modulation doping   531
modulation spectroscopy   260
modulation techniques   260, 276
molecular beam epitaxy (MBE)   314
molecular crystals   379
molecular mass   67
molecular organic chemical vapor deposition
    (MOCVD)   314
molecular vibrations   173
molecular weight   71
molecular weight M_w 68

momentum   137, 287
momentum per gas molecule   67
monochromator   130, 271
– X-ray   121
monolayer passivation   400
monolayer surfactants   401
Monte Carlo simulations   207, 291
Moore's Law   529
morphic effects   265
morphological stability   335
morphology   248
multiferroic   530
multiple reflection techniques   263, 276

n

N versus P-type barriers   474
n-GaAs   477
– Φ_{SB} 477
n-InP   479
– macroscopic Schottky barrier heights   479
nanopillars   531
nanoscale circuits   534
nanoscale CL spectroscopy   531
nanowires, nanotubes   322, 534
native point defects   456, 457, 531
NEA   528
near-surface interband transitions   170
near-surface segregation,
    metal–semiconductor reaction   524
negative electron affinity (NEA)   355, 528
negative resistance   526
negative-U centers   453, 467
nonbonding oxygen hole centers (NBOHCs)
    529
nonequilibrium energy processing   338
nonstoichiometry   436, 456
nuclear particle detector   185
nucleation   336, 341
nucleation sites   307, 323, 328, 400

o

O ion sputtering   203
"ohmic" metal–semiconductor   19
ohmic contacts   372, 502–504, 525
– III–V compound   502
One-Third Rule   62
optical absorption   142, 257, 258, 407
optical absorption coefficient   259
optical density d_{opt} 259, 260
optical excitation process   95
optical phonon modes   276
optical spectroscopies   257
optical wavelength   257

optical-reflection-sensitive force microscopy   246
optoelectronic emitter   5
orbital energies   421
orbital number   100
organic semiconductor–metal dipoles   463
organic semiconductors, heterojunction   418
orientation   306, 309, 319, 323, 512
orientational dependence   369
oscillations   316, 532
oscillator fit of   261
oscillator strength   142
outdiffusion   136, 137
overlayer growth modes   337
overlayers   331
oxide gas sensors   371
oxygen adsorption   368
oxygen vacancy   368, 369

p

parasitic resistance   28
partial yield spectroscopy   140, 141
particle–solid scattering   147
passivation   492, 514
pathogens   376
Pauling electronegativity   453
Peierls instability   50
periodic spacing   214
periodicities   219
periodicity   215, 226
permeable base transistor   335
phase   341, 345
phase changes   142
phase equilibria   335, 346
phenomena   1
– colloid chemistry   1
– corrosion   1
– electromagnetic   1
– optical interference   1
– passivation   1
– triboelectricity   1
– tribology   1
phonon   181, 264
photocapacitance spectroscopy   29
photochemical etching   323
photochemical washing   508
photodesorption   86
photoelectric   94
photoelectric effect   9, 93
photoelectrochemical etching   322
photoelectron cross section σ is   100
photoelectron spectroscopy, X-ray   112
photoemission   413

photoemission spectroscopy (PES)   93, 125, 143, 351, 410, 461
photoemission, electron energy loss   43
photoemissive cathode   5
photoexcitation, rate   96
photoionization cross section   95, 100, 105, 113
photoluminescence (PL)   83, 86, 509
photoluminescence spectroscopy (PLS)   279
photomultiplier   526, 527
photon flux   131
photons   93
photoresponse   31
photostimulated population   276
photostimulated population and depopulation   277
photovoltage transients   271, 273
photovoltage transitions   272
photovoltaic effects   411
photowashing   509
physisorption   337, 346, 348, 349
piezo-optical spectroscopies   261
piezoelectric   378
piezoelectric effect   484
piezoelectric scanner   239
piezoelectric strain   393
pillow effect   418, 464
planar-averaged potentials   430
Planck's constant   28, 93, 183
plasma oxidation   89
plasmon   169, 170, 172, 176, 179, 181
plasmon energy loss   283
plasmon frequency   174, 175
plastic deformation   323
plastic deformation, chemical reaction   457
point defects   323, 456
point group symmetries, Bravais lattices   219
Poisson ratio μ 391
Poisson solver   299
Poisson's equation   272, 433
polarity dependence   432
polarizability α 284, 354
polarization   265
polarization electric field   396
polarization fields   439
polarization vector **A**   139
polarization-dependent photoemission spectroscopy   137, 139
polarizations   266
potential barrier   19
potentiometric   372, 373, 380
primary energy E_P 170
primitive vectors   216, 217
probability density function   258

processing contacts 493
pseudomorphic growth 387, 389, 392
pseudopotential 52, 53, 420
pseudobinary alloys 399
pthalocyanines 379
pulsed laser annealing 89
pumping
– cold trap 73
– conductance 75
– cryopump 73
– desorption rate v_d 75
– diffusion pumps 73
– manipulators 76
– poppet 75
– pumping speed S_p 75
– relief valves 74
– vapor pressures 74
– vapor pumps 73
pumps 70
– rotary 70
– rough pump 70
– scroll 70
– sorption pumps 70
– titanium sublimation pump (TSP) 70
purple plague 343

q
quadrupole 202, 203
quadrupole analyzer 199
quantum "corral" 252
quantum "stadium" 252
quantum dot (QD) 535
quantum electronic 6
– carrier confinement 6
– inversion layer 6
– quantum dots 6
– quantum well 6
quantum interference patterns 253
quantum well 5
– cascade laser 6
– high electron mobility transistor 6
– laser 6
– quantum dots 6
– two-dimensional electron gas 6
quantum wires 400
quantum-scale interfaces 535
quasi-two-dimensional electron gas 531

r
radiation angular frequency 257
radiative recombination efficiency at quantum well laser 525
radio frequency (RF) cavity 129
Raman spectroscopy (RS) 83, 264

rapid thermal annealing (RTA) 529
reactions, interface 12
– chemical 12, 208, 334, 473
– heat of interface reaction 13
reactive 136
reactive metal interlayers 502
"receptor" molecules 376
reciprocal lattice 216, 217
recombination 33, 296, 322, 508
recombination centers 493
recombination velocity 26
reconstruction 12, 45, 52, 119, 219, 221, 224, 225, 227, 295, 305, 355, 357
rectifying 19
reflection high-energy electron diffraction (RHEED) 223, 227, 232, 233, 315, 317, 531
refractive index 96, 257, 258
refractory metal contacts 501
relative sensitivity factor 204–206
relaxation 219
remote oxygen plasma 487
residual gas analyzer (RGA) 200, 201
resistance 19
resonances 43
resonant frequency 371, 379
resonant Raman scattering 265
resonant tunnel diode (RTD) 525, 526
response rate 369
rods 217
ROP 489
RS cross section 264
Rutherford Backscattering Spectrometry (RBS) 86, 149, 183, 187, 339, 501
Rutherford backscattering, theory 184
Rutherford scattering cross section 149
Rydbergs 95

s
sacrificial barrier 344
scanned probe techniques 408
scanner tube 239
scanning tunneling microscopy (STM) 13, 43, 213, 237, 246, 327, 408, 513, 535
scanning tunneling spectroscopy (STS) 42, 535
scattering cross section 147, 281
scattering efficiency 264
scattering length 88
Schottky barrier 11, 19, 25, 31, 38, 48, 55, 56, 60–62, 118, 250, 251, 266, 277, 300,, 331, 333, 336, 345, 374, 380, 406, 447, 451, 462, 484, 523, 524, 528
– prediction 528
– Schottky barriers 251

Schottky barrier formation   143, 164, 458, 467, 514
Schottky barrier height   29, 32, 34, 374, 449, 456
Schottky barrier sensor   372
Schottky defects   310
Schottky–Mott behavior   471
Schottky–Mott limit   463
Schottky–Mott model   474, 491
Schroedinger's equation   41, 429
screening   376
screw dislocations   245
secondary electron threshold (SET)   408
secondary electrons   88, 97
secondary ion mass   86
secondary ion mass spectrometry (SIMS)   85, 197
secondary ion yields   203
self-consistent electrostatic calculations   467, 469
semiconductor bandgap   277
semiconductor electron affinity   22, 268
semiconductor heterojunctions   383
– broken-gap   383
– geometric structure   383
– nested   383
– staggered   383
– straddling   383
semiconductor interlayers   506
semiconductor ionicity   38, 459
sensitivity   380
sensitivity factors   161
sensor   370–372
– operating principles   370
– selectivity   376
– sensitivity   377
SEXAFS   142
shadow   191
shear modulus   389
sheet conductance $g_\square$ 366, 367, 369, 371
– self-limiting   367
sheet resistance   302
shell number   100
Shockley states   44
Si   473, 474
– Φ_{SB} 473, 474
Si reconstructions
conjugated chain model   223
– missing row model   223
– surface dimer model   223
silicide   56, 58, 345
silicide phase formation   339
silicon   11
silicon reconstructions   221

SIMS   197, 198, 200, 409
– time-of-flight   200
SIMS depth profile   207, 209, 409
SIMS depth resolution   210
SIMS equipment   199
SIMS ion images   208
site specificity   327
soft X-ray photoemission spectroscopy (SXPS)   42, 86, 132, 329, 455
soft X-rays   129
soft-X-ray absorption spectroscopy (SXAS)   142
solar cells   525
space charge-limited current (SCLC)   534
space charge region   346, 350, 368
space-charge-limited   463
spatial resolution   203, 247
spatially localized CL   301
specific contact resistance   302
spectral distribution   131
spectral resolution   132
spectroscopic ellipsometry (SE)   52, 263, 276
– imaginary   263
– real   263
spin valve   533
spintronic   336, 355, 530, 532
spiral growth   245
split-off bands   44
spring constant   247
SPS transient   272
stacking faults   392
staggered heterostructures   407
static dielectric constant   22
static secondary ion mass spectrometry (SIMS)   83
statistical weight   95
step   306–308
step edges   328
step flow   306
sticking coefficient   68, 316
STM   309, 310, 410
STM hardware   237
stoichiometry   457, 510
stoichiometry of outdiffusion   459
stokes emissions   264
stopping power   282, 291
storage ring   131
strain   308, 310, 323, 387, 393, 439, 531
strain energy σ_s 389
strain-induced polarization fields   396
strained layer superlattices (SLSs)   394
strained-layer superlattice   396
Stranski–Krastanov   337, 391
structural imperfections   456

stuffed barriers 344
superlattice band structure 420
superlattice energy bands 394
superlattices 219, 221, 244, 398
superstructures 220
surface analytic and processing chambers 89
surface and interface techniques 83
surface band bending 265
surface conductance 366
surface conductivity 24, 365, 367, 379, 380
surface depletion layer 352, 449
surface dielectric constant 261
surface dimer model 223
surface dipole 353, 404, 405
surface dipole V_M 448
surface extended X-ray absorption fine structure (SEXAFS) 141
surface lattice; surface superlattice 219
surface misorientation 400
surface phonon 170, 172
surface photovoltage 247
– surface photovoltage spectroscopy (SPS) 42, 268, 271, 274
– surface photovoltage spectroscopy equipment 270
– surface photovoltage transient 274
surface plasmons 176
surface potential 247
surface recombination velocities 510
surface reconstruction 44, 48, 191, 217, 327, 358, 439
surface reconstruction: band bending 427
surface reconstruction: interface bonding 428
surface reflectance 276
– imaginary 276
– real 276
surface reflectance spectroscopy (SRS) 42
surface relaxation 305
surface science 12, 523
surface sensitivity 114, 132
surface space charge region 20, 29, 260, 367, 402
surface states 23, 24, 267, 353, 380
– extrinsic 52
– intrinsic 45
surface work function 346
surface-enhanced Raman scattering 276
surfactants 399
SXPS 476, 478, 483
symmetry 217
symmetry-forbidden 266
synchrotron radiation 121
synchrotron radiation sources 129

t
take-off angle θ_P 114, 116, 138
Tamm states 44
"target" molecules 376
temperature sensors 379
template structures 399, 402, 439
tensile 387
tension 394
ternary III–V semiconductors 482
terrace 307, 310, 328
thermal coefficient of expansion 384
thermal oxides 509
thermal velocity 67
thermally stimulated capacitance 29
thermionic emission 9, 25, 523, 524
thermionic emission barrier 492
thermocouple 379
thermodynamic 55, 61, 330, 339
thermodynamic trade-offs 437
thermodynamically stable phases 344
thermoreflectance spectroscopy 261
thin film diffusion 341
Thomas–Fermi–Moliere 184
threading dislocation 391
threshold energy 286
tight-binding 419, 432
time-of-flight 203
time-of-flight secondary ion mass spectrometry (TOF-SIMS) 201
time-resolved charge injection (TRCI) 91
TLM 302
topographic image 244
topography 239
topping array 354
Torr 67
total internal reflection 262
transitivity relation 425
transient effects 369
transistor 10, 301, 525
– gate modulation 10
transistor action 11
transition metals 56, 329
transition-metal energies 420, 422
transition-metal impurities 422
translational vectors 220
transmission electron diffraction 227
transmission electron microscopy (TEM) 180, 181, 339, 531
transmission function 161
transmission line measurement 301
transverse dielectric constant 258
trap states 368
trap-assisted hopping 524
trapping 33

trimer   229
tunnel current   237
tunnel diodes   525
tunneling   32, 419, 502
tunneling charge transport   523
tunneling current   238, 243, 246
tunneling image   238
tunneling theory   239
tunneling transport   524
tunneling, wavefunction   12
turbopump acts   71
two-dimensional electron gas   393
– 2DEG   393
two-dimensional electron gas transistors   525

u
UHV chambers   69
ultrahigh vacuum (UHV)   12, 67
ultrahigh vacuum technology   67
ultraviolet photoemission spectroscopy (UPS)   42, 86, 354
undulators   132
unified defect model   461
unit mesh   215
unit mesh vectors   220
universal range-energy relation   288
UV   120
UV monochromator   120
UV photoelectron spectroscopy   119

v
vacancies   229, 310, 461
vacuum level E_{VAC} 86, 403
valence band   5
van der Waals bonding   347, 353
van der Waals force   252
vapor pressures   69
vapor-phase epitaxy (VPE)   314
Vegard's Law   384, 440
vibrating capacitor   268

vibrational frequencies   181
vicinal   306–308
vicinal angle   512
vicinal surface   244
virtual excitations   169
Volmer–Weber growth   337

w
wafer-scale analysis   298
wave vector dispersions   46
wavefunction decay   38, 241
wavefunction tailing   450
wavefunction tunneling   52
wavelength   93
wavelength modulation spectroscopy   261
wet chemical etching   318
– orientation effects   319
wet-chemical treatments   508
wigglers   132
Wood notation   220
work function   9, 20, 94, 237, 350, 353, 354, 356, 409, 418, 452, 459, 470, 472, 483, 523
– SET   409

x
X-ray diffraction (XRD)   501
X-ray monochromator   129
X-ray photoelectron spectroscopy (XPS)   86, 87, 489
X-ray sources   120, 122
X-ray; Auger yields   155
X-rays   280

y
yellow emission (YL)   301
yield   191
Young's modulus   389

z
ZnO   486